THE
BANKING
CRISIS
HANDBOOK

THE BANKING CRISIS HANDBOOK

EDITED BY
GREG N. GREGORIOU

CRC Press
Taylor & Francis Group
Boca Raton   London   New York

CRC Press is an imprint of the
Taylor & Francis Group an **informa** business

A CHAPMAN & HALL BOOK

CRC Press
Taylor & Francis Group
6000 Broken Sound Parkway NW, Suite 300
Boca Raton, FL 33487-2742

© 2010 by Taylor and Francis Group, LLC
CRC Press is an imprint of Taylor & Francis Group, an Informa business

No claim to original U.S. Government works

Printed in the United States of America on acid-free paper
10 9 8 7 6 5 4 3 2 1

International Standard Book Number: 978-1-4398-1853-4 (Hardback)

Visit the Taylor & Francis Web site at
http://www.taylorandfrancis.com

and the CRC Press Web site at
http://www.crcpress.com

Contents

PART II **Global, European, and Emerging Markets' Perspectives**

PART III **Preventing Banking Crises, Bank Runs,
Regulation, and Bailouts**

Foreword

"This time is different" has perhaps been the most costly error in judgment made in almost every economic crisis thus far. This assertion embodies the premature hope on the part of capital market participants that a given crisis is not a true crisis, and hence will not lead to the severe impact expected by others. As a consequence of this misconception, market participants tend to revert to bullish portfolio allocations too soon, thus carelessly destroying significant amounts of capital.

Surprisingly, a detailed analysis of the current crisis leads exactly to the same impression: "This time is different." The crisis that began specifically as a subprime crisis in the United States will likely go down in history as the event that precipitated the restructuring of the global financial system. Even now, 2 years after the first signs of the crisis began to emerge, it is still not possible to predict with any degree of certainty the mid- to long-term consequences and havoc that the ongoing crisis will bring about. It is nonetheless clear that the current crisis is more severe than any other crisis seen over the past decades and that its consequences are perhaps farther reaching than those brought about by the crisis of 1929. In order to provide the reader with a rough idea of the far-reaching consequences associated with the current crisis, the following key developments are noteworthy. The business model employed by investment banks is no longer economically viable with all major U.S. investment banks having either vanished from the market completely, adopted a commercial banking business model, taken over by other banks, or becoming insolvent. The best-documented example is perhaps the insolvency of Lehman Brothers, the impact of which was initially underestimated, thus leading to an unnecessary intensification of the current banking crisis. Citigroup, once the largest bank in the world based on market capitalization, has become a penny stock and needed to be bailed out with billions of dollars in taxpayers' money. Similar is the case with Bank of America, Fannie

Mae, Freddie Mac, AIG, etc. Private equity funds, insurance companies, and hedge funds shut down their operations by the dozens, and with them important players (investors) in the securitization market vanished, thus undermining an extremely important economic instrument for facilitating the efficient allocation of credit risk to diverse capital market participants. The rating agencies reacted by adjusting their rating methodologies, subsequently placing countless securities with negative outlooks and the high probability of a future downgrade on their respective watch lists.

The banking crisis is nonetheless not solely an American problem, with banks throughout the world either being bailed out with government guarantees, forced to merge, or forced into insolvency. The various affected governments continuously surpass one another with increasingly ambitious bailouts with a total volume that exceeds several trillion U.S. dollars. Throughout the world, central banks are buying securities from commercial banks in an effort to generate desperately needed liquidity. Even entire states such as Iceland are bankrupt, or are on the verge of becoming bankrupt, based on the default probabilities implied by the spreads of credit default swaps referenced to their debt. Investment bankers around the world have been declared as the scapegoats for the current crisis with the branches of several major banks being destroyed during the G-20 summit in London and an ever-increasing number of people demanding that performance-based bonus payments be eliminated completely. "Bank runs" have occurred both in Hong Kong and in the United Kingdom (i.e., Northern Rock). The massive and sudden withdrawal of deposits served to exacerbate the problems of the banks in question and, in the majority of the cases observed, led to their bankruptcy.

Falling equity prices, significant write-downs, diminishing liquidity, the credit crunch, rising unemployment, and dwindling consumer demand have all served to demonstrate the severe impact that the crisis has had on the "real" (i.e., nonfinancial) economy. The collapse of the ship freight index, Baltic Dry, by more than 90% and the multitude of freighters sitting idle in major ports around the world demonstrate that global trade has virtually come to a standstill. In the near term, the paralysis of market participants brought about by the initial shock may even lead to deflationary tendencies. Government aid programs have been massively expanded with a view to saving several hundred thousand jobs (i.e., in the automotive sector). However, the appeal for massive government intervention has come not only from Chrysler and General Motors. On the contrary, the appeals for government assistance/intervention have become louder in almost every sector of the economy. Money is being printed at an ever-increasing rate with concerns regarding the prospect of (hyper-)inflation already being voiced by some. Some already question the merits of capitalism and are propagating Marxism as a viable alternative economic system. Regardless of the aforementioned overreactions, it can certainly be said that the self-regulatory approach to financial markets espoused by the former FED chief Alan Greenspan does not function properly in times of systemic financial crises and that systematic bank risks can only be corrected with the state intervening in the markets as a lender of last resort.

Confidence in the interbrain market as well as the confidence of individual investors has been severely undermined, and it is to be expected that confidence will not be restored for several years to come. A quick glance at the many severe consequences stemming from the current crisis already shows the likely complexity of the answers to the key questions that have arisen: (a) What were the precipitating factors and how could it happen? (b) What needs to be done to restore confidence in the financial sector and, ultimately, find a way out of the current crisis?

The list of possible reasons for the crisis is long and discussions in the press concerning the causes are highly controversial. Aside from unjustifiable bonus payments, errors made in the valuation of individual assets, pressures to generate unrealistic returns, severe errors made by key rating agencies, ineffective risk management practices, the irresponsible use of excessive leverage for mergers and acquisitions, inefficient work on the part of regulatory bodies, inconsistencies in key regulatory frameworks and unethical behavior on the part of a few market participants, and the unquestioned faith in IT and market standard valuation models should perhaps be viewed as the most important contributors to the current market crisis. In the course of the preceding two decades, large sums of money were invested in IT systems and the development of valuation models. In this process, assumptions were made regarding correlations and the interdependency between asset classes. These assumptions were subsequently used within the framework of the aforementioned models but turned out to be unrealistic under the extreme market conditions encountered during the current crisis. In such extreme market conditions, the values of nearly all asset classes—with the exception of gold and other commodities—tend to move in the same direction. Blind faith was placed in these models and key business decisions were based on their results. The models, however, were based solely on historical data that led to unrealistic valuations. As a consequence, all models are being examined, modified, and stressed to reflect all the potential changes in the valuation of the specific asset classes caused by the crisis.

The crisis that began with the meltdown in the U.S. subprime market first expanded into a banking crisis before eventually evolving into a global economic crisis through the infection of the entire financial sector. The impacts of the banking crisis alone are so multifaceted, with many difficult aspects, that a multitude of financial experts will likely be required to isolate the underlying causes and put forth credible solutions. By compiling diverse financial articles written by established and globally active financial experts in *The Banking Crisis Handbook*, we have succeeded in highlighting the most important topics surrounding the current banking crisis. Constructive criticism is exercised, the right conclusions are drawn from past mistakes, and the relevant steps on the way into a new era for the global financial system are discussed. With a view toward the proposed solutions and changes, a quick, concerted implementation on the part of both the industrialized world and the emerging markets is the fundamental prerequisite for the restructuring of capital markets, the revival of confidence, and, thus, the prevention of a further economic downturn. Upon reading the handbook, it becomes very clear to the reader: "After this time, everything is different!"

Christian Hoppe
Commerzbank AG

Acknowledgments

We are grateful to a handful of anonymous academic referees for the selection of papers in the review process for this book. I would also like to thank Sunil Nair, Sarah Morris, Glenon Butler, all of CRC, and Suganthi Thirunavukarasu, project manager at SPi, for making this possible. Neither the editor nor the publisher is responsible for the content of each chapter.

Introduction

At the time of the writing of this introduction, the Dow Jones Industrial Average fell to 6547 on March 9, 2009, returning to its lowest level since April 25, 1997. Nobody would have ever believed that the market would go down so rapidly. Pundits around the world began comparing the current crisis to the 1929 stock market crash, placing the U.S. economy in a brief depression-like state similar to that of the 1930s. At the onset, subprime loans and the remarkable growth of collateralized debt obligations (CDOs), which peaked in the last quarter of 2006, were considered to be the culprits. However, the finger was pointed at hedge funds, and they were being blamed for destabilizing the economy and leaving the world in a deeper mess. Nobody expected the chaos to spread around the globe so quickly with large and well-established banks falling like dominoes. Recent academic research has shown that hedge funds in fact provided liquidity during the global crisis of 2008.

Well-known investment banks, such as Lehman Brothers, which became a member of the New York Stock Exchange in 1887, filed for bankruptcy in 2008. With Bear Stearns leading the pack, a plethora of banks around the world were implicated in bad home loans that were repackaged as CDOs and presented as good-quality bonds to investors. In reality, they were worthless, and the losses amounted to nearly more than \$2 trillion. The pricing of these exotic instruments (CDOs) was largely misunderstood and was too complicated for upper management and risk managers to pinpoint their real underlying risk. Senior bank officers knew that credit derivatives could be extremely profitable, amounting to massive bonuses, but were less interested and turned a deaf ear to the dark side of CDOs.

I believe that the chapters in this book highlight and shed new light on the current banking crisis. The chapters provide possible remedies as to what should have been done prior, during, and after the crisis. The exclusive, new research in this book can assist bank executives, risk management departments in banks, and Op risk teams in banking, hopefully, to get a clearer picture of the banking crisis. The chapters in this book were written by well-known academics and professionals who have published numerous peer-reviewed journal articles and book chapters.

PART I: BANKING GROSS NEGLIGENCE AND SHADOW BANKING SYSTEM

Chapter 1 highlights the events that brought the largest economy to its knees and caused the worst knockout effect since the Great Depression. While the debate is still raging, the U.S. government has finally realized that its makeshift regulatory patchwork of not only dividing responsibility between federal agencies but also between state governments is not entirely the optimum structure; however, the world continues to suffer. Many questions need to be answered not only about why warnings from the Securities and Exchange Commission (SEC) in 2003 were not heeded, but also as to why the U.S. government allowed banks not to institute the 1999 Basel II Accord. Also why did they never deregulate the real estate industry? Why did they ignore the Financial Action Task Force recommendation on money laundering? Is the very party system that accepts such large donations from lobbyists the cause? Is it the uber democracy of totally wasteful presidential elections? Is this an aberration from the Westminster System of government? This chapter examines all possible causes and explains a regulatory taxonomy that could help solve the problem—a taxonomy that was derived from Australia's near miss in 1991.

In Chapter 2, the authors examine the effectiveness of the Federal Reserve Bank's monetary interventions as a response to the deteriorating economic conditions caused by the subprime mortgage crisis. They look at the stock price reaction of financial institutions listed in the United States to interest rate cuts and liquidity injections announced by the Fed from August 2007 to April 2008. They also link their stock price reaction to a number of firm-specific factors: the size of the institutions, their exposure to the subprime crisis, and their leverage. The authors find that interest rate cuts had a stronger impact on the market value of the financial institutions than liquidity injections; on average they caused a 4.7% and 3.3% increase in their stock price, respectively. Their results offer partial support to the hypothesis that small and credit-constrained institutions with strong exposure to the crisis profited the most from the FED's interventions. Overall, the FED has been successful in restoring the confidence of the investors in the markets in the short run.

Chapter 3 discusses unreasonable mortgage lending, which implicitly relied on the never-ending housing market bubble and caused losses to many participants in the U.S. subprime mortgage market. At first glance, it is surprising that these losses, modest relative to the size of the U.S. economy, managed to cascade into one of the largest global financial crises in history. An investigation shows that numerous mechanisms magnified the initial problem. The lack of transparency in the financial markets and institutions led to the loss of trust by investors and a system-wide run on both regular banks and shadow banks (bank-like institutions subject to little regulation). The flow of funds away from all risky assets created a credit crunch for many companies that had no exposure to the subprime. The reduction in trading volumes caused a lack of liquidity and a break-down of the price discovery, which is the cornerstone of the modern financial system. In the conditions of low liquidity, highly leveraged institutions were prone to a loss spiral whereby falling asset prices forced fire-sale liquidations to meet margin calls, thereby creating additional downward pressure on the asset's value, leading to further margin calls. Poor risk management models, which used historical data without any adjustments to allow for the burst of a bubble, led to unexpectedly severe problems in the highest-rated institutions and extremely poor reliability of credit ratings. The central role afforded to the credit ratings by the financial regulators meant that even the most regulated sectors of the industry were not immune to the crisis.

In Chapter 4, the authors admit that the subprime crisis and its consequential effects highlight numerous factors and hazards that were not exactly unforeseeable. As we all know, the capital market is not perfect, correlations are not stable, and extrapolated historical volatility is a poor descriptor of future risk—to mention but a few examples. They thus raise these valid questions: Why did financial institutions not act or react much earlier? Was the combination of several coinciding shocks to the system truly inconceivable? The authors also attempt to provide appropriate answers to these questions. They critically examine whether the decision makers in financial institutions were possibly afraid of introducing overdue change in spite of knowing better, just because doing something differently could introduce personal liabilities and reduce returns in upward markets. Regulatory frameworks such as Basel II, as well as International Financial Reporting Standards and a variety of other laws, rules, and regulations are also identified to have introduced crisis-enhancing effects, inter alia, because they prescribe the use of techniques and models that do not always capture the true risk. The authors demonstrate that a combination of organizational and market-driven corrective steps are called for, including a reorientation of incentive systems, truly living up to what is called for in corporate governance, establishing enhanced and appropriate risk methodology, and impeding the use of risk methodological approaches that have demonstrably been proven wrong.

Chapter 5 suggests that while most economists and casual observers of the state of the economy have emphasized subprime lending as the central cause of the mortgage crisis, data and economic theory provide a different reading of what happened. Rather than being limited to subprime borrowers, the events that made the crisis possible concern all mortgagors. The financial fragility of all mortgagors increased to the point that their financial position became highly sensitive to changes in interest rate, amortization rate, home price, and income. Minsky provides a good framework of analysis to understand the current crisis. He shows that there are forces internal to capitalist economies that progressively push economic units in financial deals that rely more and more on refinancing and liquidation as a means to service debt commitments. He especially argues that "stability is destabilizing," i.e., that economic stability gives an incentive and forces economic units to take more risk, and progressively leads to an increase in financial fragility.

Chapter 6 analyzes the contribution of hedge funds to the crises and instability on global markets. As compared to mutual funds, hedge funds have, due to their position as private investment firms, much more freedom to act and are virtually not subject to publishing and accounting regulations. Existing hedge funds have deviated from the original idea of hedging and serve rather as leveraged derivate portfolios employed to boost equity performance. This framework leads to significant systemic risks as a consequence of instrument and balance sheet leverage. To make things clear, the authors analyze different hedge fund strategies and their contribution to the three global crises in the last decade with a strong emphasis on the financial crisis of 2007–2008. They find that existing loose regulations and opportunistic abuse of leverage instruments lead to market failure.

Chapter 7 shows that the auction rate securities (ARS) market came into being after the decline in technology stock prices in 2001–2002, reaching about $330 billion of outstanding securities by early 2008. The key to the ARS market's success involved issuing securities with long maturities, but with coupons/dividends that reset frequently, say every four or five weeks. The rates were reset at the end of each period through a Dutch auction establishing the lowest rate that would clear the securities than being sold. However, following the recent credit crisis, the ARS market failed, leaving thousands of investors stranded, unable to sell, collect on, or otherwise convert their securities. This chapter provides a comprehensive analysis of the ARS market with a particular emphasis on the origins and mechanics that caused its recent collapse.

Chapter 8 mentions that at the end of 2008, many national regulators had started to massively react to the ongoing banking and financial market crisis. One of the reasons was that the latter was increasingly affecting the "real economy." National bailout plans included the nationalization of banks and taking regulatory measures such as granting guarantees and conferring large credits to the financial sector. Taken together, the crisis has led to serious doubts on the functioning of free markets and to an unexpected high level of state interventions into the private sector all over the world. This chapter, however, argues that the crisis was not entirely triggered by failures on free markets and "bad" accounting standards per se, but that regulatory failures also significantly contributed to its emergence. This should be taken into account when discussing a new regulatory framework and tougher oversight systems, possibly at the international level. In fact, modern efficient regulation requires policy makers to understand the boundaries of national politics and the general problems of interventions into the business system.

Chapter 9 assesses the effect of the banking crisis on insurance markets and looks at the way in which events in the insurance industry have, in turn, affected the banking sector. The authors begin by considering systemic risk in banking and insurance, and conclude that the risk of structural failure is greater in the banking sector than in the insurance industry, even though there have been local "crises" in insurance markets from time to time. Nevertheless, they find that insurers have suffered considerably in the current crisis, with the greatest adverse effects in the case of financial guarantee insurers (such as the U.S. "monolines"), companies that extended their operations beyond their traditional insurance business into risk areas of structured finance (such as AIG), insurers writing lines of business that are particularly sensitive to an economic downturn (such as credit and liability insurers), and "bancassurers" (insurers having close affiliations with banks). This chapter concludes by considering how the structure of the insurance industry may change as a consequence of the current crisis, and how changes in the regulatory system may also have an impact on insurers.

Chapter 10 looks at the role that the hedge fund industry played in the recent financial crisis. The authors discuss the growth of the hedge fund industry and demonstrate that the proliferation of hedge funds was clearly a value proposition for financial intermediaries. They confront two important misgivings about the hedge fund industry. The first pertains to the regulation of the hedge fund industry, which is often misunderstood, and the second relates to the generous performance fees awarded to the managers. They demonstrate that hedge funds do not operate in a parallel lawless dimension and argue that the recent deregulation of the banking industry has provided financial institutions with considerably more latitude than that afforded to most hedge funds. Second, hedge fund managers are also often assumed to have an incentive to take on excess risk due to the particular structure of their compensation agreements. However, the authors show that although the compensation fee is asymmetric, there are several mechanisms protecting investors' interests. Finally they argue, using a couple of case studies, that the disappointing performance of hedge funds in 2008 stemmed largely from the fact that the market infrastructure collapsed beneath them and not because the "hedge fund model" was flawed.

In Chapter 11, the author states that as capitalism's latest boom goes bust, the U.S. and global financial markets are experiencing their worst financial crisis since the Great Depression. The banking system no longer operates properly, credit markets have seized up, and liquidity has completely disappeared. Fear spreads and many claim that a market solution to the crisis no longer exists. They request an immediate government intervention to avoid a financial meltdown. As Main Street blames Wall Street for the crisis, scapegoating and finger pointing abound—there can be no doubt that the usual suspects such as hedge fund and private equity fund managers will soon be arrested. Is this really the solution? Probably not. In this chapter, the author discusses the various options available to solve the current crisis. He finds evidence in particular that collaborating with private pools of capital offers an interesting alternative solution to rescue the banking system.

In Chapter 12, the author uses a de-leveraging procedure and demonstrates that the default probability of all the hedge fund strategies has started to increase since September 2008. The same procedure allows them to conclude that in 2008 the hedge find strategies did not need leverage to increase portfolio efficiency. These results could be interpreted in two different ways. The first is simpler and is based on the sudden increase of the volatility of the markets. All hedge fund strategies suffered from this dramatic increase in volatility, which caused an increase of their own asset volatility. Increasing default probability explains why "margin financing from prime brokers has been cut and haircuts and fees on repo financing have increased" (IMF, 2008). In other words, the business risk of the strategy became so high that the banks were forced to ask hedge funds for an immediate de-leveraging process. The second explanation is more intriguing. The lenders asked hedge funds for immediate de-leveraging because of their specific liquidity problems (see Adrian and Shin, 2008; ECB, 2008), independently from the concrete default risk of the hedge funds. These unjustified requests explain the sudden and correlated increase of the hedge fund's default probability.

Chapter 13 stresses that within the global economy as a whole, emerging market economies are becoming increasingly important. Any financial and banking crisis within emerging markets may well lead to rapid and widespread contagion to other financial markets and banking sectors throughout the world. Thus, a clear understanding of the dynamics of the macroeconomic and financial variables within emerging markets will be significant and valuable for the developed markets. China's equity market has grown by leaps and bounds, and foreign investment has played a key role in this expansion. Over the past year, the Chinese equity market has experienced a significant correction, partly due to the global banking and stock market turmoil, and partly due to aggressive tightening of central bank policies to address surging inflation. The author's outcome is evidence of China's recent transformation from a closed to a relatively open economy with more open capital markets. This process is likely to continue. This is positive for the long-term development of the Chinese stock markets and economy.

PART II: GLOBAL, EUROPEAN, AND EMERGING MARKETS' PERSPECTIVES

Chapter 14 examines the recent financial crisis that has erupted due to a housing boom and the subsequent inevitable bust of the housing market in the United States. The effect of this U.S. financial crisis has eventually caused substantial damage to the overall world economy. Even though the financial world experienced several episodes of financial crises in the past, none were nearly as fierce as the current crisis. During the time of this severe financial turmoil, shareholders of banking and financial companies experienced major loss of wealth. This chapter aims at investigating various issues pertinent to the financial loss sustained by global investors.

First, the authors investigate how the financial crisis has negatively impacted the share prices of U.S. and foreign banks. For this purpose, they consider a comprehensive sample of 2467 banks across 107 countries and categorize them into five portfolios based on their country of origin and the geographic distribution of their operations. They find that losses are most severe for foreign banks with substantial U.S. operations, and U.S. banks operating internationally. These are followed by U.S. banks with purely domestic operations. Interestingly, foreign banks with international operations outside the United States are severely affected as well due to the highly integrated nature of financial markets. Foreign banks with mainly home operations suffer the least damage but are not untouched by the effects of this global liquidity crisis.

Next, they test the effect of different crisis events and policy interventions on stock prices of banks. They collect various financial crisis and intervention-related dates from sources like BBC News, CNN Money, and the *Washington Post*. The crisis events that were most significantly associated with investor wealth loss were (Chapter 11) the filing of bankruptcy by Lehman Brothers, the SEC ban on short selling, the rejection of bailout legislation by the house of representatives, Paulson's announcement that TARP funds would not be used to buy illiquid assets, and the NBER declaration of formal recession. The policy interventions that had the most notable positive impact on bank stock prices were the global expansion of swap lines by central banks; the U.S. treasury's purchase of bank-preferred stocks; the FED tax cut rate to 0%; and, finally, the provision of guarantees, liquidity, and capital by the FED and FDIC to large individual banks such as Citibank and Wachovia.

Finally, the authors apply multivariate regression analysis and examine whether some firm-specific attributes such as size, leverage, and market-to-book ratio could explain the amount of loss sustained by the stock market. They find that the country of origin and operations continue to affect returns in a multivariate setting; they also find that large banks suffer the least losses. They believe that depositors, bank stock investors, as well as regulators would benefit from the findings in this chapter.

Chapter 15 demonstrates that deficient governance systems of banking firms are one of the causes of the recent financial crisis. Institutional myopia and lax constraints for self-dealings by bankers have led to the buildup of untenable risk positions in the banking industry. This chapter looks at the challenges ahead from a European perspective. The financial crisis has been largely triggered by the accumulation of bad credit risks in securitization markets, and has subsequently spread to the rest of the regulated banking system as well as the shadow banking system. In this context, the authors highlight relevant differences between the U.S. and the European securitization markets and explain why European banks were actually among those institutions that suffered the most. Focusing on the investors' perspective, this chapter analyzes different regulatory alternatives from increased transparency to strengthening financial oversight in order to shield the banking industry from similar crises in the future. It further discusses how they may contribute to the resuscitation of financial markets in the near future.

Chapter 16 investigates the fragility of the financial system as a consequence of systemic risk that has been a matter of concern for a long time (see Thornton, 1802; Bagehot, 1873). Systemic risk was thought to be caused by the irrational and subsequent herding behavior of investors who, all of a sudden, might decide to withdraw their liquid assets from an institution. The figure of a lender of last resort (LOLR) was then suggested by these authors as a way of reducing the probability that a financial collapse occurs. Since then, however, they were disturbed by the contradicting effects that an LOLR would have upon the stability of the financial system. In this regard, the literature has been divided into supporters and opponents of the LOLR. The latter would prefer arrangements such as deposit insurance contracts and/or the provision of own capital requirements, whereas those in favor of the LOLR appear to be confronted with the so-called problem of eligibility, which consists in choosing the features or criteria an institution under financial distress should fulfill in order to be eligible for the LOLR rescue.

Chapter 17 looks at the sharp slump in the economic growth forecasts and equity market prices; it is evidently not rational to assess that emerging markets are decoupled from the current global financial crisis, or at least from the U.S. economic recession. The reason appears to be simple as the United States currently accounts for about 25% of the world's import volume of goods and services and has become the first importer of the emerging countries during the last two decades. Moreover, the globalization process has rendered emerging markets more vulnerable to external shocks due to the immaturity and weakness of their financial infrastructure, and more correlated with the developed markets in times of crisis. Asking, then, how large are the impacts of the U.S. banking crisis on emerging stock markets is an important issue for academic researchers, investors, and policy makers. In this chapter, the authors focus on the finance channel of crisis shock transmission from the United States to

Argentina, Mexico, South Korea, and Thailand using a multivariate cointegration model over the period from December 1987 to January 2009. Their findings show significant but asymmetric effects of the current crisis on selected emerging stock markets due to the regional differences. Despite their efforts to reduce their financial dependences on the U.S. economy through stimulating internal demands, emerging markets seem not to be protected from the current crisis.

In Chapter 18, provisioning for loan losses plays a key role in determining the makeup, and thus the transparency and representational faithfulness, of banks' balance sheets. From a regulatory perspective, discretion in loan impairment provisioning may provide greater capacity to build up substantial buffers against deterioration in credit quality prior to the existence of actual impairment in individual loans.

However, under the approach to loan impairment and provisioning prescribed by IAS 39—*Financial Instruments: Measurement and Recognition*, any forward-looking, uncertainty-tolerant approach to loan impairment provisioning is in stark contrast to the contemporary accounting rules on the subject that emphasize the primacy of objective and verifiable evidence over future-oriented conjecture. In effect, the prudential regulatory management approach to impairment provisions is best characterized as anchored within an expected losses model, while the contemporary accounting rulemaking approach to loan impairment provisioning is anchored within an incurred loss tradition that ensures that it is historically oriented rather than future oriented.

The evidence in 2007 and 2008 suggests it was clear that substantial portions of the globe's financial and economic fabric lay in a state of severe distress; however, the financial disclosures by the Asian banks over this period show a picture at odds with this larger reality. In part, it seems strongly arguable that the impairment recognition procedures stipulated by IAS 39 represent an element of any explanation for the muted response of Asian banks to impairment recognition in the face of a gathering economic storm.

If one of the objectives of the International Financial Reporting Standards (IFRS) regime is to allow reporting entities (in this case, banks) to produce financial disclosures that are of greater assistance to users by way of being constructed on a foundation of more useful information, it may be that this objective is being poorly served by the current approach to evidence set out in IAS 39.

Chapter 19 explores the financial turmoil over the past decade that has stimulated research into various sources of vulnerability of economies around the world. In particular, both maturity and currency mismatches have been found to be associated with many of the episodes of financial fragility recorded in the past decade. This chapter addresses the following: First, the authors present empirical evidence on the extent of currency and maturity mismatches for Latin American countries using recent data from 1993 to 2007. Second, they summarize the main factors identified by the empirical literature as determinants of mismatches and shed light on the links between mismatches and financial fragility, both at the sovereign and corporate levels. Third, they discuss the roles of bond markets, financial derivatives, and capital markets, in general, in mitigating currency and maturity mismatches in developing countries. This chapter also raises issues for future research in various directions.

Chapter 20 shows that the Russian banking system is in its worst crisis since 1998: on the one hand, this is a consequence of the global financial and economic crisis; on the other hand, there are specific country factors. First of all, the Russian economy depends on a relatively small number of industries. Second, Russian firms have a large amount of foreign debt. Furthermore, when oil prices decrease, there is a decline in the ruble against the dollar and the euro. However, differently from 1998, the banking system finds itself in a better position thanks to the previous macroeconomics boom, which lasted almost 10 years. Still, the Russian banking sector may face important risks in the near future in case of a continued decrease in oil prices, a lack of stabilization in the FOREX market, and a declining quality of the collaterals, with an increase of bad loans in the banks' portfolios. Nevertheless, major improvements have been made and the Russian banks have used the current situation to improve and optimize their expenses.

Chapter 21 observes how the Australian banking system has a number of distinguishing characteristics, among which are its geographical remoteness, its uninterrupted strong growth record, and its so-called Four Pillar policy. The authors investigate the stability of the Australian banking system and analyze whether this unique set of features has kept it insulated from the 2007–2008 credit crisis. They apply Extreme Value Theory, which is particularly suitable for such a risk management analysis, and find that the Australian banks' share price return distribution functions exhibit fat tails. The risks thus exceed those indicated by the common assumption of normally distributed returns. They further find that the relatively high cocrash probabilities between the four pillar banks support the conjecture that these are "too big to fail." During the crisis, the cocrash probabilities between most Australian banks have increased markedly. Moreover, the authors show that the tail-dependence of the Australian banking sector on the American, Asian, and, to a lesser extent, European banking sectors has also been boosted.

In Chapter 22, the authors examine how Australia's banking regulation in the years leading up to the recent global financial crisis has been one of adherence to the guidelines set down by the latest and previous Basel accords requiring banks to hold adequate liquid capital to the specified percentages of risk-weighted assets as a safety valve to cover losses due to market, credit, and operational risks. The need had not been evident pre-crisis to take the additional step of insuring bank deposits. Systemic risk was deemed to have been well covered. This chapter discusses the efficacy of the Australian regulatory and institutional environments from a legal perspective and also produces empirical evidence from correlation, regression, cointegration, and causality analysis that illustrates the global positioning of the Australian banking industry in the years leading up to the crisis. Reasons why the banking system has experienced a combination of good luck and good management are put forward on the basis that the problems faced by other larger developed economies in Western Europe and North America have so far been avoided. This is not to say that Australian banking is immune from the crisis, as mining and associated companies face falling global demand for minerals and banks face higher bad and doubtful debts.

Chapter 23 reveals that in recent years, large financial institutions have expanded their operations across national boundaries. Undertaking these operations has led to stronger interconnections across institutions due to extensive interbank activities; heightened counterparty risk arising from global trading activities, inclusive of OTC derivatives contracts; and increased participation in equity, bond, and syndicated loan issuance. Such development has given rise to the "too-interconnected-to-fail" problem, which in the aftermath of the subprime mortgage crisis has become a major concern to policy makers and risk managers alike. The author introduces a methodology for assessing default risk codependence, or, in other words, how the default risk of a financial institution affects the conditional default risk of another institution. The methodology relies on market prices of default risk, so it bypasses the need to use detailed information on linkages across banks provided market prices are efficient. The methodology is applied to a sample of 25 global banks and casts some insights into the bailout of Bear Stearns and AIG, and the bankruptcy of Lehman Brothers.

In Chapter 24, the authors show that while the European financial market is experiencing a big crisis, it also has to cope with the integration of the Markets in Financial Instruments Directive (MiFID) to the legislation of all EU countries, as one step forward for succeeding as a single European financial market. Ethnographic research was recently carried out to study the first results of MiFID implementation in the Greek financial market, which reflect that most companies are not on the verge of such a big change and most of their customers are lacking knowledge on coping with such an important issue for their financial objectives. Moreover, part of the staff of these companies is still missing important aspects of this change and the implications it will bring in their everyday interaction with customers and companies' goals.

PART III: PREVENTING BANKING CRISES, BANK RUNS, REGULATION, AND BAILOUTS

In Chapter 25, the authors reveal that the 2007–2008 credit crisis not only vastly affects the financial system but is also likely to have severe consequences for the global economic development. The extent of the crisis is enormous. Due to the growing globalization and complexity of the financial system, the contagion effect of the current crisis throughout financial markets is unprecedented. The crisis clearly reveals the vulnerabilities of the financial system in its current form. Hence, it is of particular importance to understand what actually triggered the collapse of the financial system, and how such a collapse can be prevented in the future. The literature thus far on how bailout plans should be arranged is scarce. The authors take a view from the perspective of credit derivatives and explain the circumstances that led to this crisis. They describe the instruments fostering the instability of the financial system and show how the collapse was triggered. They further comment on the recent measures of short-term government intervention, which aim at limiting the acute damage to the financial and economic system. In addition, they discuss how the design of government bailout programs can influence decision making among financial institutions. Furthermore, they argue that only rescue packages including a purchase program for distressed assets create a setting where illiquid, but otherwise solvent, banks are separated from insolvent banks. Such a setting provides a valuable signal to outsiders, including investors as well as government agencies. Finally, they suggest relevant areas for improved long-term financial regulation and provide an overview of the possible consequences for the design as well as the regulation of the financial system in the future.

In Chapter 26, the authors state that a "bank run" corresponds to the phenomenon where people run to their banks to withdraw all of their deposits. This collective behavior seriously affects the bank's liquidity and often results in bankruptcies. At the present time (2008), the major Swiss private banks, UBS and Credit Suisse, are troubled due to the current "subprime" crisis. In this chapter, the authors show the main findings of a survey conducted in May/June 2008 with 363 people living in Geneva. In particular, they aim to assess individuals' confidence toward Swiss banks and attempt to recognize signals that would lead to a bank run. The authors perform this task by identifying sociological clues connected with bank run attitudes, which may be the first step in efficiently managing this type of risk. Descriptive statistics show that most people do not plan to change banks in the coming future. Moreover, Geneva inhabitants still have confidence in their banks while carefully watching the evolution of the crisis. These and other related topics (i.e., UBS president resignation, the perceived default risk of a Swiss bank, the judgment Geneva inhabitants have vis-à-vis of Swiss banks) have been analyzed, and research hypotheses have been verified on the basis of nonparametrical tests. The authors' findings highlight that people who believe their bank savings are at risk are more likely to take part in a bank run than the others. Also, confidence toward Swiss banks among Geneva inhabitants seems to be a factor that might reduce the likelihood of a bank run in the given area.

The current financial crisis places the spotlight on the ability of banks to meet their financial obligations. This chapter examines and compares changes in bank default risk in the United States and the United Kingdom over time, including the current period of crisis. A common approach used by banks to measure the probability of default (PD) among customers is the KMV/Merton structural model, which measures distance to default (DD). The authors use this same approach to measure the DD of the banks themselves. As a further measure of variation of bank risk over time, they use the value at risk (VaR) methodology to examine the banks' equity risk, as well as the increasingly popular conditional value at risk (CVaR) methodology to measure their extreme equity risk. In addition, they incorporate CVaR techniques into structural modeling to measure extreme default risk. The study finds that U.S. and U.K. banks are in an extremely precarious capital position based on market asset values, especially in the United Kingdom, where the banks are more highly leveraged.

They also find the existing credit ratings of banks are much more favorable than default probabilities indicate they should be. Movements in market asset values are currently not factored into capital adequacy requirements, and based on their findings, recommendations are made for a revised capital adequacy framework.

Chapter 27 demonstrates that the link between credit risk and the current financial crisis accentuates the importance of measuring and predicting extreme credit risk. CVaR is a method used widely in the insurance industry to measure extreme risk, and it has also gained popularity as a measure of extreme market risk. The authors combine the CVaR market approach with the KMV/Merton credit model to generate a model measuring credit risk as applied to banks under extreme market conditions.

The KMV/Merton model is a popular model used by banks to predict PD of customers based on movements in the market value of assets. The model uses option pricing methodology to estimate DD based on movements in the market value of assets. This model has been popularized among banks for measuring credit risk by KMVs who use the DD approach of Merton but apply their extensive default data base to modify PD outcomes. The authors apply this measure to the banks themselves. The current financial crisis places the spotlight on the ability of banks to meet their financial obligations. This chapter examines and compares changes in bank default risk in the United States and the United Kingdom over time, including the current period of crisis. VaR has become an increasingly popular metric for measuring market risk. VaR measures potential losses over a specific time period within a given confidence level. This concept is well understood and widely used. Its popularity escalated when it was incorporated into the Basel Accord as a required measurement for determining capital adequacy for market risk. CVaR measures extreme returns (those beyond VaR). Pflug (2000) proved that CVaR is a coherent risk measure with a number of desirable properties, such as convexity and monotonicity, among other desirable characteristics. Furthermore, VaR gives no indication on the extent of the losses that might be encountered beyond the threshold amount suggested by the measure. By contrast, CVaR does quantify the losses that might be encountered in the tail of the distribution. The authors apply CVaR in their model of DD. The study finds that U.S. and U.K. banks are in an extremely precarious capital position based on market asset values, especially in the United Kingdom, where the banks are more highly leveraged. They further find the existing credit ratings of banks are much more favorable than default probabilities indicate they should be. Movements in market asset values are currently not factored into capital adequacy requirements, and based on their findings, recommendations are made for a revised capital adequacy framework.

Chapter 28 examines how the regulators of financial institutions have identified poorly designed remuneration structures as a major contributing factor to the losses in financial institutions that precipitated the global financial crisis. Specifically, many structures encouraged excessive risk-taking on the part of individuals in these firms by paying bonuses for writing volume business in loan markets, without appropriate adjustment for the risk being incurred. The response from a number of regulators has been that financial institutions must review and "correct" their remuneration structures to prevent excessive risk-taking. In order for this to be achieved, it is fundamental that an institution identify and articulate its risk appetite. Institutions must also anticipate and establish controls to mitigate agency problems that arise with the use of risk-adjusted performance measures, as well as deal with the phenomenon of managerial overconfidence with respect to estimates of risk. These conceptual factors present significant challenges that threaten the effectiveness of risk-adjusted remuneration structures in financial institutions. There is much work to be done on the part of regulators and other relevant authorities with respect to these issues.

Chapter 29 examines the threat faced by the entire economy, both nationally and internationally, when banks get into financial trouble. Government intervention is often called for to reduce the adverse effects that would otherwise occur. Governments and the economists who work for them estimate the adverse effects that would ensue in the absence of intervention. Multiplier theory is often employed to show the secondary effects that are expected to ripple through the economy.

The problem with this approach is that policy makers focus only on the losses incurred by the banks and the adverse ripple effects that are caused by the problems in the banking sector. A good utilitarian analysis would examine the effects a policy has on all groups, both long term and short term. Rights issues are often ignored since utilitarian analyses almost uniformly disregard the existence of rights.

This chapter examines the current banking crisis and applies both the utilitarian ethics and the rights theories of Frederic Bastiat to determine when, and under what circumstances, government intervention in financial markets can be ethically justified.

Chapter 30 shows how, in the aftermath of the global financial crisis, there has been considerable controversy over the role of the state in relation to financial markets. The author demonstrates that Hong Kong's introduction of comprehensive regulation of financial services in response to repeated market failures did not deter investment or stifle innovation in this bastion of free enterprise. By the end of the last century, it had become a major international financial center, offering the highest standards of banking performance and the most open business environment in Asia. Hong Kong offers a persuasive case in favor of official measures to maintain depositors' confidence and to stabilize financial markets even when the government and the business community are deeply committed to laissez faire.

Greg N. Gregoriou and the Contributors

Editor

Greg N. Gregoriou has published 34 books, over 50 refereed publications in peer-reviewed journals, and 20 book chapters since his arrival at the State University of New York (Plattsburgh) in August, 2003. His books have been published by John Wiley & Sons, McGraw-Hill, Elsevier-Butterworth/Heinemann, Taylor & Francis/CRC Press, Palgrave-MacMillan, and Risk books. His articles have appeared in the *Journal of Portfolio Management*, the *Journal of Futures Markets*, the *European Journal of Operational Research*, the *Annals of Operations Research*, and *Computers and Operations Research*. Professor Gregoriou is a coeditor and editorial board member for the *Journal of Derivatives and Hedge Funds*, as well as an editorial board member for the *Journal of Wealth Management*, the *Journal of Risk Management in Financial Institutions*, and the *Brazilian Business Review*. A native of Montreal, he obtained his joint PhD in finance at the University of Quebec in Montreal, which merges the resources of Montreal's four major universities (McGill University, Concordia University, University of Quebec at Montreal (QUAM), and École des Hautes Études Commerciales [HEC]–Montreal). His interests focus on hedge funds, funds of hedge funds, and managed futures. He is also a member of the Curriculum Committee of the Chartered Alternative Invesmtent Analyst Association.

Contributor Bios

David E. Allen is a professor of finance at Edith Cowan University, Perth, Western Australia. He is the author of three monographs and over 70 refereed publications on a diverse range of topics covering corporate financial policy decisions, asset pricing, business economics, funds management and performance bench marking, volatility modeling and hedging, and market microstructure and liquidity.

Mohamed El Hedi Arouri is currently an associate professor of finance at the University of Orleans, France, and a researcher at the EDHEC Business School, Lille Cedex, France. He received his master's degree in economics and his PhD in finance from the University of Paris X Nanterre. His research focuses on the cost of capital, stock market integration, and international portfolio choice. He has published articles in refereed journals such as the *International Journal of Business and Finance Research*, *Frontiers of Finance and Economics*, *Annals of Economics and Statistics*, and *Finance*.

Ruggero Bertelli is an associate professor at the Richard Goodwin Faculty of Economics, University of Siena. He has been teaching banking and finance at graduate and postgraduate levels. He is responsible for the Hedge Fund and Alternative Investment Strategy Research Unit, University of Siena. He has been a member of the Financial Market Authority (FMA) board of directors as program cochair, 2005 European Conferences in Siena. He is also the 2009 FMA European Conference Meeting cochair, Torino.

Bastian Breitenfellner is a PhD student at Passau University. He holds a diploma in business administration and technology from the Technical University Munich. He also spent a visiting semester at the University of Zurich. Bastian currently works as a research assistant at the DekaBank

Chair in Finance and Financial Control at Passau University. His research is focused on credit risk valuation, credit derivatives, and risk management.

Tyrone M. Carlin is the professor of financial reporting and regulation and the chair of the business law discipline within the Faculty of Economics and Business at the University of Sydney. His current research interests lie in interdisciplinary work on corporate governance, valuation, and financial reporting. He teaches in the areas of commercial law, mergers and acquisitions, and insolvency and restructuring. He has published articles in a range of international journals, including the *Management Accounting Research*, the *Financial Accountability & Management*, the *Public Management Review*, the *Australian Accounting Review*, the *Sydney Law Review*, the *University of New South Wales Law Review*, and the *Australian Business Law Review*. He is the founding coeditor of the *Journal of Applied Research in Accounting and Finance* and the *Journal of Law & Financial Management*. He is a member of the board of management of the *Australian Accounting Review* and of the editorial boards of the *Accounting, Auditing & Accountability Journal*, the *Financial Accountability & Management*, and the *Pacific Accounting Review*.

Giuseppe Catenazzo is a research assistant at the Haute École de Gestion of Geneva, Switzerland. He holds an undergraduate degree in economics sciences and business administration, University of Aosta Valley, Italy; currently, he is a postgraduate student in applied environmental economics at the School of Oriental and African Studies, University of London, U.K. He is the coauthor of a book dealing with service management *La Gestion des Services*, Economica Ed, Paris 2008; he previously worked in the Web marketing and hospitality fields within international companies in Switzerland as well as in France.

Jorge Antonio Chan-Lau is a senior economist at the International Monetary Fund (IMF) and a fellow at the Center for Emerging Market Enterprises, The Fletcher School, Tufts University. At the IMF, he is a lead contributor to analytical and policy work on financial stability, risk modeling, and capital markets, three areas in which he has published widely. During 2007–2008 he was on leave as a senior financial officer at the International Finance Corporation, The World Bank Group, where he managed a frontier market's local currency portfolio, and for which he designed and implemented the valuation and economic capital allocation models. He also served as a special departmental advisor on credit risk modeling at the Bank of Canada in 2006 and the Central Bank of Malaysia in 2009 and is a charter member of Risk Who's Who. Dr. Chan-Lau received his PhD and MPhil in finance and economics, respectively, from the Graduate School of Business, Columbia University, and his BS in civil engineering from Pontificia Universidad Católica del Perú.

Carolyn V. Currie is a member of the Association of Certified Practising Accountants, the Chartered Secretaries Association, and a fellow of Finsia, a merger of the Australian Institute of Banking and Finance and the Securities Institute. Her experience represents almost four decades in the public and private sectors, as a merchant banker, regulator, internal auditor, and financial trainer. For the last 15 years she has been a senior lecturer in financial services at the University of Technology Sydney, as well as the managing director of her own consulting company and several private investment companies.

Dean Fantazzini is an associate professor in econometrics and finance at the Moscow School of Economics, Moscow State University. He graduated with honors from the Department of Economics at the University of Bologna, Italy, in 1999. He obtained his master's degree in financial and insurance investments at the Department of Statistics, University of Bologna, Italy, in 2000, and his PhD in economics at the Department of Economics and Quantitative Methods, University of Pavia, Italy, in 2006. Before joining the Moscow School of Economics, he was a research fellow at the Chair for Economics and Econometrics, University of Konstanz, Germany, and at the Department of Statistics and Applied Economics, University of Pavia, Italy. He is a specialist in time series analysis, financial econometrics, multivariate dependence in finance and economics, and has more than 20 publications, including three monographs. On April 28, 2009 he was awarded for fruitful scientific research and teaching activities by the former USSR president and Nobel Peace Prize winner Mikhail S. Gorbachev and by the Moscow State University rector Professor Viktor A. Sadovnichy.

Nigel Finch is a senior lecturer in accounting within the Faculty of Economics and Business at the University of Sydney, Sydney, New South Wales, Australia. Prior to this, Nigel was a lecturer in management at the

Macquarie Graduate School of Management (MGSM) in Sydney, and a director of the Centre for Managerial Finance. He specializes in the areas of accounting, financial statement analysis, and financial management. His research interests lie in the areas of, asset impairment, valuation, corporate governance, and financial reporting. Prior to joining academia, he worked as a financial controller for both public and private companies and as an investment manager specializing in Australian growth stocks for institutional investment funds. He is the founding coeditor of the *Journal of Applied Research in Accounting and Finance*.

Guy W. Ford is an associate professor of management at the Macquarie Graduate School of Management in Sydney and a director of the Centre for Managerial Finance. He teaches in the areas of strategic finance, corporate acquisitions, insolvency and restructuring, and financial institutions management. He has previously served with the Treasury Risk Management division of the Commonwealth Bank of Australia, and has published over 100 papers in a wide range of scholarly refereed journals and international conference proceedings. He is a founding coeditor of the *Journal of Law & Financial Management*. He is also the author of two books: *Financial Markets and Institutions in Australia* and *Readings in Financial Institutions Management*.

Emmanuel Fragnière, certified internal auditor (CIA), is a professor of service management at the Haute École de Gestion of Geneva, Switzerland. He is also a lecturer in enterprise risk management at the Management School of the University of Bath, U.K. He has previously served as a commodity risk analyst at Cargill (Ocean Transportation) and a senior internal auditor at Banque Cantonale Vaudoise, the fourth-largest bank in Switzerland. His research is focused on the development of risk management models for decision makers in the service sector. He has published several papers in academic journals such as the *Annals of Operations Research*; the *Environmental Modelling and Assessment*; *Interfaces*; and *Management Science*.

Xanthi Gkougkousi is a PhD student in financial accounting at the Rotterdam School of Management, Erasmus University. She completed her bachelor studies at the Athens University of Economics and Business, and then worked for two years as a freelance accountant before moving to Rotterdam, where she pursued her MSc in finance and investments at the RSM Erasmus University.

Werner Gleissner is currently the CEO of the FutureValue Group AG and the head of risk research of Marsh GmbH. He has authored more than 100 articles and more than 12 books; his current R&D activities and projects focus on risk management, rating, strategy development, the development of methods for aggregating risks, value-based management, valuation, decision making under uncertainty, and imperfect capital markets. Werner lectures at various reputable universities in the field of rating, risk management, value-based management, and entrepreneurship. He is, inter alia, the editor of the well-known loose-leave series on corporate risk management ("Risikomanagement im Unternehmen"). He holds a degree in commercial engineering (Diplom-Wirtschaftsingenieur, equivalent to a master's degree in business engineering) and a doctorate degree in economics and econometrics, both from the University of Karlsruhe, Germany.

Leo F. Goodstadt has been a chief policy adviser to the Hong Kong Government (1989–1997) and a consultant economist to leading banks in Asia. His latest book, *Profits, Politics and Panics: Hong Kong's Banks and the Making of a Miracle Economy, 1935–1985*, was published in 2007. He is also an adjunct professor at Trinity College Dublin, and a former research fellow at the Hong Kong Institute for Monetary Research.

Elena Grammenou is an experienced professional in marketing, financial management, and banking. She is a chartered marketer and has a BSc in economics from the University of Athens, a professional postgraduate diploma from the Chartered Institute of Marketing, a European Foundation Certificate in banking from the European Bank Training Network, and an MBA from the University of Bath—School of Management. Elena Grammenou is a research assistant, Haute École de Gestion de Genève", and also working in the marketing department of Piraeus Bank, Greece.

Ulrich Hommel is a professor of corporate finance and the director of the Strategic Finance Institute at the European Business School International University in Germany. He holds a PhD in economics from The University of Michigan, Ann Arbor, and has completed his habilitation at The WHU—Otto Beisheim School of Management in Vallendar. In the past, he has held visiting professorships at the Stephen M. Ross School of Business at The University of Michigan, the Krannert School of Management at Purdue University, and the Bordeaux Business School. His main research areas are corporate risk management, venture capital financing, and real options analysis.

Christian Hoppe works as the head of credit solutions in the credit portfolio management in the corporate banking division of Commerzbank AG Frankfurt. His main focus is on structured credit transactions to actively manage the corporate credit portfolio. Christian is also the cofounder and CEO of the Anleihen Finder GmbH in Frankfurt, an information platform for mezzanine and debt capital. Prior to this, he was credit portfolio manager at Dresdner Kleinwort, the investment bank arm of Dresdner Bank AG in Frankfurt. He started his career as a business and financial controller for Dresdner Bank in Frankfurt responsible for the corporate client business in Germany. He completed his economics degree at the University of Essen-Duisburg, North Rhine-Westphalia, Germany in 2003. While writing his master thesis, Christian worked in the institutional research department of Benchmark Alternative Strategies GmbH in Frankfurt. Christian is the coauthor of several articles as well as books; the author of the German book, *Derivate auf Alternative Investments—Konstruktion und Bewertungsmöglichkeiten*, published by Gabler, Wiesbaden, Germany and the coeditor of the *Handbook of Credit Portfolio Management* published by McGraw Hill, New York.

Mahmud Hossain is an assistant professor of accounting at the University of Memphis. He has published articles in various journals such as the *Journal of Accounting and Public Policy*, the *Journal of International Accounting Research*, and the *Review of Quantitative Finance and Accounting*. His research interests include auditing, banking, and capital markets.

Jason Hsu oversees the research and investment management areas at research affiliates (RA). He manages the firm's sub-advisory and hedge fund businesses. He also directs researches on asset allocation models that drives the firm's global macro and Global Tactical Asset Allocation (GTAA) products and equity strategies that underpin RA's fundamental indexation concept. He is an adjunct professor in finance at the University of California, Los Angeles (UCLA), Anderson Business School, and has served as a visiting professor at the UC Irvine Paul Merage School of Management and the School of Commerce at Taiwan National Chengchi University. Jason received his undergraduate degrees from the California Institute of Technology and his PhD in finance from the UCLA.

Pankaj K. Jain is the Suzanne Downs Palmer associate professor of finance at the Fogelman College of Business at the University of Memphis. He has previously worked in the banking industry for three years. He has published his award-winning research on financial market design in leading journals such as the *Journal of Finance*, the *Journal of Banking and Finance*, the *Journal of Financial Research*, and the *Contemporary Accounting Research*. He has been invited to present his work at the New York Stock Exchange, the National Bureau of Economic Research, and the Capital Market Institute at Toronto.

Vicente Jakas is the head of finance BAC Global Markets Fixed Income at Deutsche Bank AG, Frankfurt am Main. He holds an MSc in financial economics from the University of London (London, U.K.), a BA (honors) in business administration from the Robert Gordon University (Aberdeen, U.K.), and a BSc in business economics from the Universidad de La Laguna (La Laguna, Spain). He has more than 10 years experience in the banking industry and has worked for the "big four" audit and consultancy firms in the area of banking and finance. His main areas of research are institutions and capital markets, as well as macroeconomic policy and the financial markets.

Fredj Jawadi is currently an assistant professor at Amiens School of Management and a researcher at EconomiX at the University of Paris Ouest Nanterre La Defense, France. He received his master's degree in econometrics and his PhD in financial econometrics from the University of Paris X Nanterre. His research topics cover modeling asset price dynamics, nonlinear econometrics, international finance, and financial integration. He has published in refereed journals such as the *Journal of Risk and Insurance*, and the *Applied Financial Economics, Finance, and Economics Bulletin*.

Wilhelm K. Kross is a recognized expert in the fields of risk management and project management. Prior to starting his own business and joining the network of the FutureValue Group, he worked as a board member (inter alia as COO and CFO) of the newly founded risk consulting subsidiary of Marsh GmbH. He has previously been the head of management consulting of Value & Risk AG, a German financial services boutique consulting firm, and had formerly spent 10 years in Southern Africa and Canada. Wilhelm has held various positions as a trustee, manager, and board member (including his presidency of the PMI Frankfurt Chapter), and has worked on a wide range of projects in more than 30 countries, including rather tasks under public scrutiny, and crisis management. He has authored more than 48 publications, and has obtained an MSc equivalent in engineering from RWTH Aachen, an executive MBA from Athabasca University, and a doctorate degree in finance from the European Business School.

Alexander Kudrov is a researcher at the Higher School of Economics (Moscow, Russia), where he obtained his PhD in economics in 2008. His main area of research is extreme value theory with applications in economics and finance. He also has to his credit many publications in Russian mathematical journals.

François-Serge Lhabitant, PhD, is the chief investment officer at Kedge Capital in Jersey. He was formerly a member of the senior management at Union Bancaire Privée, Geneva, Switzerland, where he was in charge of quantitative risk management and subsequently, of the quantitative research for alternative portfolios. Prior to this, François-Serge was a director at UBS/Global Asset Management, in charge of building quantitative models for portfolio management and hedge funds. François-Serge is currently a professor of finance at the University of Lausanne, Lausanne, Switzerland and at the EDHEC Business School, Lille, France. He was formerly a visiting professor at the Hong Kong University of Science and Technology, held the Deloitte & Touch chair on risk management at the University of Antwerp, Antwerp, Belgium, and was an associate professor of finance at Thunderbird, the American Graduate School of International Management. François' specialist skills are in the areas of quantitative portfolio management, alternative investments (hedge funds) and emerging markets. He is the author of several books on these subjects and has published numerous research and scientific popularization articles. He is also a member of the Scientific Council of the Autorité des Marches Financiers, the French regulatory body.

Robert W. McGee is the director of the Center for Accounting, Auditing and Tax Studies at Florida International University in Miami. He has published more than 50 books and more than 480 scholarly papers in the fields of accounting, taxation, economics, law, and philosophy. He recently published two books on corporate governance, titled *Corporate Governance in Transition Economies* and *Corporate Governance in Developing Economies*, both published by Springer.

Max Moroz graduated from Anderson School, UCLA, Los Angeles, California with a master of science in management. He is currently the vice president, investment management and research, at Research Affiliates. His work involves the evaluation, selection, and implementation of various models related to equity alpha, risk, transaction costs, fixed income term structure, global asset allocation, etc. He also assists several institutions with the implementation of Research Affiliates' investment products. Previously, Max provided financial and risk management consulting to the energy industry at PriceWaterhouseCoopers, and valuated technology start-ups at a venture capital firm.

Sandra Mortal is an assistant professor of finance at the Fogelman College of Business. She has published and forthcoming articles at leading research journals such as the *Review of Financial Studies*, the *Journal of Financial Markets*, and the *Journal of Corporate Finance*. Her research has been featured by *Forbes* magazine, and won her a best paper award at the Financial Management Association. She also presents regularly at national and international conferences/seminars.

Stanley Mutenga is a chartered risk manager with extensive knowledge in reinsurance underwriting, structured insurance solutions, and insurance company financing. He is a senior lecturer in risk and insurance at Cass Business School, City University London. His major research interests are insurance company risk analysis and funding, and insurance company operations efficiency and strategy. He has previously served on the Finance Committee of the Butterstile Primary School Board of Governors, on the Aductus Housing Group Board as a chair of the Audit Committee, as a lecturer at Salford Business School, and as a visiting lecturer at Copenhagen Business School and at the University of Oslo. Stanley is a Children's Church leader at Celebration Churches International, London, and has also served as a cell leader, trustee, children's church leader, and sound engineer at a number of churches. He obtained his BCom (honors) in insurance and risk management with distinction from the National University of Science and Technology, Zimbabwe; his MSc in insurance and risk management with distinction from Cass Business School; and his PhD in insurance studies also from Cass Business School. His research has been published in various international journals.

Edwin Neave is an emeritus professor of finance at Queen's School of Business, Queen's University. He is the author of numerous articles and books focusing on asset pricing and on the theory of financial systems. His programs in banking education are currently used in more than 40 countries. He is also an honorary fellow of the Institute of Canadian Bankers. Neave's publications have appeared in leading financial and economic journals such as the *Journal of Economic Theory*, the *Canadian Journal of Economics*, the *Journal of Financial and Quantitative Analysis*, the *Journal of Banking and Finance*, and *Management Science*.

Duc Khuong Nguyen is a professor of finance and head of the Department of Economics, Finance and Law at ISC Paris School of Management, France. He holds his MSc and PhD in finance from the University of Grenoble II, France. His principal research areas concern emerging markets finance, market efficiency, volatility modeling, and risk management. He has published in refereed journals such as *Review of Accounting and Finance*, the *American Journal of Finance and Accounting, Economics Bulletin*, and *Bank and Markets*.

Nicolas Papageorgiou is an associate professor of finance at HEC Montreal and a director of research at Desjardins Global Asset Management. He also serves as the codirector of the DGAM-HEC Alternative Investment Research Center. He is a managing director of ReplicQuant, a consultancy firm that develops quantitative tools for the analysis of alternative investments. Professor Papageorgiou has published articles in leading academic and practitioner journals, and has been invited to present his research at numerous conferences in North America and Europe.

Christopher Parsons began his career with the Guardian Royal Exchange Assurance (now AXA) Group and joined Cass Business School in 1989. He is an officer and council member of the Chartered Insurance Institute, and a senior examiner and joint secretary of the CII Examiners' Committee. In 1998, he was awarded the CII Morgan Owen Medal for the best research paper produced by a member of the institute, which has a worldwide membership of around 70,000. He is the author of two books on insurance law, contributes regularly to insurance and legal journals, and lectures widely in the United Kingdom and overseas on the topics of law and insurance. Chris is the editor of the *Journal of Insurance Research and Practice* and a visiting professor at Warsaw University of Insurance and Banking.

Jack Penm is currently an Academic Level D at the Australian National University (ANU). He has an excellent research record in the two disciplines in which he earned his two PhDs, one in electrical engineering from University of Pittsburgh, and the other in finance from ANU. He is an author/coauthor of more than 80 papers published in various internationally respectful journals.

Robert Powell has 20 years of banking experience in South Africa, New Zealand, and Australia. He has been involved in the development and implementation of several credit and financial analysis models in banks. He has a PhD from Edith Cowan University, where he currently works as a researcher and lecturer in banking and finance, in the School of Accounting, Finance and Economics.

Julia Reichert is a research associate of the Strategic Finance Institute at the European Business School International University in Germany. She is currently on leave at the Boston Consulting Group and holds a diploma in business administration from the WHU in Vallendar. Her main research interest is on asset-backed securities with a special emphasis on governance-related aspects.

Peter Roosenboom is a professor of entrepreneurial finance and private equity at the Rotterdam School of Management, Erasmus University, and a member of Erasmus Research Institute of Management (ERIM). He holds a PhD in finance from Tilburg University. His research interests include corporate governance, venture capital and initial public offerings. His work has been published in the *Journal of Corporate Finance*; *Contemporary Accounting Research*; the *European Financial Management Journal*; *Applied Economics*; *International Review of Financial Analysis*; the *Pacific-Basin Finance Journal*; the *Journal of Accounting & Public Policy*; the *International Journal of Accounting*; and the *Journal of Management & Governance*. He is the coeditor of the book *The Rise and Fall of Europe's New Stock Markets* that has appeared in the book series "Advances in Financial Economics." He has also contributed book chapters to books on initial public offerings, mergers and acquisitions, venture capital, emerging markets, and corporate governance.

Samir Saadi is a research associate and part-time instructor of finance at the Telfer School of Management, University of Ottawa. He is currently a finance PhD candidate at Queen's School of Business. His research interests include international finance, banking management, and executive compensation. His work has been published in finance and applied economics journals including the *Journal of Multinational Financial Management*; the *Journal of International Financial Markets*; *Institutions and Money*; *Global Finance Journal*; the *Journal of Theoretical and Applied Finance*; *Review of Financial Economics*; and the *International Journal of Managerial Finance*.

Florent Salmon is a vice president, alternative management, and is overall responsible for managing the funds of hedge funds as well as the research program for the synthetic replication of investment strategies. From 1994 to 1996, he was a risk management analyst at the National Bank of Canada. In 1996, he became a pension fund analyst at Canadian National, where he specialized in the analysis and execution of securitization transactions and exotic options until 1998. He joined the alternative management team at Desjardins Asset Management in 1998 and has remained in charge since 2002. Florent Salmon also has a master's degree in finance from HEC Montréal.

John Simpson is an associate professor of finance and banking with the School of Economics and Finance at Curtin University, Perth, Western Australia. His received his PhD from the University of Western Australia on the subject of modeling international banking risk ratings. His research interests include risk and risk management in international banking and finance. He has over 30 internationally refereed publications in a growing list of reputable and respected financial economics books and journals.

Marco Sorge is an economist at the World Bank with experience in research and financial operations. Prior to joining the World Bank Group, he worked for the Bank for International Settlements, where he contributed background research for Basel II and for the Inter-American Development Bank, where he codesigned a Monte-Carlo simulation model for risk management during the Argentine crisis. Marco holds a PhD in economics from Stanford University from where he graduated in 2003 under the direction of Joseph Stiglitz.

Philip A. Stork is a visiting professor of finance at Massey University, Auckland, New Zealand. He has previously worked as a professor of finance at Erasmus University Rotterdam from where he also received his PhD.

He has worked in various roles for banks, brokers, and market makers in Europe, Australia, and the United States. His academic work has been published in major international journals such as the *European Economic Review, Economics Letters,* the *Journal of Applied Econometrics,* the *Journal of Fixed Income,* and the *Journal of International Money and Finance.*

R.D. Terrell is a financial econometrician and officer in the general division of the Order of Australia. He served as a vice-chancellor of the ANU from 1994 to 2000. He has also held visiting appointments at the London School of Economics; the Wharton School, University of Pennsylvania; and the Econometrics Program, Princeton University. He has published a number of books and research monographs and around 80 research papers in leading journals.

Eric Tymoigne is an assistant professor in the Department of Economics at California State University, Fresno. He received his PhD from the University of Missouri-Kansas City with a specialization in monetary theory and financial macroeconomics. His current research agenda includes money matters (nature, history, and theory), the detection of aggregate financial fragility and its implications for central banking, and the theoretical analysis of monetary production economies. He has published in the *Journal of Post Keynesian Economics* and the *Journal of Economic Issues,* has contributed to several edited books, and has recently published a book titled *Central Banking, Asset Prices, and Financial Fragility.*

Casper G. de Vries holds the chair of monetary economics at Erasmus University Rotterdam. He is also the vice dean of research and education at the Erasmus School of Economics. He is a fellow and board member of the Tinbergen Institute and serves as a member of the EMU Monitor group. He received his graduate training at Purdue University; he has held positions at Texas A&M University, K.U. Leuven, and he has been a visiting scholar at several European and American research institutes. His research has been published widely in leading internationally refereed journals.

Niklas Wagner is professor of finance at Passau University, Germany. After receiving his PhD in finance from Augsburg University in 1998, he held postdoctoral visiting appointments at the Haas School of Business,

U.C. Berkeley, and Stanford GSB. Academic visits also led him to the Center of Mathematical Sciences at Munich University of Technology and to the Faculty of Economics and Politics, University of Cambridge. His research interests include empirical asset pricing, applied financial econometrics, market microstructure, as well as banking and risk management. Professor Wagner has coauthored various contributions in finance, including articles in *Economic Notes, Quantitative Finance*; the *Journal of Banking and Finance*; and the *Journal of Empirical Finance*. He regularly serves as a referee for well-known journals in the area of financial economics. His industry background is in quantitative asset management with HypoVereinsbank and with Munich Financial Systems Consulting.

Jörg R. Werner is an assistant professor (Habilitand) at the Faculty of Business Studies and Economics at Bremen University (Germany). His work covers corporate governance and international accounting research, both from an empirical and a theoretical perspective.

Jennifer Westaway, BJuris LLB, MBioethics, PhD, is a lecturer with the School of Law & Taxation, Curtin University of Technology, Perth, Australia. Her research, publication, and teaching interests lie in the areas of international law, banking, finance, and trade, as well as employment and public health law. Jennifer has previously taught at the Law School at the University of Notre Dame in Perth, and has an extensive employment history in the corporate and finance sector.

Chendi Zhang is an assistant professor of finance at the University of Warwick, U.K. Prior to joining Warwick he taught at the University of Sheffield. He also served as a consultant/researcher at the World Bank and the International Finance Corporation. He has published in the *Journal of Corporate Finance*, the *Journal of Banking and Finance, Economics Letters*, and other such journals. His research interests include corporate finance, behavioral finance, socially responsible investments, and emerging economies. Chendi holds a PhD in financial economics from Tilburg University.

Andrew Zlotnik is a private asset management consultant in emerging markets investments. He started his career as an intern utilities analyst in the research department of the leading Russian investment bank, Troika

Dialog. After this, he was a leading economist and a leading risk manager at Moscow Interbank Currency EXchange (MICEX). He is a postgraduate student at the Central Economics and Mathematics Institute of the Russian Academy of Sciences. He holds a BSc in economics from Lomonosov Moscow State University, Moscow, Russia.

Contributors

David E. Allen
School of Finance and Business
   Economics
Edith Cowan University
Perth, Western Australia
Australia

Mohamed El Hedi Arouri
Faculty of Law, Economics
   and Management
University of Orleans
Orleans, France

Ruggero Bertelli
Richard Goodwin Faculty
   of Economics
University of Siena
Siena, Italy

Bastian Breitenfellner
Department of Accounting,
   Finance and Taxation
Passau University
Passau, Germany

Tyrone M. Carlin
Faculty of Economics and Business
University of Sydney
Sydney, New South Wales
Australia

Giuseppe Catenazzo
Center for Applied Research
   in Management
Haute École de Gestion of
   Geneva
Geneva, Switzerland

Jorge Antonio Chan-Lau
Center for Emerging Market
   Enterprises
The Fletcher School
Tufts University
Medford, Massachusetts

and

International Monetary Fund
Washington, District of
   Columbia

Carolyn V. Currie
Public Private Sector Partnerships
   Pvt Ltd
Mosman, New South Wales
Australia

Dean Fantazzini
Moscow School of Economics
Moscow State University
Moscow, Russia

Nigel Finch
Faculty of Economics and Business
University of Sydney
Sydney, New South Wales
Australia

Guy W. Ford
Faculty of Economics and Business
Macquarie Graduate School
    of Management
Sydney, New South Wales
Australia

Emmanuel Fragnière
Center for Applied Research
    in Management
Haute École de Gestion of Geneva
Geneva, Switzerland

Xanthi Gkougkousi
Department of Accounting
    and Control
Rotterdam School of Management
Erasmus University
Rotterdam, the Netherlands

Werner Gleissner
FutureValue Group AG
Leinfelden-Echterdingen, Stetten
Germany

Leo F. Goodstadt
Trinity College Dublin
The University of Dublin
Dublin, Ireland

Elena Grammenou
Haute École de Gestion de Genéve
Geneva, Switzerland

Ulrich Hommel
European Business School
International University Schloß
    Reichartshausen
Oestrich-Winkel, Germany

Christian Hoppe
Commerzbank AG
Frankfurt, Germany

Mahmud Hossain
Fogelman College of Business
    and Economics
University of Memphis
Memphis, Tennessee

Jason Hsu
Department of Finance
University of California
Los Angeles, California

Pankaj K. Jain
Fogelman College of Business
    and Economics
The University of Memphis
Memphis, Tennessee

Vicente Jakas
Deutsche Bank AG
Frankfurt, Germany

Fredj Jawadi
Department of Management
    and Economics
Amiens School of Management
Amiens, France

and

University of Paris Ouest Nanterre
    La Defense
Paris, France

Wilhelm K. Kross
Kross Consulting
Glashuetten, Germany

and

PMI Frankfurt Chapter
Bad Homburg, Germany

Alexander Kudrov
Department of Economics
Higher School of Economics
Moscow, Russia

François-Serge Lhabitant
Kedge Capital
Jersey, United Kingdom

and

École De Hautes Études
    Commerciales du Nord
Nice, France

Robert W. McGee
Center for Accounting, Auditing
    and Tax Studies
Florida International University
Miami, Florida

Max Moroz
Investment Management and
    Research
Research Affiliates
Newport Beach, California

Sandra Mortal
Fogelman College of Business and
    Economics
The University of Memphis
Memphis, Tennessee

Stanley Mutenga
Risk and Insurance at Cass
    Business School
City University London
London, United Kingdom

Edwin Neave
Queen's School of Business
Queen's University
Kingston, Ontario, Canada

Duc Khuong Nguyen
Department of Economics,
    Finance and Law
ISC Paris School of Management
Paris, France

Nicolas Papageorgiou
Department of Finance
HEC Montreal
Montreal, Quebec, Canada

and

Desjardins Global Asset
    Management
Montreal, Quebec, Canada

Christopher Parsons
City University of London
Cass Business School
London, United Kingdom

Jack Penm
College of Business
    and Economics
Australian National University
Canberra, Australian Capital
    Territory, Australia

Robert Powell
School of Accounting, Finance
and Economics
Edith Cowan University
Joondalup, Western Australia
Australia

Julia Reichert
European Business School
International University Schloß
Reichartshausen
Oestrich-Winkel, Germany

Peter Roosenboom
Department of Financial
Management
Rotterdam School of
Management
Rotterdam, the Netherlands

Samir Saadi
Telfer School of Management
University of Ottawa
Ottawa, Ontario, Canada

Florent Salmon
Desjardins Global Asset
Management
Montreal, Quebec, Canada

John Simpson
School of Economics and Finance
Curtin University
Perth, Western Australia
Australia

Marco Sorge
World Bank
Washington, District
of Columbia

Philip A. Stork
Department of Economics
and Finance
Massey University
Auckland, New Zealand

R.D. Terrell
National Graduate School of
Management
Australian National University
Canberra, Australian Capital
Territory, Australia

Eric Tymoigne
Department of Economics
California State University
Fresno, California

Casper G. de Vries
Erasmus School of Economics
Erasmus University Rotterdam
Rotterdam, the Netherlands

Niklas Wagner
Department of Accounting,
Finance and Taxation
Passau University
Passau, Germany

Jörg R. Werner
Faculty of Business Studies and
Economics
Bremen University
Bremen, Germany

Jennifer Westaway
School of Law & Taxation
Curtin University of Technology
Perth, Western Australia
Australia

Chendi Zhang
Warwick Business School
(Finance Group)
University of Warwick
Coventry, United Kingdom

Andrew Zlotnik
Central Economics and
Mathematics Institute
Russian Academy of Sciences
Moscow, Russia

I

Banking Gross Negligence and Shadow Banking System

CHAPTER 1

The Banking Crisis of the New Millennium— Why It Was Inevitable

Carolyn V. Currie

CONTENTS

Under the new theory of financial regulation developed by this author, collapse in the United States financial system was inevitable. Removal of protective measures accompanied by the failure to increase prudential measures, accompanied by conflicting state/federal relations, barriers to entry to the real estate industry, together with regulatory confusion resulting in regulatory arbitrage, set the scene for not only erosion from within, but also for transnational crime to destabilize the entire fabric of the world financial system. Understanding the causes helps in rectifying the financial architecture. This chapter details the enormous numbers of regulatory models that can exist and outlines formulas to assess the type of regulatory model that should be imposed on a system after assessing the stage of economic and social development.

1.1 INTRODUCTION

Today, we are greeted everyday in the media with common excuses for the global financial crisis (GFC). We admit a lack of prudential supervision and blame laissez-faire capitalism. However, we never admit to the major flaws in the regulatory structure as well as the political philosophy of the largest world economy that has almost destroyed a decade of progress.

The fact is that not only in the United States have state governments been left to administer the originators of mortgages and credit default swaps, but the federal government has never complied fully with the Financial Action Task Force money laundering provisions. By not complying they have created the ability to abort such legislation using cross-border transactions and complex financial instruments. In addition not only was the implementation of Basel II lobbied against by U.S. financial interests since 1999 and never instituted in the United States financial system, which could have highlighted the overlending to the residential mortgage market, but since 2003 SEC recommendations to control the hedge fund industry were never instituted.

A further flaw in the United States market is the fact that their real estate industry is not subject to market forces and that the tax system can encourage overborrowing. For instance, in Australia, the buyer pays no premium to a broker to find a property. All fees are negotiable and rarely go above 2%, not the 6% that is paid in the United States. In Australia, real estate agents can offer properties nationally and are subject to federal legislation. Overseas buyers can buy in subject to certain requirements—for instance, properties below A$400,000 are protected, and overseas investors in the unit/apartment market cannot buy second-hand real estate. Moreover the exemption from capital gains tax, but the non-expensing of interest on home loans against income, encourages owners to never walk away from their homes. Australian banks must prove default and cannot advertise a home merely on what is owed. They must auction and seek a fair market price from an arm's-length buyer.

This chapter discusses the types of regulatory models that exist, how they should be adapted to each individual country's stage of economic and social development, and finally to make recommendations as to the appropriate design to correct existing flaws in the United States regulatory model that have spread the crisis that developed in its core banking units worldwide.

1.2 BEHIND THE VEIL AND LESSONS THAT SHOULD HAVE BEEN LEARNED

Systemic failure in advanced not just emerging nations is now an ongoing problem. Hence we need to define regulatory failure more than ever, analyze its causes, and develop methods of measuring its severity in order to assess how much and when governments should intervene.

We also need to develop an understanding of the vast range of regulatory models through the introduction of a taxonomy, which will aid in the selection of methods for early diagnosis and prevention. It is remarkable that I should have been writing in this area for many years, have lobbied governments and regulators, was invited on June 2, 1999 to the announcement of Basel II, but then witnessed governments ignoring the regulatory literature that predicted the GFC (Currie, 1997, 1998, 2000, 2001, 2005a,b, 2006).

The lessons that should have been learned from the past financial crises (see Caprio et al., 1999, 2003 for a list of such crises) were the necessity for a staged approach to deregulation of protective measures, accompanied by an integrated, enhanced, and even approach to the development of prudential supervision and the establishment of an early warning system.

The evolution of financial services sectors in advanced and emerging nations has been marked by a change in the regulatory model from one dependent on protective measures to one more reliant on prudential supervision. It was to adjust such models for the following factors:

- The growth of financial conglomerates that demanded coordination among different regulatory bodies and levels of government and also meant that regulation, whether prudential or protective, must concentrate on ALL financial institutions. The failure of the U.S.

government to adjust its model after repealing the Glass–Steagall and McFadden Acts, meant that highly leveraged investment banks and insurance companies could bring major banks to the point of extinction, as described in Table 1.1 (Source: WSJ, 9/23/08).

- Unregulated nonbank financial institutions (NBFIs), by offering a regulatory black hole, have now been proven to provide a home to the new transnational criminal. Witness the Ponzi schemes now uncovered. Mortgage lenders, hedge funds, and unregulated finance companies have spread financial risk and promoted contagion.

Investment banks were allowed unprecedented levels of borrowings—Bear Stearns was levered at 31:1, Lehman Brothers at 34:1, Fannie Mae/Freddie Mac at 45:1, Merrill Lynch at 46:1, Goldman Sachs at 26:1, and Morgan Stanley at 20:1. None of these investment banks complied with Basel II

TABLE 1.1 NBFIs in the United States Which Triggered the GFC and Their Subsequent Fate

Bought	Failed	Transformed
A.G. Edwards		Goldman Sachs
Bought by Wachovia in 2007		Became a bank holding company in 2008
Bear Stearns		
Bought by J.P. Morgan Chase in 2008	Lehman Brothers Failed in 2008	Morgan Stanley Became a bank holding company in 2008
Donaldson, Lufkin & Jenrette Bought by Credit Suisse in 2000	(brokerage operations were bought by Barclays)	
First Boston Bought by Credit Suisse in 1988		
Paine Webber Bought by UBS in 2000		
Salomon Brothers Bought by travelers in 1997		

Note: Source was an informal lecture given by Prof. Demetri Kantarelis preceding a series of speeches by the author recorded at the Worcestor Polytechnic, Boston, MA, September 29, 2008 and Assumption College, Worcester, MA, September 30, 2008. www.socialweb.net/Events/82723.lasso. Speech available upon request to the author: www.carolyncurrie.net.

capital adequacy requirements and were geared far more than the normal on-book average of 22:1. This is why many of these investment banks disappeared, as described in Table 1.1 and the question still remains—why were they allowed to operate in such an unregulated manner given their established possibility to trigger systemic crises as proven over and over again during the 1980s with the Savings and Loans fiasco, and in the 1990s with the Long Term Credit Fund.

Why has regulatory failure occurred in 2007–2009 and spread to a global financial crisis? Causes lie not only in laissez-faire capitalism, but also in accounting rules such as mark to market, which promoted the idea that huge profits could be recognized before being realized by churning nonliquid assets. Contributing to this quagmire was regulatory confusion between state and federal legislation, lobbying against controls on residential lending, and flaws in the literature on regulatory failure that took no account of the present stage of development of an advanced economy. This translated into the lack of recognition that advanced economies had not adjusted their regulatory models as their economic and social infrastructure underwent some profound changes.

Where the system has been liberalized and globalized, where insurance, funds management, and banking activities have been merged, regulatory failure can occur in not just the traditional banking industry but in insurance, by virtue of either cross-linking guarantees or the effect on confidence. Consider also the regulatory confusion created by too many regulatory bodies and state vs. federal controls. Regulators failed to recognize the problems of insurance companies selling credit default swaps being supervised by state bodies who were also allowed to supervise without appropriate federal oversight the sale of mortgage-backed securities (MBS), not only across state borders but also across countries.

1.3 THE ROLE OF FINANCIAL INSTITUTIONS IN REGULATORY FAILURE

Causes of bank crises range from lack of investor and depositor confidence precipitated by the perception of a deterioration in asset quality. The latter is most commonly caused by excessive industry, product, country risk concentration, or intergroup lending all resulting from lack of credit control, lack of sound lending policies, and lack of internal control procedures checked upon by external auditors and the central bank supervisors. Apart from asset quality, large diversifications into new areas of business where the institution lacks expertise are reasons that financial institutions as well as corporates get into difficulties. "The risks in overtrading in banks where either the foreign exchange positions are not controlled or the option writing not fully appreciated is enormous, and spectacular losses have been made by banks in these areas. Another classic failing of financial institutions is liability mismanagement."

Within this framework of causes of bank crises, fraud is the most difficult thing for the bank analyst to predict. It is not impossible to see fraud playing a role in the entire business of toxic assets.

But the question arises of how to measure the seriousness of a financial crisis caused by regulatory failure before first prescribing the medicine, and second before taking corrective action to adjust the regulatory model to prevent a repeat.

1.4 MEASUREMENT OF REGULATORY FAILURE

The effectiveness of regulation and its corollary, failure, could be assessed against measures of stability such as exits and failures of banks, when the reference is to winding-up, or sale due to insolvency. It can also be measured by a lack of community confidence in banks or by a weakened banking system beset by poor profitability and low capitalization, and stakeholder losses (OECD, 1992).

Alternatively, a microeconomic approach could be used by regulators based on agency theory, which justifies measuring risks from balance sheet ratios. This allows assessment of how well the interests of depositors, investors, customers, creditors, and regulators have been satisfied as to the appropriate bank behavior in managing the four principal risks that can threaten the stability, safety, and structure of the financial system, namely, credit risk, liquidity risk, interest rate risk, and leverage risk. Also used is the assessment of investing and financing patterns, and dividend and executive compensation ratios.

1.5 MEANING OF REGULATION, DEREGULATION, AND LIBERALIZATION

What is clear is that we need a definition of the severity of regulatory failure, which can result in financial crises. Costs and benefits of regulatory failure or its opposite can be redefined in terms of new microeconomic indices of bank performance and macroeconomic indices of economic and social development (see Bordo et al., 1993). In this chapter, the scale proposed to measure regulatory failure is an arithmetic scale from 1 to 10 where the series represents an escalation in severity as follows:

1. Volatile security and asset prices resulting from failure by central banks to target inflation and prudentially supervise the risk behavior of financial institutions

2. Failure in major corporates, and/or a minor bank and/or a major nonbank financial institutions

3. Removal of funds from minor banks and/or nonbank financial institutions to major banks or to alternative investments such as prime corporate securities

4. Removal of funds from corporate securities to major banks, precipitating further volatility or at worst a stock market crash

5. Removal of funds from major banks to government-guaranteed securities, such as postal savings accounts in Japan

6. Flight to cash and or gold

7. Contraction in lending leading to lower economic growth, disruption to savings, and investment

8. Failure in major nonfinancial institutions, and failure in major nonbank financial institutions

9. Failure in major banks

10. All of the above together with a currency crisis leading to rapid depreciation, rescheduling of external country debt, higher inflation, and negative economic growth

Despite defining financial crises according to the OECD (1991) schema and using the above scaling of crises, a new theoretical framework is needed to aid in the prevention of such crises. The debate deregulation dates back to the 1970s—see Mitnick (1980). Deregulation does not necessarily mean the absence of regulation. Different methods of deregulation can be used. Formal deregulation, which is planned can occur in four ways: guided or unguided wind down; disintegration with transfer of programs; stripping of functions; and a catastrophic ending. In assessing how to change regulatory models, we need to distinguish between prudential and protective measures. What will assist in the discussion of why the regulatory model in the United States imploded and exploded is an understanding of the vast range of regulatory models that can exist. I have developed such a model as part of understanding why the Australian system nearly collapsed in 1991, calling for a supra regulator both nationally and internationally in an inquiry held in 1991–1992 (see the Martin Inquiry, 1992). This taxonomy is explained in Section 1.6.

1.6 TAXONOMY FOR CLASSIFYING FINANCIAL SYSTEMS

The taxonomy for classifying financial systems (Currie, 2000) distinguishes between prudential supervisory systems, which have different methods of compliance audits (strong and weak), sanctions (strong and weak) and enforcement modes (seven types), and protective measures (institutional vs. discretionary in various weak/strong combinations). These permutations and combinations give a total of 140 models ($2 \times 2 \times 7 \times 5 = 140$).

Enforcement modes range from conciliators to strong enforcers, representing a scale from weak to strong. *Conciliatory modes* are ones where law enforcement is rejected and conciliation is used to resolve disputes. *Benign Big Guns* are modes whereby enormous power is given in terms of confiscation, takeover of activities, seizure, increasing operational rules, and banning of products. Powers are rarely used—the threat is sufficient. This model has been called "regulation by raised eyebrows" or "by viceregal evasion" (Grabosky and Braithwaite, 1986). *Diagnostic Inspectorates* are modes where supervision is carried out by encouraging self-regulation by well-qualified inspectors detecting noncompliance. The goal is a cooperative relationship. *Token Enforcers* uses enforcement modes where cooperative and self-regulation is not important. *A Detached Token Enforcement mode* is more rule-book oriented, training staff, prosecuting more, seizing assets, and targeting repeat offenders. *Detached Modest Enforcement* modes involve rule-book inspections, and steady flow of prosecutions with modest penalties. *Strong Enforcers* use all forms of enforcement license suspensions, shut down of productions, injunctions, and adverse publicity as well as high penalties.

Sanction types can be industry based whereby there is consultation reappropriate preventative measures, discussion papers with written and oral input sought from industry via the Exposure Draft process, imposition of codes of conduct, imposition of direct controls, such as new prudential ratios, new interest rate ceilings, new reserve ratio rules, new capital adequacy rules, changes to Competition laws, cooperation with sister agencies to initiate prosecution to establish a precedent, changes to licensing rules, changes to banking laws, enforced divestitures or acquisitions (for instance of non bank financial institutions) on an industry basis, and finally nationalization of the banking industry. Sanction types can also be *firm based*. The broader part of the first firm-based pyramid of sanctions consists of the more frequently used regulatory sanctions coaxing compliance by persuasion. The next phase of enforcement escalation is a warning letter followed by imposition of civil monetary penalties, criminal prosecution, plant shutdown, or temporary suspension of a license to operate. Each stage is followed only if there is failure to secure compliance. At the top of the firm-based enforcement pyramid of sanctions, there is permanent revocation of licenses.

Knowledge by a firm of the enforcement pyramid actually increases the effectiveness of the enforcement. If a banking regulator only has the drastic power to withdraw or suspend licenses as the one effective sanction, it is often politically impossible and morally unacceptable to use it. Withdrawal of a license involuntarily in banking would result in that bank losing the implicit or explicit guarantee of the central bank, with a likely bank run or cessation of activities, resulting in possible contagion effects.

Hence, an incorrectly designed regulatory model can create a paradox of extremely stringent regulatory laws at times resulting in a failure to regulate. The design of the regulatory sanction pyramid should ensure that the information costs to the regulated firm of calculating the probability of the application of any particular sanction acts as a barrier and that there are sufficient politically acceptable sanctions to match escalations of noncompliance with escalations in sanctions by the state (Ayres and Braithwaite, 1992, p. 36).

Compliance audits range from weak to strong and can be applied at firm or industry level. "Firm-level compliance audits" consist of off-site examinations only using information supplied by the bank itself plus audited accounts, supplied by company-appointed auditors. External (company appointed) auditors then should supply additional data and report directly to the central bank if they are concerned regarding a bank's risk management. Note that this system includes no method of monitoring or controlling company-appointed auditors in the event of deliberate or unintentional errors in their reports. In addition prearranged on-site inspections of certain risk management aspects of a bank by regulators and/or by banking law auditors can be organized by the central bank or can also arrange surprise or spot inspections to check a bank's risk management by banking law auditors who can liaise with company law auditors. A further escalation of concern could involve surprise on-site inspections of all aspects of banks' risk management systems, when auditors are appointed by the central bank, not from the same firm as the banking law auditors, or the company-appointed auditors. A higher level could involve surprise on-site inspections but with reports only going to the bank and the central bank. The top censure could involve published reports. *Industry Compliance Audits* involve industry hearings to check on best practice and to bring banks into alignment (the best most recent example is the examination of the managing directors of the eight largest banking firms in the United States on February 11, 2009). Special reports can be commissioned by the government using outside consultants; governments can initiate inquiries at which evidence is sought form the public; Royal Commissions can be held presided over by a judge appointed by the government; and finally a commission presided over by a specialist appointed by an outside agency (e.g., the BIS or the OECD or the IMF) or presided over by another country (for instance, the Niemeyer Commission in Australia post the Great Depression and the Royal Commission into the collapse of HIH Insurance Group in 2001).

Protective measure types are usually all industry based; each type of protective measure has a range of weak to strong arrangements. *Discretionary measures* involve safety net schemes, apart from deposit insurance schemes. Safety net schemes can include one or all of the following. If all are included the regulatory model or system becomes increasingly stronger: implicit or explicit guarantees, special shareholder liability, regulatory intervention, both to ensure depositor protection and to prevent runs. Liquidity support arrangements comprise policies toward lender of last resort and toward the cheque clearing accounts that most banks hold at the central bank (known as Exchange Settlement Accounts). Activity restrictions were an old way of molding bank behavior such as restrictions on permissible activities, restrictions on branching, restrictions on equity holdings, regulations creating market segmentation, interest rate caps and floors imposed on borrowing and lending, restrictions on interlocking directors, and restrictions on banking conduct through either a voluntary or legislated code of conduct. *Institutionalized measures* include disclosure regulations such as secrecy provisions regarding client details as well as rules relating to what information a bank must publicly disclose. There are two types of information demanded: reports to regulators, such as the central bank, a banking ombudsman, and reports to shareholders, which are lodged with companies and securities regulators and also available to other stockholders, such as depositors and consumers. Institutionalized Deposit Insurance Schemes can range from weak versions using private insurance, flat fees, partial coverage, unfunded, voluntary to the strong version which is a public scheme, charging risk-related fees, offering full coverage, being fully funded and compulsory. Each type of protective measure has a range of weak to strong arrangements, but an appropriately designed regulatory model is thought out in advance and designed to match the economic and social development of a nation (see below). The worst mistakes are made by suddenly changing the model in response to an emergency. For instance, in Australia the sudden introduction of a deposit guarantee on all deposits and borrowings by Australian banks caused not only a run on mutual funds, but a flight of foreign banks from Australia as they could not raise funds as cheaply as their Australian counterparts.

When such controls are removed whether protective or prudential, and whether by intent or neglect, such a phenomenon is known as deregulation. It can and has had some different outcomes ranging from

- Return, risk, and efficiency measures remaining the same

- Return measures increase but so do risk levels but not to a level to threaten systemic stability, while efficiency measures decline but not to a dysfunctional level

- Return and risk decline, while efficiency measures improve

- Return measures decline, risk level increase while efficiency deteriorates

- Obviously the fourth outcome is suboptimal, which is why we need a new theory of financial regulation

1.7 THE NEED FOR A NEW THEORY OF FINANCIAL REGULATION

Why do we need regulation? Advanced nations have experimented with a total loosening of all types of protective and prudential measures and have learned the old lesson that regulation is required to promote a stable economic structure in order to prevent the price and output volatility that can lead to financial crises. Regulation is multifaceted and also involves establishing a regulatory model the role of which is to contain and mold the risk taking and management behavior of both financial and nonfinancial institutions as well as market participants through prudential supervisory systems appropriate to the strengths or weakness of the protective measures.

How then to design a model appropriate to a particular country's circumstances? My suggestion is "a four factor interactive system," which assumes a central role of financial institutions. The design of the regulatory model, called the M factor, must be taken as a starting point to the promotion of an economy and society up the development scale. The exact components of that model in terms of protective and prudential measures must be specified. It is postulated that there are three other factors: C, O, and H. These are necessary to the achievement of economic and social development in an emerging, transition, or advanced nation, but they are not sufficient without interaction which requires a feedback mechanism or adjustment process between the components of the envisaged system. This feedback mechanism promotes the advancement up a scale representing the degree of development of all four factors.

In the model output, Y1 represents economic development, defined as sustainable growth, which can be measured at the level of the individual by the increase in a maintainable and stable level of income per capita, at the corporate or institutional level by the increase in a maintainable and stable accumulated earnings per capita, and at the country level by improvements in the ratio of external debt and current account balance to gross domestic product (GDP), as well as increases in the level of maintainable and stable GDP per capita. A further output Y2 represents social development, defined as growth in the equitable distribution of wealth, which can be measured by the dispersion and distribution of per capita income, and participation in institutions, which could be measured by a scale ranking the democracy of the government of a country.

Inputs are the legal infrastructure—competition, bankruptcy, and commercial and criminal laws with legislative barriers to entry and exit to the finance sector to promote stability, measured by the Compendium of Standards. This is the C factor. Another input is the structure and pattern of ownership of both financial and nonfinancial institutions—the O factor, which requires a knowledge basis for understanding the rights and obligations of ownership, and leads to demand-led investment in education. The quality of human capital, in terms of education, knowledge, and skills is also vital. In this theory, human capital or the H factor involves not just increasing the capacity to learn, but enabling the individual to participate in a financial system so that social and organizational capital, or the interrelationships and systems for mediation and dispute resolution, can be adapted to increasing stages of development.

The importance of ownership structure to the financial system cannot be overemphasized. Widespread ownership creates synergy between shareholders, bondholders, and depositors interests. Concentration of the banking sector is also proven to promote stability. Changing ownership structures can be achieved through employee share ownership trusts, privatization with shares issued to employees, and through the increasing role pension or superannuation funds, publicly listed funds, and community and family groups. The necessity for the O factor lies not only in the improvements to efficiency and removal of political intervention (OECD, 1993), but also in the promotion of participation in adapting a development strategy to the needs and capacities of the underlying economic and societal systems (World Bank, 1995). Participation is required to build consensus and induce change from within (Stiglitz, 1998).

Within this model, there are certain "constraining factors." Government goals (G) influence the development of M and O. That is, they are policy-dependent factors. There is also the E factor, or the starting set of economic resources and infrastructure, both developed and undeveloped, that influences the attainment of economic growth and social development by limiting, influencing, and constraining C and H.

The new theory of financial regulation thus conceives a national economy as a set of interrelating systems and subsystems, which can be described in terms of a set of equations as follows:

1. If government goals or $G = X1 = f (S, S, S, Co, Cf)$ where the terms safety (S), stability (S), structure (S), convenience (Co) of users of financial services (such as access to a product or service) and confidence (Cf) are general public confidence in the financial system. These terms are adapted from Sinkey's (1992) theory of regulation and

2. Economic resources or $E = X2 = f (D, FDI, K, A)$, then

3. Economic development or $Y1 = M (X1). C (X2)$, and

4. Social development or $Y2 = O (X1). H (X2)$

The availability of economic resources devoted to economic and social development will depend on government funding, which may require deficit spending (D), direct foreign investment (FDI), private capital formation (K), and foreign aid (A).

Can this model be used to describe what is happening now and whether there are leakages from the system? The convention against Transnational Organized Crime (2001), also known as The Palermo Convention nonexclusively defined transnational crime to include money laundering and computer-related crime. However, the use of derivatives and nonregulated institutions is an area into which organized crime has and can move. Subprime criminality in the subprime crisis included fraud by brokers or borrowers; unscrupulous underwriters, brokers, and financial planners, who copped a sling from pushing through questionable mortgage applications; originators who exaggerated assets and income or concealed liabilities and outgoings and a proliferation of Ponzi schemes. Many of the lenders who have produced toxic assets or subprime loans were NBFIs whose success was measured by loan throughput.

Consider also the fact that many of the hedge funds, and other NBFIs use tax havens as their corporate base means that the checks and balances on them to ensure they are not being used by transnational criminals is deficient. In the Caribbean, 46.7% of funds are domiciled according to the Global Data Feeder.

Despite legislation prohibiting its bankers from assisting in the flight of capital and tax evasion, Switzerland still rates as a safe haven for ill-gotten gains—in every crisis funds leaving a country for Switzerland will cause currency depreciation of the exit country and appreciation in the tax haven.

1.8 CONCLUSION: A REGULATORY BLACK HOLE

Actions taken to date to relieve the crisis apart from stimulus packages and direct aid to banks, include open market operations to ensure member banks' liquidity; central banks reducing interest rates charged to member banks for short-term loans; reexamination of the credit rating process; muted regulatory action on lending practices, possible reforms to bankruptcy protection, tax policies, affordable housing, credit counseling, education, and the licensing and qualifications of lenders; future amendments to disclosure rules and promotion investor education and awareness and public debate.

However, in 2003 the SEC recommended when reviewing hedge funds certain actions that are still not implemented. They included

- Registering investment advisers under the Advisers Act, with strict disclosure requirements

- Introducing standards on Valuation, Suitability, and Fee Disclosure Issues relating to registered fund of hedge funds (FOHFs) such as those that placed monies with Madoff

- Limiting general solicitation in fund offerings to qualified purchasers

- Ensuring the SEC and the NASD monitor capital introduction services provided by brokers

- Encouragement of unregulated industries to embrace and further develop best practices

So there is obviously an urgent need to review why the SEC not only was ignored in its recommendations for investors advisors, but why the Federal Reserve did not object to state governments supervising mortgage originators and the writers of credit default swaps such as AIG. Also why the number of regulatory bodies that allowed regulatory arbitrage was never reviewed and integrated supervisors established as in Canada, Europe, Australia, and the United Kingdom. Why was Basel II lobbied against when it would have brought to an abrupt standstill the excessive concentration on mortgage lending in the United States financial system. Why also was the banking industry but not the real estate industry allowed to deregulate. Why has the U.S. market (FAS 157) never been reviewed when it allowed both incredible write ups and disastrous write downs. The list is endless but there is an obvious need to redesign the U.S. regulatory model from scratch.

So how should this be done? A detailed method is described in Currie (2005b). In essence this involves assessing the existing system in terms of inputs and outputs given constraining factors and then considering changes not only in terms of the regulatory model and its components, but also in terms of the existing human expertise and how it is motivated (note the dysfunctional effect of incorrectly designed executive compensation was one of the conclusions of my thesis reaffirmed by Bank of England studies—Currie, 1998), and also in terms of the legal infrastructure, government goals, and starting economic resources. Just as Australia found a flawed federal system responsible for regulatory black holes, so too should the new U.S. government reexamine its entire regulatory model in terms of this new theory of financial regulation.

REFERENCES

Ayres, I. and Braithwaite, J., 1992. *Responsive Regulation, Transcending the Deregulation Debate*. Oxford University Press, New York.

Bordo, M. D., Rockoff, H., and Redish, A., 1993. A comparison of the United States and Canadian banking systems in the twentieth century: Stability vs efficiency. National Bureau of Economic Research, Cambridge, MA. Working Paper 4546.

Caprio, G. and Klingebiel, D., 1999, 2003. Episodes of systemic and borderline financial crises. World Bank, Washington, DC. Policy Research Working Paper.

Currie, C. V., 1997. An analysis of changes in regulatory models governing the Australian financial system, 1973–1993. Australian Institute of Banking and Finance Conference Proceedings, Australia, Melbourne.

Currie, C. V., 1998. Reform of the Australian financial system—Will the Wallis proposals jeopardise systemic stability? Economic Papers, September.

Currie, C. V., 2000. The optimum regulatory model for the next millennium—Lessons from international comparisons and the Australian–Asian experience. In Gup, B. (Ed.), *New Financial Architecture for the 21st Century*. Quorum/Greenwood Books, Westport, CT.

Currie, C. V., 2001. Is the Australian financial system prepared for the third millennium? In Kantarelis, D. (Ed.), *Global Business and Economics Review—Anthology 2001*. Business and Economics Society International, Worcester, MA.

Currie, C. V., 2005a. The need for a new theory of economic reform. *Journal of Socio-Economics*, 34(4), 425–443, 1053–5357, Elsevier, August.

Currie, C. V., 2005b. Preventing financial crises: The need for a new theory of financial regulation. *ICFAI Journal of Financial Economics*, p. 6, ICFAI University Press, Vol. III.

Currie, C. V., 2006. A new theory of financial regulation: Predicting, measuring and preventing financial crises. *Journal of Socio-Economics*, 35(1), 1–170, (Elsevier), February.

Grabosky, P. and Brathwaite, J., 1986. *Of Manners Gentle—Enforcement Strategies of Australia Business Regulatory Agencies*. Oxford University Press, Melbourne, FL.

Martin, S. 1991, 1992. (Ed.). A pocket full of change: Inquiry into the Australian financial system. Report of the House of Representatives Standing Committee on Finance and Public Administration. Australian Government Publishing Service, Canberra.

Mitnick, B. M., 1980. *The Political Economy of Regulation*. Columbia University Press, New York.

OECD (Organisation for Economic Co-operation and Development), 1991. Systemic Risks in Securities Markets, OECD Publication, Paris.

OECD (Organisation for Economic Co-operation and Development), 1992. Banks under Stress. OECD Publication, Paris.

OECD (Organisation for Economic Co-operation and Development), 1993. Financial Conglomerates. OECD Publication, Paris.

Sinkey, Jr. J. F., 1992. *Commercial Bank Financial Management*. Maxwell MacMillan, New York.

Stiglitz, J., 1998. Must financial crises be this frequent and this painful? McKay Lecture, Pittsburgh, PA, September 23.

World Bank, 1995. *Structural and Sectoral Adjustment: World Bank Experience, 1980–1992*. A World bank Operations Evaluation Study, Washington, DC.

CHAPTER 2

The Effect of Monetary Policy on Stock Prices: The Subprime Mortgage Crisis

Xanthi Gkougkousi and Peter Roosenboom

CONTENTS

We examine the effect of the Fed's monetary interventions as a response to the subprime mortgage crisis. More specifically, we look at the stock price reaction of 129 financial institutions listed in the United States to cash injections and interest rate cuts and link this reaction to a number of firm-specific characteristics: size, exposure to the crisis and leverage. We find that interest rate cuts produced on average a 4.7% increase in the market value of the financial institutions while cash injections a 3.3% increase. Our results also offer partial support to the hypothesis that small and credit-constrained institutions with strong exposure to the crisis profited the most from the Fed's interventions.

2.1 INTRODUCTION

On January 22, 2008, the Federal Reserve Bank (Fed) cut the target rate by an almost unprecedented 0.75%, the largest cut since 1982. This was preceded by four interest rate cuts and followed by another four; in less than 8 months the federal fund rate target was reduced from 5.25% to 2% and the discount rate from 6.25% to 2.50%. The abovementioned interest rate reductions, along with liquidity injections have been part of a monetary intervention plan of the Fed, as a response to the continuously deteriorating economic conditions in the United States, caused by the subprime mortgage crisis.

The purpose of this study is to examine the effect of the monetary actions of the Fed during the subprime crisis on stock prices of financial institutions in the United States and to examine whether the Fed has been successful in its efforts to restore the confidence of investors in the market. We will also link the stock reactions of the financial institutions to company-specific variables, in order to see which institutions profited the most from the Fed's interventions.

Several papers have studied the effects of monetary policy on stock prices (Ganapolsky and Schmukler, 2001; Bernanke and Kuttner, 2005). However, to the best of our knowledge, there has been no study on the effects of the latest Fed's interventions on stock prices. The interest of this topic lies in the size of the interest rate reductions (up to 0.75%) and the fast and decisive manner in which the Fed slashed the interest rates as a response to the crisis; in a short period of time, the target and discount rates were reduced by almost 400%. Moreover, the Fed employed both cash injections and interest rate cuts in its effort to enhance liquidity, thus allowing us to make direct comparisons between these two types of monetary intervention.

The rest of the study is organized as follows. In Section 2.2, we shortly describe the subprime mortgage crisis and the reasons that lead to it as well as the central banks' interventions. Section 2.3 contains a comprehensive overview of the papers relevant to our topic. Next, we develop the hypotheses of our research in Section 2.4. Section 2.5 provides a description of the sample and the methodology and Section 2.6 presents the findings of the research. In Section 2.7, we summarize our results and make some suggestions for further research on this area.

2.2 CRISIS

The roots of the subprime mortgage crisis can be found in the beginning of the millennium. After the dot-com bubble burst in early 2000 and the attacks of 9/11, the U.S. economy was on the verge of recession. Central banks all around the world tried to fuel the financial system by creating capital liquidity and by lowering the interest rates. U.S. citizens were explicitly encouraged to support their economy by increased consumer spending, or as former president, Bill Clinton put it to "get out and shop." On top of that, securitization and the deregulation of lending further facilitated the expansion of the subprime mortgage market.

All the abovementioned coupled with rising house prices caused the number of mortgages and more specifically, the number of subprime mortgages to rise. Borrowers depended heavily on house price appreciation, which would allow them to refinance their debt at lower rates and take the equity out of the home for consumer spending. Subprime mortgages rose from $190B in 2001 to $625B in 2005, an increase of more than 300%, while by 2006 a fifth of all new mortgages were subprime.

Federal fund target rates started rising since mid-2004, while the inventory of unsold houses created by overbuilding caused house prices to start declining in 2006. The combination of declining house prices and the kick in of market-based interest rates when the teaser rates* ran out spurred an increase in the number of subprime defaults and foreclosures in the same

---

* Teaser rate is an initial rate on an adjustable rate mortgage that is typically below the market rate. Teaser rates are in effect only for a few months, after which a full indexed rate applies.

year. The liquidity in the market started drying up and the stock prices both in the United States and in Europe started dropping. The financial institutions were the ones hardest hit by the crisis due to their strong exposure either in the form of mortgage-backed security holdings or in the form of mortgage loans. The first casualty was American Home Mortgage Corporation, which filed for bankruptcy on August 6, 2007, followed by more bankruptcies, write-downs and bank runs both in the United States and in Europe.

As a response to the rapidly deteriorating economic conditions, the Fed and its counterparts across the Atlantic intervened, by either injecting money to the markets or by lowering interest rates, in order to boost the dried liquidity and to encourage borrowing. The two central banks that mainly reacted were the Fed and the European Central Bank (ECB). They followed different strategies; ECB being less aggressive mainly injected cash in the market, while the Fed along with cash injections slashed the federal fund rate target and the discount rate.

In order to understand the monetary choices of the Fed, it is important to recognize the role of the federal fund and the discount rate and their effects on the economy. The federal fund rate is the interest rate at which depository institutions lend balances at the Federal Reserve to other depository institutions overnight and the discount rate is the interest rate charged to banks on loans they receive from the Fed. The federal fund target rate cannot be imposed on financial institutions; however, a change in the target rate is a clear signal of a change in the Fed's monetary stance.

Previous studies have shown that both the target and the discount rate changes have a strong effect on the real economy (Bernanke and Blinder, 1992). Their influence can be transmitted through the following channels: the interest rate and the wealth channel. According to the interest rate channel theory, the reductions in the discount and target rates will eventually cause an adjustment in the real interest rates, thus affecting borrowing and investments. The adjustments in real interest rates are part of the financial institutions' portfolio management; they have to adjust their interest rates, in order to maintain competitiveness and to generate profit. For example, a decrease in policy rates will cause a downward adjustment in market rates, which will lower the cost of borrowing and investments and will finally boost investing and consumption activities. The wealth channel proposes that interest rate changes cause an opposite change to long-lived assets (stocks, bonds, and real estate), hence affecting the wealth of the households and influencing spending and investments (Kuttner and

Mosser, 2002). Thus, a reduction in the interest rates is expected to cause an increase in the value of assets such as stocks, which will increase the availability of money and hence will boost spending and investments. The same predictions hold for the existing study. We expect that monetary interventions in the form of interest rate cuts and cash injections will cause an increase in stock prices of the financial institutions in the short run, which can be interpreted as a sign of increased investors' confidence. We will also compare the effects of interest rate cuts to those of liquidity injections and will try to explain the reaction of the financial institutions based on variables such as their size or their exposure to the crisis.

2.3 LITERATURE REVIEW

This section will focus on a presentation of the most important articles related to monetary interventions and subsequent stock reactions. We will thus gain a better understanding of the topic under consideration and we will be able to better formulate the hypotheses of our research.

One of the first articles on the effects of monetary policy changes on stock returns is that of Rozeff (1974), who reports a strong relationship between changes in the growth rate of money and stock returns. Sellon (1980) refers to the way discount rate changes affect stock prices. Decreases in the discount rates are found to be linked to increased borrowing provided by the Fed. Moreover, even when the lending on behalf of the Fed remains unchanged, discount rate reductions have a strong impact on banks' expectations and as a consequence on their level of lending. Smirlock and Yawitz (1985) study the stock reaction to discount rate changes and find that increases (decreases) lower (raise) stock prices, while Cook and Hahn (1989) and Thorbecke and Alami (1994) report similar results when studying the effect of federal fund rate target changes. Bernanke and Kuttner (2005) find that a hypothetical unexpected 25 basis point cut in the federal fund rate target is associated with a 1% increase in the price of a broad stock index, and attribute these excess stock returns not to changes in the real interest rates but to changes in expected future excess returns or expected future dividends. Rigobon and Sack (2003) take a different perspective on the same issue. They study how the stock market affects the short-term interest rates and find that a 5% rise (fall) in the S&P 500 index rises the probability of a tightening (easing) of the monetary policy by about a half.

There are a number of articles that study the stock price reaction of firms to government interventions during crises. Klingebiel et al. (2001)

report positive short-term stock reactions of banks in four East Asian countries to government guarantees of bank liabilities during the East Asian crisis. According to the authors, the stock reaction is a sign of the restored confidence of the market participants, but not necessarily evidence of an improvement of the health of the financial sector. Ganapolsky and Schmukler (2001) also find positive stock reactions of the Argentinean market to policies announced by the Argentinean government aimed at preventing the spillover effects of the Mexican crisis, while Kho, Lee, and Stulz (2000) reach similar results when examining the stock price reaction of U.S. banks during financial crises other than the East Asian one. The existing paper adds to the abovementioned stream of literature by studying the effects of monetary intervention during the subprime mortgage crisis. The bold actions on behalf of the Fed and the simultaneous cash injections and interest rate cuts offer a unique setting, which will allow us to get a better understanding of the monetary intervention plans.

2.4 HYPOTHESES

Next, we develop the hypotheses of our study, which we will later test empirically. In line with previous research (Smirlock and Yawitz, 1985; Cook and Hahn, 1989; Thorbecke and Alami, 1994; Bernanke and Kuttner, 2005), we expect that decreases in federal fund target and discount rate will raise stock prices. We cannot estimate separately the effect of target and the effect of discount rate decreases because of the simultaneous occurrence of these two events in six out of nine cases.

Hypothesis 1: *Decreases in the federal fund rate target and decreases in the discount rate will have a positive impact on the stock price of financial institutions.*

The Fed's announcements of cash injections are expected to raise the stock price of the financial institutions (Klingebiel et al., 2001). Given the fact that the main side-effect of the subprime mortgage crisis was the drying up of liquidity due to panic and loss of confidence in the markets, we expect that liquidity injections will benefit the financial sector as a whole and not only the banks that will choose to make use of the Fed's emergency credit lines.

Hypothesis 2: *Announcements of cash injections will have a positive impact on the stock price of the financial institutions.*

We expect that unanticipated changes in the target and discount rate will have a larger impact on the stock price of financial institutions than anticipated ones (Kuttner, 2001). According to the semi-strong efficiency hypothesis, markets are efficient and share prices reflect all publically available information. Thus, only unexpected changes, which convey new information to the market, can have an impact on stock prices. Our estimate of market's expectations is based on extensive research in the financial press in the weeks preceding each Fed's intervention.

Hypothesis 3: *Unexpected changes in target and discount rates will have a stronger impact on the stock price of the financial institutions than expected ones.*

Due to agency costs induced by information asymmetries, firms need to use collateral when raising money in the credit markets. Smaller financial institutions typically do not have as much collateral as larger ones and thus do not have the same ability to raise external funds. In addition, there is less publically available information for smaller firms, making it even more difficult for them to access the credit market. Therefore, the stocks of smaller institutions are expected to react stronger than the stocks of larger ones, both to interest rate reductions and to cash injections, since interest rate cuts and cash injections will improve their liquidity constraint situation (Thorbecke, 1997; Perez-Quiros and Timmermann, 2000).

Hypothesis 4: *Changes in monetary policy will have a stronger impact on the stock price of smaller financial institutions in comparison to the stock price of larger ones.*

Not every financial institution is equally exposed to the subprime crisis. We expect that financial institutions that are more exposed to the crisis will profit more from interest rate cuts and cash injections and hence are expected to present higher cumulative abnormal returns. The ratio of mortgage to total loans is used as a proxy for their exposure to the crisis.

Hypothesis 5: *Changes in monetary policy will have a stronger impact on the stock price of financial institutions with higher mortgage to total loans ratios.*

In line with the reasoning above, financial institutions that are more exposed to the subprime mortgage crisis, that is, they have higher mortgage-backed to total securities ratios, are expected to profit more from interest rate cuts and cash injections and thus are expected to present higher cumulative abnormal returns.

Hypothesis 6: *Changes in monetary policy will have a stronger impact on the stock price of financial institutions with higher mortgage-backed to total securities ratios.*

The financial literature suggests that credit constrained firms react stronger to changes in interest rates than firms that are less constrained (Ehrmann and Fratzscher, 2004). Following the example of Lamont, Polk, and Saa-Requejo (2001) and Kaplan and Zingales (1997), we use the debt to total assets ratio as a proxy for credit constraints of the financial institutions. Consequently, we expect that firms with higher debt to total assets ratios will react stronger to interest rate cuts and liquidity injections.

Hypothesis 7: *Changes in monetary policy will have a stronger impact on the stock price of financial institutions with higher debt to total assets ratios.*

2.5 DATA AND METHODOLOGY

In order to test our hypotheses, we conduct an event study and then regress the abnormal returns on financial institutions' characteristics. First, we will describe our sample and following we will present the methodology that we use.

2.5.1 Data

Our sample consists of 129 financial institutions that are listed in the United States and are included in the Compustat Global database. We excluded from our sample 13 financial institutions: 3 financial institutions that did not trade continuously during the period under consideration and 10 for which historical stock prices were unavailable. The financial institutions included in our sample belong to one of the following industries: financial services (2%), commercial banks (64%), savings institutions/Fed-chartered (12%), savings institutions/not Fed-chartered (3%), security brokers and dealers (9%), personal credit institutions (2%), finance lessors (1%), mortgage bankers (2%), federal credit agencies (2%), and investment advice and life insurance (3%). Our sample consists exclusively of financial institutions, since they were the ones most exposed to the subprime crisis.

The accounting variables were collected from Compustat Global and Compustat North America, while some missing items were hand-collected from the annual reports of the companies, which were downloaded from Thomson Research. Our accounting variables relate to 2006, the year before the crisis.

Our sample covers 14 Fed-induced events starting on August 9, 2007 till the April 30, 2008. The event dates were collected after research in the *Financial Times* and the *Wall Street Journal* using Factiva. Our events include eight decreases in the discount rate, seven decreases in the federal fund rate target, and five cash injections. Six out of fourteen events concern simultaneous cuts of the discount and the target rate.

2.5.2 Methodology

2.5.2.1 Event Study

For the event study we use the market model. The event window is equal to 3 days and the estimation period is equal to 250 trading days. We use daily stock price returns from Datastream and use the North America Datastream price index as our market index in the event study. There are five contaminated events taking place on 10/31/2007, 12/11/2007, 12/12/2007, 3/17/2008, and 3/18/2008. This happens either because their event windows coincide (12/11/2007 and 12/12/2007, 3/17/2008 and 3/18/2008) or because other events that had a strong influence on the stock prices of the financial institutions took place at the same time (10/31/2007 and 3/17/2008). However, we choose to run the event study including these five dates—so as not to lose possible valuable results—and to interpret them with special attention.

The expected returns of the financial institutions are calculated as follows:

$$E(R_{it}) = a_i + \beta_i R_{mt} + \varepsilon_{it}$$

where $E(R_{it})$ and R_{mt} are the expected rate of return of the financial institution i and the actual return of the North America Datastream price index at date t, respectively. The error ε_{it} is assumed to have a mean equal to zero and to be uncorrelated with both R_{mt} and across the financial institutions. An ordinary least squares (OLS) regression between the index and the stock prices of each financial institution is run to estimate the a_i and β_i. The abnormal return of the financial institution i at date t, AR_{it}, is calculated as the difference between the realized return (R_{it}) and the expected return:

$$AR_{it} = R_{it} - (a_i + \beta_i R_{mt})$$

To determine the abnormal returns produced by each event j, we calculate the cumulative average abnormal returns (CAAR$_j$) by summing up the average abnormal returns of the days −1 to +1 of the event window, as follows:

$$CAAR_j = \sum_{t=-1}^{1} AAR_{jt}$$

where AAR_{jt} is the average of the abnormal returns of all the companies in our sample at date t for the event j.

As Brown and Warner (1985) propose, it is necessary to control for event-induced variance. However, the common technique of using the cross-sectional variance of the event period instead of the variance of the estimation period (Boehmer et al., 1991) cannot be used in our case; the cross-sectional variance rises dramatically due to the nature of the events under consideration. We account for this problem by using the variance of the estimation period of the first event only. Thus, to test for statistical significance, we divide the cumulative average abnormal returns of each event by the standard deviation of the average abnormal returns of the first event ($SAAR_1$) times the square root of the number of days included in the event window, which is in our case equal to three. The formula we just described is

$$t\text{-stat}_j = CAAR_j/(SAAR_1 \times \sqrt{3}) \tag{2.1}$$

Next, we calculate the cumulative abnormal returns of each financial institution i to the event j (CAR_{ij}) by summing up the abnormal returns of the financial institution i during the event window (AR_{ijt}):

$$CAR_{ij} = \sum_{t=-1}^{1} AR_{ijt}$$

In order to control for outliers, we also calculate the cumulative median abnormal returns ($CMAR_j$) and test them for statistical significance. The CMARs are the sum of the median abnormal returns of the days -1 to $+1$ (MAR_{jt}):

$$CMAR_j = \sum_{t=-1}^{1} MAR_{jt}$$

2.5.2.2 Regressions

In order to explain the cumulative abnormal returns, we run the following cross-sectional OLS regression for each event separately:

$$CAR_{ij} = \alpha + \beta_1 Loan_i + \beta_2 Sec_i + \beta_3 LnSize_i$$
$$+ \beta_4 Debt_i + \beta_5 Commercial_i + u_{ij}$$

TABLE 2.1   Descriptive Statistics

Variable	Mean	First Quartile	Second Quartile	Third Quartile	Maximum	Skewness	Kurtosis
Loan	0.38	0.12	0.35	0.55	1.00	0.60	−0.52
Sec	0.52	0.32	0.54	0.75	1.00	−0.34	−0.90
Size	121.16	5.96	12.89	58.00	1884.32	3.68	14.70
Debt	0.21	0.17	0.17	0.24	0.93	2.37	6.83

Notes: This table presents the descriptive statistics of our sample. Loan is the ratio of mortgage to total loans; Sec is the ratio of mortgage-backed securities to total securities; Debt is the ratio of debt to total assets; Size refers to total assets in billions of dollars. Our sample consists of 129 financial institutions in the United States.

where

CAR_{ij} is the cumulative abnormal return of financial institution i produced by the event j

$Loan_i$ is the ratio of mortgage to total loans of financial institution i

Sec_i is the ratio of mortgage-backed to total securities of financial institution i

$LnSize_i$ is the natural logarithm of total assets of financial institution i

$Debt_i$ is the ratio of debt to total assets of financial institution i

$Commercial_i$ is a dummy control variable that takes the value 1 if the financial institution is registered as a commercial bank and 0 otherwise

The independent variables are only weakly correlated (Pearson correlation is less than 0.3 in every case). We use White's t-statistics to control for heteroscedasticity. Table 2.1 presents some descriptive statistics of the independent variables.

2.6 RESULTS

2.6.1 Event Study

Tables 2.2 and 2.3 summarize the results of the event study. All of the results reported, apart from the events on 3/17/2008 and 3/18/2008, are statistically significant at the 95% confidence level. On 3/17/2008 and 3/18/2008, the results are insignificant, which can be attributed to the fact that the events are contaminated.

The results offer support to Hypotheses 1 and 2, since both interest rate cuts and cash injections have a positive effect on the stock price of the financial institutions. On average, monetary policy actions cause a

TABLE 2.2   Event Study Results: Results per Event

Event Date	CAAR (%)	CMAR (%)	Significant Reaction (%)	Positive Reaction (%)
8/9/2007	3.64***	3.10***	47	71
8/17/2007	5.66***	5.23***	75	91
9/18/2007	2.28***	2.02***	33	84
10/31/2007	−3.02***	−2.64***	42	15
12/11/2007	−2.66***	−2.55***	40	19
12/12/2007	−4.46***	−4.09***	59	13
1/10/2008	3.11***	2.39***	42	76
1/22/2008	9.61***	9.68***	87	93
1/30/2008	3.68***	2.67***	47	80
3/7/2008	2.65***	3.76***	67	81
3/11/2008	3.65***	3.66***	59	88
3/17/2008	−1.31	1.09	38	65
3/18/2008	1.38*	3.36***	57	82
4/30/2008	2.02**	1.78**	29	76

Notes: This table presents the results of the event study for each event separately. Column 1 presents the dates of the event occurrence; column 2 presents the cumulative average abnormal returns produced by each event; column 3 presents the cumulative median abnormal returns produced by each event; column 4 presents the percentage of the financial institutions that had a statistically significant stock reaction to each event at the 5% level; and column 5 presents the percentage of the financial institutions that had a positive stock reaction to each event. ***, ** and, * represent statistical significance at the 1%, 5%, and 10% levels, respectively (two-tailed *t*-test).

TABLE 2.3   Event Study Results: Results per Event Type

Event Type	N	CAAR (%)	CMAR (%)
All	14	1.87*	2.1**
All excluding contaminated	9	4.03***	3.81***
All cash injections	5	1.72	1.76
Cash injections excluding contaminated	4	3.26***	3.23***
All interest rate cuts	9	1.96	2.29*
Interest rate cuts excluding contaminated	5	4.65***	4.27***
Unanticipated interest rate cuts excluding contaminated	3	5.85***	5.64**
Anticipated interest rate cuts excluding contaminated	2	2.85***	2.22***

Notes: This table presents the results per event type. Column 1 presents the type of event; column 2 presents the number of events for each event type; column 3 presents the cumulative average abnormal returns for each event type; column 4 presents the cumulative median abnormal returns for each event type. ***, ** and, * represent statistical significance at the 1%, 5%, and 10% levels, respectively (two-tailed *t*-test).

4.03% abnormal increase in their stock price. The interest rate cuts produce abnormal returns equal to 1.96% and the cash injections abnormal returns equal to 1.72%. When we correct for contamination, the interest rate cuts produce abnormal returns equal to 4.65% and the cash injections present abnormal returns equal to 3.26%. We notice that interest rate cuts produce, on average, higher abnormal returns than cash injections. A possible explanation is the widespread impact of interest rate cuts; all financial institutions are to profit from interest rate reductions, whereas cash injections mostly affect the financial institutions that will make use of the extended credit lines of the Fed. Note however that this difference is not statistically different from zero (t-stat = 1.41).

We separate the interest rate cuts into two subcategories, excluding the contaminated events: unanticipated interest rate cuts (8/17/2007, 9/18/2007, and 1/22/2008) and anticipated ones (1/30/2008 and 4/30/2008). Consistent with hypothesis 3, the unexpected changes in interest rates have a stronger impact on the stock price of the financial institutions than expected ones. Anticipated interest rate reductions produce on average a 2.85% increase in the stock price of the financial institutions, while unanticipated ones cause 5.85% increase. However, the difference in the two averages is again not statistically different from zero (t-stat = 1.32).

Tables 2.2 and 2.3 also report the CMARs. The use of CMARs changes only slightly the results of our research, hence we will not refer to them further. Finally, the percentage of companies that produced statistically significant abnormal returns for each event is reported in the fourth column of Table 2.2.

2.6.2 Regressions

According to the results of our regressions, the size of the financial institutions has stronger explanatory power when compared to all the other variables of our model; in 7 out of 14 events, the relationship between the cumulative abnormal returns and the size of the financial institutions is negative and statistically significant. In 3 out of 14 events, the exposure of the companies to mortgage loans and the degree of credit constraint is statistically significant and positively related to their stock price reaction, while only in one occasion the exposure of the companies to mortgage-backed securities is statistically significant and positively related to their equity value. The coefficients of the regressions do not depend on the type of monetary intervention. Table 2.4 presents the coefficients of the variables for each event.

TABLE 2.4 Regression Coefficients

Variable	8/9/ 2007	8/17/ 2007	9/18/ 2007	10/31/ 2007	12/11/ 2007	12/12/ 2007	1/10/ 2007	1/22/ 2008	1/30/ 2008	3/7/ 2008	3/11/ 2008	3/17/ 2008	3/18/ 2008	4/30/ 2008
							Date							
Loan	0.03**	0.02	0.00	−0.01	−0.02	−0.02	0.04*	0.05**	0.03	−0.03	0.04	0.11	0.11	−0.03*
	(2.42)	(0.90)	(0.16)	(−1.06)	(−1.33)	(−1.05)	(1.95)	(2.23)	(1.56)	(−1.49)	(1.37)	(1.00)	(1.35)	(−1.94)
Sec	−0.02	0.03*	−0.01	−0.02	−0.01	0.00	−0.02	0.02	−0.03	0.01	0.04	0.02	0.02	0.01
	(−1.18)	(1.82)	(−0.77)	(−1.05)	(−0.28)	(−0.28)	(−0.86)	(−0.35)	(−1.47)	(0.55)	(1.27)	(0.60)	(0.45)	(0.74)
LnSize	−0.04***	−0.02***	−0.01***	0.01	−0.01**	−0.01**	0.02***	0.01	−0.01	−0.02***	−0.02***	−0.02	0.00	0.00
	(−6.79)	(−3.98)	(−4.49)	(1.53)	(−2.03)	(−2.09)	(4.48)	(−1.01)	(−1.19)	(−4.12)	(−2.62)	(−0.85)	(0.04)	(−0.80)
Debt	0.11***	0.07**	−0.01	−0.05*	−0.02	−0.02	0.02	0.05	0.10**	0.01	0.00	−0.21	−0.18	0.04
	(2.68)	(2.23)	(−0.38)	(−1.77)	(−0.53)	(−0.49)	(0.53)	(0.06)	(2.05)	(0.28)	(−0.05)	(−0.83)	(−0.93)	(1.10)
Commercial	−0.01	−0.01	0.00	0.01	−0.01	0.00	−0.02	0.01	−0.01	0.05***	0.01	0.07	0.07	−0.02*
	(−0.52)	(−0.63)	(0.63)	(0.56)	(−0.77)	(−0.37)	(−1.36)	(1.17)	(−0.68)	(4.71)	(0.30)	(1.08)	(1.27)	(−1.82)
Adjusted R^2 (%)	26	14	11	4	3	2	17	3	3	35	6	2	2	1

Notes: This table presents the coefficients of the multivariate regressions and the corresponding White's t-statistics in parentheses. Loan is the ratio of mortgage to total loans; Sec is the ratio of mortgage-backed to total securities; LnSize is the natural logarithm of total assets; Debt is the ratio of debt to total assets; and Commercial is a dummy variable that is equal to 1 when the financial institution is a commercial bank and zero otherwise. ***, **, and, * represent statistical significance at the 1%, 5%, and 10% levels (two-tailed t-test), respectively.

The results offer partial support to Hypotheses 4, through 7, with stronger support being offered to Hypothesis 4. More specifically, Table 2.4 indicates that small and credit-constrained financial institutions as well as institutions that had a strong exposure to the crisis either in the form of mortgage loans or in the form of mortgage backed securities were the ones that profited the most from the Fed's interventions. The size of the financial institutions seems to have the stronger explanatory power; however, the results are weak and should be interpreted with caution. Finally, the explanatory power of the dummy variable "Commercial" is low, so we conclude that the reaction of the commercial banks is not what drives the results.

2.7 CONCLUSION

Our findings show that the monetary policy actions of the Fed as a response to the subprime mortgage crisis had a positive short-term influence on the stock price of the financial institutions. This suggests that investors perceived the Fed's interventions to be successful at least in the short run. More specifically, a decrease by 50% in the discount and/or target rate caused on average a 4.65% increase in the stock price of our sample. Liquidity injections or announcements of future liquidity injections also drove up the stock prices of the financial institutions; on average, liquidity injections were accompanied by a 3.26% increase in the equity value.

The relevant literature offers a number of different explanations for the stock price increases as a response to the interest rate cuts. Bernanke and Kuttner (2005) claim that stock prices increase due to expected future excess returns or expected future dividends and not due to expected decreases in the real interest rates, while Klingebiel et al. (2001) propose that what drives up the stock prices is the restored confidence of the market participants. The answer to this question is something that we cannot infer from our results and is beyond the scope of this study. However, in line with Kuttner (2001), we find that unanticipated interest rate cuts have a stronger effect on the stock price of the financial institutions, when compared to anticipated ones. On average, an unexpected interest rate decrease produces a 3 percentage points higher stock price reaction than an expected one. Finally, our results offer partial support for the hypothesis that small and credit-constrained institutions with strong exposure to the crisis have profited more from the Fed's interventions.

There is a lot of room for further research on this topic. First of all the sample could be enlarged from the 129 financial institutions of our sample to all the financial institutions listed in the United States. Moreover, quarterly instead of yearly accounting data could be used and more explanatory variables could be introduced. Examples of explanatory variables that could be incorporated in our model are the measures of liquidity developed by Berger and Bouwman (2008), the credit ratings of the financial institutions, the degree of wholesale funding of the financial institutions (Yorulmazer, 2008), and the ownership structure of the financial institutions. In addition, the federal funds future rates (Kuttner, 2001) could be used, in order to distinguish between expected and unexpected monetary interventions. Last but not least, future research could expand the data set to other sectors, apart from the financial sector, to examine the effects of monetary interventions on a larger part of the U.S. economy.

REFERENCES

Berger, A.N. and Bouwman, C.H.S. (2008) Bank liquidity creation. Working Paper, available at http://ssrn.com/abstract=672784.

Bernanke, B.S. and Blinder, A. (1992) The federal funds rate and the channels of monetary transmission. *American Economic Review*, 82(4): 901–921.

Bernanke, B.S. and Kuttner, K.N. (2005) What explains the stock market's reaction to Federal Reserve policy? *Journal of Finance*, 60(3): 1221–1257.

Boehmer, E., Musumeci, J., and Poulsen, A.B. (1991) Event study methodology under conditions of event-induced variance. *Journal of Financial Economics*, 30(2): 253–272.

Brown, S. and Warner, J. (1985) Using daily stock returns: The case of event studies. *Journal of Financial Economics*, 14(1): 3–31.

Cook, T. and Hahn, T. (1989) The effect of changes in the federal funds rate target on market interest rates in the 1970s. *Journal of Monetary Economics*, 24(3): 331–351.

Ehrmann, M. and Fratzscher, M. (2004) Taking stock: Monetary policy transmission to equity markets. *Journal of Banking Credit and Money*, 36(4): 719–737.

Ganapolsky, E.J.J. and Schmukler, S.L. (2001) Crisis management in Argentina during the 1994–95 Mexican crisis: How did markets react? *World Bank Economists' Forum*, Vol. 1, World Bank, Washington, DC, pp. 3–30.

Kaplan, S.N. and Zingales, L. (1997) Do investment cash flow sensitivities provide useful measures of financing constraints. *Quarterly Journal of Economics*, 112(1): 169–215.

Kho, B., Lee, D.W., and Stulz, R.M.M. (2000) U.S. banks, crises, and bailouts: From Mexico to LTCM. *American Economic Review*, 90(2): 28–31.

Klingebiel, D., Kroszner, R., van Oijen, P., and Laeven, L.A. (2001) Stock market responses to bank restructuring policies during the East Asian crisis. World Bank Policy Research Working Paper Series, available at http://ssrn.com/abstract=632640.

Kuttner, K.N. (2001) Monetary policy surprises and interest rates: Evidence from the Fed funds futures market. *Journal of Monetary Economics*, 47(3): 523–544.

Kuttner, K.N. and Mosser, P. (2002) The monetary transmission channel: Some answers and further questions. *Economic Policy Review*, 8(1): 15–26.

Lamont, O., Polk, C., and Saa-Requejo, J. (2001) Financial constraints and stock returns. *Review of Financial Studies*, 14(2): 529–554.

Perez-Quiros, G. and Timmermann, A. (2000) Firm size and cyclical variations in stock returns. *Journal of Finance*, 55(3): 1229–1262.

Rigobon, R. and Sack, B. (2003) Measuring the reaction of monetary policy to the stock market. *The Quarterly Journal of Economics*, 118(2): 639–669.

Rozeff, M. (1974) Money and stock prices, market efficiency, and the lag in the effect of monetary policy. *Journal of Financial Economics*, 1(3): 245–302.

Sellon, Jr. G.H. (1980) The role of the discount rate in monetary policy: A theoretical analysis. Federal Reserve Bank of Kansas City, *Economic Review*, 65: 3–15.

Smirlock, M. and Yawitz, J. (1985) Asset returns, discount rate changes, and market efficiency. *Journal of Finance*, 40(4): 1141–1158.

Thorbecke, W. (1997) On stock market returns and monetary policy. *Journal of Finance*, 52(2): 635–654.

Thorbecke, W. and Alami, T. (1994) The effects of changes in the federal funds rate target on stock prices in the 1970s. *Journal of Economics and Business*, 46(1): 13–19.

Yorulmazer, T. (2008) Liquidity, bank runs and bailouts: Spillover effects during the Northern Rock episode. Available at http://ssrn.com/abstract=1107570.

CHAPTER 3

Shadow Banks and the Financial Crisis of 2007–2008

Jason Hsu and Max Moroz

CONTENTS

Historically, commercial banks have served as mechanisms for providing financial leverage to the economy. Banks took in deposits and made loans, and thus, facilitated the pooling and sharing of risks. However, the explosive growth of hedge funds and the supporting investment banking activities have led to a rapid growth also in nontraditional financing as well as complex derivatives for leveraging, which are not regulated and, in terms of risks, are poorly managed. In this chapter, we explore this shadow banking system which has contributed significantly to the current banking crisis.

3.1 INTRODUCTION

Year 2005 marked the first time since reliable history was kept that the United States suffered no bank failures (Figure 3.1).

The prospects looked good too, with all the major indicators above historical averages, and well in excess of the requirements for well-capitalized banks (Table 3.1).

As late as the fall of 2006, the regulators in their quarterly report focused on the strong earnings and revenues, the fast asset growth, the robust capital indicators, and the all-time low achieved by the noncurrent loan rate (FDIC, 2006). Industry analysts concurred. "The U.S. banking industry is entering 2006 in a relatively healthy condition," noted the rating firm A.M. Best in its annual review (Best, 2006). And indeed, no banks failed in 2006 either, for the second time ever.

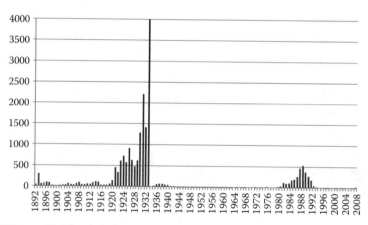

FIGURE 3.1   U.S. bank failures 1892–2008.

TABLE 3.1   Banking Industry Financial Health Indicators

	2006	2005	2004	2003	2002	2001	2000	Well-Capitalized
Core capital (leverage) ratio (%)	8.23	8.25	8.11	7.88	7.86	7.79	7.71	5
Tier 1 risk-based capital ratio (%)	10.52	10.66	10.76	10.47	10.43	9.90	9.42	6
Total risk-based capital ratio (%)	12.99	12.98	13.19	13.00	13.00	12.72	12.13	10
Net charge-offs to loans (%)	0.38	0.50	0.56	0.78	0.97	0.83	0.59	N/A

But just outside the solid fortress of the heavily regulated banking industry was hiding a beast labeled by Paul McCulley of Pimco the "shadow banking system" (McCulley, 2007). It attracted little attention from either the regulators or the media, until suddenly it emerged as one of the main reasons for "the worst financial crisis since Depression" (IMF, 2008).

A financial crisis occurs when multiple financial intermediaries fail simultaneously, thus disrupting the operation of a certain financial market. Barring an unlikely coincidence, this requires one or both of the following conditions:

1. Detrimental *external factor* affects many institutions at once

2. Problem in one institution spreads to others through an internal *contagion mechanism*

Viewing this from the perspective of individual institutions, each of them can fail due to

1. Investment risk (losses from directional bets)

2. Hedging risk (large variance between a position and the risk it is supposed to offset)

3. Counterparty risk (default of trading partners)

4. Liquidity risk (inability to sell assets at fair value due to time constraints)

In this chapter, we investigate the role that the shadow banking system played in the present crisis.

3.2 BANK RUNS

The past two years have seen the reemergence of bank runs, once a recurrent phenomenon causing enormous welfare destruction, but by now largely eliminated through various policy measures such as deposit insurance. In this section, we describe the theory behind bank runs.

A bank run is a fast loss of deposits or other short-term financing by a financial institution, which causes it to fail from the lack of liquidity.

While a traditional bank run has depositors queuing outside a retail bank branch, the concept of a bank run applies to many other situations. For example, mutual fund investors may decide to redeem their shares all at once, a hedge fund may find no lenders willing to roll over its repos, a conduit may find itself unable to sell short-term commercial paper to refinance the expiring obligations, and so on.

In their seminal paper, Diamond and Dybvig (1983) explain why banks typically have liquidity mismatch in their balance sheets, and how this mismatch enables bank runs. The Diamond–Dybvig model assumes that risk-averse consumers look for investment opportunities but do not know in advance when they might need their money back. The investment projects of the production sector are illiquid in the sense that selling them early (before they mature) can only be done at a penalty. The role of banks is to satisfy investor demand for liquidity by converting illiquid assets (loans to the production sector) into liquid liabilities (demand deposits). In this model, all banks are mutual banks (i.e., have no equity), and create liquidity just by pooling risks among its depositor-owners.

The main finding of the paper is that this setup allows multiple equilibria. In the "good" equilibrium, only the depositors experiencing idiosyncratic need for liquidity withdraw their money early. The bank sells *some* of its illiquid assets at a penalty to generate the cash for those withdrawals. However, due to risk pooling that penalty is partially borne by depositors who withdraw at maturity. As a result, individual investor payoff in the two scenarios ("need liquidity" and "don't need liquidity") is smoothed out. This creates value because investors are risk-averse.

In the "bad" equilibrium, a bank run occurs. All depositors, regardless of their liquidity needs, withdraw their money early fearing bank failure. The bank is forced to sell *all* its assets at a penalty. Risk pooling does not work, since no one waits until maturity. The bank cannot meet its contractually promised payoffs and has to declare bankruptcy.

In the bank run equilibrium, investors redeem early because they suspect (rightly or wrongly) that others might want to do the same, and (correctly) concluding that this would cause the bank to default. In the case of impending default, it is better to redeem early, while the bank is still attempting to meet its obligations. The fear of a bank run thus becomes a self-fulfilling prophecy.

Thus, in Diamond–Dybvig model, a bank run occurs when a number of investors come to believe that many other investors are going to withdraw their funds (or, equivalently, deny refinancing). For example, suppose an (unbiased) rumor spreads on Wall Street that a certain bank finds it hard to obtain overnight loans. A bank that suddenly cannot roll over its overnight liabilities is likely to run out liquidity and default. There is an incentive to abandon the suspect bank as early as possible as the default would hurt those who provided the very last loans. Therefore, the rumor can scare all potential lenders away, and the bank can fail from a run by its counterparties in a self-fulfilling prophecy.

The strength of the Diamond–Dybvig model is that it can explain such sporadic runs, in which depositors or counterparties leave mostly out of fear to be the last ones on the sinking ship, and would have stayed if they knew others were staying.*

On the other hand, in case of a real economic problem (or a rumor thereof), investors do not care whether others are staying or leaving: no one is willing to keep their funds tied up in a failing institution regardless of what other investors are doing. In such scenario, the Diamond–Dybvig mechanism should not be considered the primary cause of failure although it can certainly speed up the demise (a failing institution meets its end even faster when it loses liquidity). To distinguish the two scenarios, we will refer to the former as a Diamond–Dybvig bank run, and to the latter as a fundamental bank run.

Bank panic is a simultaneous run on many banks. It can consist of many independent bank runs, caused by an external economic problem such as a spike in bad loans. More often, however, it is fueled by the contagion mechanism: a default of a few banks causes direct losses at counterparties, lowers prices on assets held by other banks due to fire-sale liquidations, and reduces the amount of liquidity in the system (since creditors usually do not know exactly which particular banks are in trouble and reduce overall lending).

Measures against bank runs include

- Deposit insurance

- Lender of last resort facility

- Suspension of convertibility

- Maintenance of sufficient liquidity

---

* The extreme case is that of a *sunspot run*, where no economic factor is present whatsoever. The sunspot refers to any random variable that has no independent economic significance and yet influences the choice of the equilibrium. The term originated from an obscure economic research by Jevons (1875) that attempted to predict corn prices using sunspot activity. The empirical research on sunspot runs is controversial. Some authors (Gorton, 1988; Allen and Gale, 1998) argue that bank panics are caused by actual economic problems. Others (Ennis, 2003) claim that sunspot theory is a likely explanation for many bank panics. It remains beyond doubt that a very minor economic issue (e.g., one with no perceivable impact on default probability) can cause a bank run through the sunspot mechanism. The news itself can serve as the coordination device that selects the "bad" equilibrium.

The first two measures are under the control of an external authority (e.g., the government needs to provide the funds and can set requirements for the deposit insurance program). The last two are largely under the control of individual institutions.

Today, most countries address the problem of bank runs by providing government guarantees for deposits (at least up to a certain amount). According to the International Association of Deposit Insurers, deposits are currently guaranteed by the governments of nearly 100 countries, including 28 of the 30 OECD member states (the only exceptions being Australia and New Zealand). Once depositors do not bear the risk of a loss, they lose the incentive to withdraw at the first sign of trouble. This is by far the most effective way to avoid bank runs. However, deposit insurance cannot be extended beyond the (nearly risk-free) bank deposits.* Insuring a risky asset is economically infeasible because it destroys its raison d'être: to transfer risk to those investors who are willing to bear it in return for a higher return.

The lender of last resort can provide a loan to a financial institution when the market refuses to do so, possibly avoiding default caused by the lack of liquidity. In the United States, this function is performed by the Federal Reserve (Fed). Until recently, it was limited to commercial banks, and rarely used (for fear of a stigma associated with the need for emergency funds). During the recent crisis, however, the Fed greatly expanded this function both through individual loans (such as the $85B initial loan to AIG) and by the introduction of new programs (such as Term Auction Facility).

Until early twentieth century, U.S. banks often suspended depositors' withdrawal rights without going through bankruptcy. This provided extra time to liquidate assets in a more orderly manner, and thus avoid a fire-sale liquidation that might cause insolvency. This approach is still used today by hedge funds, mutual funds, and others who have implied but not legally binding obligation to redeem their own shares on demand. It does not work for any debt obligations, so if buyers cannot be found for new commercial paper, the option to simply delay the payment on the expiring paper does not exist (except through a bankruptcy).

---

* As we discuss below, the US government did offer temporary insurance for money market fund investments in 2008. However, money market fund shares are the safest securities other than bank deposits. Furthermore, the insurance only covered existing investments as of the announcement date, and not a single dollar of new investments after that. This prevented the otherwise inevitable moral hazard effect.

Finally, holding sufficient liquid assets reduces the chance that investors would become concerned about liquidity to begin with. Each institution makes its own decision about what amount of liquidity is sufficient. In the United States, there are regulatory requirements for banks to hold sufficient reserves but those requirements have not been binding for at least a decade (Bennett and Peristiani, 2002).

All the measures listed above are very effective against a Diamond–Dybvig run (whether on an individual bank or on the whole industry). However, no amount of liquidity (whether injected by the Fed, obtained by delaying payments on liabilities, or from internal reserves) can stop investors running away from a failing institution. This leaves deposit insurance as the only solution for fundamental bank runs. Due to the limitations discussed above, it is generally feasible only for actual bank deposits. Thus, fundamental bank runs will happen unless all financial institutions have such a conservative risk profile that they are never in trouble—an unrealistic constraint on the modern financial industry.

3.3 SHADOW BANKS

Rephrasing McCulley, shadow banks are entities that fund illiquid assets with short-term liabilities and yet remain outside of the banking regulation. The liquidity mismatch makes shadow banks prone to bank runs, while the traditional safeguards embedded in the bank regulation are missing. However, as we argue below, most of the bank runs experienced by shadow banks were of the fundamentals rather than the Diamond–Dybvig variety, and as such would not have been prevented by any existing regulation.

Shadow banks include broker-dealers, hedge funds, private equity groups, structured investment vehicles, conduits, CDO structures, money-market funds, nonbank mortgage lenders, and other similar entities. Note that the liquidity mismatch does not necessarily mean that the assets have longer term than liabilities. The litmus test is the following. Assume that none of the liabilities can be rolled over, and hence the assets have to be sold off to meet the obligations. Such scenario is likely to result in a fire-sale price discount borne by the holder. The higher this price penalty, the larger the liability mismatch.

Demand deposits and investments in mutual funds have an effective maturity of a single business day, since redemptions can be made without notice. If an asset-backed commercial paper (ABCP) money-market mutual fund had to sell its assets within one day, the fire-sale price penalty would be devastating if it happened when the ABCP market was strained by subprime mortgage fears. On the other hand, a fixed-income mutual fund investing in long-term Treasury bonds could dispose of its assets in one day with almost no price penalty. Therefore, the liquidity mismatch is quite high in an ABCP money-market fund, but virtually nonexistent in the Treasury bond fund despite the latter's much longer asset duration.

3.4 ECONOMIC PRECONDITIONS FOR THE CRISIS

The root of the financial crisis was in the home price bubble that began somewhere around 2000. Following (Cecchetti, 2008), we look at the ratio of home prices to rents for the residential real estate (see Figure 3.2). It is similar in concept to the equity *P/E* ratio, and as such should be driven by the expected growth rate of rents and the expected return required from the investment in one's own house. Despite various complicating factors in the housing market economics,* this ratio stayed in the 9–11 range for half a century, before skyrocketing from 11 to 16 at an accelerating pace during 2000–2005. At the same time, the inflation-adjusted rent[†] was growing at a relatively steady rate (averaging 0.5% per year, and never exceeding 2.1% per year) since the data became available in 1983 (see Figure 3.3). There was no obvious reason for the expected rent growth

FIGURE 3.2   Ratio of home prices to rent.

---

* For instance, the required return on one's own house is lower than on other investments with a similar risk profile due to the substantial nonmonetary utility derived from living in one's own house. This fact should increase the home price/rent ratio beyond a similar ratio for rental properties. Furthermore, an increase in the desire of consumers to own a home, or in their wealth (hence their ability to afford such nonmonetary utility), should increase this ratio.

† We used owners' equivalent rent of primary residence as reported by the Bureau of Labor and Statistics, and seasonally adjusted CPI excluding energy and food.

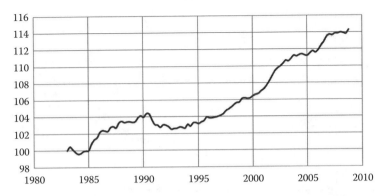

FIGURE 3.3   Real rent growth.

to spike so high above the historical level; in fact, rent growth had been slowing down since 2003. As a result, it would be reasonable to conclude that home prices to rent ratio would have to revert back to its historical levels, which as of late 2005 required a 30%–35% fall in the home prices. However, some researchers suggest that it was not obvious at the time that home prices were due for a significant fall (Gerardi et al., 2008). Whether or not the housing market meltdown was expected, it certainly became the origin of the crisis.

As can be seen in Figure 3.4, which depicts the S&P/Case–Shiller seasonally adjusted U.S. national home price index, the first signs of the slowdown could be seen in 2005. In Q2 2006, home prices fell for the first time in 14 years.

Many subprime borrowers could only afford mortgage payments by continuously refinancing the property at fast-growing valuations. As a result, the past-due (30–89 days) mortgage loan rate jumped by 0.13%

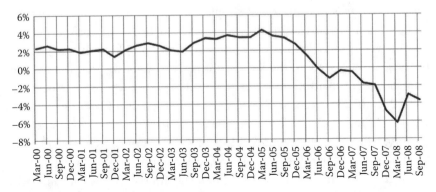

FIGURE 3.4   Quarterly change in home prices.

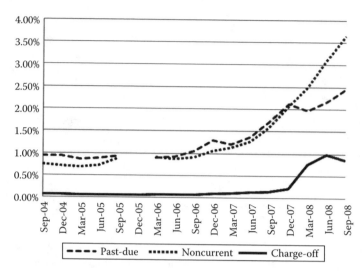

FIGURE 3.5   Residential mortgage loan performance.

points in Q3 2006. It took another quarter for this to translate into noncurrent (over 90 days) mortgage loan rate and the charge-off rate (Figure 3.5). We removed the data point for Q4 2005 due to the impact of a one-time accounting rule change.

3.5 CRISIS TIMELINE

On February 8, 2007, HSBC surprised analysts by announcing a much larger than expected loan-loss provision for 2006, blaming deteriorating U.S. mortgage business. This was the beginning of a constant stream of reports of major losses and bankruptcies among subprime lenders, home-builders, and other institutions exposed to those markets. At this point, the problem was not limited to simple investment losses for institutions involved. First, the trading in securities tied to the subprime market became very slow, which caused severe liquidity problems for the holders of those assets. Second, the lack of transparency meant that the market punished some institutions that had no subprime exposure whatsoever.

In March 2007, a Bear Stearns hedge fund, High-Grade Structured Credit Strategies (HGSCS), experienced its first down month since its inception in 2003, with a modest loss of 3.7%. As of March 31, 2007, the fund had $925 M in capital and was $9.7B long and $4B short various subprime mortgage market securities. Upon learning of the loss in late April, investors, scared by unexpected losses, redeemed $100 m. The requisite asset sales from this tiny redemption were probably in the range of $1B due to leverage. This was enough to move the illiquid market in the mortgage-backed CDOs causing further damage to the fund's valuation.

In mid-April, Bear's counterparty Goldman Sachs revalued the subprime securities it was involved in trading with Bear. The marks went from ¢98 on the dollar to 50–60 effectively destroying the HGSCS and its sister fund, High-Grade Structured Credit Strategies Enhanced Leverage (Cohan, 2009). By early June redemptions were frozen, but lenders increased margin requirements and marked down the collateral. Despite calls from Bear for a negotiated orderly liquidation, one of the lenders, Merrill Lynch, offered for sale about $400 m of the seized collateral, but found bids only at significant discounts. Any additional sales would further depress the prices, and force more margin calls and more sales (not just from this fund, but also from any other investors who carried exposure to subprime mortgages). This loss spiral, described in detail by Brunnermeier (2008), threatened to cause more defaults among highly leveraged investors in subprime assets as well as among institutions exposed to them. The latter would include Bear Stearns, the largest prime broker to hedge funds.

Finally, in late June, Bear Stearns offered to bail out the less leveraged HGSCS with a $3.2B facility into which the fund's repo loans were rolled. The other fund would be liquidated by the lenders. The partial rescue did little to stop the contagion effect; the securities valuations continued to plummet, and trading in any subprime-related securities was virtually nonexistent. The initial losses in the two hedge funds were regular (and quite modest) investment losses. The subsequent run by investors was prevented by a freeze on redemptions. The damage was done by the loss spiral fueled by the lack of liquidity in the markets and by the high leverage of the funds.

On August 9, 2007, BNP Paribas froze redemptions in three of its investment funds citing its inability to value the assets due to a total lack of trading activity in the asset-backed securities market. BNP Paribas reopened the funds less than 3 weeks later, with virtually no change in the NAV. In other words, the funds were solvent and liquid; the problem was purely the failure of the market price discovery mechanism. However, BNP Paribas' announcement triggered a panic in global financial markets: freezing interbank lending and short-term funding markets (see Figure 3.6 that compares the total outstanding amounts of ABCP and commercial paper issued by financial and nonfinancial companies).

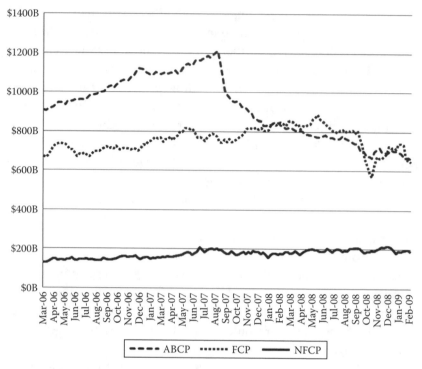

FIGURE 3.6   Commercial paper outstanding.

The first major victim was Northern Rock, a U.K. retail bank that used brokered deposits to finance a significant part of its balance sheet. The lack of demand for commercial paper meant that it could not refinance its maturing obligations. The bank reported liquidity problems to the U.K. regulators on August 13, 2007, and for a month the Bank of England attempted in vain to find a solution (such as a takeover). Finally, on September 14, 2007, the Bank of England announced emergency measures to support the bank, triggering the widely publicized run on the bank. By then, Northern Rock had already been doomed by the bank run at the institutional level; the retail bank run was an afterthought rather than the cause of the crisis (Shin, 2009). The only difference between Northern Rock and other U.K. retail banks was the unusually high reliance on the wholesale capital market and hence the lack of deposit insurance on that part of its liabilities (brokered deposits were not insured). The run on Northern Rock was part of an overall fundamentals run on the capital markets, driven by a combination of a subprime mortgage meltdown and the lack of information about exposures and valuations in that market.

The next wave of victims consisted of issuers of ABCP that could not find buyers when they tried to roll it over. Cheyne Finance and Rhinebridge were among those that succumbed to the run on the commercial paper in October 2007. By March 2008, Bear Stearns, the most leveraged of the investment banks, had already been weakened by exposure to agency bonds, subprime mortgages, and the failed mortgage fund, Carlyle Capital. By March 11, its liquidity pool started to deteriorate. On March 11, a rumor was started that Goldman Sachs was unwilling to face Bear Stearns as counterparty.* The rumor caused a run on Bear Stearns by hedge fund clients and repo counterparties. By Sunday, the bank was bailed out in a takeover by JP Morgan with a $29B loan from the Fed. According to the SEC, Bear Stearns' liquidity pool fell from the perfectly adequate $18.1B on March 10 to $11.5B on March 11 and to $2B on March 13. However, throughout the week, Bear Stearns remained firmly in compliance with the net capital requirements for its broker-dealers and the capital ratio standards for its holding company (Cox, 2008). The SEC blamed the demise of the bank on the "crisis of confidence," which was just another name for a bank run. This was one of the rare examples where a Diamond–Dybvig bank run destroyed a major institution in the current crisis.

On July 11, 2008, IndyMac was seized by FDIC in one of the biggest bank failures in the U.S. history. The bank struggled due to a large amount of mortgage loans that it was unable to securitize and instead kept on its books. On May 12, the bank reported in its 10-Q filing that it barely met the "well-capitalized" regulatory guidelines; indeed, accounting for the April downgrades on the mortgage-backed securities it was holding, it was already below the threshold. A controversial public letter by Senator Schumer on June 26 warned of the bank's imminent demise, and caused a small run on IndyMac (quite irrational since most depositors were fully FDIC-insured); the loss of $1.3B in deposits over the 11 day period became the final straw. The main cause of failure was a simple investment loss and not a bank run.

On September 7, 2008, Fannie Mae and Freddie Mac were taken over by the federal government for fear that their falling capitalization and

---

* According to Goldman Sachs, late in the day on March 11, Hayman Capital requested a trade whereby Goldman would insure Bear Stearns' obligations in a derivatives trade. Goldman Sachs replied that they would consider it. Hayman Capital misinterpreted this as refusal. Even though the next morning Goldman Sachs actually accepted the trade, and continued trading with Bear Stearns for the rest of the week, the damage to Bear Stearns was irreversible.

inability to raise new funds might cause further disruption to the U.S. housing market. These agencies were badly mismanaged and had a history of accounting problems. The crisis only made their problems more visible.

On September 15, 2008, Lehman Brothers announced bankruptcy after huge losses from subprime mortgage-related securities on its books. Lehman was not taken down by a liquidity crisis or a bank run: unlike Bear Stearns half a year earlier, Lehman had access to the Fed's discount window, and major counterparties including Goldman Sachs, Merrill Lynch, and Citigroup continued doing business with it until the end. Lehman Brothers failure was a result of the slow erosion of capital in a stream of write-downs. This time the Fed did not come to the rescue, and the shareholders were wiped out, while the bondholders recovered less than ¢10 on the dollar.

The next day, September 16, 2008, one of the largest insurers of mortgage-backed securities, AIG had to be rescued by the Fed's credit line of $85B (expanded by March 2009 to $163B). The crisis at AIG came about after the company's debt rating was downgraded on September 15. The rating agencies cited mounting losses and extremely limited ability to raise cash due to falling stock price and widening yields on debt, as well as difficult capital market conditions. The downgrade allowed AIG's credit default swap counterparties to demand up to $14.5B of additional collateral, which AIG did not have. Without the rescue package, AIG would have likely defaulted within 24 hours; it had already hired the law firm to draw up bankruptcy papers. There was no run on AIG; on the contrary, several of its U.S. counterparties made an effort to prevent AIG's collapse, fearing market panic far worse than that caused by Lehman's collapse. AIG's near-failure was caused by the lack of liquidity that prevented it from deleveraging as required by the market conditions. The contagion effect was avoided by the costly rescue by the U.S. government based on "too large to fail" concept.

Also on September 16, the Reserve Primary Fund, the nation's oldest money-market fund, announced that its NAV fell to $0.97. The fund, which had $62B in assets as of the morning on September 15, became the second in the industry's 37 year history to "break the buck." The cause was the $785 m face amount of Lehman Brothers commercial paper and medium-term note securities in the fund's portfolio. These securities were revalued from par to $0.80 per dollar on the morning of September 15, but the NAV remained at $1.00. Nearly $39B of redemption requests were submitted at $1.00 per share by investors who learned about the fund's exposure to Lehman debt (this information was publicly available). At 3 p.m. on September 16, Lehman debt was revalued to zero. By then, the fund shrank to $23B, and the revaluation pushed the NAV to $0.97. Note that the $0.03 difference in price received before and after 3 p.m. on September 16 was caused not by the run on the fund but by the fall in the valuation of the Lehman debt.

Of course, redemptions kept coming (all but $3B in assets had been redeemed by the end of the week). Given the extreme illiquidity of the market at that time, the run could have caused enormous losses to investors. The Primary Fund dealt with this problem quite simply: it did not sell any of its assets. The first $10.8B in redemptions was paid out on September 15 with an overdraft facility from its custodian, and the rest was left unpaid: at first under a 7 day delay on redemption payments, and then under an SEC-approved suspension of all redemption rights. The fund later announced that any assets that it cannot sell at or above cost would be held until maturity (3–12 months). Face with a choice between capital preservation and liquidity, the Board of the Primary Fund went with capital preservation.

The problems with the Primary Fund scared investors in other money-market funds, even those without any exposure to Lehman Brothers. The $10B Reserve Government Fund, which invested in agency securities, never broke the buck, but faced $6B in redemption requests for the week. Due to illiquidity of the agency markets, The Reserve followed precisely the same approach with this fund (it was later liquidated at $1).

On September 17, in a then unprecedented move, Putnam Investments announced the liquidation of its $12B money-market fund after significant redemption pressure. Liquidation was another technique to avoid selling assets when the market could not provide liquidity. The fund maintained $1 NAV and was acquired the following week by Federated Investors at $1 per share. In total, investors in institutional nongovernment money-market funds redeemed a record $193B (or about 14% of the total assets) during the week; $82B of that money flowed into funds that invested in treasury bonds.

The runs on the money-market funds contained elements of both Diamond–Dybvig and fundamental bank runs, as investors feared both the loss of liquidity due to redemptions and the loss of NAV due to poor asset credit quality. Such runs threatened irreparable damage to the sectors of the economy that relied on short-term capital market, as money flowed out of the money-market funds that invested in any kind of commercial paper. As a result, on September 19, 2008, the Treasury Department announced a temporary guarantee program to protect shareholders of money-market mutual funds from loss of principal. The program, offered at an annualized cost of 0.04%–0.06% of the fund's total NAV, is currently scheduled to last until at least April 30, 2009. The insurance was only offered to funds that consistently maintained NAV at $1, and only covered positions as of September 19, 2008 (not new purchases). The intention was to prevent the run on mutual funds by existing shareholders, rather than use the government backing to attract new flows. All major fund families joined the program between September 29 and October 8. The following week, the negative asset growth in the most fragile sector, the institutional nongovernment money-market funds, was reversed. Nevertheless, as of February 25, 2009, the total assets in this sector barely approached the September 17, 2008 level, and remained far below the pre-Lehman bankruptcy level (see Figure 3.7).

On September 25, 2008, Washington Mutual Bank (WaMu) was closed by the Office of Thrift Supervision (OTS) in the largest bank failure in U.S. history. In a way, it was another victim of Lehman's collapse: the OTS announced that its action came as a result of a run on WaMu that began on the day of Lehman's bankruptcy, and drained it of $16.7B in deposits over the 10 day period. As with IndyMac, most WaMu depositors were fully FDIC-insured, and they ran because they were poorly informed about deposit protection and scared by Lehman's collapse. This "run of misunderstanding" clearly contributed to the bank's collapse by stripping it of liquidity. Whether WaMu was solvent at the time it was taken over is

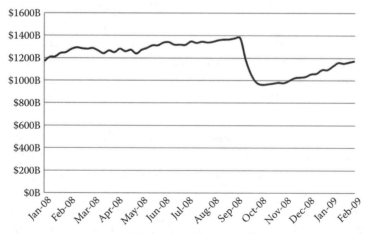

FIGURE 3.7   Money-market mutual fund net assets.

far from clear. On the one hand, the shareholders and subordinated debt holders were wiped out, which suggests insolvency. On the other hand, the regulators' incentive was not to protect shareholders but rather to secure the deal as soon as possible to minimize the potential cost to taxpayers (indeed, no FDIC funds were spent). FDIC might have sold WaMu at fire-sale prices to safeguard taxpayers' money.

3.6 CONCLUSION

Bank runs on the shadow banking system have been a significant factor in the spread of subprime losses to the overall financial system. However, most bank runs could be classified as *fundamental* (i.e., driven by the lack of trust in the institution by each investor individually, rather than by the coordination mechanism characteristic of the Diamond–Dybvig runs). Therefore, traditional bank run safeguards would have been relatively ineffective.

More importantly, highly leveraged shadow banks with illiquid assets suffered from the loss spiral effect whereby they were forced to deleverage due to higher margin requirements and falling asset prices. This deleveraging increased margin requirements and reduced asset valuations, thus fueling the next round of the loss spiral (Brunnermeier, 2008). In addition, this inherent instability was exacerbated by a variety of contributing factors.

First, informational problems reached unprecedented levels. Investors could not trust security ratings, price discovery was not functioning due to the lack of trading, banks' exposure to toxic assets was hidden from everyone (including the banks' management), and so on. As Fed Governor Mishkin (Mishkin, 2008) pointed out, "financial innovations often have flaws and do not solve information problems as well as markets have hoped they would." When these flaws become evident, financial markets seize up.

Second, numerous agency problems distorted incentives for market participants. For example, loan originators did not have to provide any warranty about loan performance beyond a short initial guarantee, and hence were motivated by fees rather than loan quality.

Third, the reliance on historical data to estimate future risk has been discredited numerous times, but was nevertheless widely practiced at the highest level of decision making, both by the industry and the regulators. As a result of this poor risk modeling, investment losses and defaults were far larger than expected. This scared many investors away from any risky assets altogether.

Fourth, the lack of a multilateral settlement mechanism in such markets as CDS created gridlocks where trading partners could not cancel out offsetting positions because of concerns about counterparty credit risk (Brunnermeier, 2008).

REFERENCES

Allen, F. and Gale, D. (1998) Optimal financial crises. *Journal of Finance*, 53(4): 1245–1283.

Bennett, P. and Peristiani, S. (2002) Are U.S. reserve requirements still binding? *Economic Policy Review*, 8(1): 53–68.

Best, A. M. (2006) U.S. banking trends for 2005. Special Report, A. M. Best, Oldwick, NJ.

Brunnermeier, M. (2008) Deciphering the liquidity and credit crunch 2007–08. Working Paper, National Bureau of Economic Research, Cambridge, MA.

Cecchetti, S. (2008) Monetary policy and the financial crisis of 2007–2008. Working Paper, Centre for Economic Policy Research, Washington, D.C.

Cohan, W. D. (2009) *House of Cards: A Tale of Hubris and Wretched Excess on Wall Street*. Doubleday: New York.

Cox, C. (2008) Sound practices for managing liquidity in banking organizations. Available at http://www.sec.gov/news/press/2008/2008-48_letter.pdf.

Diamond, D. and Dybvig, P. (1983) Bank runs, deposit insurance, and liquidity. *Journal of Political Economy*, 91: 401–419.

Ennis, H. (2003) Economic fundamentals and bank runs. *Economic Quarterly*, 89(2): 55–71.

FDIC (2006) Quarterly banking profile. Report, Federal Deposit Insurance Corporation, Washington, DC.

Gerardi, K., Lehnert, A., Sherland, S., and Willen, P. (2008) Making sense of the subprime crisis. Working Paper, Federal Reserve Bank of Boston, Boston, MA.

Gorton, G. (1988) Banking panics and business cycles. *Oxford Economic Papers*, 40(4): 751–781.

IMF (2008) World Economic Outlook. Report, International Monetary Fund, Washington, D.C.

Jevons, W. S. (1875) Influence of the Sun-spot period on the price of corn. *Nature*, 13, 15.

McCulley, P. (2007) Global central bank focus. Available at http://www.pimco.com/LeftNav/Featured+Market+Commentary/FF/2007/GCBF+August-+September+2007.htm.

Mishkin, F. (2008) On "Leveraged losses: Lessons from the mortgage meltdown." Proceedings, U.S. Monetary Policy Forum, New York.

Shin, H. S. (2009) Reflections on modern bank runs: A case study of Northern Rock. *Journal of Economic Perspectives*, 23(1), 2008.

CHAPTER 4

Ineffective Risk Management in Banking: Bold Ignorance or Gross Negligence?

Wilhelm K. Kross and Werner Gleissner*

CONTENTS

* The authors wish to thank Mrs. Judith Süssmeier-Kross and Mr. Thomas Berger for their
  valuable support in helping to compile this chapter and to perform background research.

An analysis of the subprime crisis highlights that numerous factors and hazards were not exactly unforeseeable. So, why did financial institutions not act or react much earlier? Why have outdated valuation and portfolio optimization models not been abandoned? Have regulatory frameworks such as Basel II been useful? What was learned from this recent crisis as well as earlier financial crises? This chapter is an attempt to provide qualified answers to these questions. The authors demonstrate that a combination of organizational and market-driven corrective steps are called for including a reorientation of incentive systems, truly living up to what is called for in corporate governance, establishing enhanced risk methodology, and impeding the use of risk methodological approaches, which have demonstrably been proven wrong.

4.1 INTRODUCTION

A thorough analysis of the subprime crisis triggers and their consequential effect on the financial sector and ultimately on the world economy highlights that numerous factors and hazards were not exactly unforeseeable. Hence, inter alia, the following questions can and should be raised:

1. Why did financial institutions not act or at least react much earlier?

2. Have regulatory frameworks such as Basel II been useful or have they added to the problems at hand?

3. Why did regulatory agencies not react and act much earlier?

4. Why have valuation and portfolio optimization models that are based on perfect capital market theories not been abandoned many years ago?

5. Why were corporate governance frameworks effectively ignored?

6. How far has poor risk communication added insult to injury?

7. Were the extensive state guarantees and loans truly necessary?

8. What truly are the lessons learned from this recent crisis as well as from earlier financial crises?

This chapter is an attempt to provide qualified answers to these questions, without overemphasizing the theoretical foundation and the extensive empirical background research in the field, the details of which may be provided on request to interested patricians and researchers. Some readers may be surprised to learn how long some rather significant defects and inefficiencies in banks' risk management systems have remained uncompensated and how far value was hence destroyed irresponsibly by those who should have known better. Moreover, it is demonstrated that a combination of organizational and market-driven corrective steps are called for, including but not limited to a reorientation of incentive systems, truly living up to what is called for in corporate governance, and rendering it compulsory to use at least those aspects of enhanced risk methodology that have been established for numerous years, while on the other hand impeding the use of risk methodological approaches that have demonstrably been proven wrong.

4.2 REVIEW OF CRISIS COVERAGE IN MEDIA AND BUSINESS JOURNALS

For a more in-depth appreciation of the inherent inefficiencies of risk management systems in banks, it is appropriate to first briefly reflect on the recent subprime crisis and the direct and indirect effects it had on the world economy. To set the stage we should summarize a variety of facts and interpretations from the recent media coverage, given that the subprime crisis and its consequential effects happened too recently to have enjoyed a sufficiently widespread coverage in academic literature. In fact, the book in which this chapter is published has a good chance of being one of the first edited books on the recent banking crisis.

As the media have demonstrated, a few fundamental aspects have been misinterpreted in North America and elsewhere. Reportedly, the boom in the United States property market in the early and mid-2000, caused overconfidence and rather biased interpretations of property prices. Banks started to issue mortgage bonds for private and commercial properties for 100% of the purchase price, or even more. Mortgage financiers started to offer structures such as 100% financing including transfer fees, and in some cases allowed that the repayment of funds was only going to start with a delay of several months after having taken occupation of the property. Needless to say, this equated the financing of well in excess of 100% of the original market price of the property.

A distinct misconception was the belief that property investments are safe and secure, and always on the rise. Historically this may have been the case, as is evidenced by the fact that neither the burst of the new economy bubble nor the various energy crises truly affected the property market by orders of magnitude. The most recent banking crisis has shown, however, that property prices are not necessarily uncorrelated. As the media reported, changing interest rates and certain changes in the economic climate triggered the increased number of defaulting borrowers. Free-falling property market prices then led to the observation made by more home owners that it was possibly easier to simply vacate the property and hand it back to the bank than to pay off a mortgage bond that by far exceeded the true value on the market. With more and more home owners copying that pattern, and commercial property becoming more speculative and less profitable due to lower occupation rates, a downward spiral evolved with a much higher magnitude of impact on the world economy than would be have been expected from the singular set of event or stress factors—the mortgage loan crisis—in the United States, as shown in Figure 4.1.

Falling property prices alone could not have had the ripple effect that has been observed across the globe in recent months. The problem at hand was more complex. A fundamental issue, which was discussed quite openly in the media, related to the fact that banks had started to combine the traditional mortgage bond financing business with more sophisticated capital market transactions. This issue of mortgage bond financing, which has been somewhat simplified for this chapter, was no longer the business of simply issuing funds to a borrower, refinancing this on the capital market with the inherently better rating that a bank possesses, and living off the difference between the bank's refinancing rate and the borrower's mortgage bond interest conditions. Rather, the lending was repackaged into longer term asset-backed securitization (ABS). These were sold to conduits, which were then refinanced through commercial papers. These

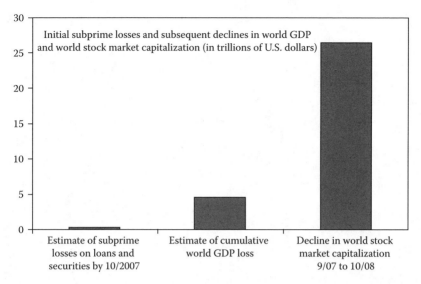

FIGURE 4.1 Initial subprime losses and subsequent declines in world GDP and world stock market in trillions of U.S. dollars. (From Blanchard, O., The crisis: Basis mechanisms, and appropriate policies. Working Paper 09-01. Massachusetts Institute of Technology (MIT)—Department of Economics; National Bureau of Economic Research (NBER), Cambridge, MA, 2008. With permission.)

commercial papers, asset backed, were perceived to be very secure investments, inter alia given that the financial system had earlier survived such shocks as 9/11. Liquidity risk was at the same time perceived to be close to zero and was secured by a liquidity line that in most cases was well underdimensioned. In some cases, this coincided with a lack of diversification in the business model, as the case of Bear Stearns indicates (Figure 4.2).

To make things worse, even more financial products (i.e., certificates and the like) were restructured, in most cases reflecting derivative structures, which, it is fair to say, were not truly understood by the large majority of investors. Not understanding risk profiles implies that risk management systems become unproductive or even counterproductive. In turn, when operating without effective risk monitoring and analysis systems, organizations became overextended, overleveraged, and undercapitalized without even knowing what they are really in for. It is analogous to flying in bad weather without radar.

Government intervention, and in some cases the lack of government intervention, rendered things worse as the case of Lehman Brothers may indicate. The reported financial losses of almost $100 billion in 1 year as

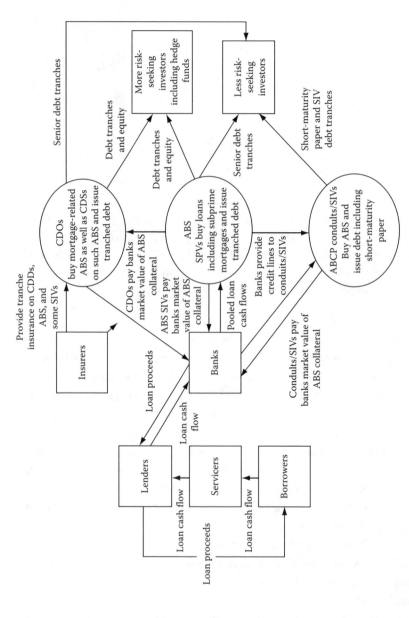

FIGURE 4.2 Mortgage market flows and risk exposure. ABS, asset-backed securities; ABCP, asset-backed commercial paper; CDO, collateralized debt obligations; CDS, credit default swap; SIV, structured investment vehicle; SPV, special purpose vehicle. (From International Monetary Fund, *Global Financial Stability Report*, Washington, DC, 2007, p. 11. With permission.)

the case of AIG reflects (including the astonishing 2008 fourth quarter loss in excess of $67 billion that was apparently caused by credit insurance payments that had been negotiated to secure CDOs) certainly exceeds the order of magnitude that almost any financial analyst would have ever expected, or that a risk radar system would have assumed as a shock scenario. And the reader may be aware that these tremendous losses were incurred in spite of the U.S. government's cash injections to AIG totaling some $150 billion throughout 2008.

Incidentally, the abovementioned case example of Lehman Brothers indicates that even further misconceptions and crisis amalgamating factors must have prevailed. Some of these have been also identified in common literature and have been discussed in the media. One such aspect is related to the level of independence of larger banks and the impact that their failure might have at home and abroad. The media have speculated that the U.S. government decided not to intervene in the failure of Lehman Brothers because a significant part of their involvement and in particular their liabilities were outside the United States. Reportedly, at least within the United States, the U.S. economy appears to have been interpreted for many years as being largely independent, thereby not taking cognizance of the fact that the U.S. contribution to the worldwide gross domestic product had declined to the current level of approximately 27%, as reported by the World Bank.

Hence a much higher degree of interaction between the U.S. economy and those of other parts of the world has evolved than had historically been assumed. Furthermore, investments in certificates and similarly structured financial products were perceived to be as safe as investments in other investment vehicles such as mutual funds, stocks, and bonds, this in spite of the fact that certificates in most cases simply are a structured loan that is provided to the issuer of the certificate, usually backed and secured by a minute portion of the investment vehicle. At the same time, savings deposits were believed to be safe and commodity prices were believed to be on the increase from now until eternity. And yet another fundamental error was the belief that very large banks would survive, no matter what happens, and that neither the U.S. real estate price slumps nor the banking crisis would have a significant effect on other industry sectors such as automotive, transport, electronics, and retail.

Another severe misconception is related to the United States' financial reporting standards, which were for many years believed to be superior to those implemented elsewhere because they enabled financial institutions to reflect unrealized profits fast, and of course derive attractive personal bonus payments in upward markets. The rather detrimental and crisis-accelerating effects of such standards in a downward-oriented market were ignored or not appreciated. It appears that more than a few of the former proponents of the current financial reporting standards are now calling for different solutions, which would enable them to in some regards reflect the valuation of market values as they will be achievable over the medium to longer term, in particular when it is clearly not their intention or when there is no other good reason to have to dispose of a currently undervalued asset in a hurry.

Recent media coverage provides insight into the fact that shocks to the financial system have been there before, many times in the last 100 years. The issue seems to be, however, that in 2008, a combination of two shocks coincided, the one of a banking bubble triggered in turn by a property price bubble, and that of a serious deterioration of investor confidence. It is fair to say that the implications to the financial system are rather far-reaching because these coincide with other effects that have the tendency to make things worse, these including but not limited to rating agencies who tend to overextrapolate upward and downward trends, an international financial reporting system that tends to drive investors and other players in the financial market toward short-termism, and significantly overex-tended and cash-strapped governments. The new liquidity crisis which has evolved in recent months and weeks is clearly of an extent that most active players in the financial market and most governments had never expected. By now, refinancing has become very expensive, and the volume of funds that is loaned to the private sector has dropped sharply. Banks had to raise their equity positions in order not to be penalized by business analysts with a downgrading of the rating, a crisis-enhancing strategy which coin-cided with the raising of minimum requirements in order to qualify for financing, and the hesitation or nonpreparedness of bureaucrats to engage in any decision at all. Governments provided guarantees to the major players in the financial market, thereby in some cases jeopardizing the stability and the rating of their own currencies. Speculation on defaulting governments has recently added to the lack of investor confidence, as the cases of Iceland, the Baltic States, Hungary, and other countries in Eastern Europe may indicate. As *The Economist* reported in its February 29, 2009 edition, the biggest risk in the emerging world today comes not from sovereign borrowing, but from the debts of firms and banks. As foreign

	If one green bottle should accidentally fall...			
Country	Current-account as % of GDP*	Short-term debt as % of reserves*	Banks' loan/ deposit ratio	Overall risk ranking†
South Africa	−10.4	81	1.09	17
Hungary	−4.3	79	1.30	16
Poland	−8.0	38	1.03	14 =
South Korea	1.3	102	1.30	14 =
Mexico	−2.5	39	0.93	12 =
Pakistan	−7.8	27	0.99	12 =
Brazil	−1.5	22	1.36	10 =
Turkey	−2.3	70	0.83	10 =
Russia	1.5	28	1.51	9
Argentina	0.2	63	0.74	8
Venezuela	0.8	58	0.75	7
Indonesia	1.2	88	0.62	6
Thailand	0.3	17	0.88	5
India	−2.4	9	0.74	4
Taiwan	7.9	26	0.87	3
Malaysia	11.3	15	0.72	2
China	5.2	7	0.68	1

Sources: HSBC; Economist Intelligence Unit   *2009 Forecast   † Higher score implies higher risk

FIGURE 4.3   Current economic liquidity indicators. (From N.N., Domino theory, *The Economist*, Feb. 29, 2009. With permission.)

capital dries up, they will find it harder to refinance maturing debts or to raise new loans. Figure 4.3 introduces three indicators to judge how vulnerable economies are to the global credit crunch. Among the countries in the table, Pakistan, South Africa, and Poland are tipped to run current-account deficits of 8% or more of GDP this year—the size of Thailand's deficit before its crisis in 1997.

Based on the background research that was published by *The Economist*, in contrast, the Asian emerging markets generally look the safest, the main exception being South Korea, which, thanks to its large short-term foreign debts and highly leveraged banks, is deemed to be as risky as Poland. Vietnam, though not included in the table, reportedly scores high on the risk rating too. It is worth mentioning that the overall score in the table only ranks countries' relative risks. To assess the absolute risk of a crisis, one needs to estimate external-financing needs (defined as the sum of the current-account balance and the stock of short-term debt) over the next 12 months. As reported by *The Economist*, UBS has reportedly calculated the gap between this and the stock of foreign-exchange reserves for 45 countries. While only 16 of these countries have a financing "gap," all others reflect reserves that are more than sufficient to cover a year's worth of payments, even if there were no new capital inflows (Figure 4.4). Virtually all of those 16 countries are in Central and Eastern Europe. Luckily, most emerging economies' large reserves will help to keep them out of danger. However, the longer the credit crunch continues, the more those reserves will start to dwindle. Further speculation on what the next shock to the financial system will be, and when this will likely occur (e.g., a commodity crisis or a significant change in inflation rates) adds insult to injury. The insurance broker AON reflected in its 2009 political risk map (see www.aon.com) that sovereign default risk is exceptionally high in numerous countries (i.e., the list being topped by Zimbabwe, Kyrgyzstan, and 11 other countries considered rather critical) and that numerous countries are particularly vulnerable to a commodity crunch.

Reportedly, the financial crisis is now quite clearly visible in other market sectors too, and may reach an overall impact that goes well beyond the losses that were experienced by banks. Beyond financial institutions, the automotive sector appears to have been the first to experience the shock, as the recent negotiations between the large automotive manufacturers and the U.S. government demonstrate. The fast reaction is no surprise given the level of sophistication that is inherent in the optimized just-in-time management approaches and the rather complex logistics networks in that sector. But it appears that cross-sector economic forecasts and in particular unemployment figures are recognizing now that the impact of the financial crisis will be far greater than had been assumed just a few weeks ago.

Of course, the rather significant deterioration of value in the order of trillions of dollars has provided opportunities too. Governments investing in financial market participants, while their stock is traded extremely low, may have found a means to reduce overall government debt by orders of magnitude in the medium to long term. Investment bankers specialized in the field of mergers and acquisitions have become busy, given that more and more organizations are unable to pay their debt (as assessed in conformity with the rather counterproductive financial reporting standards),

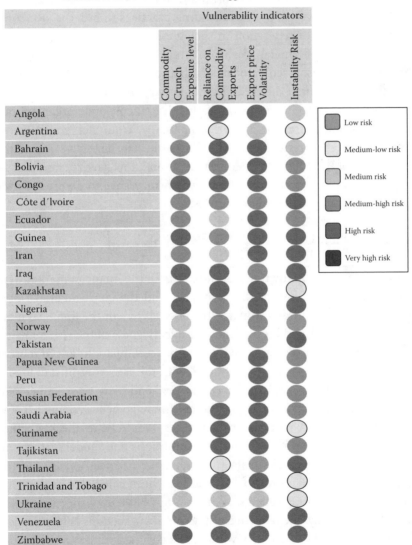

Commodity crunch exposure matrix

Volatility in global commodity prices in the 1970s and early 1980s contributed to political and economic instability in a number of countries. What countries are vulnerable in 2009 if commodity prices continue to fall, as some forecasters suggest?

FIGURE 4.4   Commodity crunch matrix. (Extracted from AON's 2009 political risk map, see www.aon.com. With permission.)

or given that for whatever reasons their credit lines have been terminated now seek a suitable purchaser instead of going into liquidation.

Needless to say, while this chapter was written, the end of the financial crisis was and is not foreseeable. Some media are speculating whether investment bankers and those working to support their glory might currently engage in overly restrictive lending practices in order to overcompensate their loss of reputation in society at large, this being in spite of their remarkable financial wealth. Certain signs are presently visible, though, and are discussed in the media that some sectors might have managed themselves out of the crisis, particularly in the United States, as long as (other) banks and large economically relevant corporate organizations obtain sufficient guarantees and bridging finance from their governments to survive the short term. Hence, the U.S. banks may have experienced the worst parts of the 2008/2009 financial crisis and now have reasons to look forward to a brighter future; a trend which may have to be proven over time to be paralleled in other countries.

4.3 A CLOSER LOOK AT MODERN RISK MANAGEMENT SYSTEMS

It is fair to say that many if not most players in the financial market were taken by surprise. The severity of impact and the consequential ripple effects clearly were underestimated by most players in the financial market, and beyond. A good test of the true effectiveness of the risk management systems that were employed, would be to compare financial market participants' forecasts for the year 2008, with the rather significant losses, in some cases, that were experienced at year-end. Any unexplained gap reflects a defect or inadequacy of the risk management system that was employed. A slightly more sophisticated test would be to check whether the integration of certain additional assumptions and model features, to reflect the learning results from the 2008/2009 financial crisis, and the re-simulation of the figures that should have been calculated as the 2008 budget, in combination reflects the impact of what truly happened.

Be this as it may, the objective of the following sections is to provide a discussion of further factors, beyond media coverage, that can be demonstrated to have had an effect on organizations performance in the recent banking crisis. While this discussion is not exhaustive and all-encompassing, it does provide food for thought and sensible concepts and ideas on what to improve in the not so distant future. The reader is advised to paint his or her own picture of whether the severity of failures of risk management systems, and the rather long time which has passed since these effects were first detected, must result in the summary conclusion that some players in the market have been grossly negligent as opposed to having been just boldly ignorant.

4.3.1 Lending Practices

Even the superficial coverage that the subprime crisis and its consequential ripple effects have enjoyed in the media, has demonstrated that banks' lending practices have been problematic. Markets can go up, and down, and an upward or downward move can have an implication on property price regimes. The same applies to the field of vehicle financing, which— as is visible in the media—pushes car sales by means of offering discounts and financing mechanisms that, as it appears in some cases currently, are worse than the practices that had been prevalent in property financing. Some countries reportedly are now struggling to provide sufficient spare for the (temporary) storage of repossessed cars.

Unless banks are keen on experiencing similar or even worse financial crises again, lending practices need to be scrutinized to be limited to a realistically achievable residual value of the property, assuming that a borrower defaults and that backup pledges or guarantees at that stage cannot be realized or liquidated in the near future.

It is conceivable too that the way in which individuals or companies are rated, is in need for improvement. In particular, it is questionable how good pledged securities depots, personal guarantees, etc., truly are in periods of economic downswing, and whether common haircuts are sensibly calibrated. As the discussion below on liquidity risk outlines, it has been common knowledge for many years that if an asset must be sold in a hurry, the realizable market value may be far less than the one to be achieved if time is not an issue.

While banks and regulatory agencies will need to ensure that credit practices will need to become a little more robust and restrictive, both governments and regulatory agencies will need to ensure that lending practices will not become as overly restrictive as the case has been in many countries in recent weeks. Refinancing has become rather expensive, which renders lending less profitable and ultimately leads to an increase in lending rates to be passed on to borrowers, a good recipe to enforcing the downside impact of a recession. At the same time, while jobs in the banking industry are being slashed by the thousands, bank branch staff, credit specialists, and some of the many bureaucrats in the financial systems have become overly sensitive in order to certainly make no mistake ever, thereby possibly overcompensating their fear that they might loose their jobs too at some stage. Needless to say, in the interest of most people and organizations, this downward spiral needs to be stopped; the sooner the better.

4.3.2 Ineffective Operational Risk Management

Widely published case examples such as the recent trading scandal at the Société Générale have demonstrated that operational risk management systems can be rather ineffective, in spite of the fact that regulatory regimes such as the recent transformation of Basel II into European Community legislative and regulatory frameworks for banking had attempted to address such organizational defects rather intensely. Needless to say, a more stringent and robust real-life implementation of corporate governance is called for, in combination with the enhanced market discipline and the more proactive involvement of regulatory agencies as is called for under the second and third pillars of Basel II (Figure 4.5).

And as submitted elsewhere, real-life implementation should start with the focusing of risk management strategies (Figure 4.6), not like the commonly observable "tail wagging the dog" (Kross 2006).

In addition, it appears that known facts and scientific evidence on risk-taking behavior has not been reflected as boldly as it should have in the said regulatory framework commonly referred to as Basel II and its respective transformations into local-level regulatory frameworks on banking supervision. To mention an example, Shapira in his 1995 book on the managerial perspective of risk-taking behavior, demonstrated that individuals and managers quite consistently deviate in their risk-taking behavior as a function of perceived success or failure. While the tendency to take risk is relatively low, as decision makers experience a little success, a little failure or simply a break-even, risk-taking behavior increases rather considerably when considerable or significant success is experienced (i.e., decision makers are prepared to bet). Moreover, as things evolve toward the negative side, that is, considerable failure is experienced, personal risk taking increases tremendously, far higher than on the positive side. Rather significant discrepancies evolve between what decision makers believe they would do, versus what others would do, versus what they should do.

Needless to say, risk management systems and their ability to cope with subjective estimates and personal judgment need to be sufficiently robust to be able to compensate the negative extremes. Given that control mechanisms in the banking sector already are rather stringent, and procedural,

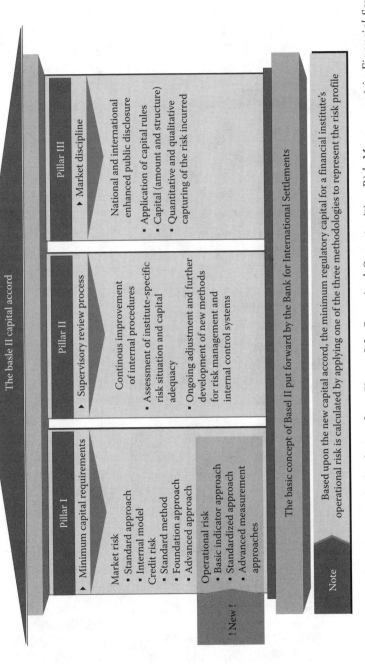

FIGURE 4.5 The Basel II framework. (Redrawn from Kross, W., *Organized Opportunities: Risk Management in Financial Services Operations.* Wiley-VCH: Weinheim, Germany, 2006.)

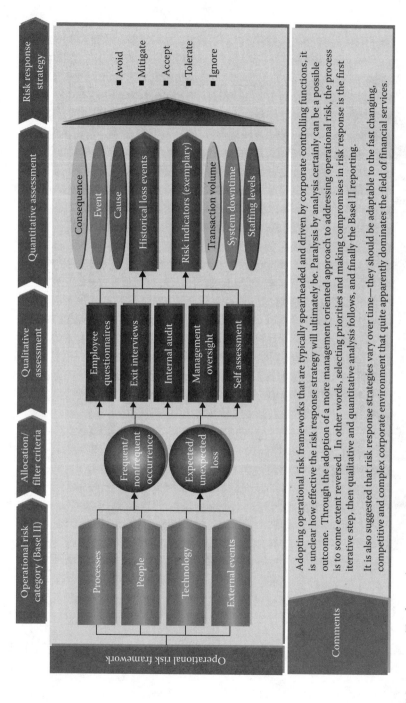

FIGURE 4.6 Traditional approach to implementing Basel II OpRisk. (Redrawn from Kross, W., "Integrating management" into "OpRisk management". In: Gregoriou, G. N. (Ed.), *Operational Risk Toward Basel III*. John Wiley & Sons: New York, 2009.)

it is submitted that adding more elements of control to this environment should not be the predominant mechanism to achieve the desired results. Rather, a cultural alignment of decision makers is called for, which in all likelihood can only be achieved if reward and recognition systems take cognizance of risk factors in addition to being linked to the key indicators of sales volume and profitability. Risk-adjusted performance should be rewarded, an element that unfortunately is not even reflected in the bulk of real-life implementation of balanced scorecards (Figure 4.7).

4.3.3 Appropriate Measures of Risk

It has been recognized for a long time that risk can be described using a diversity of key indicators and risk measures (Figure 4.8). Similarly, it has been recognized that defining risk as the volatility in future cash flows, is an inappropriate and incomplete reflection of what truly needs to be addressed in risk management.

While it is recognized that modern banks do employ a number of approaches to measuring risk and assessing the likelihood of failure, it is submitted that there is room for improvement as Berger and Gleissner demonstrated in 2006 for German publicly traded corporations. When estimating the market risk inherent in securities, for example, or employing the Markowitz approach to asset portfolio optimization, volatility is the predominant risk key indicator while on the other hand certain downside risk indicators such as the value-at-risk (VaR), the conditional VaR, the deviation VaR or what is commonly referred to as the maximum drawdown are widely ignored. In fact, such downside risk metrics are easier to understand intuitively, than volatility is. At the same time, it is submitted that sensible key indicators to measure risk must take into consideration what clearly is at risk, and how this risk could best be described. For example, for ABS transactions, it likely is inappropriate to simply reflect the extrapolated volatility of historic cash flows over the last 250 days, if on the other hand credit risk is a predominant factor in the overall future performance.

The earlier notion that banks and investors were faced with a lack of transparency and with rather limited capabilities to assess the true nature of risk factors inherent in structured and exotic financial products raises some further fundamental concerns. In particular, it appears that key risk indicators were assessed at a level that was far too abstract and too high-level to serve the truly intended purpose. Also, the number of key performance indicators used was very small (Figure 4.9).

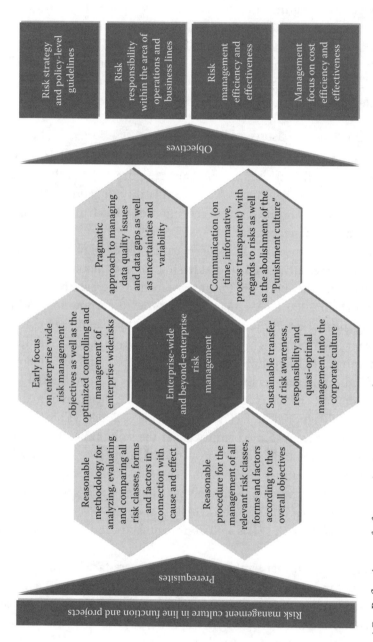

FIGURE 4.7 Reflecting soft factors in enterprise risk management. (Redrawn from Kross, W., *Organized Opportunities: Risk Management in Financial Services Operations.* Wiley-VCH: Weinheim, Germany, 2006.)

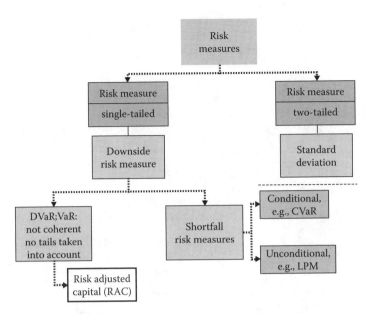

FIGURE 4.8   Risk metrics.

The wealth of funds and hedge funds that evolved in recent years has further augmented this set of problems. Instead of simply investing in stocks or bonds, rather significant volumes of money were invested into intermediary structures. In turn, to assess the level of risk inherent in a fund, performance analyses and risk valuation approaches were typically focused on the last 250 days' performance as opposed to the inherent risk profiles of those stocks and bonds investments (and synthetic, interme-diary, or derivative structures), which were the results of asset portfolio management optimization within the fund. This coincided with perfor-mance and risk-reporting standards which were minimal compared to the level of effort that a bank has to engage in. To render things worse, some funds decided to invest in yet other funds, which in turn made it nearly impossible for investors to perform drill-down analyses into the truly underlying risk factors. It has been widely discussed in the media that most if not all larger players in the investment banking arena, and beyond, directly or indirectly overinvested in the U.S. property market which had been performing so well for years—without truly recognizing the extent of their exposures.

As a "lesson learned" from the recent financial crisis, it is submitted that in future, risk metrics must take into consideration what truly is at risk, and what the major risk drivers are and will be. A simple extrapolation

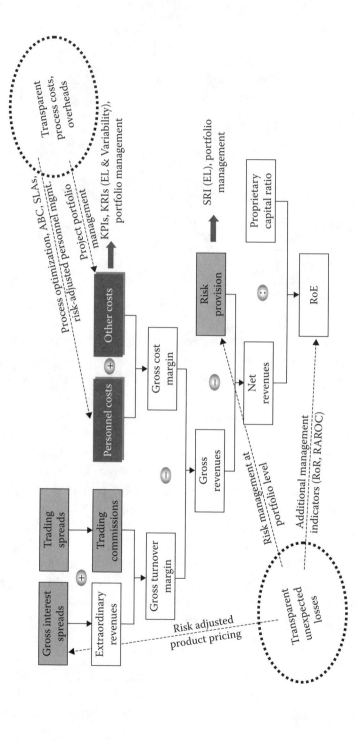

FIGURE 4.9 Introducing sensible key risk indicators and OpRisk management measures. (Redrawn from Kross, W., *Organized Opportunities: Risk Management in Financial Services Operations*. Wiley-VCH: Weinheim, Germany, 2006.)

of the past 250 days' volatility appears not to have been a good indicator of risk. Additional risk metrics are required including but not limited to the VaR, or what is commonly referred as the maximum drawdown. Economic models are required to truly explain what the impact of individual risk factors (with their inherent uncertainty) and their interaction will mean for future performance (Kross and Gleissner 2009).

4.3.4 Meta Risk

As submitted earlier, risk should not be ignored if, in spite of the inherent regulatory constraints, return on investment is the main focus. The above sections have highlighted that certain things can go wrong in the way in which risk is identified, analyzed, quantified, evaluated, and managed. Clearly, a failure in quantifying risk appropriately and reflecting the right assumptions, both relevant factors in the so-called meta risk, are issues that can no longer be neglected.

To highlight what the extent of model errors can be, recent analyses demonstrated that by means of employing a normal distribution to the stock exchange crash of 1987, one is dealing with an event that might only happen once in 10^{87} years. Empirical observations have demonstrated, however, that such crashes can happen once in 38 years (see Romeike and Heinecke 2008). Moreover, the recent financial crisis in combination with the extent of the financial downturn after 9/11 leaves room to the interpretation that rather extreme shock events, such as these recent stock market crashes, might be on the increase. Also, Chopra and Ziemba in 1993 (based on empirical evidence) found that estimation errors in the expected values of assets are of approximately 10-fold importance when compared to the respective variances, which in turn reflect double the effect of errors in correlations. Implicitly this demonstrates why the expected returns of an investment portfolio are quite relevant when the investor reflects a relatively low or moderate risk aversion. However, with increasing risk tolerance and decreasing risk aversion the sensitivity of individual assets within a portfolio increases.

It has been submitted by numerous authors that far too many risk management systems in practice place too little emphasis on the empirical evidence that the extent of risk in itself is volatile (i.e., GARCH process) and that it reflects extreme market readjustments (i.e., crashes), which are rather poorly incorporated in the standard distribution-based approaches to measuring risk that are predominant in the financial sector today. It is submitted that the rather overdue adoption of more enhanced approaches to describing risk has not been realized in practice to the extent that it should have. The reader may be aware of the fact that recent empirical evidence has inter alia provided the basis for more appropriate risk distribution functions, which in effect are a combination of normal or lognormal distributions, and Pareto distributions, for capital market data in the extreme regions that are commonly referred to as "fat tails."

With respect to the various uncertainty factors that should be reflected in a modern risk management system, one usually distinguishes estimation errors in input parameters (i.e., regression analyses of historic data), parameter uncertainty (i.e., coefficients of quantitative prognosis and regression analysis models), and uncertainty inherent in the prognoses (i.e., the qualitative assumptions in the prognoses obtained from asset managers or business analysts). Of course, only exclusively future-oriented simulation techniques will suffice to quantify risk at the portfolio level while historical simulation will likely prove to be increasingly inadequate and misleading.

One further aspect of meta risks applies to the time horizon for which risk quantification models are suitable. It has long been recognized that common practice (e.g., the Markowitz approach for Portfolio Optimization) employs single period models. In other words, usually these were originally conceived to provide forecasts for a period of maximum 1 year. In practice, however, these models have been applied to derive recommendations and support decisions for long-term investment strategies. Whether or not such models have been enhanced to include the effect of dividend reinvestments and the like plays a minor role in this regard. Much more important is the fact that such models usually reflect stable correlations, stable investment environments, and static circumstances, which in the true sense cannot be extrapolated from now until eternity.

4.3.5 Compensating the Inappropriate "Perfect Capital Market" Hypothesis

While the above discussions on the implication for risk management have considered the environment in which operational risk management was performed, it has long been recognized that a number of methodological errors have existed and have not been corrected in most real-life organizations to this day. One fundamental issue relates to the fact that the valuation models that are applied in practice rely on the perfect capital market hypothesis. This observation is still valid inter alia for the capital asset pricing model (CAPM), the Markowitz portfolio optimization approach, the arbitrage pricing theory (APT), and several other established valuation paradigms and approaches. The problem is the capital market is not perfect. Anomalies exist and these in fact drive financial market participants' performance. Information advantages exist, which differentiate the bearers of such privileged information and those who can benefit from such information or speculation thereon from other capital market participants. Liquidation costs, taxes, transaction costs, and other factors play a role too, as has been obvious for decades.

At the same time, the recent financial crisis and its consequential effects are a clear indication that many players in the financial market underestimated what is commonly referred to as liquidity risk. This can of course imply an upside and a downside. The downside is simply explained when considering the scenario of having to sell a property or an investment in a private entity at rather short notice. The shorter the permissible time span, the lower the realizable price, but there can be an upside too. Liquid assets that are currently valued on a mark-to-market basis to assess the volatility range over a maximum 24 h, could potentially be valued significantly higher if the current owner of these liquid assets or financial products does not intend to sell them within the next 24 h.

It has furthermore been known too for years if not decades that one of the most important factors in financial risk management are the correlations between individual investment targets and their respective risk/return profiles. Adding to the abovementioned observation on the perfect capital market hypothesis which in itself has been a problem, it is submitted that the way in which correlations are reflected in today's valuation models is largely incorrect. While it is recognized that some players in the financial market do assess and adjust correlations regularly to reflect current knowledge, typically, correlations are reflected in a risk system once and forever, as a single value between −1 and +1, and are deemed to be stable from now to eternity. The recent financial crisis is a good case example to demonstrate that correlations in real life are unstable and uncertain. Correlations and the portfolio diversion effects that are calculated on this basis, may of course be more or less stable in stable upward markets. However, on the other hand, correlations between most securities tend to approach +1 in shock scenarios similar to the one that was experienced in the recent financial crisis.

It is submitted that in order to address these problems sensibly and suitably, it is necessary to recognize that it will no longer be possible to ignore unsystematic risk all together. Moreover, the risk profiles of organizations and hence those factors that should be reflected in the weighted average cost of capital (WACC) must reflect the true focus, the extent, and the quality of risk management systems. Today's software applications render it realistically achievable to reflect correlations as an uncertain variable that is influenced by outside factors and trends, and that evolves over time as a result of certain economic factors, the interdependencies of which are largely understood. It is submitted that in future, the reflection of correlations as a stable and certain parameter can no longer be accepted by regulatory agencies as an appropriate means of measuring market or credit risk at portfolio level. And of course, the various underlying assumptions in such approaches as Markowitz-type portfolio optimization, CAPM, and APT must be corrected and compensated as a matter of urgency.

4.3.6 Compensating the Impact of Rating Agencies and Financial Reporting

In recent months, the media have rather clearly highlighted that rating agencies have contributed strongly to the financial crisis and in some cases clearly misrepresented asset valuations and the extrapolation of historical upward or downward trends. It has been submitted that rating agencies must start taking into consideration what truly is at risk in an organization, and what the range of possible impacts of such risk factors could be. It has long been submitted, and was explained in the context of the CAPM model further above, that simply using a capital market derived beta factor to express risk is a gross misrepresentation given that unsystematic risk has been proven for decades to play a role. Unsystematic risk cannot be fully compensated through diversification; anomalies exist and in fact are what drives extraordinary returns on investment.

It is submitted that rating agencies will need to adopt empirical evidence in the field of risk research that was gained in the last three decades, and inter alia adopt more enhanced economic forecast models to avoid the overemphasized and overextended extrapolation of upward and downward trends. Qualifying statements on risk/return profiles should complement recommendations on historic or future over- and under-performance. It should be taken into consideration in such qualifying statements that the so-called momentum effect often does not exist as a standalone effect. For example, some economic research has demonstrated that historical over-performers may under-perform at some stage in the future, while on the other hand historical under-performance is compensated with turnaround strategies that in turn yield considerable over-performance at some stage in the future.

Needless to say, the above observations for rating agencies are of similar validity for the development, implementation, and calibration of internal rating models, as they are applied more and more often in larger banks at this point in time. Rather than simply extrapolating historic trends, forecasts of future performance need to be explained on the basis of robust and sensible economic models as opposed to being assessed on the basis of some historic figures that bear little if any relevance in future (Kross and Gleissner 2009).

It has been recognized too and was mentioned earlier in this chapter that the international financial reporting standards (IFRS) are rather problematic in many regards. Briefly, the IFRS currently are a rather rigid mixed model which combines a mark-to-market valuation approaches as prescribed for capital market instruments and liquid assets, with the amortized residual values for certain illiquid assets such as fixed property. This was an attractive model for those who wanted to reflect current potentials quickly, and reflect the impacts of unrealized profits from certain capital market instruments as quickly as possible in their quarterly or annual reports. While this approach does not sound overly problematic at first sight, it is logical that problems occur when an investment in illiquid assets such as fixed property (e.g., an office building) is refinanced through capital market instruments and financial derivatives that have to be valued mark-to-market. Depending on market trends, a rather volatile balance sheet can evolve, which in turn might yield the observation by business analysts who assess organizations' risk positions based on historic volatility, that the organization is performing in a more volatile environment than it truly is. As an interim compensation in the absence of tangible market data for illiquid assets, it was agreed that hedge accounting relationships are defined under which the refinancing of certain objects with derivatives are treated as one unit as long as certain correlation corridors (i.e., 80%–125%) are demonstrated not to have been exceeded. Needless to say, such analyses can be rather complicated when structured financial products are involved, and the compromise is fairly weak.

To add to the above, in recent months, the IFRS have been observed to have been more problematic than had been recognized earlier on. In particular it has now been proven that through the mark-to-market approach, investors and analysts have been driven toward an unhealthy level of short-termism. Tremendous asset value write-downs had to be reflected in financial institutions' financial statements, which effectively reflect the assumption for the respective assets that the financial institutions have to dispose of them at short notice. Particularly when assets are valued lowly, however, it makes very little sense to dispose of them, unless of course certain liquidity gaps cannot be covered through other means. These, however, as the media reported, were in contrast resolved via government guarantees. In other words, for financial reporting purposes, many investments in financial instruments had to be cut down to an entirely unrealistic value, far below the true value that will ultimately be realized when they are truly disposed of. It remains to be seen whether the mark-to-market approach, due to its crisis-enhancing effects, will be abolished all together or whether yet another set of compromises will be invented.

4.4 PROBLEMS AND SOLUTIONS

The above discussion has highlighted quite clearly that in the field of applied risk management in financial institutions, rather significant flaws remained uncompensated. It is hoped that the following discussion of methodological approaches and true solutions to the problems at hand will be used as guidance both by practitioners and academia.

4.4.1 Management on the Basis of Sensible Key Risk Indicators

First and foremost, it is appropriate and well overdue that the overall approach to addressing risk is looked at rather critically. A sufficient number of fatal flaws have been highlighted in this chapter, and further issues such as the explicit decision not to include strategic risk as a relevant factor under the Basel II framework have been identified as being quite problematic. Adding the "management perspective" and rendering OpRisk initiatives operationally manageable—to if possible generate net value and competitive advantages from successfully implemented OpRisk management investments—is not a trivial undertaking, particularly when the initiative was originally conceived for regulatory compliance or risk-controlling purposes only. Numerous interlayered aspects need to be taken into consideration, and a rather complex overall master plan needs to be followed, which reflects that OpRisk management initiatives must ultimately cope with what is commonly referred to as major-scale, cultural change in an organization. It is suggested that at least the following topics need to be addressed:

- Scope, target, and the generic approach to methodology, limits, and risk retention

- Responsibility allocation, that is, the definition of risk owners/managers vs. risk controlling and process models for the line/project organization

- Implementation of a temporary central risk management function to enhance envisaged/planned/targeted/anticipated cultural change

- Setting up an OpRisk management implementation project, coping with known and unknown implementation challenges, and raising management effectiveness priorities above documentation priorities

- Moderated interviews to identify the delta between responsibilities and capabilities and functional authority, to define and negotiate escalation triggers, to provide training as needed, and to design reporting and escalation mechanisms

- A documented conclusion that risk owner responsibilities can (once trained and integrated into a reporting and escalation process structure) from now on be "lived"

Neither of these success-critical factors is explicitly mentioned in the framework referred to as Basel II, or the support documentation that was provided by the Bank for International Settlements or the regional and local-level regulatory agencies who worked on the transformation into local-level laws and regulations. The results in terms of enabled, efficient, and effective risk management, with an independent risk-controlling function, furthermore needs to be both manageable over the short, medium, and long term with respect to the explicit efforts to be undertaken by each individual and team, and with respect to the data and key indicators, which are collected and analyzed to support risk-based decision making. A horror vision commonly voiced by several larger banks in Europe highlights that capturing and collecting several 10,000 data points on a daily basis, quality-enhancing them and maintaining them over a period of 10–30 years, certainly is not manageable realistically.

Taking into consideration that any financial institution has implemented risk management in the one or other way already, it is hence appropriate to introduce a graphical flowchart of sensible steps for the reorientation and refocusing operational risk management (Figure 4.10)—thereby

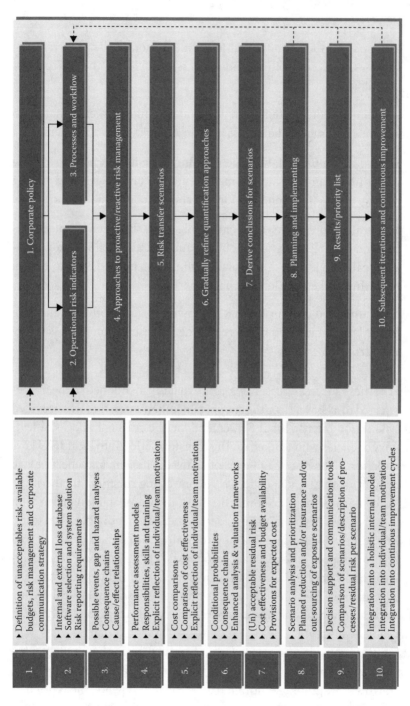

FIGURE 4.10 Recommended approach to refocusing OpRisk management. (Redrawn from Kross, W., *Organized Opportunities: Risk Management in Financial Services Operations.* Wiley-VCH: Weinheim, Germany, 2006.)

focusing on the management perspective instead of just brainless regulatory compliance reporting.

The above discussions on appropriate measures of risk, and on the treatment of meta risk factors, do not need to be repeated here except to say that volatility as the predominant measure of risk does not suffice. Downside risk measures such as the VaR, the conditional VaR, the deviation VaR, or what is commonly referred to as the maximum drawdown, as well as "safety first" approaches should be used as complementary approaches. Of course, the perfect capital market assumption and the various abovementioned features that go along with it must from now onward be treated as being outdated, wrong, and in the true sense unnecessary given today's computing power of enhanced simulation modeling software. At the same time, parameters need to be identified and assessed to describe the true risk aversion of an investor in order to derive optimized suggestions and recommendations for the structuring of investment portfolios and the reflection of residual risk restrictions (i.e., safety-first approach, guaranteed return, and portfolio insurance concepts).

As discussed elsewhere, risk management should conceptually be shifted from the predominant regulatory compliance focus toward a true value-based management. Besides the more cultural issues mentioned above, on the technical level, this calls for risk management practices that eventually support the management with better risk information. To be able to fulfill this task, especially the role of risk modeling and aggregation must be emphasized (Figure 4.11).

Risk factors affecting overall goals should all be systematically identified and then evaluated. Nowadays it is common to evaluate risk using binomial distributions for outcomes, that is, a standardized and analytically robust function to capture exactly quantified consequences. As is seen in practice, however, risk in most cases cannot be modeled using just one possible outcome, but rather to reflect a range of possible outcomes. This implies that besides the commonly used normal distribution in models, and the binomial distribution for event-driven risks, alternative distributions must be taken into consideration. This includes triangular distributions, Pareto distributions, or even uniform distributions. Closely linked to this, risk exposures should be derived from an aggregation of risk factors with the help of simulation techniques, rather than continuing to rely on the commonly established approaches such as variance–covariance models, which are based on normal distributions. Also, correlations between risk factors have to be analyzed individually, and reflected explicitly.

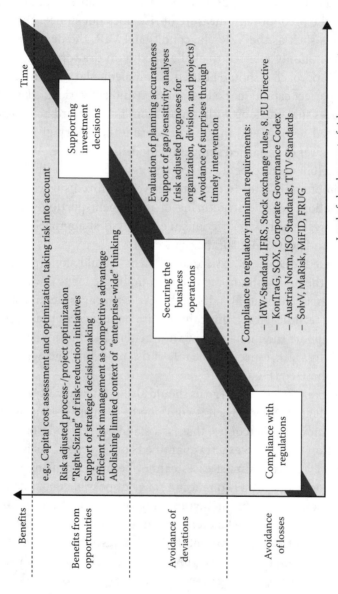

FIGURE 4.11   Refocusing the goal and objectives of risk management.

This calls for economic models to map interrelationships of economic factors and commonly used key indicators, as a robust prognosis.

The most suitable risk aggregation procedure—and the most flexible one—is the simulation of individual (and combined) risk factors using Monte Carlo simulation. Such a simulation technique uses random numbers to derive modeling results (and respective inherent sensitivities) such as the distribution of earnings before taxes, as displayed in Figure 4.12.

A separate discussion that was published elsewhere, furthermore demonstrates that insurance coverage had historically been somewhat neglected in real-life OpRisk initiatives, partially due to the fact that the early versions of the Basel II framework did not accept insurance as a permissible means of minimum regulatory capital reduction. Moreover, proponents of the CAPM to date believe that insurance implies no net value generation, given that only capital market–related aspects captured in the beta factor truly count in the description of the risk position of an enterprise. Change is bound to happen in this regard too, however, for good reasons. Higher exposures to risk generally reduce the enterprise value due to a higher need for (expensive) proprietary capital. Hence it is sensible to specifically

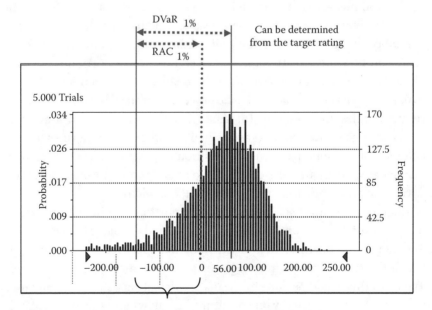

FIGURE 4.12   Sample output of a Monte Carlo simulation.
... and to assess the risk adjusted capital (RAC) and the certainty of the planning.
DVaR = deviation value at risk
RAC = risk adjusted capital
(Courtesy of FutureValue Group AG, Leinfelden-Echterdingen, Stetten, Germany.)

work on risk transfer strategies that reduce the overall risk position efficiently and effectively, for example, as demonstrated through a reduction of the total cost of risk (TCOR). Insurance and other risk transfer mechanisms are, therefore, no longer understood as a cost factor that simply adds no conceivable value to the value of an enterprise (because it does not affect the beta factor), but rather as a set of suitable instruments that can (through a reduction of the proprietary capital required) deliver a positive net contribution to the enterprise value. In turn, the optimization of the residual risk position of an enterprise through appropriately designed and implemented risk transfer mechanisms permits the focusing on the true core business of the enterprise, and the devotion of proprietary capital to those initiatives that best enforce the core strategy and the sustained competitive advantages of the enterprise (Kross and Gleissner 2009).

4.4.2 Enhanced Approaches to Asset Portfolio Risk Management

Performance is often defined and measured in a way that does not truly reflect the risk perception of an investor, as the above confirms. Additional complementary performance measures are required (e.g., the Modigliani–Modigliani measure), and further metrics and key risk and performance indicators will likely prove to be sensible. This in turn yields the comparability of alternative portfolio allocations with their respective risk/return profiles and sensitivities. Moreover, investment portfolios are usually structured and optimized with the focus of 1 year only. Stochastic, dynamic models as well as stochastic processes for the description for risk/return profiles are required. Information on short-term and long-term investment goals are required as an input in order to derive rules for the adoption of portfolio planning, in addition to portfolio structure recommendations. Further intelligent portfolio securitization mechanisms such as time invariant portfolio protection (TIPP) could be used.

Illiquid assets are usually neglected in portfolio optimization, which in turn leads to a suboptimal allocation of funds in liquid asset classes. The integration of simulation-based valuation approaches is required, in which the respective risk/return profiles of illiquid assets are described. Typically, this would be done by means of calculating a stock vs. bond mix that reflects the expected value of future cash flows and the expected risk inherent in future cash flows. Further risk/return key indicators may be required, in addition to or complementary to market prices. These may include revenue, capital turnover, EBIT margin, debt ratio, etc. The introduction of the risk/return profiles of illiquid assets in turn yields a better

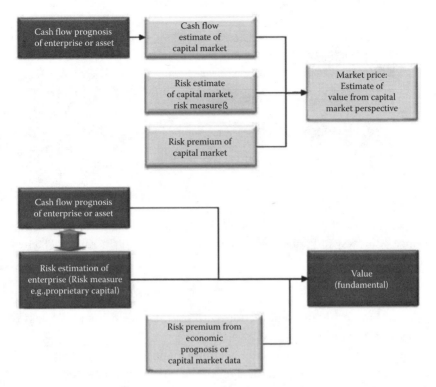

FIGURE 4.13 Traditional vs. sensible approaches to quantifying enterprise risk positions. (From Gleissner, W. and Wolfrum, M. Cost of capital and valuation with imperfect diversification and unsystematic risks, Finexpert publications, available at http://www.finexpert.info/fileadmin/user_upload/downloads/pdf/notes_members/Gleissner_2009_Cost_of_Capital_and_Valuation_an_finexpert_260109.pdf, 2009.)

understanding of the true value drivers and valuation changes in the overall investment portfolio (Figure 4.13).

Traditionally, correlations have captured the interrelationships between historic investment returns, but did not explain them economically. An economic model is required to explain the interrelationships of economic factors and commonly used key indicators. Stochastic processes as described above need to be defined for future expected performance and interest, share and bond investment, inflation, and economic growth. Causal interdependencies need to be defined. The result is a future-oriented explanation of risk and return profiles of individual assets and asset classes as well as the (uncertain) correlation of these asset classes. This in turn enables the analysis of the impact of special crisis scenarios, for example, the modeling

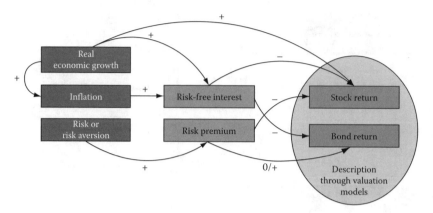

FIGURE 4.14 Reflecting the relevance of economic indicators for asset performance. (Courtesy of FutureValue Group AG, Leinfelden-Echterdingen, Stetten, Germany.)

of a banking crisis in combination with an oil price crisis and a loss of investor confidence crisis (Figure 4.14).

Hence it is fair to say that asset portfolio management will likely experience some rather significant changes in the not-so-distant future. Now as in the future, the objective is to optimize the asset allocation across asset classes in order to benefit from diversion effects (i.e., reduce the impact of unsystematic risk factors) and to yield a near-optimum investment strategy based on a given investor's risk tolerance, which may vary over time. But almost everything else will change. Asset portfolio management will need to explicitly identify and compensate the effects of extreme negative crises and shocks to the system. The predominant key risk indicator will be the probability and the extent of losses (e.g., expected shortfall). Given that risk and return are not freely interchangeable; restrictions need to apply with respect to the maximum risk-bearing capacity and the ability to cover capital requirements with third party equity or debt.

Robust, context-dependent prognoses suitable for longer term investment decisions and the respective (uncertain) correlations between the expected risk and return profiles of individual assets and asset classes will need to be derived from economic forecasting models. Uncertainty will need to be reflected to describe the likely inherent extents of estimation errors, prognosis errors, model errors, and in particular the impact of subjective estimates and decisions inherent in each of these. The optimal asset portfolio allocation will also need to reflect the extent and the characteristics of illiquid assets and asset classes (e.g., immovable property, investment holdings in non-listed companies, etc.), which requires specific risk and return valuation know-how that has historically not been used by practitioners in the field of asset portfolio management. Portfolio insurance strategies can in some cases be used to (temporarily) optimize risk/return profiles of investment portfolios, whereby of course a procyclical investment behavior should be avoided through a thorough observation of fundamental valuation levels. And capital market anomalies such as the abovementioned "momentum effect" and short-term value anomalies exist, and should be explicitly incorporated in the above.

4.5 RECOMMENDATIONS AND OUTLOOK

As described in this chapter, initiating an OpRisk management initiative without reinventing the wheel is a complex undertaking with numerous inherent challenges. The far-reaching ripple effects of the recent financial crisis demonstrate, however, that major change in the positioning, realization, and operation of risk management systems is urgently called for (Kross et al. 2009). As the findings presented in this chapter reflect, however, a rather large body of knowledge and experience has evolved in recent years, which renders most of the key steps manageable predictably. Lessons learned have been published to an increasing extent in recent years, and generic frameworks and fairly capable standard software solutions are starting to make risk managers' lives easier.

Implementing risk management systems does not necessarily imply that a tremendous amount of money is wasted. Of course this would be the typical expectation of a decision maker, who has historically invested in risk-reporting systems such as the ones required under Basel II. To provide some further insight into the reasons why one would truly be able to derive net value from such initiatives, an analogy is appropriate, which refers to a model first put forward by Kerzner (2000), who is often referred to as one of the leading thinkers and practitioners in the field of project management. Kerzner's model introduced the observation that knowledge workers' intensive drive for increasing levels of maturity cannot be the ultimate solution. In the context of products introduced into a market, maturity may be the stage from which onward everything goes downhill. Rather, Kerzner submitted, top management should start to recognize their knowledge workers willingness to change, and provide further direction to develop their project management competencies beyond maturity, to excellence. The following graphical representation was modified in that

FIGURE 4.15   Migrating from maturity to OpRisk management excellence.

Kerzner's expression "project" was replaced by the word "risk," which follows the inherent feeling (in the absence of tangible scientific evidence) that the same reference model is similarly valid for the evolution of risk management maturity and the further net benefits and competitive advantages, if risk management excellence were to be established (Figure 4.15).

Kerzner was able to substantiate in his publications and presentations that working toward project management maturity helps in reducing the likelihood of failure, which is usually achieved by and large with in a time period of 1–2 years. Of course this is a value proposition on its own. The link to strategy crafting and to the development of competitive advantages through excellent project management would introduce a more complex set of opportunities, which may not be as easy to capture. It could hence be an interesting research project at postgraduate level to assess in how far Kerzner's model is truly convertible to the field of operational risk management, and the tracking of which key indicators would best help to substantiate for deductive thinkers that advanced levels of operational risk management, evolving far beyond regulatory compliance, have the potential to add substantial levels of net value to an organization.

Hence reflecting back to the objectives of this chapter, as formulated in the introduction, it is appropriate to conclude that a thorough analysis of the subprime crisis triggers and its consequential effect on the financial sector and ultimately on the world economy highlights that numerous factors and hazards were not exactly unforeseeable. Financial institutions, and regulatory agencies for that matter, could and should have reacted much earlier.

Commonly practiced valuation and portfolio optimization models that are based on perfect capital market theories, should have been abandoned many years ago. Regulatory frameworks such as Basel II have not truly helped, and in some regard have been misleading, inter alia through the separate treatment of risk factors that cannot truly be treated separately.

The much more significant problem seems to be, however, that the way in which the implementation of Basel II and similar regulatory frameworks in real-life organizations has driven the net impact into a much more negative spectrum, while on the other hand the promises of vs. the net result derived from corporate governance frameworks left a lot to be desired. Risk communication strategies were everything but well organized or well implemented, and added to the losses in investor confidence. A combination of organizational and market-driven corrective steps are called for, including but not limited to a reorientation of incentive systems, truly living up to what is called for in corporate governance, rendering it compulsory to use at least those aspects of enhanced risk methodology which have been established for numerous years, while on the other hand impeding the use of risk methodological approaches, which have demonstrably been proven wrong. And whatever the various players in the financial market truly learned from this financial crisis remains to be seen.

REFERENCES

AON Group (2009) Commodity crunch exposure matrix, in: 2009 political risk map. Available at: http://www.aon.com/risk-services/political-risk-map/images/2009_PE_Risk_Map_Small.pdf.

Berger, T. and Gleissner, W. (2006) Risk reporting and risks reported—A survey of German DAX-listed companies. Working Paper Presented at the International Conference on Money, Investment and Risk, Nottingham, U.K.

Blanchard, O. (2008) The crisis: Basis mechanisms, and appropriate policies. Working Paper 09-01. Massachusetts Institute of Technology (MIT)—Department of Economics; National Bureau of Economic Research (NBER), Cambridge, MA.

Chopra, V. K. and Ziemba, W. T. (1993) The effect of errors in means, variances and covariances on optimal portfolio choice. *The Journal of Portfolio Management*, 19(2):6–10.

Gleißner, W. (2009) Kapitalmarktorientierung statt Wertorientierung: Volkswirtschaftliche Konsequenzen von Fehlern bei Unternehmens- und Risikobewertungen. *WSI Mitteilungen* 6:310–318.

Gleißner, W. and Wolfrum, M. (2008) Eigenkapitalkosten und die Bewertung nicht börsennotierter Unternehmen: Relevanz von Diversifikationsgrad und Risikomaß. *Finanz Betrieb* 09:602–614.

Gleißner, W. and Wolfrum, M. (2009) Cost of capital and valuation with imperfect diversification and unsystematic risks. Finexpert publications, available at http://www.finexpert.info/fileadmin/user_upload/downloads/pdf/notes_members/Gleissner_2009_Cost_of_Capital_and_Valuation_an_finexpert_260109.pdf

International Monetary Fund (2007) *Global Financial Stability Report*, Washington, DC.

Kerzner, H. (2000) *Project Management: A Systems Approach to Planning, Scheduling and Controlling*. John Wiley & Sons: New York.

Kross, W. (2006) *Organized Opportunities: Risk Management in Financial Services Operations*. Wiley-VCH: Weinheim, Germany.

Kross, W. (2009) "Integrating management" into "OpRisk management". In: Gregoriou, G. N. (Ed.), *Operational Risk Toward Basel III*. John Wiley & Sons: New York.

Kross, W. and Gleissner, W. (2009) OpRisk insurance as a net value generator. In: Gregoriou, G. N. (Ed.), *Operational Risk Toward Basel III*. John Wiley & Sons: New York.

Kross, W., Hommel, U., and Wiethüchter, M. (2009) Plausible operational value-at-risk calculations for management decision-making. In: Gregoriou, G. N. (ed.) *VaR Implementation Handbook*. McGraw-Hill: New York.

N.N. (February 29, 2009) Domino theory. *The Economist*, Economic Focus Section.

Romeike, F. and Heinicke, F. (2008) Schätzfehler "Moderner" Risikomodelle. *Finance* 2: 32–34.

Shapira, Z. (1995) *Risk Taking: A Managerial Perspective*. Russell Sage: New York.

CHAPTER 5

U.S. Mortgage Crisis: Subprime or Systemic?*

Eric Tymoigne

CONTENTS

This chapter argues that the current U.S. mortgage crisis and the events leading to it are a textbook example of Minsky's analysis of capitalist economies. After briefly reviewing Minsky's framework of analysis, this chapter illustrates his theory with the trends recorded since the early 1990s. Rather than being mainly the result of granting mortgages to individuals with less-than-perfect creditworthiness or of the greed of home speculators, the crisis was caused by the fact that, as Minsky used to say, "stability is destabilizing." This reading of the crisis leads to specific policy recommendations that are presented briefly at the end of this chapter.

* This project benefited from the financial support of the Ford Foundation. A more extended analysis can be found in Tymoigne (2009b).

5.1 INTRODUCTION

Most commentators have pointed at subprime lending and home speculators to explain the current financial crisis. While those trends contributed to the latter, they are only part of the story. There were much deeper forces at play that resulted in a progressive transfer toward financial deals relying more and more on liquidation and borrowing as a means to meet financial commitments. This process was at work in all parts of the mortgage industry, prime and nonprime, as well as in consumer finance and leveraged buyouts.

Minsky's theoretical framework is a good point of reference to understand the current financial crisis. He provided a detailed explanation of how "stability is destabilizing," that is, how a period of enduring economic prosperity creates a favorable environment for the emergence of what he called a Ponzi process. The forces that generate instability are not based on irrationality, greed, or market imperfections; they are intrinsic to the way the capitalist economic system works. Thus, rather than providing financial literacy, imposing a tax on financial transactions, or improving disclosure of information, Minsky emphasized the importance of orienting financial reforms toward understanding and managing systemic risk from a cash-flow perspective.

Section 5.2 reviews Minsky's framework by presenting some of the systemic forces that progressively lead to economic instability. Section 5.3 illustrates Minsky's theory by analyzing some of the trends in the financial industry that have led to the crisis. Section 5.4 describes the immediate causes of the crisis and its unfolding. Section 5.5 concludes and provides some recommendations for financial reforms.

5.2 MINSKY'S FINANCIAL INSTABILITY HYPOTHESIS

The "financial instability hypothesis" was first put forward by Hyman P. Minsky in 1964. Stated concisely, it claims that "stability is destabilizing," that is, capitalist economies, even though very innovative and productive, are intrinsically unstable (Minsky, 1986). The instability of a capitalist economy is not mainly the result of irrational choices, asymmetries of information, or other "imperfections" of markets and individuals. Instead, it is the result of psychological, sociological, economic, and policy factors that a capitalist economy tends to exacerbate. Those are presented briefly in this section; an extended explanation is provided in Tymoigne (2009a).

Minsky first notes that we live in an uncertain world (in Knight's and Keynes's sense of the term). In this context, he notes that individuals' decision-making process is quite different from the one assumed in risky environments. Indeed, probability calculus is not as reliable (or not reliable at all) and people tend to complement it (or to replace it) by simple rules like anchoring and adjustment (e.g., extrapolation of the recent past) and representativeness (e.g., chartist approach in finance). These psychological factors have been studied extensively by Kahneman and Tversky (e.g., Tversky and Kahneman, 1974). Psychological factors are complemented by sociological factors because, under uncertainty, individuals tend to look for the approval of others. This leads to the creation of economic conventions, that is, mental models commonly agreed upon by a group of individuals. Individuals more or less know that conventions do not represent the true model of an economy; however, they are used because they are essential in an uncertain environment. Indeed, conventions create a focus point by providing a reading of economic events and a view of the future, which helps to rationalize (sometime ex post) economic decisions. The "New Economy" of the 1990s was a famous convention.

The previous types of behavior are "irrational" if one follows the strict jacket of the New Consensus economic framework. However, the hypotheses underlying rationality in the latter framework are either those of pure and perfect competition (which implies perfect and costless information), or of risky and imperfect environments (probability distributions are well established or can be established over time with additional, albeit costly, information). In an uncertain world, in which the future is unknowable, behaving as if everything is known, or could be known, leads to dangerous consequences for individuals. In addition, in an uncertain world, more information does not necessarily lead to a better decision-making process. In fact, psychologists have shown that the quality of decisions declines above a certain amount of information and that only confidence is positively related to more information (Wärneryd, 2001: 168). Finally, information is subject to interpretation (through the psychological and social factors cited above) and the latter is what matters rather the information itself.

These psychological and sociological factors promote instability over time in at least two ways. First, the longer a period of expansion, the more people think recessions are a thing of the past and so the more indebted and the less liquid they are willing to be. Individuals forget the lessons from the past and become more confident to perform riskier financial deals. A second way is the ignorance of important information, or the interpretation of negative information as positive information.

Some economic factors also promote instability. First, competition for monetary accumulation pushes economic agents to try to guess the uncertain future in order to obtain a bigger monetary profit relative to their competitors. This race toward the future is the source of the productivity of the capitalist system, but also of its instability. Indeed, it forces individuals to forget about the big picture concerning where the economy is heading, and to narrow their effort on beating the competition by all means (sometime illegal) because their own economic survival is at stake. One of this means is the use of debt; for example, managers are not rewarded for managing a stable business but for an aggressive expansion.

Second, competition is an essential ingredient in the formation of conventions and their wide use by economic agents. Indeed, given the fast pace, "in-the-present" world of entrepreneurial leadership, the previous sociological and psychological factors tend to be followed more closely. Also, competition pushes competitors to follow those who perform best, and to ignore information that is too costly to obtain or that could threaten a competitive position.

A third economic factor that promotes instability is the fact that the macroeconomy operates under rules different from the microeconomy (e.g., the paradox of thrift); therefore, things that make sense at the individual level may be fallacious for the economy as a whole. This generates positive feedback loops peculiar to the capitalist system (Minsky, 1986: 227) that individuals are unaware of, or unwilling to take into account, because they are too complex to analyze. Minsky especially emphasizes the Kalecki equation of profit, which shows that aggregate monetary profit is determined by the investment expenditures of entrepreneurs. Thus, the more entrepreneurs invest, the more they earn, and so the more their expected earnings rise which encourages additional investment. This leads to the "paradox of debt" (Lavoie, 1997) for which a higher willingness to go into debt leads to lower aggregate debt-to-equity ratio. This positive "frustration" of expectations reinforces entrepreneurs' optimism and so increases their willingness to use leveraged positions.

A fourth economic factor that promotes instability is the shortening of the maturity of debts. According to Minsky, the proportion of short-term debts (short relative to the maturity of the operations they fund) tends to grow over a sustained economic expansion because they are less expensive and because refinancing operations grow. Shorter maturity compounds the effect of higher interest rates on debt-service payments by increasing the speed of repayment. Maturity mismatch also creates a need to refinance and so make an economic unit more vulnerable to disturbances in the financial sector.

A final economic factor that may promote instability is financial innovations. The latter are essential to maintain the profitability of financial institutions because, like for any other industry, the market for a given product always ends up saturating. Over a period of enduring expansion, innovations involve extending the use of existing financial products to more risky enterprises and the creation of financial products with higher embedded leverages. In addition, new financial products are marketed as sophisticated products that are better able to measure and/or to protect against risks associated with leverage, which tends to let people believe that the use of debt is safer than in the past (Galbraith, 1961).

Finally, at the policy level, there are, for Minsky, several factors that may promote instability. Once a capitalist economy is booming and close to full employment, it tends to promote inflation. This pushes the central bank to raise its rate of interest and so generates an increase in interest payments. This is all the more the case that debts are based on flexible interest rates and short maturity, and that a lot of refinancing operations are performed. While the central bank raises its interest rates, fiscal policy also becomes more restrictive, either for political reasons (reaching a budget surplus usually is part of the political agency of any Administration) or to limit inflationary pressures. This reduces the net incomes received by private entities and so further decreases their capacity to service debts, which leads to an increasing dependence on refinancing sources. Finally, given the existing economic set up, inappropriate regulatory and supervisory institutions also will tend to promote instability. Notably, a regulatory system that is not flexible enough to account for new innovations and changes in behaviors may create competitive disadvantages and perverse incentives (e.g., regulation Q in the 1970s).

All these factors may promote economic instability (both upward (inflation and speculation) and downward (deflation and recession)) because they tend to increase the financial fragility of economic agents over an enduring expansion. Given everything else, progressively there is growing reliance on external funds and position-making operations (i.e., liquidation and refinancing). Thus, economic agents come to rely increasingly on the expected availability of some refinancing sources to repay their outstanding debts; this represents what Minsky called "speculative finance."

Over time, this speculative financing worsens into Ponzi financing for which the servicing of a *given* amount of outstanding debts is expected to require a *growing* amount of refinancing operations or liquidation at *rising* prices. When refinancing unexpectedly sources dry out, massive forced liquidations occur and the system heads toward a debt-deflation process (Fisher, 1933).

It is important to note that for Minsky, the boom phase of a business cycle is only one component of the story. Financial fragility does not need mania and irresponsible lending practices to emerge even though they compound the previous tendencies. A long period of sustained expansion (during which economic results are excellent and so optimism is justified) is the main driver of financial fragility, the boom is only there to give the *coup de grâce*. Stated another way, by acting according to the rules of the economic system and making intelligent and rational decisions within the prevailing conventions (which define the norms of behavior), economic agents may still promote economic instability.

5.3 THE ROAD TO THE CURRENT CRISIS: AN APPLICATION OF MINSKY'S ANALYSIS

Since the end of 2006, the housing sector has been in a state of limbo. As illustrated in Figures 5.1 through 5.3, foreclosure started and delinquency rates have been rising steadily and have reached historical high levels, and house prices have been dropping at an accelerating rate.

Nevada, California, and Florida lead the nation in terms of foreclosure rate according to RealtyTrac and, with Arizona, they are the states that have recorded the biggest drop in house prices with declines superior to 30%

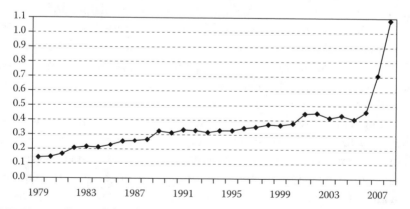

FIGURE 5.1 Rate of foreclosure started, all mortgages (percent). (Data from Mortgage Bankers Association, Washington, DC; www.mbaa.org.)

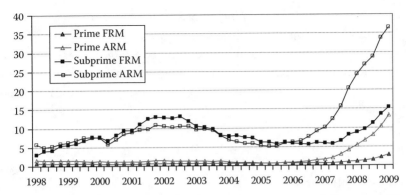

FIGURE 5.2 Percentage of single-family mortgages in serious delinquency. (Data from Mortgage Bankers Association, Washington, DC; www.mbaa.org.)

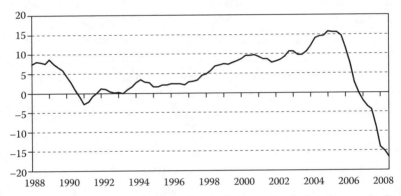

FIGURE 5.3 Annual growth rate of U.S. home price index (percent). (Data from Standard and Poor's, S&P/Case-Shiller Home Price Indices; http://www2. standardandpoors.com/portal/site/sp/en/us/page.topic/Indices csmahp/0,0,0,0,0, 0,0,0,0,1,3,0,0,0,0,0.html.)

from the second quarter of 2006 to the third quarter of 2008 (Tymoigne, 2009b). The drop has concerned all types of single homes but is more pronounced for low-tier homes.

Most commentators have stated that the mortgage crisis is a subprime crisis but as the delinquency figures suggest, the subprime area is only part of the story. All mortgages have been performing much worse, and adjustable rate mortgages (ARMs) and other nontraditional mortgages have been performing horribly whatever the creditworthiness of the mortgagors. Rather than being confided to the subprime mortgage business, the events that led to the possibility of a crisis were generalized to all sectors of the mortgage industry and were compounded by people who truly wished to stay in their home. In order to understand why, one needs to remember that since the mid-1980s, the U.S. economy had been experiencing a long period of stability with only two minor recessions. Thus, the financial community and the public progressively became used to relatively stable default rates and rising home prices, which had three major consequences.

A first consequence was that financial institutions became willing to create complex financial products that involve higher leverage and that gave the impression that credit risk and liquidity risk can be managed more efficiently than in the past. Securitization has been the main driver of those innovations that were extended progressively to more and more esoteric activities. In addition, through subordination and other credit-enhancement methods, financial institutions were able to create investment-grade securities that contain very high embedded leverage and that are backed by nonprime mortgages and noninvestment-grade securities.

Securitization first started in the mortgage industry in the early 1970s and, from the mid-1980s until today, was progressively extended to other economic activities like auto loans, student loans, and, more recently, carbon emission allowances and nonperforming loans.

However, it is likely that there is a finite amount of illiquid receivables that can be used to create asset-backed securities (ABSs), simply because there is a limited number of economic activities (or their number grows too slowly to meet the demands of financial institutions in terms of market expansion). To counter this problem, resecuritization, re-resecuritization, and so on, have been developed and have consisted in securitizing securities themselves rather than nontradable financial claims. However, pilling up levels of securitization also has limits because modeling becomes extremely complex and data to evaluate the risks involved in the new securities is inexistent, which creates valuation problems that became acute during the crisis. To counter this barrier to market growth, synthetic securitization was developed in 1997 and it has grown rapidly since 2001. In 2000, the growth in synthetic securitization was boosted by changes in the regulatory framework described below, and it has led to a change in the main motive of securitization (Tymoigne, 2009b), which, since 2001, has been driven by portfolio arbitrages rather than by the removal of credit and liquidity risks from the balance sheets of loan originators.

The crisis has put a halt to the innovative frenzy but, surely, more innovations will emerge, by widening the number of structured products and by increasing the appeal for existing securities issued by special purpose entities (SPE), in order to counter regulatory barriers and other limits to market growth. Das provides a nice insider view of the need for the financial sector to innovate constantly.

> We need 'innovation', we were told. We created increasingly odd products. These obscure structures allowed us to earn higher margins than the cutthroat vanilla business. The structure business also provided flow for our trading desks. [...] New structures that clients actually wanted were not that easy to create. Even if somebody came up with something, everybody learned about it almost instantaneously. [...] Margins, even on structured products, plummeted quickly. (Das, 2006: 41)

Basel II imposes very high weights (much higher than 100%) on structured securities with a credit rating below BBB in order to provide an incentive to issue investment-grade securities (Renault, 2007: 394ff.). However, mezzanine and rated junior structured notes have been very popular among financial institutions because of the higher spread they provide relative to traditional securities with similar ratings. Thus, it is likely that financial institutions will find ways to counter this barrier put on their profitability.

A second consequence of the long period of expansion, combined with the previous developments in the financial system, has been a willingness to let financial institutions self-regulate* their activities and to let more of them participate to the securitization process. First, the Financial Modernization Act of 1999 legalized the increasing diversification of financial activities undertaken by financial companies. One of the main consequences is that financial companies have become involved in activities in which they have had limited experiences and that may not be coherent with their core business. This increases the potential financial fragility of a company as well as systemic risk. AIG and Citi are perfect examples of the danger of too much diversification toward unfamiliar activities and/or activities with a risk level incoherent with the core business. Second, the

---

* In addition, given the strong unpopularity of government regulation since the early 1980s, regulators like the SEC and the FDIC have been seriously understaffed and undertrained for years (if not decades), which has resulted in the incapacity to detect and to prosecute properly dangerous financial practices. The S&L crisis (Black, 2005) and the recent Madoff and Co. scandals (Wayne, 2009), which are only the tip of the iceberg given the reckless behaviors of all major financial institutions in the 2000s, are good illustrations of this state of affairs.

Commodity Futures Modernization Act of 2000 left credit default swaps (CDSs) and, later, equity default swaps (EDSs) completely unregulated by federal agencies, which, given their exclusion from state gambling laws (on the basis that they are derivatives and so are not considered gambling activities (Adelson, 2004: 5)), led to a huge boom in the CDS market from 2001. This has allowed synthetic securitization, especially for arbitrage purposes, to grow tremendously. Third, in November 2000, the Employee Retirement Income Security Act was amended to permit pension funds to buy investment-grade structured securities (with a grade independent from the rating of the underlying collateral), which was essential to widen the demand for structured securities. This, in turn, was essential to sustain the growth of mortgage lending because mortgage products, especially subprime residential mortgages, were the main source of collateral for structured securities (Bank of International Settlements, 2008: 5).

As the result of enduring expansion, innovative frenzy, and deregulation, the conditions became ideal for the emerge of a boom in the mortgage industry, which has consisted in widening lending to riskier borrowers, as well as progressively loosening underwriting procedures and consumer protection for all types of mortgage (prime and nonprime). This was essential to maintain the profitability of financial institutions and to counter limits to market growth.

From 2001 to 2003, there was a large refinancing process going on in the prime mortgage business, but, as the market dried, mortgage originators turned to nonprime borrowers, as shown in Figure 5.4. One may

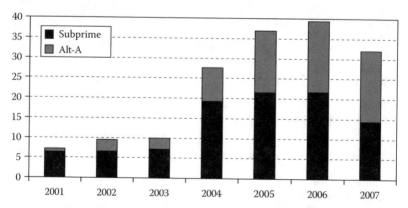

FIGURE 5.4   Share of nonprime mortgages in securitized purchase mortgage originations (percent). (Data from Rosen, K.T., Anatomy of the housing market boom and correction, Fisher Center for Real Estate and Urban Economics, Working Paper No. 306, University of California at Berkeley, Berkeley, CA, 2007.)

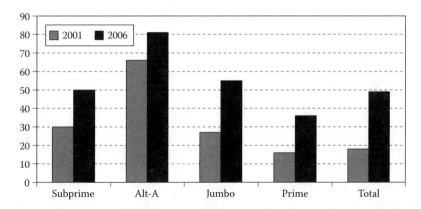

FIGURE 5.5   Distribution of low/no-doc share of purchase origination (percent of origination dollars of securitized loans). (Data from Zelman, I.L., Prime conforming mortgages have 0% of no-doc mortgages, 13, 2007.)

note that from 2005, the source of growth of nonprime mortgages was alt-A lending rather than subprime lending. As shown in Figure 5.5, this was accompanied by the proliferation of mortgages with low or no documentation (to verify borrower's income, assets, etc.) in the nonprime and prime categories. In fact, the growth of low-doc/no-doc mortgages was much higher in the prime-mortgage business from 2001 to 2006, while this was already a well-established practice in the nonprime business from 2001 (especially for alt-A mortgages).

According to Zelman (2007), in 2005 and 2006, at least 50% of all new mortgages backing private-label mortgage-backed securities (PL MBS) had a low documentation; and this proportion climbed to 77.9% in 2006 for alt-A mortgages purchased to back PL MBSs. Among PL MBSs issued in 2006, 5.8% of the jumbo mortgages backing them did not have any documentation, 3.3% for alt-A mortgages, and 0.2% for subprime mortgages.

The boost provided to mortgage lending, by extending it to nonprime borrowers and by loosening underwriting requirements was reinforced further by a large increase in frauds. The lenders compounded those problems by not verifying the information provided by borrowers even though it was very easy to do so (Morgenson, 2008). In fact, the Financial Crimes Enforcement Network finds that mortgage brokers, appraisers, and borrowers were all central parts of fraudulent schemes that consisted mainly in misrepresentation of income/assets/debts and forged/fraudulent documents (over 70% of all frauds in the mortgage industry) (Financial Crimes Enforcement Network, 2008). Fraud became a quasi-institutionalized way of operating for all eyes to see, with Web sites advertising software that allows to print fake pay stubs and fraudulent methods to raise credit scores (Creswell, 2007).

Financial institutions further pushed back limits to market growth by advertising "low-cost" mortgages that seemed attractive at first but were highly toxic for borrowers. As a consequence, the chance of default increased for any level of creditworthiness. In addition to the advertisement of teaser interest rates, the rise of nontraditional mortgages has taken several forms. One form, shown in Figures 5.6 and 5.7, has been an increase in interest-only (IO) mortgages and payment-option mortgages. Another form, shown in Figure 5.8, has been an increase in the proportion

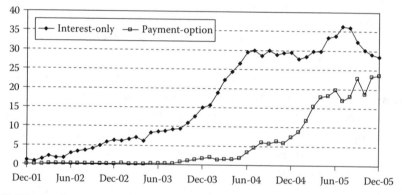

FIGURE 5.6   Proportion of nonprime mortgage originations with nontraditional characteristics (percent). (Data from *FDIC Outlook*, Summer 2006, http://www.fdic.gov/bank/analytical/regional/index.html.)

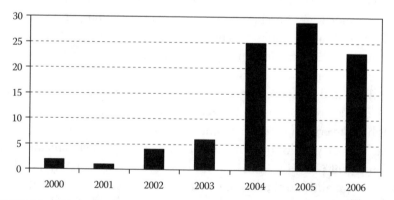

FIGURE 5.7   Share of interest-only and payment-option mortgages as a percent of all mortgage originations purchased. (Data from Zelman, I.L., Mortgage liquidity du jour: Underestimated no more, Credit Suisse, Equity Research, New York.)

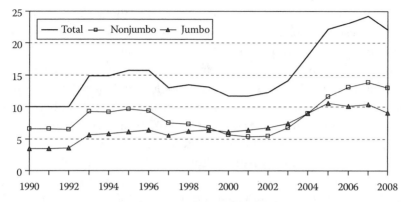

FIGURE 5.8 Proportion of conventional single-family mortgages with adjustable rate (percent). (Data from Federal Housing Finance Agency, Washington, DC, Average of the first three quarters for 2008; http://www.fhfa.gov/Default.aspx?Page=70.)

of ARMs because of their lower monthly cost relative to fixed-rate mortgages (FRM), at least until the end of 2006 (Tymoigne, 2009b).

The low-cost and low-doc effects were reinforced by a higher loan-to-value ratio, piggyback lending, and short-term interest reset (Zelman, 2007: 4) in order to create a large increase in origination volumes by lending to new borrowers and by encouraging existing borrowers to refinance. All this was compounded by the willingness of households (which was itself initiated and nurtured by aggressive marketing strategies) to go for those "low-cost" mortgages even though rates on traditional mortgages were already at historical low in 2003:

> Finally, "we thought, why not live a little bit dangerously and do the interest only?" Mr. James said. "Why pay the principal if we don't know how long we're going to be there?" (Bayot, 2003).

As Bayot notes candidly about IOs:

> Leaving the principal balance on a mortgage untouched, however, carries risks. [...] This is not to worry as long as home prices are appreciating [...] and owners can simply sell for more than they paid. (Bayot, 2003)

Thus, by 2003, the Ponzi process was working at full speed in the mortgage industry. As the previous quote shows, this Ponzi process was not only sustained by speculators but also by individuals who did not think that they would stay in the house for a long time, even though they were genuinely interested in staying in the house. However, the underwriting process faulted by qualifying them on the basis of interest payments or of the introductory payments, and by taking the "long" history of rising home prices as a given. The latter trend created a feeling that selling a home is a normal and safe way to repay a mortgage, so much so that house prices were used to judge the intrinsic creditworthiness of borrowers.

The Ponzi process was not limited to the mortgage industry but was present also in consumer finance (Brown, 2007; Tymoigne, 2007), as well as in corporate leveraged buyouts that went on at a frenetic pace with extremely high (and rising) leverage (International Monetary Funds, 2008a: 10, n. 21.). A 2006 *New York Times* article by Duhigg clearly shows the mood of moment:

> With Wall Street caught up in a wave of acquisitions, normally cautious bond investors are living like Las Vegas high rollers [...]. And for some companies, the more they borrow, the safer they are deemed. [...] The wave of recent purchases by private equity groups has pushed issuance of high-risk debt to record levels. [...] But instead of worrying that defaults will increase, debt prices are signaling that bondholders believe that companies will have few problems paying off new loans. (Duhigg 2006)

This is a classic example of the boom period analyzed by Minsky.

Overall, therefore, there is more to the story than just subprime lending, which stagnated from 2005. This threatened the continuation of the Ponzi process by making it more difficult to qualify a growing number of people and to sustain home price growth. The latter was essential because liquidation came to be seen as a normal way to service mortgage payments. In order to counter the stagnation of subprime lending and some policy events described below, mortgage brokers shifted lending toward alt-A borrowers and, as shown in Figure 5.6, toward more exotic forms of mortgage relying less on interest-rate incentives and more on payment options. In addition, lending requirements on prime and nonprime mortgages loosened significantly and fraud grew rapidly. All this allowed the Ponzi process to continue but increased the financial fragility of mortgagors. As shown later, this also increased the financial fragility of financial companies by increasing counterparty risk and by raising wishful thinking through a higher reliance on automated underwriting programs, and level 2 and level 3 valuation models.

Because of the previous trends, the broader access to homeownership and consumer lending has been associated with growing financial fragility and higher repossessions. Therefore, the gains on the real side have been short lived, and homeownership has been declining since 2005 while the homeowner vacancy rate has been rising steeply to historical highs since 2006.

5.4 IMMEDIATE CAUSES OF THE CRISIS AND ITS UNFOLDING

The possibility of a crisis was generated by the previous trends. The immediate causes of the crisis involve financial events as well as two policy events. On the financial side, principal payments started to kick in for IOs and teaser interest rates began to be reset upward, which has led to an increase in delinquency from the third quarter of 2006. The number of interest-rate resetting is expected to continue to rise until the end of 2011; therefore, defaults will continue to rise sharply if nothing is done (International Monetary Fund, 2007: 8). On the policy side, first, the Federal Reserve started to raise its interest rate rapidly from early 2004, which affected ARMs very rapidly and led to a significant slowdown in ARM originations from 2006. Second, from the end of 2003, the Treasury decided to reduce its deficit and planned to reach a surplus by 2012. These two policies progressively squeezed the income of the private sector (Tymoigne, 2007). Given the increasing financial fragility of the private sector induced by the previous tendencies, only a relatively small decline in income was necessary to unwind the Ponzi process in the mortgage industry. The process had been stretched to its maximum to counter those policy actions and limits to market growth, and started to unwind at the end of 2006.

As a consequence of the rise in delinquency, defaults on mortgages have started to rise sharply as the rate of foreclosure suggests. Thus, loan originators and other SPE sponsors did no receive debt service payments and so could not service the securities issued by SPEs, which greatly affected the profitability of financial institutions. This led to a decline in lending activities, which, given that the growth of home prices was based on a Ponzi process, has led to a sharp decline in house prices. The decline has been so steep that originators could not recover the outstanding principal of mortgages as they expected. All this has led to a large decline in the value of all tranches of SPE securities, senior or subordinated, and, in July 2008, BBB-rated CDSs on ABSs traded on average at a 90% discount (60% for AAA ABCDS) (International Monetary Funds 2008b: 13).

Defaults and large declines in the value of securities triggered the unwinding of swap contracts and other securities affected by them (e.g., leveraged super-senior notes and credit-linked notes). The unwinding has been so large that super senior tranches, that were supposed to be extremely safe, have been affected. Given the losses, monolines, pension funds, and hedge funds, who are the net sellers of credit protection (International Monetary Fund, 2008a), could not meet payments on CDSs and other securities, and their financial problems have been transmitted rapidly to others. Indeed, for example, net buyers of CDSs who thought that they had hedged their short CDS positions by buying CDSs figured out that the counterparty could not pay; therefore, those institutions could not make good on the contingent payments required by the CDSs they sold. A "long" period of stable default rates and rising home prices had given the impression that selling protections on credit risk was a safe and easy way to make money. Thus, financial institutions did not put aside any funds to meet contingent payments; or if they did, the amount put aside was too small to meet the large required payments given the size of the drop in the value of securities.

All these developments in the CDS and other markets were compounded by additional factors, which, all combined, have led to massive liquidations and spectacular failures in the financial sector. First, the crisis made it very difficult for SPEs to refinance their positions (especially asset-backed commercial paper conduits and special investment vehicles that fund their positions in long-term assets with short-term and medium-term securities),* and the automatic unwinding triggers of SPEs forced the latter to liquidate their positions. The financial difficulties of SPEs have led to a return to the originators' balance sheet of the credit risk and liquidity risk that they wanted to avoid either through credit lines granted to SPEs or through other forms of guarantees provided to the latter. As a consequence originators' equity capital and cash reserve were rapidly depleted, which reinforced the refinancing and liquidation pressures on the financial system. Second, monolines (especially Ambac and MBIA) were downgraded in the middle of 2008, which contributed to massive liquidations and write-downs of structured products. Indeed, their downgrading affected the strength of the third-party insurance they provided, and so

---

* The creation of a maturity mismatch in the funding of securitization was "encouraged by the implicit belief that ready access to [short-term] financing would always be there" (Counterparty Risk Management Policy Group III, 2008: 38). This is a clear example of the use of what Minsky called speculative finance.

affected the credit rating of securities relying on this insurance; therefore, pension funds and others required to buy only investment-grade securities had to sell some of their positions. Third, the complexity of structured notes and lack of data to value them have led to a large increase in level 3 valuation methodology, for which homemade models are created to value a security (International Monetary Funds, 2008a). Of course, level 3 can lead to large abuses, and, sometimes, financial firms used different pricing models to value financial assets for their customers and internally (Counterparty Risk Management Policy Group III, 2008: 88). With level 3, the assessment of solvency and liquidity by regulators and private companies becomes extremely difficult. The Troubled Asset Relief Program has shown how difficult it is to find a way to value those securities (probably most of them are worthless).

Combined with further actual and potential threats (e.g., the rise of margin requirements (International Monetary Funds, 2008a)), all these events brought the U.S. financial system (and with it the whole U.S. economy) on the brink of complete destruction. As a consequence, in addition to trillions of dollars of advances provided by the Federal Reserve to meet short-term liquidity needs, the federal government had to intervene in an unprecedented manner through massive lending programs, capital injections, and purchases of toxic securities for a committed amount of 23.7 trillions of dollars as of June 30, 2009 according to TARP Special Inspector General. In October 2008, losses concerned mainly U.S. banks with write-downs totaling around 350 billions of dollars. Worldwide the financial sector has lost about 750 billions of dollars, including 600 billions by banks and 100 billions by insurers; the losses are mostly in mortgage and leveraged-loan products and there are growing rapidly (International Monetary Fund, 2008b: 17). Given the trend of home prices, interest-rate resets, foreclosures, and delinquencies, more losses and government interventions are to be expected.

5.5 CONCLUSION

Rather than being the result of lending to less creditworthy borrowers and home speculation, the crisis is the result of a widespread loosening of underwriting criteria and of the quality of all mortgages, which has increased the financial fragility of all mortgagors. This has resulted in historical high levels of delinquency and foreclosure for all mortgages.

The decline in underwriting standards and consumer protection were the result of a long process of deregulation and unchecked financial innovations that was driven by (1) an enduring period of economic stability (which pushed to find new ways to make money as markets saturated, and which gave the confidence to increase leverage in financial innovations and existing economic activities); (2) cutthroat competition (which pushed to innovate frenetically and promoted sloppy underwriting and rating standards); and (3) beliefs that market mechanisms and profit motive always lead to social optimal and that government should get out of the way.

This period of time was a classic example of the transfer from hedge to Ponzi financing that Minsky analyzed in detail. Once established the Ponzi process in the mortgage business has required more and more daring financial practices, which ultimately have consisted in letting borrowers choose what income to state and how much debt to service, and letting financial companies use esoteric methods to price and to rate structured securities.

All this has several implications in terms of policy and only a few of them are briefly presented here (Tymoigne, 2009a,b provide a more in-depth analysis). First, nonprime lending is not synonymous with "bad" lending and may be perfectly normal as long as the financial terms are adapted to the needs of borrowers *and* are related to their normal sources of cash inflow. Of course, default rates are much higher on nonprime loans but that is a given, and lenders who decide to enter into this business ought to be able to protect themselves. Second, the estimation of creditworthiness needs to be based on a cash-flow approach instead of just looking at credit history. Indeed, rising home prices and a long period of good credit history lower default probabilities and raise credit ratings (which allows more customers to qualify for a mortgage and so sustains house-price growth), even though the underlying creditworthiness of borrowers may be weakening and may be relying heavily on a Ponzi process (Tymoigne, 2009b). Rather than asking "will you repay on time?", one should ask "*how* will you repay on time?" and the liquidation of a home, or the expectation that refinancing will be available, should never be a criteria to judge the *intrinsic* capacity to repay. In addition, customers should be qualified only on the basis of the full debt service payments, even if they plan to repay fully before principal servicing begins. Third, better financial literacy and better disclosure of information to financial investors will not help to prevent similar future crises. Not only because more information does not mean better decisions, but also because economic agents are willing *and are forced* to take more financial risks as the economy performs better and as financial innovations give confidence that risks can be hedged more efficiently. Fourth, the main purpose of regulation and supervision is currently to uncover isolated "bad behaviors" and fraudulent Ponzi schemes à la Madoff, without recognizing that the system itself encourages Ponzi, albeit legal, financial practices.

Everybody may behave "wisely/cleverly" according to the norms of behavior but may generate a great deal of systemic instability. Thus, in addition to a prudential approach to regulation and supervision, we need a systemic approach that involves an aggregate analysis of cash flows and position-making channels. Fifth, financial innovations (i.e., new financial products or new ways of using existing financial products) need to be monitored by a government agency before and after they enter the economy to make sure that they are safe and do what they claim. Ponzi-inducing financial innovations should be forbidden to enter the economy even if financial companies claim this is the only way they can maintain their competitiveness, and even if it looks like those innovations improve standards of living. The new mortgage contracts and securities were praised for allowing low-income households to become homeowners. However, given their structure, those financial innovations also led to the emergence of a national Ponzi scheme, and the welfare gains *predictably* were short-lived. Some of those innovations should not have been allowed to exist and higher low-income homeownership may not be sustainable without further enhancing government programs. Sixth, competitive pressures should be alleviated in the financial industry. Many economists already have noted the destabilizing effects of compensations based on relative short-term performances. In addition, in combination with a government oversight of financial innovations, a form of patent should be provided to financial inventions that are certified safe. This would encourage financial companies to take the time to develop reliable financial products that meet the needs of their costumers and society as a whole.

REFERENCES

Adelson, A. (September 13, 2004) CDOs in plain English, Nomura Fixed Income Research, New York.

Bank of International Settlements (BIS) (2008) *Credit Risk Transfer*. Basel, Switzerland.

Bayot, J. (2003) Interest-only plans catch on as curb to mortgage payments. *New York Times*, August 16, accessed February 20, 2008 at http://query.nytimes.com/gst/fullpage.html?res=9C05E4DF1530F935A2575BC0A9659C8B63.

Black, W.K. (2005) *The Best Way to Rob a Bank Is to Own One: How Corporate Executives and Politicians Looted the S&L Industry*. University of Texas Press: Austin, TX.

Brown, C. (2007) Financial engineering, consumer credit, and the stability of effective demand. *Journal of Post Keynesian Economics*, 29(3): 427–453.

Counterparty Risk Management Policy Group III (2008) *Containing Systemic Risk: The Road to Reform*. Counterparty Risk Management Policy Group III, New York.

Creswell, J. (2007) Web help for getting mortgage the criminal way. *New York Times*, June 16, accessed December 10, 2008 at http://www.nytimes.com/2007/06/16/technology/16fraud.html.

Das, S. (2006) *Traders, Guns and Money*. Prentice Hall: New York.

Duhigg, C. (2006) Fast credit, easy terms, buy now. *New York Times*, November 21, accessed September 5, 2007 at http://www.nytimes.com/2006/11/21/business/21place.html.

Financial Crimes Enforcement Network (April, 2008) *Mortgage Loan Fraud*, Washington, DC.

Fisher, I. (1933) The debt-deflation theory of great depressions. *Econometrica*, 1(4): 337–357.

Galbraith, J.K. (1961) *The Great Crash*, 3rd ed. Riberside Press, Cambridge, MA.

International Monetary Fund (October 2007) *Global Financial Stability Report*. Washington DC.

International Monetary Fund (April 2008a) *Global Financial Stability Report*. Washington, DC.

International Monetary Fund (October 2008b) *Global Financial Stability Report*. Washington, DC.

Lavoie, M. (1997) Loanable funds, endogenous money, and minsky's financial fragility hypothesis. In: A.J. Cohen, H. Hagemann, and J. Smithin (eds.) *Money, Financial Institutions, and Macroeconomics*. Kluwer Nijhoff: Boston, MA.

Minsky, H.P. (1986) *Stabilizing an Unstable Economy*. Yale University Press: New Heaven, CT.

Morgenson, G. (2008) A road not taken by lenders. *New York Times*, April 6, available http://www.nytimes.com/2008/04/06/business/06gret.html.

Renault, O. (2007) Cash and synthetic collateral debt obligations: Motivations and investment strategies. In: A. de Servigny and J. Norbert (eds.) *The Handbook of Structured Finance*, pp. 372–396. McGraw-Hill, New York.

Rosen, K.T. (2007) Anatomy of the housing market boom and correction. Fisher Center for Real Estate and Urban Economics, Working Paper No. 306, University of California at Berkeley, CA.

Tversky, A. and Kahneman, D. (1974) Judgment under uncertainty: Heuristics and Biases. *Science*, 185(4157): 1124–1131.

Tymoigne, E. (2007) A hard-nosed look at worsening U.S. household finance. *Challenge*, 50(4): 88–111.

Tymoigne, E. (2009a) *Central Banking, Asset Prices and Financial Fragility*. Routledge: London, U.K.

Tymoigne, E. (2009b) Securitization, deregulation, economic stability, and financial crisis: Do we need more of the same type of regulation? Working Paper, Levy Economics Institute, Annandale-on-Hudson, NY.

Wärneryd, K.-E. (2001) *Stock-Market Psychology*. Edward Elgar: Northampton, U.K., New York.

Wayne, L. (January 27, 2009) Troubled times bring mini-madoffs to light. *New York Times*, http://www.nytimes.com/2009/01/28/business/28ponzi.htm.

Zelman, I.L. (March, 2007) Mortgage liquidity du jour: Underestimated no more. Credit Suisse, Equity Research, New York.

CHAPTER 6

Regulation and Financial Stability in Laissez-Faire Hong Kong: A Reassuring Record

Leo F. Goodstadt

CONTENTS

After 50 years of bank runs and financial scandals, Hong Kong became Asia's second largest international financial center and China's biggest source of foreign direct investment (FDI) in the 1980s once laissez faire had given way to a modern regulatory system. Thereafter, its monetary institutions successfully withstood dramatic political change and external financial turmoil. This analysis identifies the relevance of Hong Kong's experience to the current global financial crisis.

6.1 WHY HONG KONG

With predictions of a looming recession for most economies in 2009 to match the Great Depression of 80 years ago, there has been a search for historical parallels to put this extraordinary crisis into a manageable context and to derive from past experience some credible indicators of likely future trends in the absence of a firmer basis for rational predictions (most notably in Reinhart and Rogoff, 2009). At the same time, despite the obvious case for radical measures to ward off worldwide deflation, a recurrent regret has been the perception that the era of free market economics and liberal regulatory regimes has come to an end (e.g., King, 2008). In this context, Hong Kong offers a source of reassurance. As the United States and Western Europe were dismantling their supervisory constraints on banking from the 1970s onward, Hong Kong chose to strengthen its regulatory arrangements. But tighter controls over financial institutions did not check Hong Kong's emergence as Asia's freest and most innovative financial center. Nor was the government forced to abandon its general commitment to noninterventionism.

Despite Hong Kong's anomalous status as a Chinese city under British colonial rule until 1997, its historical experience is of major interest in any review of Asia's development in the last 50 years. Hong Kong has long possessed an independent, sophisticated, and international financial system and has been among Asia's top three financial centers for almost a century. It has been China's largest single source of external finance since the nation embarked on the economic reforms that generated an annual GDP growth rate of 9.8% between 1978 and 2007. Hong Kong's prosperity has come to depend increasingly on the efficiency of its financial services sector because since the 1980s, the economy has switched from exporting light industrial products to reliance on the services sector, which now accounts for 93% of GDP.

Laissez faire has always been Hong Kong's preferred economic ideology. The U.S. Heritage Foundation and the Canadian Fraser Institute have labeled

Hong Kong the world's "freest economy" in every year since the mid-1990s. The constitutional blueprint enacted by Beijing for postcolonial Hong Kong binds the administration to a free port, balanced budgets, minimal taxation, free movement of capital and currency, and the full noninterventionist philosophy inherited from the former British rulers. Banking and financial services, however, have long been an exception to the almost total freedom from state interference. Between 1935 and 1985, the government was forced, reluctantly but repeatedly, to rescue banks from collapse and to expand regulation because market forces, moral hazard, and self-regulation proved increasingly incapable of protecting the financial system from disaster.

This chapter starts with a summary of the regulatory and policy failures, particularly in the United States and the United Kingdom, that are currently identified as major factors in the global financial crisis that started in 2007. Next, Hong Kong's record of recurrent banking crises and unenthusiastic government reforms is reviewed, highlighting the marked propensity of banks to engage in self-destructive behavior if left unsupervised.* Throughout the analysis, attention is drawn to the parallels from Hong Kong's historical experience that match developments in the United States and United Kingdom over the last decade and form the background to the current global crisis.

6.2 ANATOMY OF A CRISIS

The anatomy of the financial crisis that has swept the world since 2007 can be summed up for the purposes of this book in the following terms:

Reluctant regulators

The crisis was made possible by regulatory failures. Officials declined to use either legal powers or moral suasion to check imprudent and improper business practices that had emerged over the previous decade.

- In the United States, regulators were convinced of the capacity of financial markets to regulate their own affairs and to enforce prudent behavior on participating firms, a confidence that was symbolized by the end of the Glass–Steagall Act in 1999. Complacency within the Federal Reserve was accompanied by a refusal to recognize the need

---

* Unless other wise indicated, details of Hong Kong's economic record, its political arrangements and their commercial implications, its monetary and fiscal policies, and its banking history are taken from Goodstadt (2005, 2007) and Schenk (2008).

for supervision of new sources of credit risk arising from the proliferation of commercial mortgage-backed securities (CMSs), collateralized debt obligations (CDOs), credit default swaps (CDSs), and similar products that were to prove so toxic.

- In the United Kingdom, regulators have claimed that the crisis was imported from the United States and had been unforeseeable. But they had at least as much advance warning of potential instability as their American counterparts, which was ignored. As in the United States, the British government wanted to retreat from responsibility. In 1997, for example, the United Kingdom's central bankers had been given the duty of the ensuring stability for the entire financial system. By 2007, this responsibility had been diluted to a duty of making a contribution to overall financial stability.

Property's perils

The crisis began with failures of individual financial institutions whose property exposure was grossly excessive in relation to their funding capacity, which has proved to be a global phenomenon (Sentence, 2008). Here, the regulators' indifference to self-destructive tendencies was a crucial contributing factor.

- U.S. regulators refused to respond to evidence of dangerous mortgage practices that were proliferating despite warnings that can be traced back to 2001. In particular, they had failed to comprehend how securitization encouraged lower standards of screening for mortgage applications. (See Demyanyk and Van Hemert, 2008; Keys et al., 2008.)

- In the United Kingdom, the 2007 collapse of Northern Rock revealed a similar reluctance by the regulatory authorities in the past to act on clear evidence of an unsustainable mortgage strategy. As the global crisis intensified during 2008, the regulators were found to have previously ignored similar levels of imprudent lending behavior and reckless disregard of obvious market risks across most of British banking.

Cross-contamination

The losses created in the property sector were aggravated by shoddy banking practices.

- In the United States, securitization of loans was supposed to allow financial institutions to transfer their mortgage risks in particular into "nonbanking" portfolios. In practice, banks became the major purchasers of these products, which spread the contagion worldwide.

- In the United Kingdom, commercial banks added to their property woes by taking equity positions in their major borrowers, thus increasing their potential losses in the event of default.

Fraud and incompetence

Extensive malpractice and mismanagement proved an inevitable side effect of the global financial crisis.

- The collapse of the Madoff "Ponzi" scheme with losses initially estimated at $50 billion was notable both for its scale and for the gullibility displayed by individuals and institutions whose financial reputations were previously of a very high order. The Madoff affair was paralleled by creative accounting at Fannie Mae and Freddie Mac, which masked the true and alarming extent of their involvement in subprime lending.

- Mismanagement on a staggering scale was a global phenomenon. For example, Société Générale lost $7 billion through unauthorized transactions by a junior trader, Jérôme Kerviel. At Credit Suisse, "pricing errors" led to profits for 2007 being overstated by $2.85 billion in relation to asset-based securities.

6.3 HONG KONG'S RELUCTANT REGULATORS (1948–1964)

Even before World War II, the Hong Kong government had come under considerable pressure to introduce a modern regulatory system that would end the chronic instability of local banking institutions. Officials resisted these calls for reform. They claimed that Chinese clients regarded their bank deposits very much as speculative investments, which they were prepared to lose without complaint in return for the prospect of high interest rates. Officials in this period also insisted that regulation would prove not only futile and expensive but would also mislead the public by creating a false sense of security about the soundness of individual banks. These two misconceived assertions were to be recycled by the colonial administration over the next 40 years whenever it needed to justify its rejection of responsibility for the stability of deposit-taking institutions.

In 1947, the colonial administration came under renewed pressure to modernize its financial and monetary arrangements, this time from the Chinese Mainland where the ruling Guomindang party was struggling for survival. The national economy was crumbling, and hyperinflation paralyzed all normal commerce and investment. The Guomindang regime accused Hong Kong with its free port and unrestricted currency markets of facilitating rampant smuggling, black markets, and a protracted capital flight from the Mainland. The British colonial rulers recognized that, despite the growing strength of Mao Zedong and the Chinese Communist Party, the Guomindang authorities were still powerful enough in southern China to organize a blockade that would bankrupt Hong Kong. As a result, a diplomatic agreement was signed that was supposed to impose strict regulation on Hong Kong's banks and compel their compliance with the Mainland's financial and currency laws.

Hong Kong duly drafted legislation to bring all banking activities within a licensing system that would enable the government to impose a measure of control. However, the original proposals were diluted by senior Hong Kong officials who were still strongly attached to laissez faire (in marked contrast to the rest of the United Kingdom's colonial empire). They also saw no serious political costs in leaving the fate of the banks entirely to market forces because officials still believed that Chinese depositors would not protest if they lost all their savings in a bank collapse. In addition, colonial officials had a genuine desire to protect what they regarded as "traditional" Chinese business practices, and they felt that it would be unfair to force "Western" corporate structures and modern business behavior on an industry that was dominated by family firms.

In consequence, the new legislation was interpreted very liberally. London's recent directive that all colonial governments should protect bank depositors was ignored, and the new law's minimalist provisions for official supervision of the newly licensed banks were not enforced energetically. Licenses were given indiscriminately to virtually all applicants even when an individual firm was known to be operating unlawfully. The government did not even insist on local banks complying strictly with the commercial legislation applied to the rest of the business world. The consequences in terms of bank runs and failures were predictable enough. Of the 132 banks licensed in 1948, 55 had gone out of business by 1955.

The colonial administration's indulgence toward "traditional" banking practices meant that the smaller local banks were under no pressure to put the protection of the depositor ahead of profits. These firms stuck to familiar business operations—mainly foreign currency, gold dealing and real estate—although these were much riskier activities than the manufacturing sector (whose exports grew at an annual 115% during the 1950s). Criminality was almost impossible to avoid because the strict foreign exchange controls then in force meant that all gold and much of the currency transactions were technically unlawful, while bullion and currency shipments involved smuggling and, therefore, contact with criminal (triad) organizations.

6.4 PROPERTY'S PERILS

Property, however, proved the largest threat to the stability of the banking system. Industrial takeoff in the 1950s led to a building boom to house the growing population and to provide the factories for sustained industrial expansion. This business was particularly attractive to local banks, which were often part of family business groups that also owned property and construction companies. They tended to treat the public's deposits as the family's working capital, and much of their real estate loans inevitably went to related parties. In the early 1950s, building regulations were relaxed, increasing significantly the redevelopment value of building sites. The number of housing projects rose dramatically as landowners sought to capitalize on this legislative windfall. By the late 1950s, the property sector had been transformed by new engineering techniques that allowed the construction of large-scale, multistory housing projects that were offered for sale in advance of completion.

Either directly or through associated companies, a bank could be involved in every stage of a real estate project: site purchase, building design and construction, and the long-term mortgages offered to the public. As demand for property-related loans surged, there was little to restrain the expansion of lending to the real estate sector on a reckless scale. The tolerance shown by the colonial administration toward misconduct and mismanagement by bankers served as an encouragement for illegal and imprudent behavior, and problems of default and fraud started to surface. It was revealing that, at this stage, the government agency responsible for corporate and land affairs—and not the officials overseeing banking affairs—introduced the first measures to protect the public.

The near collapse in 1961 of a family-owned bank linked to dubious land deals, followed by revelations of outright theft by its chief executive, made it hard for the government to resist demands for reforms. The colonial administration now accepted the case for a modern system of regulation. But there was still no sense of urgency about drafting the necessary legislation. A property bubble continued to undermine the basic rules for prudent lending until 1964 when property prices started to fall, leaving banks overexposed to property with loans secured on real estate assets whose value was slumping catastrophically.

Hong Kong officials admitted after the event that they had foreseen the likelihood of the 1960s property crash. They justified their failure to intervene on the grounds that the government would have been wrong to take any steps that might impede a badly needed increase in the supply of housing. Currently, American officials are being attacked for adopting a similar approach and accused of laying the foundation for the subprime crisis because of their political decisions to give priority to the expansion of home-ownership (Kregel, 2008).

The business model adopted by much of Hong Kong's banking in this era had been based on the misconceived assumption that a bank's risks from exposure to the property sector were reduced if the lending could be compartmentalized among different (though frequently, in practice, related) entities: funding to purchase the redevelopment site; building and construction loans; and mortgage facilities for the home-owner. This approach was not so different in concept and consequences from the emergence later in the century of the "originate-to-distribute" model in the United States property market where banks originated loans, generated a variety of fees and then shared the revenues with other investors, while disclosure to both their customers and the regulators was minimal (Berndt and Gupta, 2008).

6.5 SEARCH FOR STABILITY (1965–1970)

As the end of the property boom in 1964 was undermining the foundations of the "traditional" banking model, legislation was at last coming into force to introduce modern standards of regulation and establish a professional supervisory agency. But the reforms were kept as modest as officials and banking interests could achieve in the face of public anxiety about potential bank failures. Thus, for example, proposals designed to prevent conflicts of interest on the part of bank owners and directors were deleted from the draft law. Nevertheless, depositors with licensed banks were now assured of protection against mismanagement. Strict requirements were set for a bank's liquidity, which ensured that its lending was tied directly to the bank's deposit base and that there would be a reasonable level of reserves to meet sudden surges in withdrawals. And for the first time, the authorities had the power to obtain comprehensive and timely data from each licensed institution.

Unfortunately, the legislative process had been so protracted that three banks failed outright between 1961 and 1965. That year saw widespread erosion of depositor confidence as the public became aware of the banking industry's property problems, and a further six banks, including Hang Seng Bank, then Hong Kong's second largest bank, were only saved through government guarantees or cash injections.

The 1965 crisis destroyed the colonial administration's assumption that somehow Chinese depositors would docilely accept the loss of their savings when banks failed. Hong Kong now had "mass" banking that relied on the marketing of savings accounts to the community at large in order to boost the deposit base that set the statutory ceiling for the individual bank's loans and advances (except for foreign banks, as explained below). Street protests followed the colonial administration's decision not to bail out a bank that had been hit by rumors of fraud and mismanagement. (These reports were later shown to have been true and long known to officials.) Depositors had become a political force that the colonial administration could no longer ignore. Henceforward, the government's policy would be to pump liquidity into any bank in danger of collapse.

The crisis had a profound effect on Hong Kong's public finances. It convinced the government of the need to build up large fiscal reserves through under-spending its annual budgets. These funds were accumulated not for use in a Keynesian reflation of the economy during recessions (which, in any case, Hong Kong was not to suffer until 1998). They were solely for the defense of the currency and for the rescue of banks in trouble through injections of emergency liquidity, just as they had been used successfully during 1965 (and again in 1967 when Maoist extremism imported from the Mainland undermined business confidence with a campaign of strikes and bombs in protest against colonialism).

Officials found it convenient to blame the 1965 crisis on excessive competition among the banks rather than on the colonial administration's prior negligence. They, therefore, endorsed a cartel set up by the industry to fix interest rates for retail deposits. At the same time, new applications for new licenses, whether from foreign or local firms, would not be granted (Schenk, 2006). Once again, banking was to be exempted from the normal rules of laissez faire, this time by curtailing competition through direct government intervention.

There was one further departure from the normal rules of free and fair markets that this British colony usually proclaimed. The government assumed that banks with headquarters outside Hong Kong could be trusted to manage their businesses prudently with minimal official supervision of their funding arrangements. As a result, they were allowed to meet their statutory liquidity requirements through window-dressing transactions with their head offices. Thus, these banks could expand their loan books well beyond their deposit base, unlike their locally owned competitors. This "liquidity loophole" was to have serious consequences in later decades through facilitating imprudent lending.

6.6 FOREIGN CONTAMINATION (1971–1978)

In 1965, the colonial administration was able to prevent a general collapse of public confidence in the banking system. The regulatory reforms introduced in the previous year had proved their worth and provided ample justification for abandoning laissez-faire preconceptions. But this experience did little to change the deep-seated distaste for regulatory responsibilities among officials. When in the 1970s, banking stability came under threat once more, it seemed that the lessons of the recent past had been forgotten.

The new challenge was the emergence of deposit-takers outside the scope of banking legislation. A stock market bubble started in 1971 during which an estimated 1500 "finance houses" sprang up mainly acting as intermediaries willing to fund stock market investors whom licensed banks did not regard as acceptable credit risks. In 1973, share prices fell by 75%, bringing down a large number of the financial intermediaries. (The stock market was not to recover fully until the end of the decade.) The better capitalized and more professionally staffed firms that survived now formed a new and increasingly important class of financial institution: the "deposit-taking company" (DTC)—unlicensed, unregulated but able to perform most banking functions. In many instances, these DTCs were set up by licensed banks as well as by foreign banks that could not obtain licensed status because of the government's anticompetition policies.

The bank-related DTCs were to play a special role as Hong Kong was transforming itself into Asia's premier business center. While domestic exports grew during this decade by an annual average of 19% (even faster than the annual 17% in the 1960s), the annual rate of growth for exports of services had accelerated from 12% in the previous decade to almost 17%. By 1980, financial and business services, together with real estate, accounted for 23% of GDP, the same share as manufacturing.

In New York and London, more sophisticated financial services were coming into fashion, including loan syndication, project finance and term lending, and currency swaps (Jao, 1979). The potential market in Hong Kong for such facilities was expanding in the 1970s. Chinese family enterprises that had made their first fortunes from manufacturing were seeking modern corporate structures as they expanded into new ventures (particularly property) and looked for stock market flotations.

In serving this new generation of Hong Kong tycoons, foreign banks enjoyed special advantages. Their staff had experience and expertise acquired in the much more advanced U.S. and U.K. financial markets. They had ample funding to finance corporate clients because the "liquidity loophole" allowed foreign banks to escape from prudential statutory liquidity requirements. They could thus lend well beyond their deposit base, enabling them to finance a significant share of the stock market speculation in this decade. If they could not obtain a banking license, a DTC subsidiary allowed them to function as investment bankers without the inconvenience of official oversight. Indeed, even for licensed banks, a captive DTC provided an opportunity to engage in business practices that would be unacceptable to the banking regulators.

Leading bankers publicly warned the government of a growing threat to the prudential management and the stability of the financial sector as a whole. But the colonial administration decided to accept the risks involved. In the early 1970s, the government believed that if statutory supervision were imposed on DTCs, many of them would be unable to meet the regulators' requirements. The forced closures that would follow could slow down Hong Kong's development as Asia's financial center, officials feared. They were determined, in any case, not to expand the government's responsibility for protecting the public's deposits, and the colonial administration publicly repudiated any duty to secure the prudent conduct by DTCs of their business.

Even after DTCs became subject to registration in 1975, there was no suggestion of inspecting them to ensure compliance with the minimalist rules that were supposed to govern their lending. Not until the excesses of a new share and property boom in 1978 was the government forced to accept that DTCs should no longer be left to their own devices, and they were made subject to regulation on the same basis as licensed banks. But again, as with the banks in the 1950s, the resources required for effective supervision of DTCs were not provided. Official statistics revealed very clearly the hole in the regulatory system that the under-supervised DTCs represented. Of Hong Kong's 124 licensed banks, 28 had DTC subsidiaries, and a further 187 DTCs—almost 80% of the total number—were controlled by overseas banks unable to obtain licensed status.

The 1970s showed how deeply entrenched was the antipathy of the government toward intervention in financial markets despite the persuasive evidence from the troubled 1960s of the benefits of effective regulation. Officials closed their eyes to the dubious credentials of many DTCs, just as they had done when licensing local banks in the 1940s and 1950s. The absence of effective regulation and the government's lack of enthusiasm for policing DTCs fostered a burst of misconduct and mismanagement, in which leading international banks were to be involved for the first time in Hong Kong. As in earlier decades, officials in the 1970s displayed a stubborn refusal to enforce existing legislation regardless of ample proof of imprudent and even illegal behavior. Much the same regulatory attitudes have been identified as facilitating the start of the global financial crisis in 2007 by allowing banks to operate "outside the bounds of established regulatory controls," introducing "a new element of volatility and instability into the model of many U.S. banking institutions" (Whalen, 2008).

6.7 FRAUD AND INCOMPETENCE (1979–1986)

A heavy price was to be paid for this official negligence in the following decade as the economy came under severe political as well as economic pressure. These problems were aggravated by the influence on official regulators' attitudes of the colonial administration's determination to minimize the government's oversight of financial institutions. The regulators' response was to focus narrowly on technical compliance with statutory requirements. They refused, for example, to report the criminal activities of owners and executives to law enforcement agencies on the misconceived grounds that financial inspections were totally confidential. In this environment, fraud flourished on a grand scale. For example, after the collapse of a DTC with a loan book of only $150 million, it was discovered that it had achieved an apparent world record for "check-kiting" frauds. It had used its American bank accounts to obtain $21.7 billion by passing worthless checks. No alarm bells went off in either Citibank or HSBC about the legitimacy of the daily flood of transactions from this minor financial corporation. (The parallels with the Madoff and Société Générale revelations of 2008–2009 are obvious.)

Investors had started to lose confidence during 1982 in the asset bubble financed by unconstrained lending via the DTCs. The more vulnerable lenders, desperate to disguise their losses, turned to fraud and false accounting. In 1986, the share of GDP generated by financial and business services, together with real estate, fell back to 17%. By this date, the government had been forced to rescue seven licensed banks to avoid a general collapse of public confidence in the banking system, and some 90 DTCs had closed their doors. Not even HSBC—then Hong Kong's quasi-central bank—was immune to the financial losses and corporate scandals caused by DTC subsidiaries.

Then, abruptly, the entire financial services industry was cleaned up. Outrage had mounted in the early 1980s at the financial scandals and the corporate collapses and, even more, at the cost to the public of rescuing banks from crises attributable to official incompetence. The entire regulatory system was overhauled, and the colonial administration now accepted a duty to protect the customers of all deposit-taking institutions. It helped that the law-enforcement agencies were enthusiastic about investigating misconduct and prosecuting executives no matter how high their rank or well-connected their employers.

6.8 LESSONS OF THE PAST

The lessons had finally been learned, and there were to be no bank failures either in the 1998 Asian financial crisis or during the current global crisis. A half century of trauma triggered by property bubbles had convinced the banks themselves that overexposure to property would create unacceptable risks as uncertainty mounted during the run-up to the end of British rule and the return to Chinese sovereignty scheduled for 1997. In 1991, banks decided that the loan-to-value ratio for mortgages should not exceed 70%, and the regulatory authority subsequently adopted this ceiling as an official benchmark. Between 1994 and 1998, the barrier against excessive lending to real estate was raised still higher when the regulators set a 40% limit to property exposure within a bank's total Hong Kong loan portfolio. As a result, although property values slumped by 70% between 1998 and 2003, the delinquency rate never rose above 2.5% of total bank mortgage lending.

Officials too had learned from past experiences of crisis management. In 1998, the currency came under severe pressure in tandem with share prices. Speculation was driving down currencies almost everywhere else in the region, and there seemed a high probability that the Hong Kong dollar, which had been linked since 1983 to a fixed rate against its U.S. counterpart, might be forced to devalue. The postcolonial government was in a strong position to meet this challenge because it had inherited from its British predecessors virtually no public debt, substantial fiscal reserves, and total backing in USD assets for the entire note issue. With these resources, it was able to commit $15 billion to large-scale intervention in the financial markets (on which it made a profit of 68%). Although the government acquired a substantial share portfolio in the process, officials were not tempted to become actively involved in corporate management or to induce firms to adopt anti-deflationary measures during Hong Kong's first recession in almost 40 years. Throughout the severe deflationary pressures of the early years of the new century, the government rejected Keynesian remedies and remained loyal to the economic policies of the previous century. Its primary goal was to avoid budget deficits and to curb welfare spending (Goodstadt, 2009).

The current global crisis has not left Hong Kong completely unscathed but the pressures have come from the downturn in world trade rather than financial instability. As in 1998, defense of the exchange rate link with the USD has been the primary problem for Hong Kong since the start of the current global financial crisis. The challenge this time, however, has been to prevent upward pressure on the currency because of a sustained inflow of funds in search of stability.

In common with the United States and the European Union, Hong Kong bankers have found it difficult to avoid public complaints about a credit squeeze and the increased difficulties that firms face in securing loan facilities. The constraints on lending have not been the result of problems within the banking system, however, but reflect deteriorating business conditions. None of the banks has been at risk of insolvency since holdings of subprime mortgage and related assets have been very limited. Thus, the government was not forced to inject liquidity into the banking system (despite standby arrangements to do so if necessary). The main official initiative in 2008 was a $13 billion revolving guarantee to underwrite bank lending to small and medium-sized enterprises. The public seemed to take for granted that the regulatory arrangements put in place after the scandals of the 1980s ensured the safety of financial institutions, so that when formal government guarantees for bank deposit-holders were announced, the move seemed almost unnecessary.

The one serious scandal centered on the sale by banks of an estimated $2 billion worth of Lehman "mini-bonds" to 20,000 retail investors. The damage to the banks involved has been reputational rather than financial since the complaint is that bank staff misled customers about the nature of these products. The only serious test of public confidence came from another sector of the financial services industry. As AIG seemed about to topple in New York, its two Hong Kong subsidiaries were hit by a wave of cancellations among its 1.95 million local policy holders. The panic promptly stopped when the regulatory authority announced that the subsidiaries were operating as separate and financially independent businesses that would be restrained from transferring assets to the parent company without prior official approval.

6.9 CONCLUSIONS

Perhaps the most striking conclusion to be drawn from the above analysis is how similar have been the behavior patterns of the main parties involved in the management of financial crises–the policy makers, the regulators, and the financial and banking executives—and how predictable the outcome of their decisions has been, whether in the current financial crisis or in the light of Hong Kong's financial history. This finding is in marked contrast to other research findings about financial crises that suggest their "anatomy" is largely dictated by "local" conditions (Sheng, 1998) and that the present global crisis is without modern parallels (Sentence, 2008).

Hong Kong's banking history offers persuasive evidence that neither financial markets nor financial institutions can be left to police themselves. After each set of bank runs and corporate failures, Hong Kong officials grudgingly implemented incremental but minimalist expansions of their responsibility for financial stability. The next property boom would then provide new opportunities for illegal and imprudent banking activities that highlighted the need for more regulation. The cumulative impact of this long reform process was impressive, nevertheless. By the end of the century, Hong Kong had achieved what was arguably the most robust banking system in Asia operating in what was indisputably the most open and competitive financial environment in the region.

The trend to comprehensive regulation of financial services did not smother enterprise or deter innovation. Furthermore, the government's readiness to deploy its reserves to reflate banks during a run and to rescue financial markets from panic did not lead inexorably to deficit finance and profligate spending at the taxpayers' expense. Hong Kong thus represents an important example of the merits of government intervention in financial affairs even when a government and the business community are deeply committed to laissez faire.

REFERENCES

Berndt, A. and Gupta, A. (November 2008) Moral hazard and adverse selection in the originate-to-distribute model of bank credit, available at SSRN-id1290312[1].pdf.

Demyanyk, Y. S. and Van Hemert, O. (December 5, 2008) Understanding the subprime mortgage crisis, available at SSRN-id1020396[1].pdf.

Goodstadt, L. F. (2005) *Uneasy Partners: The Conflict between Public Interest and Private Profit in Hong Kong.* Hong Kong University Press: Hong Kong.

Goodstadt, L. F. (2007) *Profits, Politics and Panics: Hong Kong's Banks and the Making of a Miracle Economy, 1935–1985.* Hong Kong University Press: Hong Kong.

Goodstadt, L. F. (2009) A fragile prosperity: Government policy and the management of Hong Kong's economic and social development. HKIMR Working Paper No. 1/2009, January 2009.

Jao, Y. C. (1979) The rise of Hong Kong as a financial centre. *Asian Survey,* 19(7): 674–694.

Keys, B. et al. (December 2008) Did securitization lead to lax screening? evidence from subprime loans 2001–2006, EFA 2008 Athens Meetings Paper available at SSRN-id1093137[1].pdf.

King, M. (2008) Banking and the bank of England. *Bank of England Quarterly Bulletin* 2008 Q3, available at www.bankofengland.co.uk/publications/speeches/2008/speech347.pdf.

Kregel, J. (April 2008) Changes in the U.S. financial system and the subprime crisis, Levy Economics Institute Working Paper No. 530, available at SSRN-id1123937[1].pdf.

Reinhart, C. M. and Rogoff, K. S. (2009) The aftermath of financial crises. Paper presented to the American Economic Association, January 2009, available at http://www.economics.harvard.edu/faculty/rogoff/files/Aftermath.pdf.

Schenk, C. R. (July 2006) The origins of anti-competitive regulation: Was Hong Kong 'Over-banked' in the 1960s? HKIMR Working Paper No.9/2006, Hong Kong.

Schenk, C. R. (Ed.) (2008) *Hong Kong SAR's Monetary and Exchange Rate Challenges: Historical Perspectives.* Palgrave Macmillan: Basingstoke, U.K.

Sentence, A. (2008) The current downturn—A bust without a boom? MPR Monetary Policy and the Markets Conference, December 2008, available at http://www. bankofengland.co.uk/publications/speeches/2008/speech370.pdf.

Sheng, A. (1998) Bank restructuring revisited, in G. Caprio Jr. et al. (Eds.), *Preventing Bank Crises: Lessons from Recent Global Bank Failures.* The World Bank: Washington, DC.

Whalen, R. C. (March 1, 2008) The subprime crisis: Cause, effect and consequences, Networks Financial Institute Policy Brief No. 2008-PB-04, available at SSRN-id1113888[1].pdf.

CHAPTER 7

Auction Rate Securities: Another Victim of the Credit Crisis

Edwin Neave and Samir Saadi

CONTENTS

The auction rate securities (ARS) market came into being after the decline in technology stock prices in 2001–2002. During the heyday of the scheme, investors were attracted to ARS because they represented high-grade short-term paper with a higher yield than Treasury bills. However, following the recent credit crisis, the ARS market failed leaving thousands of investors stranded, unable to, sell, collect on, or otherwise convert their securities. This proposed chapter provides a comprehensive analysis of the ARS market with a particular emphasis on the origins and mechanics that caused its recent collapse.

7.1 INTRODUCTION

First introduced in 1984, auction rate securities (ARS) are municipal bonds, corporate bonds, and preferred stocks with interest rates or dividend yields that are reset regularly through auctions. Marketed as safe and cash-equivalent investment, ARS attracted wealthy investors as well as corporate treasurers. However, the growth of the ARS market was hampered in 1980s and 1990s due to multiple auction failures as well as evidence of auction manipulation by ARS brokers/dealers. Nevertheless, the ARS market grew rapidly after the decline in technology stock prices in 2001–2002, after which it experienced a wave of failures in 2007 and 2008. Its growth and decline were both stimulated by an interaction of factors.

In late 2002, the U.S. federal funds rate had been reduced to a low of 1%, and because of the historically low yields on high-quality short-term bonds, investors were seeking higher rate investment opportunities. At the same time long-term interest rates remained considerably higher, and long-term borrowers were seeking to reduce the interest rates they were paying. Investment bankers sought to profit from serving both types of client, and proposed ARS as a means of achieving their goals. During the next several years, the ARS market worked well, but was still flawed because it represented an attempt to finance long-term assets through short-term borrowings. When it ultimately failed in 2008, additional attempts to raise funds using ARS failed.

During the ARS' heyday, investors were attracted to them because they represented high-grade short-term paper with a higher yield than Treasury bills. Long-term borrowers (municipalities, hospitals, utilities, port authorities, housing finance agencies, student loan authorities, and universities) were equally attracted to the market because, for a time, ARS allowed them to issue 20 year term debt at rates much lower than 20 year fixed rates.

The scheme proved popular with both investment bankers and borrowers, and by early 2008 about $330 billion (U.S.) of ARS had been issued. Even collateralized debt obligations (CDOs) were sold in the ARS market to finance long-term (subprime) assets. When failures began to occur in mid-2007, it was suspected that many CDOs were backed by low-quality assets. At the same time, bond insurers were leveraged at triple-digit multiples of their capital and investors lost confidence in bond insurers' ability to redeem the liabilities created by the default insurance. As a result, investor interest in auctions fell sharply, and in some auctions there were no bids. The dealers who were to provide liquidity in the paper were unable to do so because of other difficulties they were facing, and the ARS market was frozen.

This present chapter provides a comprehensive analysis of the ARS market with a particular emphasis on the origins and mechanics that caused its recent collapse. First, we provide a general background on the ARS market in Section 7.2. Section 7.3 describes the auction process and Section 7.4 describes the evolution of the ARS market. Section 7.5 discusses the reasons behind its collapse while Section 7.6 concludes the chapter.

7.2 AUCTION RATE SECURITIES

ARS are long-term instruments with interest rates that are reset at periodic intervals through Dutch auctions establishing the lowest rate that would clear the securities then being sold. The auctions, which run by the underwriters, typically occur every 1, 7, 14, 28, 35, and 49 days, or 6 months. Since they were first created in 1984, ARS were an attractive investment for investors. Thanks to the auction process and given that ARS are traded and callable at par, ARS market allow investors to sell their securities and hence get back the face value of their investment. ARS underwriters market ARS as safe and liquid cash alternatives that provide higher returns than other several comparable investments such as certificate of deposits, money market funds, and variable rate demand obligations (VRDOs). These features help explain the growth of the ARS market over the last two decades to a total of some $330 billion in early 2008. At the same time, the ARS market would not have grown so large if it were not also an attractive source of funding for issuers. Indeed, ARS allow issuers to obtain long-term financing at short-term rates. ARS are generally more attractive than similar long-term financing alternatives such as the VRDO for which interest rate is also reset periodically. In fact, in contrast to VRDO, ARS do not requires letter of credit, "put" or tenders features and annual short-term bond ratings.

There are two types of ARS: bonds with long-term maturities and perpetual preferred equity with a cash dividend. Auction-rate bonds represent the major portion of the auction-rate market. The approximately $165 billion in bonds outstanding at the end of 2007 was nearly 76% of the market. These bonds, often tax-exempt, are issued by municipalities (hence the name "munis"), student-loan authorities, states and state agencies, cities, museums, and many others. Sometimes, several small borrowers get together to make a large issue. By far the most dominant groups of issuers are municipalities (roughly 50% of the market) and student-loan authorities (nearly 34% of the market). While the latter only issue auction-rate bonds to fund student loans, municipalities issue auction-rate bonds to help fund, among other projects, roads, public school construction, and improvements to public infrastructure. Auction-rate bonds issued by municipalities are not (default) risk-free, but the fact that they are insured and have top credit rating (e.g., AAA/Aaa) makes them highly marketable. This, however, makes them also highly sensitive to changes in credit rating and to the ability of the insurer (very often a monoline) to cover the losses associated with these bonds. Indeed, the main role of monolines has been to insure mortgage bonds and CDOs.

Unlike auction-rate bonds, perpetual preferred equities are not issued by government-related issuers, rather by closed-end funds. Closed-end funds' shares are typically traded on a stock exchange. The assets of a closed-end fund may be invested in stocks, bonds, or a combination of both. As of end 2007, the total value of auction preferred stocks issued by closed-end funds is estimated at about $63 billion with nearly 48% of it is tax-exempt. It is noteworthy that unlike auction-rate bonds, preferred stocks are not insured and generally have higher default risk. Over the past decades the leading underwriters of ARS have been Goldman Sachs, Morgan Stanley, UBS, Lehman Brothers, JP Morgan, Citigroup, and Bank of America. The usual issue size of ARS is $25 million. With a minimum investment of $25,000 typical buyers of ARS are wealthy individuals and corporations. Following rule 2a-7 of the Securities and Exchange Commission (SEC), money market funds are prohibited from holding ARS given that final maturity of such investment typically exceed 397 days. The rate of returns earned on ARS is determined by auction process through brokers/dealers known as "remarketing agents." Remarketing agents can be a single underwriter or syndicate of multiple brokers/dealers. In Section 7.3, we describe the ARS auction process.

7.3 ARS AUCTION PROCESS

Unlike regular long-term bonds and preferred stocks, rates on ARS are reset at periodic intervals through Dutch auctions. The auction process is the tool that makes the ARS attractive, particularly because it allows ARS investors to liquidate their investment by selling their ARS to potential buyers. Each potential buyer indicates to brokers/dealers the quantity of ARS he/she wishes to buy along with the rate she requires in order to hold the securities until next auction. Holders of existing ARS can either hold their securities regardless of the clearing rate (known as "hold-at-market"), hold their securities only at a specific minimum rate otherwise sell at the clearing rate (known as "hold-at-rate"), or just sell their ARS at any clearing rate. In summary there can be four types of bid: Buy, hold-at-market, hold-at-rate, and sell. If all holders of existing ARS decide to hold their securities regardless of the clearing rate they will earn an "all hold" rate, which determined in the security's official statement and is usually based on Bond Market Association Index (BMA) rates or commercial paper rates.

Brokers/dealers collect all the bids and convey them on behalf of potential buyers to the auction agent, who is usually a third-party bank selected by the issuer. The auction agent then sort the bids on ascending order based on bid rate to determine the clearing rate. If the buyer demand is higher than the seller supply, then the clearing rate is the lowest bid rate at which all ARS can be sold. Bids that are higher than the clearing rate are not filled, while bids that are at or below the clearing rate are filled in the following order of priority: (1) Hold bids, (2) hold-at-rate and buy bids below clearing rate, (3) hold-at-rate at the clearing rate, and (4) buy bids at the clearing rate. The issuer of the ARS will have to make interest payments (until the subsequent auction) to winning bidders based on the clearing rate.

To illustrate how the clearing rate is determined let's assume that there 2000 ARS outstanding. Let's also assume that holders of 800 ARS hold their securities regardless of the clearing rate, meaning that the total supply of ARS is equal to 1200 ARS. That is $30 million (i.e., 1,200 × $25,000) worth of ARS is available for sale at the auction. Suppose that the participants at the auction have placed the following orders (or bids) through their corresponding brokers/dealers:

Potential buyers of available ARS:

> Bidder # 1: Buy 300 at any rate
> Bidder # 2: Buy 400 at 2%
> Bidder # 3: Buy 100 at 4.5%
> Total buy bids: 800

Remaining holders of existing ARS:

> Bidder # 4: Hold 350 at 3.5%
> Bidder # 5: Hold 250 at 4%
> Bidder # 6: Sell 700 at any rate
> Total sell and hold-at-rate bids: 1300

So the total demand for ARS is equal to the buy bids (i.e., 800) plus the hold-at-rate bids (i.e., 350 + 250): 1400. For now, let's see what the wining bids are in our hypothetical example. After brokers/dealers have conveyed their clients' bids to the auction agent, the latter assemble the bids from smallest to highest based on bids' corresponding rates to determine the clearing rate. Table 7.1 illustrates how the clearing rate is established. With an ARS supply of 1200, and once orders for bidder 1, 2, 4 are fully filled at the clearing rate of 4%, only 150 ARS were left for bidder 5, which explain its partial fill. Order for bidder 3 is not filled not because there were no ARS left but because the ARS rate required by bidder 3 exceeds the clearing rate. The auction agent advises the brokers/dealers, of the auction results who then record and settle the trades for the next business day settlement.

Fortunately for bidders 1, 2, 4, and 5 the total demand for ARS exceeds the total supply, otherwise they would have met the same fate as bidder 5. In fact, the condition that the demand for ARS should be at least equal to the supply is necessary for the success of the ARS auction. If the demand

TABLE 7.1 ARS Auction Process and Clearing Rate

Bidder	Placed Bid	Bid Type	Bid's Rate	ARS Allocated	Remaining ARS	Bids Filed (at the Clearing Rate)
1	300	Buy	Any	300	900	Filled
2	400	Buy	2.0%	400	500	Filled
4	350	Hold-at-rate	3.5%	350	150	Filled
5	250	Hold-at-rate	4.0%	150	0	Partially filled
3	100	Buy	4.5%	0	0	Not filled

was less than the supply, the auction would have failed and ARS would have become illiquid. Now let's introduce some modification in our example to show how a lack of demand leads to failed auction. Suppose instead of 400 ARS units, bidder 2 informs his/her broker that he/she wishes to buy 100 units. This will lead to a drop in the demand for ARS from 1400 to 1100 units. Since the supply side is unchanged (i.e., 1200) the auction will fail and none of the orders will be filled. Table 7.2 depicts the new situation. It is noteworthy that we could have faced an action failure also if the supply went up by more than 200 ARS units. In both cases, the auction agent is incapable to determine a clearing rate.

Unlike an auction failure that is due to lack of supply, a failure that is due to lack of demand can have serious consequences for the holders of ARS, issuers of ARS, underwriters of ARS, and to the ARS market as a whole. The first form of failure occurs when all holders of ARS decide to continue to hold (hold-at-market). Although holders of ARS will usually earn a rate that is below the market, the perception of ARS being a safe and cash-equivalent investment is not altered.* However, when the auction fails because there are too few or no potential investors wishing to buy ARS, then holders find themselves forced to hold their ARS until next auction, if not longer. A failed auction does not mean that the ARS go into default, but it can put holders of ARS, especially those of an emerging need for cash, in a troubling situation. To meet their cash flow needs, some investors may consider selling other securities in their portfolio or borrowing at margin at an interest rate that may exceed the yield earned from the underlying security.

TABLE 7.2   ARS Auction Failure

Bidder	Placed Bid	Bid Type	Bid's Rate	ARS Allocated	Remaining ARS	Bids Filed (at the Clearing Rate)
1	300	Buy	Any	300	900	Not filled
2	100	Buy	2.0%	100	800	Not filled
4	350	Hold-at-rate	3.5%	350	450	Not filled
5	250	Hold-at-rate	4.0%	250	200	Not filled
3	100	Buy	4.5%	100	100	Not filled

* ARS issuer, California Housing Finance Agency (CHFA) had to pay an interest rate of 1.4% on its ARS when auction failed on February 26, 2008 because all holders of CHFA's ARS decided to continue holding their security.

According to ARS regulation, when auction fails, the issuer has to pay a "penalty rate" known also as "maximum rate" or "fail rate." The penalty rate is often a multiple of a reference rate, such as the London Interbank Offering Rate (LIBOR) or an index of Treasury securities, but does not exceed a fixed cap. The multiple may depend on the credit rating of the ARS issuer. Penalty rate is usually very higher but it is much higher for auction-rate bonds than for preferred equity. For instance, the interest rate paid by Port Authority of New York and New Jersey had gone up from 4.25% to a penalty rate of 20% when auction of its ARS failed on February 12, 2008, increasing the borrowing cost by $0.3 million per week. Yet, the penalty rate for closed-end funds (issuers of preferred equity) is, on average, as less as 3.5%. To hedge against penalty rate, some issuers of auction-rate bonds may use Interest Rate Cap (IRC) or enter into Interest Rate Swaps (IRSs). IRC is an insurance policy that protects the ARS issuer against the risk of paying high interest rate. Using IRS, the ARS issuer enters into a contractual agreement with a counterparty to exchange a series of fixed rate–interest payments for a series of floating rate–interest payments at predetermined intervals for a stated period of time. Hence, both IRC and IRS allow the ARS issuer to keep its borrowing cost at a preestablished limit.

While they are not legally required to do so, underwriters of ARS can provide a "clearing bid" to circumvent the event of an auction failure. As expressed above, ARS were marketed by underwriters as highly liquid investment, however, when ARS holders find themselves unable to liquidate theirs positions, they are likely to lose confidence in the underwriters, the auction process, and the ARS market as a whole. Repetitive auction failures cause potential investors to lose interest in ARS leading to further auction failures.

7.4 EVOLUTION OF THE ARS MARKET

In early 1980s, when financial institutions were striving to engineer new financial products that would lower their borrowing cost caused by high inflation, an investment banker at Lehman Brothers, Ronald Gallatin, invented ARS.[*] According to Carow, Erwin, and McConnell (1999), during that period, financial institutions in the United States invented about

---

[*] ARS were not the only brainstorm of Ronald Gallatin. In fact, he invited a series of other financial instruments such as Zero Coupon Treasury Receipts, Money Market Preferred Stock, and Targeted Stock.

26 new financial instruments. With exception of a few, including the ARS, most of these new and complex instruments were short lived. The ARS main objective was to allow financing long-term debt at short-term interest rates that are reset weekly or monthly through an auction process. In 1984, American Express issued the fist ARS: A $350 million issue of money-market-preferred shares with a minimum purchase of $500,000. The dividend rate was reset at auction every 49 days. Following American Express, other banks such as Citicorp (currently Citigroup) and MCorp started selling ARS. Although the first ARS were auction-rate-preferred shares, the concept soon spread to auction-rate bonds with the first being issued on 1985 by Warrick County to finance the Southern Indiana Gas and Electric Company (McConnell and Saretto, 2008).

A year after the emergence of the ARS, U.S. Steel issued the first insured ARS. The insurance company provides coverage to ARS holders against default risk associated with the issue. In case of default, the insurance company pays principal and interest to holders of the defaulted ARS. While ARS insurance boosted both issuers' credit rating as well as investors' confidence, it did not however avert auction failure. Indeed, on 1987 the auction of the MCorp's $62.5 million preferred stock issue failed when not enough investors showed interest, marking the first auction failure in ARS market. As a result, the dividend rate on MCorp's ARS soared to a penalty rate of 13%. Subsequent auction failures in the 1990s led several large ARS issuers such as Citigroup and JP Morgan Chase & Co to retire their ARS as their borrowing costs surpassed the 14%. These failures made investors realize that neither an AAA bond rating nor the insurance company can prevent an auction failure. In 1995, investors' confidence in ARS market was wobbled again by the SEC accusation to Lehman of manipulating 13 American Express auctions. The company agreed to pay an $850,000 fine but without admitting or denying any misconduct.

The market took off after 2000 as public sector started using ARS as a major source of funding. In fact, the annual sales of municipal auction-rate bonds grew from $9.56 billion to more than $40 billion in 2003 and 2004. The percentage of insured ARS has also increased fueling ARS market growth. On 2006, however, the annual sales of auction-rate municipal bond shrunk to about $30 billion following a SEC investigation on how ARS auctions are conducted. The critics complain that brokers/dealers controlled all information and influenced the bids, which of course violated the basic rules of auction process. Moreover, given that brokers/dealers were not required to guarantee against an auction failure, investors may have been oblivious of the true liquidity risk associated with the ARS. Brokers/dealers argued that they were trying to prevent auction failure, which will potentially impair all market participants (i.e., investors, issuers, and underwriters). The SEC investigation, however, uncovered serious "violative practices" that go beyond the aforementioned practice.

SEC investigation showed that about 15 brokers/dealers (Respondents) engaged in one or more of the following violative practices in connection with certain auctions*:

Completion of open or market bids: "Some investors placed open bids and/or market bids in auctions. When an investor placed an open bid, it allowed the Respondent to designate some or all of the bid's parameters, such as the specific security, rate, or quantity... After viewing other orders in the auction; certain Respondents supplied the bid parameters missing from open bids and/or the rate for market bids."

Intervention in auctions:

Bids to prevent failed auctions: "Without adequate disclosure, certain Respondents bid to prevent auctions from failing...Respondents submitted bids to ensure that all of the securities would be purchased to avoid failed auctions and thereby, in certain instances, affected the clearing rate"

Bids to set a "market" rate: "Without adequate disclosure, certain Respondents submitted bids or asked investors to change their bids so that auctions cleared at rates that these Respondents considered to be appropriate "market" rates."

Bids to prevent all-hold auctions: "Without adequate disclosure, certain Respondents submitted bids or asked investors to submit bids to prevent the all-hold rate, which is the below-market rate set when all current holders want to hold their positions so that there are no securities for sale in the auction."

---

* SEC Settlement, May 31, 2006, pp. 5–8.

Prioritization of bids: "Before submitting bids to the auction agent, certain Respondents changed or "prioritized" their customers' bids to increase the likelihood that the bids would be filled."

Submission or revision of bids after deadlines: "… these practices, except when solely done to correct clerical errors, advantaged investors or Respondents who bid after a deadline by displacing other investors' bids, affected the clearing rate, and did not conform to disclosed procedures."

Allocation of securities: "Certain Respondents exercised discretion in allocating securities to investors who bid at the clearing rate instead of allocating the securities pro rata as stated in the disclosure documents."

Partial orders: "When an auction is oversubscribed, investors may receive a partial, pro rata allocation of securities rather than receiving the full amount of the securities for which they bid. When this occurred, certain Respondents did not require certain investors to follow through with the purchase of the securities even though the bids were supposed to be irrevocable."

Express or tacit understandings to provide higher returns: "… certain Respondents provided higher returns than the auction clearing rate to certain investors."

Price talk: "Certain Respondents provided different "price talk" to certain investors. In certain instances, some investors received information that gave them an advantage in determining what rate to bid, thereby displacing other investors' bids and/or affecting the clearing rate."

The SEC commission fined the 15 banks and brokers/dealers a total of $13.8 million for market manipulation.* Though, SEC did not impose new regulations, the SEC settlement led to an improvement in information

---

* They are Bear, Stearns & Co. Inc.; Citigroup Global Markets, Inc.; Goldman, Sachs & Co.; J.P. Morgan Securities, Inc.; Lehman Brothers Inc.; Merrill Lynch, Pierce, Fenner & Smith Incorporated; Morgan Stanley & Co. Incorporated and Morgan Stanley DW Inc.; RBC Dain Rauscher Inc.; Banc of America Securities LLC; A.G. Edwards & Sons, Inc.; Morgan Keegan & Company, Inc.; Piper Jaffray & Co.; SunTrust Capital Markets Inc.; and Wachovia Capital Markets, LLC.

disclosure and auction practices. For instance, in an effort to address the issues raised by SEC investigation, the Securities Industry and Financial Markets Association (SIFMA) issued on April 3, 2007 *Best Practices for Broker–Dealers of Auction Rate Securities* or *Best Practices*.*

Despite all of these events, the ARS market as a whole continued to expand until February 2008, at which time it froze and collapsed, as described in the following section.

7.5 THE PERFECT STORM: COLLAPSE OF THE ARS MARKET

Although the ARS market experienced some downturns between 1984 (when the first ASR was issued) and the end of 2006, the collapse of the market in 2008 was completely unexpected. While 13 auction failures only were recorded during the 1984–2006 period, over 1000 occurred on the first week of February 2008 alone. Indeed, the market broke down during the week of February 13, 2008, when ARS brokers/dealers simultaneously withdrew their support for the auctions, leaving thousands of investors stranded, unable to, sell, collect on, or otherwise convert their securities.

While there are different views on what caused the collapse of the ARS market, there is a consensus that the recent credit crisis was the main stimulus, as ARS market conditions have changed dramatically since the middle of 2007. In its March 2008 Statement on Financial Market Developments, the President's Working Group on Financial Markets (PWG) described the current situation as follows[†]:

> Since mid-2007, financial markets have been in turmoil. Soaring delinquencies on U.S. subprime mortgages were the primary trigger of recent events. However, that initial shock both uncovered and exacerbated other weaknesses in the global financial system. Because financial markets are interconnected, both across asset classes and countries, the impact has been widespread.

In late 2008, regulators in the United States began to negotiate with issuers to buy the paper back. UBS agreed to repurchase $19 billion, Merrill and Citigroup about another $20 billion. But the buyback arrangements

---

* *SIFMA's Best Practices for Broker–Dealers of Auction Rate Securities.* The guide is available online at http://www.sifma.org/services/pdf/AuctionRateSecurities_FinalBestPractices.pdf.

† The President's Working Group on Financial Markets, Statement on financial market developments, March 2008, pp. 1–9. The report is available online at http://www.iasplus.com/crunch/creditcrunch.htm#0803pwg.

reached at this writing cover just a little less than 15% of the $330 billion ARS outstanding, and the market itself remains frozen.

In what follows, we present the different factors that have been proposed to explain the collapse of the ARS markets. We believe, however, that the simultaneous rising of these factors (due to credit crisis) has led to the formation of a "perfect storm" that caused the market to fail. In other words, the ARS collapse should not be attributed to one factor or another, but to the accumulation of different factors.

7.5.1 Trouble with Monoline Insurers

As default on subprime mortgage soared and started spreading into the mortgage securitization market by late 2007, investors became wary about the ability of insurers to cover losses associated with the auction-rate bonds. Despite the fact the credit ratings of ARS issuers have remained high, investors were concerned about the potential credit downgrading of monoline insurers. Indeed, the demand for ARS started to decline sharply because of the concerns of credit downgrading of the two largest insurers in the auction-rate market (MBIA, Inc. and Ambac Financial Group, Inc.) which at that time had AAA ratings. The situation got worse starting from July 2007 when both monoline insurers reported lower profits due to subprime losses. As a result, an increasing number auction failure occurred starting from August 2007. Interestingly, even uninsured ARS experienced similar failures. This showed the importance of the role played by monoline insurers in the success and stability of the ARS market. In the same line, commenting on the factors that contributed to the credit crisis, the PWG's statement added[*]:

> More generally, uncertainty about asset valuations in illiquid markets and about financial institutions exposures to asset price changes left investors and markets jittery. Of particular concern were the exposures of financial guarantors to protection they had provided on supersenior CDO tranches. Downgrades of these institutions credit ratings could disrupt financial markets in which investors rely heavily on their guarantees, notably the U.S. municipal bond market. In addition, failure of one of the guarantors would impose additional losses on financial institutions that had purchased CDO protection from the guarantor.

---

[*] The President's Working Group on Financial Markets, Statement on financial market developments, March 2008, pp. 9–10.

7.5.2 Brokers/Dealers Liquidity Problem

Prior to the credit crisis, brokers/dealers (which are mainly large banks) often bid themselves in ARS auction to avert auction failure although they are not obligated to do so. However, after losing hundreds of billions of dollars on subprime mortgages, brokers/dealers stopped bidding at auctions for their own accounts. This led to more auction failure and rise of the borrowing cost for ARS issuers as ARS rates jumped to the maximum level. While the high interest rates compensated the ARS holders for their loss of liquidity, they did not lessen investors' fear of losing their entire investment. Analysts and media started blaming brokers/dealers for creating an artificial market that became illiquid the moment these dealers withdrew from the market*.

7.5.3 Increase in Demand and Decrease in Supply

As the word of the auction failure spread, ARS were perceived as risky and illiquid investment leading to further decline in demand for ARS. Moreover, the supply of ARS increased as holders raced to sell their ARS creating even greater pressure to find buyers to make the auctions succeed. This has widened further the gap between demand and supply for ARS, making the likelihood of auction success just about zero.

7.5.4 Deceptive Practice by Brokers/Dealers and Lack of Sound Information Disclosure Mechanism

Selling pressure from ARS holders was trigged not only by the fear of getting stuck with an illiquid security, but also by the fear that brokers/dealers would engage in deceptive practices. Although 2006 SEC investigation has uncovered some serious violative practices, it did not lead to implementation of a legal framework that regulate the ARS market and in particular the auction process. Analysts and media believe that brokers/dealers continued to engage in violative practices after the SEC investigation such as prioritizing some clients in filling the bids on the detriment of others.

Just after collapse of the ARS market, several investors started putting pressure on issuers to buy back the securities and suing the brokers/dealers for misleading them by marketing ARS as safe and cash-equivalent investment. Notably, in mid-2007, when the ARS market started to show

---

* G. Morgensen, It's a long, cold, cashless siege, *New York Times*, April 13, 2008. G. Morgensen, How to clear a road to redemption, *New York Times*, May 4, 2008.

alarming signs, certain dealers increased the marketing of their ARS to potential investors as safe and liquid investment while choosing not to disclose the failed auctions in an attempt to fill the gap of corporate and other institutional investors, who had fled the market. The SEC started investigating whether dealers informed their investors about all the risks associated with these securities and also how they persuaded issuers to sell ARS.

Some brokers/dealers who sold ARS started writing their ARS down from their balance sheets. On the first week of April, Palm, Inc. for instance recorded a $25 million write-down related to ARS it cannot sell.

7.5.5 Level of Embedded Interest Rate Caps

McConnell and Saretto (2008) criticized the view that by not biding on the auctions at any rate, investors had "irrationally" caused auction failure. The authors attribute, instead, the failures to embedded interest rate caps that limited the returns on the bonds. According to McConnell and Saretto (2008), to compensate for the increasing risk associated with ARS, investors "rationally" bid for higher rate of returns than the auctions generally allowed. At its peak in a sample of 793 bonds analyzed by McConnell and Saretto, the overall auction failure rate was 46%, and the authors produce evidence suggesting that in the failed auctions market clearing yields lay above the level of the embedded caps.

7.6 CONCLUSION

Allowing for low-cost financing with the most flexibility, ARS were an attractive source of funding for public and private sectors. Investors were also attracted to ARS as they were considered safe, liquid and cash-equivalent investments providing higher yield than Treasury bills. The market worked relatively smoothly until late 2007, as default on subprime mortgage soared and started spreading into the mortgage securitization market. The collapse of the ARS market by mid-February of 2008, left thousands of investors stranded, unable to, sell, collect on, or otherwise convert their securities. Different factors and explanations have been put forward to explain the collapse of the ARS markets: (1) inability of insurers to meet their obligations; (2) failure of brokers/dealers to support the auctions due to liquidity problem; (3) sharp decrease in demand and increase in supply following greater selling pressure as ARS were perceived risky and illiquid investment; (4) lack of sound information disclosure mechanism in the ARS market and deceptive practice by brokers/dealers; and (5) relatively low level of embedded interest rate caps that limited the returns on the bonds. We believe, however, that the interaction of these factors (due to credit crisis) has led to the formation of a "perfect storm" that had blown the ARS market away.

With credit crisis persisting along with investors and issuers' lack of confidence, the future of the ARS market is uncertain. Ronald Gallatin, the inventor of ARS, holds little hope: "The back of the market is broken... I think the market's problem started with credit, but now credit isn't the problem."* This said, there are to date some attempts by The Municipal Securities Rulemaking Board (MSRB) and the SIFMA to restore the restore ARS market.

REFERENCES

Carow, K., Gayle, E., and McConnell, J. (1999) A survey of U.S. corporate financing innovation: 1970–1997. *Journal of Applied Corporate Finance*, 12(1): 55–69.

McConnell, J. and Saretto, A. (2008) Auction failures and the market for auction rate securities. Working Paper, The Krannert School of Management. Purdue University, West Lafayette, IN.

---

* M. Quint and D. Preston. Auction-rate collapse costs taxpayers $1.65 billion, available online at http://www.bondsonline.com/News_Releases/news05190801.php.

CHAPTER 8

The Banking Crisis and the Nation-State

Jörg R. Werner

CONTENTS

In this chapter, the hypothesis is critically discussed that the regulatory actions after the banking crisis indicate a return of "strong" nation-states. It is argued that the crisis is not only entirely triggered by failures on free markets and "bad" accounting standards but also that regulatory failures significantly contributed to its emergence. Following from that, it seems questionable whether state interventions are a reasonable answer.

8.1 INTRODUCTION

In autumn 2008, many nation-states began to massively react to the current banking and financial market crisis. National bailout plans included the nationalization of banks* and regulatory measures such as granting guarantees and conferring large credits to the financial sector. The crisis originated from the so-called subprime crisis in the United States that spilled over to many other countries (Sanders, 2008). The structural background of the crisis can be traced to relatively low interest rates around the world since the early 2000s and, connectedly, real estate bubbles emerging mainly in the United States and also in the United Kingdom, Spain, and other countries (Barrell and Davis, 2008). With banks collapsing and the crisis increasingly affecting the "real economy," regulators responded to public demands to resolve the crisis by means of state interventions. The latter were intended to stabilize and rescue the national economies and to prevent from a total collapse of national banking systems and financial markets. Interventions were justified by the argument that it is a state's duty and responsibility to react to excesses and crises on markets that were supposed to be triggered by irresponsible and selfish managers. In general, nation-states followed a two-step approach: First, regulatory actions were undertaken to curtail the national impacts of the crisis. In a second step (which is currently going on), it is intended to create new rules for the financial system—possibly on the international level—to avoid a recurrence of the crisis.

This chapter addresses the question whether the regulatory actions are indicative of a return of a "strong" (i.e., interventionist) nation-state, which was until recently supposed to be an outdated model (Genschel and Zangl, 2008). In the following, it will be argued that the banking crisis is not entirely triggered by failures on free markets and "bad" accounting standards but that also regulatory failures significantly contributed to the emergence of the crisis. Hence, increased state interventions seem unlikely to be a reasonable answer to the crisis. Moreover, the return of strongly interventionist nation-states might not only be a wrong answer to the current crisis but in the long run even further the weakening of nation-states.

---

* The first nationalization—also heralding the financial crisis—was that of Northern Rock in the United Kingdom, see Shin (2009) for a discussion.

8.2 EMERGENCE OF THE CRISIS

The banking and financial market crisis mainly originated from the United States. It was a usual practice there to confer mortgage loans to borrowers with a low degree of creditworthiness. This was not at all an irrational practice as the houses bought by the borrowers were supposed to massively increase in value; in case of a mortgage default, banks would then have the possibility to sell the houses above their historical costs. Of course, the expected increase in house prices can now be explained by a market bubble which was, however, at that time not identified as such. The burst of the real estate pricing bubble in 2007 led to rapidly decreasing house prices and created a problem for lending banks: As defaults increased massively, mortgage loans had to be depreciated. This affected not only the banks that directly conferred the loans but also a large number of other institutes. The reason for this is that the risky mortgage loans had been bundled (securitized) to so-called asset-backed securities (ABS) and sold to banks all over the world. The buying banks usually refinanced the purchase of these financial instruments by issuing revolving bonds with short maturities (usually designed as off-balance-sheet transactions). When the real estate market crisis in the United States revealed that the underlying loans were uncollectible, the issued bonds also lost in value. As maturities were short, the respective banks faced massive liquidity problems: With the bond market collapsing, banks had to refinance the redemption of the matured bonds by other means than issuing new bonds. With many banks facing similar liquidity problems, it got, however, nearly impossible to obtain funds from other sources such as the interbank market. Securitization of mortgages thus made a regional crisis in the real estate sector a global crisis of the financial system. Does this indicate a failure of "free" financial markets? And if there is such a failure, does it point to a need for more (or better) regulation? At first glance, the answer must be affirmative; a closer look, however, raises doubts about such a conclusion.

8.3 THE ROLE OF EXCESSIVE RISK TAKING AND BANK CONCENTRATION

First, it has to be said that conferring loans, the securitization of receivables and bets on future market trends are nothing spectacular on financial markets. Such transactions are virtually essential for the functioning of (efficient) markets. However, a serious problem is excessive risk taking. Indeed, there was excessive risk taking on several levels: Relying on ever-increasing house prices, mortgage loans were conferred without serious checking of creditworthiness; buying the securitized receivables without adequate risk assessment also created severe liquidity risks. The (near-) collapse of some banks and the call for state interventions thus can be traced to excessive risk taking by private banks. Does this justify state interventions?

From a liberal point of view, this question has to be abnegated. Following this regulatory approach, it is the state's responsibility to support the functioning of markets, but not to rescue failed or inefficient private companies. In fact, elimination of inefficient companies is an important feature of the market process. This insight is trivial but sometimes abnegated for the banking sector. The argument is made that the collapse of some financial institutions leads to a loss of confidence in the whole financial system and to massive negative consequence for the real (nonfinancial) economy. This concern is indeed justified but it also has to be asked why elimination of some banks would have such tremendous economic consequences. The answer lies in a *too big to fail* problem that can be traced to the relatively high concentration in the banking system (see Figure 8.1). Concentration is defined as the ratio of the three largest banks' assets to total banking sector assets.

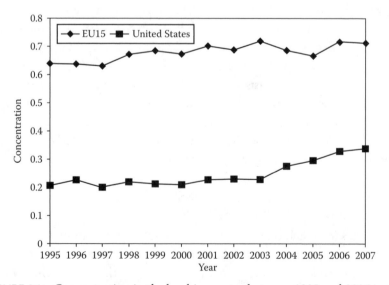

FIGURE 8.1   Concentration in the banking sector between 1995 and 2007 in the EU-15 countries (average) and the United States. (Data from Beck, T. et al., *World Bank Econ. Rev.*, 14, 597, 2007, own calculations.)

Indeed, there is empirical evidence for an increasing concentration in the banking sector in the last decades (see e.g., Nissan and Niroornand, 2006). Interestingly, ongoing concentration in the banking sector was usually appreciated by national politicians. For instance, the merger of *Commerzbank* und *Dresdner Bank* in Germany was welcomed by the German finance minister who also was (and still is) in favor of further concentration. It seems that many politicians aim at creating "national champions," which are able to compete in the global financial system (see Carbó Valverde et al. (2007) for the European context and Schnabel and Hakenes (2006) for the German context). Concentration in the national banking sector is seen as a means to the end of strengthening the national financial market places. The downside is that increasing concentration leads to less competition and the emergence of a *too big to fail* problem. Particularly the latter makes it impossible that big banks collapse. Collapses of companies are, however, an important—if not a necessary—feature of free markets: If banks cannot fail, why should they prevent from excessive risk taking and why should they be engaged in creating efficient governance structures? There is some empirical evidence for banks assuming that regulators adopted a *too big to fail policy* and for a positive relationship between bank concentration and financial distress (see Cipollini and Fiordelisi, 2009).* Hence, when regulators trace the financial crisis to a failure of free markets, this might draw off the attention of own failures mainly in competition policy.

8.4 DID "BAD" ACCOUNTING STANDARDS TRIGGER THE CRISIS?

Accounting standards were also criticized for having been contributed to the banking and financial crisis. For instance, Sinn (2008) argues that the main reason for the banking crisis can be seen in a combination of moral hazard problems at the company level and bad accounting standards. Criticism is particularly raised against International Financial Reporting Standards (IFRS) and their broad reliance on fair values. Indeed, IFRS require regular revaluations of certain financial instruments

---

* It has, however, to be mentioned that empirical evidence on the relationship between concentration and financial system stability is mixed. For instance, Beck et al. (2003) find that crises are less likely in economies with more concentrated banking systems. Angkinand (2009) suggests that the output cost of crises is higher for countries with deposits concentrated within a few large banks. In his empirical study, the coefficient has the expected sign but turns out as being insignificant.

and other assets and liabilities (see e.g., Zimmermann and Werner, 2006). Revaluations partly can lead to realization of "paper profits" in earnings. Critics point to the risk that revaluations increase the volatility of earnings and share prices (Penman, 2007). In this regard, historical cost accounting is supposed to be superior as positive revaluations will not be realized; thus, depreciations will be lower in the case of a crisis.

This argument seems plausible at first sight. However, it goes in hand with the contra-intuitive notion that accounting is better if it does not depict current values in the balance sheet. Even more surprising, criticism against fair value accounting in the context of the banking crisis often ignores that entry prices are almost equal both under historical cost and fair value accounting schemes. This implies that the purchase of ABS falling below historical costs would also have created a need for depreciation under a historical cost accounting system; the only difference is that the amount to be depreciated is higher under fair value accounting if—and only if—there were positive revaluations before the crisis. Interestingly, it has never been proved that such positive revaluations took place.

A probably more important problem is that positive revaluations also lead to an increase in equity. According to the Basel II scheme, banks have to underlay all assets with a certain fraction of equity depending on the risk connected to the respective assets. In turn, a higher value of equity also allows for increasing the overall business activity. The problem is that the activity level cannot possibly be reduced quickly when equity is decreasing as a consequence of impairments (see Penn (2008) for a discussion of IFRS's impact on determining regulatory capital in the United Kingdom). However, this is not a problem of accounting rules per se that are designed to provide a true and fair view of a firm's financial situation but rather one of linking regulatory capital requirement to financial statements. This again seems to be a regulatory failure. Interestingly, IFRS financial statements—designed for providing information to investors— are on one hand used for regulatory purposes here while they are, on the other hand, not regarded as being apt to determine corporate dividend payouts or as a basis for calculating corporate taxes, both due to IFRS's featuring of fair value accounting.

In fact, as far as accounting standards are concerned, there are only two issues that need to be resolved: (1) Buying collateralized debt obligations (CDOs) and refinancing the purchase price by issuing revolving bonds usually was treated as an off-balance-sheet transaction (Lander et al., 2009). From the viewpoint of banks, off-balance-sheet transactions are very attractive. The main advantage is that no regulatory capital has to be assigned to such transactions. The main problem is that there is no transparency about the risk of such transactions. This problem was already identified in the context of the Enron crisis (see Healy and Palepu, 2003) and is not satisfactory resolved until now. (2) In case of down-correcting balance sheet numbers (either in form of an impairment test under historical cost accounting or as a negative revaluation under fair value accounting) a problem emerges when markets are, temporarily, inactive. This is, however, not a severe problem of the accounting standards as the International Accounting Standards Board's (IASB's) quick reaction to resolve this problem has proven.

8.5 THE ROLE OF EXECUTIVE COMPENSATION

In the public debate, performance-based compensation of bank managers was also criticized. For instance, Sharfman et al. (2009) report that employees at Goldman Sachs, Merrill Lynch, Morgan Stanley, Lehman Brothers and Bear Sterns received $39 billion in bonuses in 2007 which is $3 billion more than the year before. The authors conclude: "If only these companies had retained the bulk of these large bonuses over the last two years (…), perhaps the magnitude of the financial crisis would have been significantly less." (p. 5). Reacting to the criticism against excessive executive compensation, some nation-states conditioned bailouts on banks' willingness to curtail executive compensation. Even though this is a comprehensible reaction, regulators should be cautious with generally criticizing variable parts of compensation. According to agency theory, the logic of compensation contracts and variable payments is to set incentives that managers act in the best interest of the owners. This implies a general usefulness of variable payments that contribute to economic welfare by reducing agency costs. Even in an economic crisis, variable payments can make sense to reward agents that acted in the interest of their principals. However, practical implementation of efficient contracts suffers from CEOs having power over their boards and, as a consequence, compensation contracts not being negotiated at arm's length (see Bebchuk and Fried, 2006). As a result, compensation might be too high, and, even more crucial, there might be a low correlation between managers' effort and variable payments (this point was already discussed in a seminal paper by Jensen and Murphy, 1990). The problem of compensation thus can be regarded as a problem of weak corporate governance structures. In fact, the crisis has shown that these structures are particularly weak in public banks (or those who at least have close relationships with the state). The regulatory answer thus should rather be one of strengthening corporate governance structures than a direct intervention into private companies' compensation policies.

8.6 CONCLUSION

It is, by no means, a new phenomenon that crises lead to a reconsideration of regulations at place and interventions by the state (Stead and Smallman, 1999; Zimmermann et al., 2008). However, the current banking and financial market crisis has led to serious doubts on the functioning of free markets and to an unexpected high level of state interventions into the private sector all over the world.

The discussion above revealed that reality is more subtle than conventional wisdom suggests. From a liberal point of view, markets will only function if there is competition. However, in the last decades, nation-states backed concentration in the banking sector leading to *too big to fail* problems. Facing such problems, nation-states now indeed have to intervene into their financial systems. Costs of these interventions are tremendous and some states are, as a consequence, near national bankruptcy. National budgets are likely to suffer from the interventions for a very long time. Hence, the observed return of the strong state might, in the long run, lead to further curtailing of nation-states' possibilities to fulfill their core functions.

In this sense, history might iterate: The expansion of state functions in the 1970s, mostly observable in the Continental European countries, was argued to have weakened the nation-states in the long run as they turned out as not being able to meet the excessive demands aroused. If new regulatory frameworks and tougher oversight systems, possibly on the international level, are discussed, the possible failure of such regulatory actions should be anticipated. Modern efficient regulation requires policy makers to understand the boundaries of national politics and the general problems of interventions into the business system. Interventions should aim at enhancing the functioning of markets and at creating incentives for desired economic behavior. In fact, such regulation is hard to achieve for mainly two reasons: First, it requires refraining from giving a populist answer to the crisis. Second, it requires overcoming resistance against forestalling off-balance-sheet transactions and against enforcing more competition. Of course, even though protected by deposit protection systems at place, voters will probably also dislike the possibility of bank collapses—but only letting some banks collapse will make them pay for wrong economic decisions and excessive risk taking and prevent other banks from doing so in the future.

REFERENCES

Angkinand, A.P. (2009) Banking regulation and the output cost of banking crises. *Journal of International Financial Markets, Institutions & Money*, 19(2): 240–257.

Barrell, R. and Davis, E.P. (2008) The evolution of the financial crisis of 2007–2008. *National Institute Economic Review*, 206(1): 5–14.

Bebchuk, L.A. and Fried, J.M. (2006) Pay without performance: Overview of the issues. *Academy of Management Perspectives*, 20(1): 5–24.

Beck, T., Demirgüç-Kunt, A., and Levine, R. (2000) A new debate on financial development and structure (updated May 2009). *World Bank Economic Review*, 14(3): 597–605.

Beck, T., Demirguc-Kunt, A., and Levine, R. (2003) Bank concentration and crises. NBER Working Paper No. W9921. Available at SSRN: http://ssrn.com/abstract = 437490.

Carbó Valverde, S., Humphrey, D.B., and López del Paso, R. (2007) Do cross-country differences in bank efficiency support a policy of "national champions"? *Journal of Banking & Finance*, 31(7): 2173–2188.

Cipollini, A. and Fiordelisi, F. (2009) The impact of bank concentration on financial distress: The case of the European banking system. SSRN Working Paper. Available at SSRN: http://ssrn.com/abstract = 1343441.

Genschel, P. and Zangl, B. (2008) Transformations of the state—From monopolist to manager of political authority. TranState Working Papers, No. 76, University of Bremen, Germany.

Healy, P.M. and Palepu, K.G. (2003) The fall of Enron. *Journal of Economic Perspectives*, 17(2): 3.

Jensen, M. and Murphy, K. (1990) Performance pay and top-management incentives. *The Journal of Political Economy*, 98(2): 225–264.

Lander, G., Barker, K., Zabelina, M., and Williams, T. (2009) Subprime mortgage tremors: An international issue. *International Advances in Economic Research*, 15(1): 1–16.

Nissan, E. and Niroornand, F. (2006) Banking bigness: Concentration of the world's 50 largest banks. *Multinational Business Review*, 14(1): 59–78.

Penman, S.H. (2007) Financial reporting quality: Is fair value a plus or a minus? *Accounting & Business Research*, 37(Special Issue): 33–43.

Penn, B. (2008) The whole sorry story. *International Financial Law Review*, 27(5): 52–55.

Sanders, A. (2008) The subprime crisis and its role in the financial crisis. *Journal of Housing Economics*, 17(4): 254–261.

Schnabel, I. and Hakenes, H. (2006) Braucht Deutschland eine starke private deutsche Bank? Über die Notwendigkeit nationaler Champions im Bankwesen (Is there a need for a 'strong private German bank'? On the desirability of national champions in the banking sector). *Kredit und Kapital*, 39(2): 163–181.

Sharfman, B.S., Toll, S.J., and Szydlowski, A. (2009) Wall Street's corporate governance crisis. *Corporate Governance Advisor*, 17(1): 5–8.

Shin, H.S. (2009) Reflections on Northern Rock: The bank run that heralded the global financial crisis. *Journal of Economic Perspectives*, 23(1): 101–119.

Sinn, H.-W. (2008) Why banking crises happen: The role of bad accounting and moral hazard missteps. *The International Economy*, 22(3): 60–61.

Stead, E. and Smallman, C. (1999) Understanding business failure: Learning and un-learning from industrial crises. *Journal of Contingencies & Crisis Management*, 7(1): 1–19.

Zimmermann, J. and Werner, J.R. (2006) Fair value accounting under IAS/IFRS: Concepts, reasons, criticisms. In: Gregoriou, G.N. and Gaber, M. (Eds.), *International Accounting—Standards, Regulations, and Financial Reporting.* Elsevier: Amsterdam, the Netherlands.

Zimmermann, J., Werner, J.R., and Volmer, P.B. (2008) *Global Governance in Accounting: Rebalancing Public Power and Private Commitment.* Palgrave Macmillan: Basingstoke, U.K.

CHAPTER 9

The Banking Crisis and the Insurance Markets

Christopher Parsons and Stanley Mutenga

CONTENTS

In this chapter, we assess the impact of the current banking crisis on insurance markets and the role of the insurance industry in the crisis itself. After briefly considering the ways in which the barriers between banking and insurance operation have diminished in recent years, we examine some previous insurance "crises" before comparing the nature and degree of systemic risk in insurance and banking, noting the part that reinsurance might play in increasing systemic risk. We then go on to discuss the ways in which the insurance industry has become involved in the current upheaval in world banking. We conclude by considering the structural changes in the insurance industry that might result from the crisis, including possible effects on "bancassurance" activity, and offer some thoughts on changes in the regulation of insurance markets that might ensue.

9.1 INTRODUCTION

Historically, there has been a marked separation between banking and insurance markets in many countries—not least in the United States—so that events in one sphere usually had little effect on the other. However, in recent years, the barriers between insurance and banking operations have been partly dismantled, resulting in much closer affiliations between banks and insurance companies and more linkages and overlaps in their activities. This phenomenon (which is partly, but not wholly, encapsulated in the term "bancassurance") began at an earlier date in Europe than in the United States, where depression-era legislation, which largely separated the activities of banks and insurers was repealed only at the end of the Clinton era, and earlier than in major Asian markets such as Japan and Korea, where similar legislation separating banks from insurers has existed until very recently. However, given that the barriers referred to above have been at least partly removed in many countries, we may well ask whether the current banking crisis has spilled over into an "insurance crisis" and whether this, in turn, has made the banking crisis worse. As we shall see, the banking crisis has indeed affected insurance markets to a significant degree and events in insurance markets have, in turn, deepened the crisis to some extent. However, most of these effects have been relatively muted, largely owing to some key differences in the structure of the two markets, the nature of their products, and the relationships between the players in them.

9.2 PREVIOUS INSURANCE "CRISES"

Has there ever been an "insurance crisis" which parallels current events in the banking world? Indeed, there have been a number of upheavals in insurance markets which have been described, at least by some, as "crises": that is, periods characterized by the failure (or near failure) of one or a number of insurance firms, reduction in the supply of insurance and consequent disruption of economic activity. A notable example is the 1984–1986 U.S. "liability insurance crisis." During this period U.S. property/casualty insurers made huge losses and insurer insolvencies became commonplace. For most of the time reinsurers made even greater losses than direct insurers, with the result that reinsurance firms in Europe and North America reduced their lines on U.S. risks, increased rates, and imposed restrictive conditions. Direct insurers were obliged to follow suit and liability insurance capacity dwindled alarmingly as insurers and reinsurers attempted to restore profitability. The causes of this "crisis" are still disputed, but reported effects included, *inter alia*, the closure of some children's playgrounds when insurance for them became unobtainable and rocketing premiums for medical malpractice insurance, which allegedly diverted physicians away from these branches of medicine (such as obstetrics) where premiums were most punitive.

The near collapse of the 300 year old Lloyd's insurance market in the early 1990s provides another interesting example of a major upheaval in the world of insurance. A combination of many years of injudicious underwriting, an exceptional number of major catastrophes in a short space of time, and internal management problems caused Lloyd's collectively to make huge losses over the period in question (around GB£8 billion between 1988 and 1992) bankrupting many underwriting members and bringing Lloyd's close to insolvency. In the end, Lloyd's did not fail. Rather, it recovered, restructured, and reinvented itself, regaining its leading place in the world insurance market for large, unusual, extra-hazardous, and internationally traded risks. However, had it occurred, the failure of Lloyd's would have dealt a hard blow to the world insurance system by curtailing the supply of cover for the sorts of risks described above, which would have caused significant economic disruption.

At this point it is worth remarking an interesting parallel between one facet of Lloyd's problems and a key source of the current banking crisis. This was the London Market Excess (LMX) reinsurance (or retrocession) "spiral" which developed at Lloyd's in the early 1990s. Retrocession is the further transfer of risk by a reinsurer to another (re)insurer, the latter being known as a retrocessionaire. This enables risks to be spread more widely but also creates a credit risk for the reinsurer concerned which could, in turn, impact upon (re)insurers from which the reinsurer itself has accepted risks. The LMX spiral developed from the continued retrocession by Lloyd's syndicates of much of their business, so that many syndicates underwrote again the very risks they had transferred, sometimes without knowing it. Commission was shaved off the premiums each time the risk was passed on, threatening to leave insufficient capital to pay for losses, thus contributing to the overall Lloyd's crisis. This situation, with multiple participants parceling up and selling on risks to the point where the players lose track of their own and others' exposures—effectively magnifying risk rather than managing it—is obviously echoed in the frenetic trading of credit derivatives that a played major part in the current banking crisis. In fact, reinsurance has been identified as one part of the insurance market where there is, at least potentially, some systemic risk. This is discussed below.

Besides the U.S. liability crises and the Lloyd's near-debacle, there have been many other upheavals, sometimes affecting one line of insurance only (such as the product liability crisis of 1976–1977 and the shortage in terrorism cover following the events of September 11, 2001). However, these "crises" have been relatively short in duration and, more important for the purpose of this study, mostly local in terms of their impact. Furthermore, we should not confuse a sharp rise in insurance prices with a failure of supply. The cyclical nature of insurance markets is such that sharp price "corrections" (or overcorrections) are quite frequent, but it is very rare indeed for any common form of insurance to be unobtainable at any price. We should further bear in mind that the direct economic disruption caused by insurance "crises" such as those described is unlikely to be as severe as the devastation which can flow from a banking crisis. We know only too well that a credit crunch, triggered by a banking crisis, can severely damage the real economy, leading to a worldwide recession. On the other hand, even a massive rise in the cost of insurance is unlikely in itself to cause much damage to the real economy. This is so because the cost of insurance, relative to other costs, is quite small for most businesses.

9.3 SYSTEMIC RISK IN INSURANCE AND THE ROLE OF REINSURANCE

In any event, the fact that there has never been a major worldwide "insurance crunch" suggests that there are some fundamental differences between banking and insurance markets, with less risk of systemic failure in the case of the latter. Indeed, this is the case. For one thing, clients cannot withdraw money they have paid to an insurer at the time of their choice. Rather, they are entitled to be paid only if and when a loss occurs in the case of a non-life insurance contract or on death of the insured or maturity of the plan in the case of life insurance. It is true that life insurance policyholders might be tempted to cash in life policies with a savings element prematurely if they lost confidence in the insurer concerned, but penalties for early exit usually provide a considerable disincentive against precipitate action of this sort. Thus, there is no real insurance equivalent of a "run on the bank," where loss of confidence in the institution's ability to meet its obligations brings about the very thing that depositors fear.

A further key distinction between banks and insurance companies lies in the fact that insurers do not trade with each other to a significant extent. Thus, there is little possibility that misjudged risks assumed by one insurer will find their way into, and contaminate, the portfolios of other firms or that the failure of one insurer will cause other companies to falter. However, there is one qualification to this rule, which lies in the practice of reinsurance and retrocession, mentioned earlier. Reinsurance is the process by which insurers lay off some of the risks they assume with other firms (reinsurers) which may, in turn, transfer some of the risk to yet more insurers (retrocessionaires). Insurers do this (especially in relation to very large, unusual, and unbalanced risks) in order to increase their capacity to write insurance business, spread risk more widely, and stabilize their portfolios against the sharp fluctuations that might result from large random losses. Most insurance firms buy reinsurance for at least some risks. Worldwide, insurers transfer around 6% of the premiums they receive to independent reinsurers, though non-life insurers transfer much more (around 12%) than life insurers (around 2%). In turn, around 20% of premiums received by reinsurers are transferred to retrocessionaires, that is, firms that accept risks which the reinsurers themselves lay off. Sellers of reinsurance include so-called professional reinsurers (such as Swiss Re, Munich Re, and Berkshire Hathaway Re) which write only reinsurance business and ordinary insurance companies which write both (direct) insurance and reinsurance. The former provide around 75% of the world insurance capacity and the latter around 25%. We should bear in mind, however, that many insurance groups (such as the Allianz, Generali, Swiss Re, and Munich Re groups) comprise both professional reinsurers and ordinary insurance companies.

The net effect of this activity is that insurers do trade risks to some extent by sharing them through reinsurance networks. We might then ask whether this creates a potential systemic weakness in the world insurance market, with the possibility that ill-judged risks assumed by one or more insurance firms might fatally infect the reinsurers to which they are transferred. In turn, if a major reinsurer were to fail for this (or any other) reason, might this create a domino effect, bringing down insurance firms that rely upon the reinsurer? In fact, evidence suggests that the systemic risk posed by reinsurance is quite small. The perils which underlie insurers' liabilities are mostly unlike the exclusively financial risks which banks assume. Insurers' liabilities mainly arise from so-called pure risks (such as fires, storms, explosions, law suits, earthquakes, and the like), which are mainly localized and not correlated to a significant extent. This means that they can be effectively diversified across the globe by reinsurers who accept transfers from very large numbers of insurance firms, across many different lines of insurance business and many different national markets. What if, nevertheless, a large insurer was to collapse? Swiss Re (2003) argue that the failure of even a very large reinsurer is unlikely to pose a threat to other (direct) insurers that use its services. It reports that while claims on reinsurers make up 10%–36% of balance sheet total for non-life insurers (and between 2% and 12% for life insurers), direct insurers spread their risk across a number of reinsurers so that even the largest reinsurer rarely represents more than 4% of an insurer's assets. It further notes that only 24 reinsurers failed in the period 1980–2002 and these failures affected on average only 0.02% of the total premiums transferred to reinsurers in this period. Reinsurance problems are a significant factor in only about 5% of insurance company failures (Swiss Re, 2003) and actual reinsurer failure (as distinct from the following of an inappropriate reinsurance strategy) accounts for only about 3.5% (see Figure 9.1).

It would seem therefore, that the whole insurance/reinsurance system could only be threatened by an unanticipated exogenous shock far greater than any that yet have yet occurred.* All this does not mean that the insurance industry has proved immune to the effects of the current banking crisis, not does it mean that insurance markets and mechanisms have played no part in enhancing its effects. The insurance industry has been affected by the current financial crisis in two main ways. The first is through the underwriting activities of insurers and the deterioration in the risk environment which the financial crisis has brought about. As we shall see, some insurers have moved away from the "traditional" insurance risks described earlier and started to assume, by way of insurance, financial (and especially credit-related) exposures which are much more prone to systemic risk, thus aligning their fortunes more closely with firms in the banking sector. The second way in which insurers have been affected

---

* Reinsurance markets have proved resilient enough to absorb quite easily major shocks from such events as 9/11 and Hurricane Katrina.

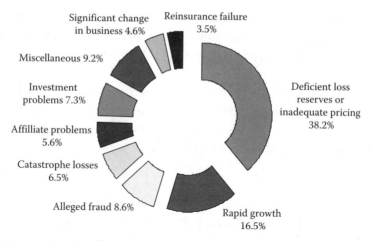

FIGURE 9.1 Reasons for U.S. property casualty insurance company failures between 1969 and 2005. (From A.M. Best, Property/casualty impairments hit near-term lows despite surging hurricane activity, available at http://www.bestweek.com, Oldwick, NJ, 2005. With permission.)

is through erosion in the value of their assets, including their investments in credit-linked assets. Each of these is considered in turn.

9.4 EFFECT OF THE CRISIS ON INSURANCE RISK

In the simplest of terms, insurers make profits by agreeing to pay for defined losses which their clients might suffer in exchange for premium payments, and by investing these premiums to generate extra income. Profitability will suffer if claim payments rise unexpectedly, so that the premiums collected are inadequate, or if expected investment income is not realized—or, of course, if operating expenses are higher than anticipated. In this section, we are mainly concerned with the first problem—unexpected rises in claim payments.

In the case of non-life insurance, claim levels tend to rise generally in an economic downturn. Almost inevitably, there will be a rise in fraudulent or exaggerated claims as the recession bites and more claims will be made for relatively trivial losses, which would tend to go unclaimed in better times. Demand for non-life insurance is relatively inelastic, because few substitutes for insurance exist and because some major lines (such as motor) are mandatory. However, the financial crisis will inevitably reduce demand to some extent, forcing insurers to cut costs (e.g., by shedding staff) in order to maintain profitability. In the case of life insurance, claim payments as such do not increase in a recession, but the cashing in of policies, and the drop in demand for new ones (particularly for saving plans and mortgage-linked products), is likely to be more severe, creating a greater imperative to cut costs.

Quite apart from these general effects, there are some specific lines of non-life insurance where the impact of the banking crisis is likely to be more direct and more intense. These are liability insurance and insurance which covers credit risk. We look at these next.

9.4.1 Liability Insurance

Besides insuring property and various forms of financial loss, non-life insurers also write liability (or casualty) insurance, covering losses which individuals or firms incur when they are sued for negligence or other legal wrongdoing. Banks, financial firms, and their professional advisers (such as lawyers and accountants who operate in financial markets), real estate agents, mortgage brokers, and the like are at risk of compensation claims by clients who believe themselves to be victims of negligence. Stakeholders (such as shareholders) in financial businesses who suffer loss as a consequence of mismanagement by the firms' own directors or senior management may also target the individuals concerned. The first of these risks is insured under Errors and Omissions insurance (E&O) and the second under Directors' and Officers' insurance (D&O). Leading insurers in these markets include AIG (which writes around 35% of D&O insurance by premium volume), Chubb (which writes around 15%), and XL and Lloyd's of London.

A major financial downturn always brings an increase in liability suits of this kind as firms and individuals look around for people to blame for their losses—the "dotcom" crash of 2000 being a recent episode which triggered such a spike in claims. Most business leaders in the United Kingdom believe that the current financial crisis will generate even more claims than the punctured dotcom bubble (Lloyd's, 2008), and all insurers in this market anticipate a sharp rise, with one U.S. analyst suggesting that claims will reach US$9.6 billion, comprising US$5.9 billion for D&O and US$3.7 billion for E&O (Advisen, 2008). However, estimating the ultimate cost to the insurance industry is difficult. Many—probably the majority—of these claims will fail, because proving negligence or positive wrongdoing on the part of professional firms or corporate directors is difficult: they are not liable for mere errors in judgment or losses caused by market fluctuations beyond their control. Nevertheless, insurers are generally bound to defend their clients, even when the suits they face are unmeritorious, and such defense costs always constitute a significant portion of insurers' outlay. Furthermore, liability insurance of this type is "long-tail" business, with significant delays in the notification of claims and long settlement periods. This means that the final cost of the credit crunch and banking crisis will not be known to liability insurers for many years. However, despite the inevitable rise in claims cost, there is no sign of any shortage of supply for these lines of insurance at the time of writing. Premiums will certainly rise in the short or medium term, and this will have some impact for insurance buyers on the cost of doing business, but there seems to be no immediate likelihood of insurance becoming unaffordable, let alone unobtainable.

9.4.2 The Underwriting of Credit Risk by Insurers

Several types of insurance cover credit risk, including mortgage protection insurance (MPI) and payment protection insurance (PPI). However, the most important ways in which insurers and reinsurers underwrite credit risk are through credit insurance and, in recent years, though credit default swaps (CDSs).

In essence, credit insurance covers losses which insured firms suffer when their clients fail to pay for goods and services. The main risks covered are insolvency of the client or protracted default (failure of the client to pay within a set time period). In addition to these risks of commercial default some insurers also cover short-term political risks—for example, failure of a foreign buyer to pay as a consequence of war or revolution. The world market for credit insurance is dominated by four groups: Euler Hermes, Atradius, Coface, and Credito y Caucion (CyC) which have around 37%, 23%, 18%, and 7% of the market, respectively (Swiss Re, 2006). Western Europe accounts for around 74% of credit insurance buyers, with North America and Asia at just 6% and 3%, respectively. Important insured sectors include constructions (16%), metals and engineering (15%), agriculture and food (13%), services (12%), and electronics (9%).

Claims against credit insurers do not arise from purely financial transactions, such as a borrower defaulting on a loan, but from the default of buyers who fail to pay for goods or services. Therefore, credit insurers are not directly affected by the credit crunch and banking crisis but rather by the increase in bankruptcies in the real economy that these have generated. Currently, credit insurers are experiencing a sharp rise in claims as the recession bites. The market leader reported a sharp deterioration in the claims environment in the fourth quarter of 2008 and expected the position to deteriorate rather than improve (Euler Hermes, 2008). However, credit insurers are well used to managing substantial peaks and troughs in their business that coincide with the business cycle. Swiss Re (2006) notes that the loss ratio for German credit insurers has varied from 33% (in 2004) to 106% (in 2006) and that dynamic management of the credit limits (in simple terms, a process whereby the credit insurer has a right to reduce or remove the credit limits for future transactions of a specific buyer if its financial positions deteriorates) enables credit insurers to ride these cycles effectively and reduce their loss ratios well before bankruptcies start to decline. Credit insurers can also refuse to renew their policies and decline new business. Evidence suggests that all this is happening, with credit insurance cover becoming increasingly difficult to obtain (Jetuah, 2009).

Withdrawal of credit insurance can have serious consequences. First, it can create a domino effect along the supply chain in which companies that cannot buy insurance cut ties with suppliers who are judged high risk. Furthermore, companies in difficulty become less attractive to potential buyers if they cannot get credit insurance, making it harder for insolvency practitioners to find a buyer for the business. Clearly, the reduction in the supply of credit insurance is having a significant effect in reinforcing the economic downturn that has been triggered by the banking crisis. This has raised concern at the government level. For example, the French government has promised that the state-owned Caisse Centrale de Réassurance will provide a guarantee for additional credit insurance when an insurer decides to reduce its cover for nonpayment of bills (Financial Times, 2008a) and, in the United Kingdom, the Business Minister has met with the big three credit insurers to discuss the possibility of a similar state guarantee to underpin the sector (Financial Times, 2008b).

The role of CDSs or pure credit swaps in the banking crisis is well understood. A CDS is like a multi-period credit option, that is, buying credit insurance. The lending financial institution pays an annual fee to the counterparty, and if the borrower defaults the swap counterparty pays a default payment to the lending financial institution that is equal to the difference between the par value of the loan and its market value. These instruments played a major role in the demise of Bear Stearns and Lehman Brothers and the troubles of Merrill Lynch and AIG, to name but a few firms. A CDS is often described as, or at least compared to, an insurance policy. However, it is not insurance in the traditional sense: there is no requirement to actually hold any asset or suffer any loss, so CDS are widely used simply to speculate on market changes. Many firms that write them do not engage in traditional insurance business. This means, crucially, that they have not been subject to regulation in the way that insurance firms have: indeed, much of the CDS market has hardly been regulated at all. Note, however, that in the case of long-term swaps the out-of-the-money party would have an incentive to default to defer current and future losses. Thus, Basel II imposes capital requirements. Having said that, banks can reduce their capital requirements by obtaining guarantees from insurers, which in turn leaves banks undercapitalized and exposes insurers to CDS defaults. Thus, one might argue that insurers appeared to offer banks a way to get around the Basel rules, via these unregulated insurance contracts.

Swiss Re, using data from the British Bankers' Association (BBA), suggests that the global market volume for CDS was just US$1.189 trillion in 2001 (Swiss Re, 2003). The main buyers at that time were banks, securities firms, and hedge funds, but the main sellers also included monoline (financial guarantee) insurers, reinsurers (with a market share of 20%), and primary insurers (13% share). Since then most insurers and reinsurers have reduced their exposure to CDS relative to banks and hedge funds, accounting for just 17% of protection sellers in 2006 according to the BBA. However, since 2001 the market for CDS has grown massively (to US$57.9 trillion in 2007 according to the Bank for International Settlements, US$62.17 trillion according to ISDA), leaving some insurers heavily exposed to fall out from the subprime and subsequent banking crises. Notable examples include the U.S. monoline insurers. These differ from most insurers in that they are not permitted to write any line other than financial guarantee insurance (hence their name). In recent years, these companies have moved away from their traditional role of guaranteeing bonds issued by U.S. municipalities toward writing business related to structured asset-backed finance deals, such as mortgage-backed bonds and collateralized debt obligations (CDOs). Losses on this type of business, mainly through exposure to subprime mortgages led to rating agency downgrades of several monolines in 2008, including AMBAC, MBIA, and CIFG Guaranty. This, in turn, triggered simultaneous downgrading of hundreds of thousands of bonds issued by municipalities and financial institutions, thus deepening the banking crisis.

One major reinsurer, Swiss Re, has reported very substantial losses on its CDS portfolios which, combined with other write-downs, resulted in the firm posting a significant net loss of around SFr1 billion (US$860 million) for 2008 and requiring it to raise to raise fresh capital, including SFr3 billion (about US$2.6 billion) from Berkshire Hathaway, Warren Buffett's holding company. Of course, the other main "insurance" case is AIG, whose huge exposure to CDS resulted in a notorious US$85 billion government bailout in late 2008, effectively taking the group into public ownership. Weakness at AIG could have forced financial institutions worldwide which bought these swaps to take write-downs or losses. Furthermore, banks and mutual funds are major holders of AIG's debt and many would have taken a major hit if AIG had defaulted. We must emphasize that the downfall of AIG was not attributable to losses made in its relatively healthy core insurance operations but rather to the imprudence of its subsidiary, AIG Financial Products, effectively a hedge fund that wrote the CDS contracts out of its London offices. In simple terms, AIG was brought down not by its core insurance activities but what many have described as its "dabbling" in a risky and unregulated area of finance.

9.5 THE IMPACT OF THE CRISIS ON INSURERS' ASSETS AND ON THEIR VALUATIONS

In this section, we evaluate the extent to which the current credit crisis has affected insurance markets across the globe. Changes in returns in the periods 2007–2008, 2008–2009, and 2007–2009, computed from annual stock market indices data obtained from Thompson DataStream and Yahoo Finance, are used to measure the extent of the impact. These periods were chosen because they represent the timeline of the current credit crisis. We compute changes in returns by taking the natural logarithms of index values for year t, and $t-1$. Indices chosen are those that had complete data for the period's analyzed. The underlying assumption is that insurer equity valuations and returns are determined by investment, underwriting, and capital management policies adopted by each market. The effects of underlying economic variables associated with the current credit crisis on these key insurance value drivers should be reflected in movements in equity valuations. This is captured by the changes in returns across the globe reported in Table 9.1.

The impact of the crisis across world insurance markets has clearly been uneven. The lowest decline on the regional indices of 29.95% on the DJ STOXX index is observed in Asia (−31.94% on the Best's Asian/Pacific Insurance Index), for the whole period 2007–2009. The greatest decline of 77.67% was observed on the OSE4030 Insurance Index in Norway. This is comparable only to the United States, where the declines on the S&P

TABLE 9.1 Changes in Returns for World Major Indices

Index		% Change for Period 2007–2009	% Change 2007/2008	% Change 2008/2009
FTSE-350 Insurance Index	United Kingdom	−45.29%	−17.24%	−33.90%
FTSE GBR—BANKS	United Kingdom	−65.98%	−29.04%	−52.06%
FTSE GBR—INV BANK	United Kingdom	−41.12%	36.20%	−56.77%
SWX SP Insurance TR	Switzerland	−42.02%	−12.88%	−33.44%
SWX SP Insurance PR	Switzerland	−45.80%	−15.39%	−35.94%
DJES Insurance P	Zurich	−56.55%	−24.01%	−42.82%
DJES Insurance R	Zurich	−53.59%	−21.99%	−40.50%
S&P Insurance Index	United States	−62.60%	−12.29%	−57.37%
DJ STOXX Americas 600 Insurance Index	United States	−53.72%	−6.72%	−49.42%
DJ US Insurance IND	NYSE	−57.68%	−11.21%	−52.34%
FTSE USA—BANKS	United States	−50.29%	−28.39%	−30.59%
OSE4030 Insurance	Oslo	−77.67%	−39.25%	−63.24%
Best's Global Non-Life Insurance Index	Global	−25.16%	2.56%	−27.03%
Best's Global Life Insurance Index	Global	−60.93%	−12.13%	−55.54%
FTSE GLOBL BANK	Global	−48.98%	−19.79%	−36.39%
Best's Asian/Pacific Insurance Index	Asia	−31.94%	−7.34%	−26.55%
DJ STOXX Asia/Pacific 600 Insurance Ind.	Asia	−29.95%	−8.50%	−24.90%
FTSE EU—BANKS	EU	−58.02%	−22.29%	−45.98%

Insurance Index and DJ US Insurance Index were 62.60% and 57.68%, respectively. In Europe, declines on the DJES Insurance P index were also high at 55.69%, while those on other Swiss insurance indices were less than 50%. Of all the developed markets analyzed the United Kingdom had the lowest decline of 45.29% on its FTSE-350 Insurance Index. These varying market performances can be explained by differences in

- The stage which each market has reached in the global recession

- The extent to which equities markets and banking stocks have declined

- The extent of insurance companies' involvement in CDOs, asset-backed security (ABS) and subprime exposure (see Table 9.2)

TABLE 9.2   Insurance Companies' CDO, ABS, and Subprime Exposure

	CDO and ABS Exposure		Subprime Exposure		Write-Down and Value Adjustments
	Exposure in €BN	As % of Total Assets	Exposure in €BN	As % of Total Assets	In €BN
Aegon	20.8	6.6%	2.7	1.0%	0.82
Allianz	35.0	3.4%	1.7	0.2%	1.83
Aviva	2.6	0.6%	0.2	0.0%	—
Axa	15.4	2.1%	2.2	0.3%	1.5
Fortis	45.3[a]	7.5%	2.9	0.0%	3.7
ING	92.3	7.0%	3.2	0.2%	4.6
Munich Re	7.9	3.6%	0.6	0.3%	0.5
Swiss Re	26.6	9.1%	2.0	0.7%	3.96
ZFS	18.8	7.3%	0.2	0.1%	0.858
AIG	72.0	9.0%	21.8	2.7%	10.8

Source: Oliver Wyman (2008)/Bloomberg, www.creditwritedowns.com. With permission.
[a] This is the combined bancassurance exposure; the insurance business exposure was €3.1 billion in 2007.

A direct comparison between life and non-life global returns during the period 2007–2009 shows that life business, which had a negative return of 60.93%, has been more affected than non-life business, which had a more moderate negative return of 25.16%. We can also see that for both indices, most losses came in the period 2008–2009 when the credit crisis intensified. In fact, the global non-life index had positive gains of 2.56% during the period 2007–2008 and the global life index lost only 12.13% of its value.

High negative returns in the United States and European insurance markets and moderate negative returns in Asia are a reflection of the length and depth of recessionary trends in these markets. Furthermore, advanced western markets are more integrated, which means that the performance of their equity markets is highly correlated. The volume of toxic assets traded across the Atlantic is also relatively high compared with those bought by Asian companies. The high volume of toxic assets within the financial system has been cited as the main reason for the depth and entrenchment of recessionary trends in world economies. This is reflected in declines in the banking indices in Table 9.1. The FTSE global bank index fell by 48.98% in the 2007–2009 period with EU banks experiencing a 58.02% decline, while U.K. banks suffered the largest fall of 65.98%. However, only the U.K. banking index performed more badly than the global life insurance index. It is also striking that U.S. banks have fared rather better than U.S. insurance companies in this period, with negative returns of 50.29% compared with 62.60% for the latter. For the other two periods, 2007–2008 and 2008–2009, changes in returns were greater during the latter, but greater swings were recorded for insurance companies than for banks. This is consistent with a study which shows that insurer failures tend to lag downturns in economic activity by 1 or 2 years (Best, 2008). Based on this study, we might expect to see higher impairment rates for insurers 2009 and 2010.

We noted earlier that the impact of the crisis on non-life insurer's liabilities is limited mainly to those companies that directly underwrite perils associated with the current crisis—that is, credit-related risks. Our focus now turns to the role that assets have played in affecting insurance company valuations. Most of the assets in which insurance companies invest their funds have been significantly affected by the economic downturn. Under IFRS 7, an asset is impaired when there is objective evidence that loss events have impacted the estimated future cash flows of the asset. Figure 9.1 (Section 9.3) shows that asset impairment (or overstatement in some cases) contributed to only 7.3% of insurance company failures between 1969 and 2005. Certainly, very few insurance companies have failed in the course of the current crisis despite record slumps in equity markets across the globe. However, the 2008 study by A. M. Best mentioned above, showing time lags in insurance company failures following economic downturns, suggests that any optimism in this respect may be premature. The extent to which asset impairments affects insurance companies depends also on the nature of the contracts underlying their liabilities. Under life insurance contracts such as variable annuities and participating business that offers guarantees, the performance of the assets backing them is crucial, as this will determine whether the insurer will be able to meet its obligations. The recent major decline in equities and real estate returns (where life insurers generally hold most of their assets) helps to explain the huge negative swings in life insurer valuations discussed earlier. In a volatile economic environment, hedging these contracts is very difficult, so the quality of returns generated by these assets is crucial to the financial performance of the life companies offering them. Furthermore, since life insurers sell long-term products there is no scope for mid-term price adjustments, and any impairment on assets will be recognized directly on equity. Non-life insurers offer short-term contracts, which provide scope for quick adjustment in pricing if they incur realized investment losses.

Although investment strategies adopted by non-life insurers are generally thought to be more conservative than those of life insurers, U.S. life and non-life insurers in fact invested 58.4% (Insurance Information Institute, 2009a) and 61.2% (Insurance Information Institute, 2009b) in credit market instruments, respectively, in 2007. Most regulatory authorities across the globe also levy more capital on non-life liabilities than on life liabilities. The relative stability of non-life insurers can therefore be explained by the current behavior of their liabilities. On the other hand, life insurance company solvency capital is mostly levied on assets rather than the liabilities. With the turbulence that has been seen in markets recently, we can infer that the valuations in Table 9.1 reflect the greater burden that investment risk has on life solvency capital.

In the crisis so far, relatively few insurance companies have had their solvency margins breached when compared with banks. Insurance companies (and especially non-life companies) hold significantly more capital than banks, which might provide a plausible explanation for lower declines in the case of the former. A further difference between insurance companies and banks, especially in Europe, lies in the financial instruments which they can use as substitutes for equity in their capital structures. Insurance companies in the EU, like banks, use a tier system to allow them to leverage their capital, by permitting certain types of hybrid instruments and debt to be treated as capital (CEA and Mercer Oliver Wyman Limited, 2005). Capital constituents under the current U.K. risk based system (Financial Services Authority, 2004) and the proposed Solvency II regime,* as under Basel II, are composed of two tiers. Under Solvency I, companies should hold a minimum of 50% of tier I equity capital including certain approved additions. Under the proposed Solvency II regime, companies would be allowed to use other new innovative and non-innovative capital allowances, which further reduce the quantity of equity capital an insurance company should hold under tier I and tier II to a minimum of 25% total capital. These structures of debt that qualify as equity have the benefits of improving return on equity, but can only stand if core equity is high enough. If insurance companies under Solvency II follow the same route as banks in reducing core equity to 25% under both tiers, they might be more vulnerable to catastrophic events in the future.

As discussed earlier, and as we can see from Figure 9.2, insurance company valuations have declined significantly in the course of the crisis,

---

* The CEIOPS call for Advice No. 19 proposes a tiering system of capital similar to European banks.

yet this does not reflect the degree of solvency they have displayed so far. Company valuations are mainly based on conjectures made by outside investors. While managers and regulators can observe both core and transient components to insurer cash flows, outsiders bundle both components of risk together when determining insurer cash flow volatility. This opaqueness of insurance cash flows makes insurers the most difficult institutions to value. Pottier and Sommer (2006) observed that smaller insurance companies, stock insurers, those with a history of reserving error, those that use less reinsurance, those with high investments in shares and low grade bonds, and those that are geographically diversified are difficult to assess. Furthermore, failure to create insurance portfolios free of moral hazard, information asymmetry, and correlation contamination make it very difficult to accurately observe cash flows encumbered with private information. This is also exacerbated by the use of different accounting practices, even among those countries that have adopted international financial reporting standards (IFRS). This means that financial statements used in valuations are neither reliable nor transparent, hence the need for fair value accounting for insurance companies under IFRS Phase II. Volatility exhibited by insurance companies is due to both components of risk mentioned above and incorporates information asymmetry premiums (Myers, 1984; Myers and Majluf, 1984). From Table 9.1 and Figure 9.2, we can see that valuations of life business have declined significantly,

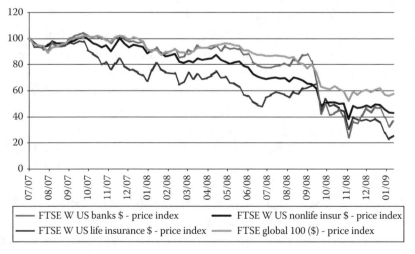

FIGURE 9.2 Performance of global insurance and banking stocks during the credit crunch.

tracking the banking indices and faring much worse than the global share index. The reason for this may be that life insurer valuations are being driven mainly by outsiders' perception as to the quality of insurers' asset portfolios. While banks and monoline insurers have done much to disclose the quality of their assets, this has not been the case with insurance companies. The extent of insurance companies' exposure to credit-related assets and liabilities nearly doubled from US$106 billion in September 2008 to US$205 billion in January 2009 (Insurance Information Institute, 2009c), while those for the banking sector only increased from US$600–$780 billion. The beginning of this period saw the collapse of AIG and Lehman Brothers. This period also saw a significant dip in both insurance and bank valuations, as shown in Figure 9.2. We could interpret this as market reaction to what was perceived to be significant hidden information concerning the quality of insurers' asset portfolios. Until all information is released, investors will not be able to observe the true value of insurance companies. Investors need to have all pre-loss information pertaining to a portfolio if they are to make an informed decision on the price to charge for their capital. Therefore, investors may continue to charge information asymmetry discounts on insurance stocks until all information on their assets is known, and more insurers, like Swiss Re, may need to seek additional capital from well-informed investors.

9.6 CONCLUSION

In January 2009, the IMF projected total losses from the credit crisis of around US$1.4 trillion, of which US$780 billion was expected to be paid by banks and US$205 billion or 20.8% by insurers (Insurance Information Institute, 2009c). The figure for insurers is almost double the September 2008 projected loss of US$106 billion, in marked contrast to the banking sector increase of only 30% from US$600 billion. This new projected figure may have a further significant impact on insurer valuations. Clearly, the effects of the banking crisis on insurance markets are by no means insignificant. We have seen how, in turn, upheavals in insurance markets have exacerbated the banking crisis and amplified its effects. The impact of the crisis on individual insurers has been uneven. The firms most directly affected are the specialist financial guarantee insurers (such as the U.S. monolines) and companies (such as AIG and Swiss Re) that extended their operations beyond traditional insurance business into risky areas of structured finance, and, to a lesser extent, firms writing lines of insurance business that are particularly sensitive to an economic downturn, such as credit and liability insurers. Unsurprisingly, bancassurers—integrated financial services providers and insurers which have close affiliations with banks—have been among the major insurance market victims. Of course, all insurance firms have suffered an erosion in the value of their assets and, for the reasons given above, life insurers have suffered more than non-life firms. We have seen that at some points in the emerging crisis, insurance stocks have fallen as sharply as banking stocks—notably in September 2008 when the news first broke on the parlous state of AIG. Insurance stocks have since recovered to some degree, no doubt because investors have started to believe that the contagion of insurance markets might be less severe, and less general, than they first imagined.

How will insurance markets change as a consequence of all this? First, we are likely to see a rise in M&A activity amongst insurance firms, with smaller insurers that have been weakened by the crisis becoming attractive targets for larger and better-capitalized companies. A question mark also hangs over future developments in the field of bancassurance. Until recently, bancassurance activity—where insurers use banks to distribute their products, or banks and insurance firms integrate more fully within one corporate structure—has been attractive to banks and insurers alike, generating potential economies of scale and scope, opportunities for cross selling and new income streams. In the current climate some of these benefits look questionable. Some banks, for example, may be obliged to sell off their insurance subsidiaries and withdraw from insurance activities in order to preserve capital and concentrate resources on core activities. Conversely, banks with well established, stable, and profitable insurance businesses may seek to retain them at the very least until their insurance businesses can be sold on better terms. From the point of view of insurance firms themselves, the benefits of close affiliations with banks look increasingly dubious. For one thing, insurers have in the past used bank distribution channels in order to gain from the stronger brand, image and reputation of their banking partners. These reputational assets are now badly tarnished and, in any event, the market for typical bancassurance products (e.g., savings plans and mortgage or loan related insurance policies) is stagnant. Again, several insurance groups (including Allianz, Europe's largest non-life insurer) have suffered more from problems with their banking subsidiaries than from their core insurance activities, which weakens the case for an insurer retaining or acquiring banking interests. For these reasons, insurers are now likely to focus more closely on traditional insurance business and, like Swiss Re, dispose of or cut down their structured credit and capital markets activities, as well as limiting their involvement with banks.

Finally, we can speculate briefly on how the regulation and supervision of insurance markets might change as a consequence of the crisis. A spate of failures among financial firms always leads to calls for tighter regulation, and this seems inevitable in the case of the banking sector. On the other hand, most senior figures in the insurance industry see no need for wholesale changes in insurance regulation. For example, the Chief Risk Officers of major European insurers have expressed their confidence in the adequacy of the current European project—Solvency II—to ensure that the industry is adequately regulated in the aftermath of the financial crisis (CRO, 2008) and the U.K. financial regulator, the Financial Services Authority, does not believe that the financial crisis has raised any doubts about the appropriateness of the capital adequacy regime for insurers (Financial Services Authority, 2009). Nevertheless, it seems likely that at least some extra regulation will spill over from the banking and securities sectors into insurance markets, especially at points where the activities of banks and insurers overlap, such as in relation to investment products sold by life insurers. In fact, there have already been developments in one such area of overlap. This is the market for CDS, discussed at various points above. Concern at the unregulated nature of this market has already led New York State, with effect from January 1, 2009, to regulate CDS as financial guarantee insurance in those instances where the buyer of the CDS owns the underlying security for which the instrument provides protection. Thus, all buyers who hold or reasonably expect to hold a "material interest" in the reference obligation will need to be licensed as financial guarantee insurers and only financial guarantee insurers will be allowed to issue such CDS. This is intended to ensure that the protection seller has sufficient capital and surplus, and a suitably robust risk management system to protect their buyers. It will prevent firms (such as, in the past, AIG and the monolines) from issuing CDS out of noninsurance subsidiaries.

REFERENCES

A.M. Best (2005) Property/casualty impairments hit near-term lows despite surging hurricane activity. Available at http://www.bestweek.com.

A.M. Best (2008) Best's impairment rate and rating transition study—1977 to 2007, Best's rating methodology—structured finance. Available at http://www.bestweek.com.

Advisen (2008) The crisis in the subprime mortgage market and the global credit markets: The impact on E&O insurers. Report available at http://corner.advisen.com/Subprime_E_O_final_3.pdf.

CRO (2008) Chief risk officer forum response to the financial crisis. Report available at http://www.croforum.org/publications.ecp.

Euler Hermes (2008) Press release available at http://www.eulerhermes.com/en/financial-news/financial-news_20081202_00100049.html.

Financial Services Authority (2004) Policy statement 04/16. In *Integrated Prudential Sourcebook for Insurers*. Financial Services Authority: London, U.K.

Financial Services Authority (2009) *Financial Risk Outlook 2009*. Financial Services Authority, London, U.K., available at http://www.fsa.gov.uk/pubs/plan/financial_risk_outlook_2009.pdf.

Financial Times (2008a) *The Financial Times*, London, U.K., November 28, 2008 at http://www.ft.com/cms/s/0/9e83517c-bcec-11dd-af5a-0000779fd18c.html.

Financial Times (2008b) *The Financial Times*, London, U.K., December 11, 2008 at http://www.ft.com/cms/s/0/4b9f8c6e-c728-11dd-97a5-000077b07658.html.

Insurance Information Institute (2009a) Life/health insurer financial asset distribution, 2003–2007. Available at http://www.iii.org/media/facts/statsbyissue/life/.

Insurance Information Institute (2009b) Property/casualty insurer financial asset distribution, 2003–2007. Available at http://www.iii.org/media/facts/statsbyissue/industry/.

Insurance Information Institute (2009c) A report on the State of the US insurance market. Available at http://server.iii.org/yy_obj_data/binary/804919_1_0/connecticut.pdf.

Jetuah, D. (January 8, 2009) Accountancy age, London, U.K., Available at: http://www.accountancyage.com/accountancyage/analysis/2233495/credit-insurance-triple-whammy.

Lloyd's (2008) Directors in the dock—Is business facing a liability crisis. Report by Lloyd's of London in association with the Economist Intelligence Unit, London, U.K.

Myers, S. (1984) The capital structure puzzle. *Journal of Finance*, 39(3): 575–592.

Myers, S. and Majluf, N. (1984) Corporate financing and investment decisions when firms have information that investors do not have. *Journal of Financial Economics*, 13(2): 187–122.

Oliver Wyman (2005) Essential groundwork for the Solvency II project. Report by Comité European des Assurances (CEA) and Mercer Oliver Wyman Limited. Available at: http://oliverwyman.com.

Oliver Wyman (2008) Beyond the crunch: Insurers should see the silver lining. Report available at http://www.oliverwyman.com/ow/pdf_files/OW_Eng_FS_Publ_2008_beyondthecrunch.pdf.

Pottier, S.W. and Sommer, D. W. (2006) Opaqueness in the insurance industry: Why are some insurers harder to evaluate than others? *Risk Management & Insurance Review*, 9(2): 149–163.

Swiss Re (2003) Reinsurance—A systemic risk. Sigma No. 5/2003, Swiss Reinsurance Company, Zurich, Switzerland.

Swiss Re (2006) Credit insurance and surety: Solidifying commitments. Sigma No. 6/2006, Swiss Reinsurance Company, Zurich, Switzerland.

CHAPTER 10

The Role of Hedge Funds in the Banking Crisis: Victim or Culprit?

Nicolas Papageorgiou and Florent Salmon

CONTENTS

This chapter will study the role of hedge funds in the recent financial crisis and look at how the massive deleveraging process that has taken place will affect the hedge fund industry going forward. Does the hedge fund industry deserve all the blame that is being directed toward it for the recent upheaval in the markets? We will attempt to disentangle the role of the different agents involved in this fiasco (banks, brokerage firms, fund managers, etc.) and provide a clearer picture of their share of the responsibility for the recent market collapse.

10.1 INTRODUCTION

The 2008 credit crisis has decimated financial markets and the negative repercussions on the global economy have been swift and devastating. Central Banks and governments have taken unprecedented steps to shore up a very fragile banking system, with interest rates being slashed to historic lows, and already over $1 trillion of taxpayers' money being used to bailout financial institutions. We have witnessed the fall from grace of some of the world's leading financial institutions, with a number of them being nationalized, others restructured, and several filing for bankruptcy. Although we have not yet turned the corner and the possibility of a deep and drawn-out recession is increasingly likely, economists and financial analysts have already started a postmortem examination of this latest crisis. Nobody questions the pivotal role that securitization played in the build up to the credit crisis; however, the responsibility of different financial markets' participants is still very much up for discussion.

We will first examine the role that hedge funds played in the recent financial debacle. The media and many financial experts have attempted to place a disproportionate share of the blame on hedge funds. Needless to say, recent scandals such as the one involving Bernard Madoff have clearly not done much to aid their reputation. These supposedly unregulated investment vehicles present an easy target as they are considerably less transparent than more traditional funds, and undertake much more sophisticated and complex investment strategies. These "bad boys" of modern finance also provide an ideal scapegoat for other financial intermediaries who seek to deflect the blame for their role in the present debacle. In order to provide a clearer evaluation of their responsibility, we will first look at the exceptional growth of the hedge fund industry, highlighting the important role that financial institutions played in the hedge fund boom of the last 5 years. We will demonstrate the importance of hedge funds both as a source of returns as well as an unparalleled source of fees for the large brokerage firms. The proliferation of hedge funds was clearly a value proposition for these financial intermediaries.

Next, we will confront two important misgivings that the general public has about hedge funds and their managers. The first pertains to the regulation of the hedge fund industry, which is often misunderstood and the second relates to the generous performance fees awarded to the managers. We will demonstrate that in fact, hedge funds are subject to a significant amount of regulation and legislation and that they do not operate in a parallel lawless dimension. In fact, in certain aspects, the recent deregulation of the banking industry provided financial institutions with considerably more latitude than that afforded to most hedge funds. Hedge fund managers are also often assumed to have an incentive to take on excess risk due to the particular structure of their compensation agreements. We will show that although the compensation fee is asymmetric, there are several mechanisms protecting investors' interests; furthermore, the remuneration scheme is fully disclosed upfront and is contractual. In contrast, bonuses and directors' compensation in financial institutions are considerably more discretionary and are not always linked to positive performance, as we have clearly witnessed in 2008.

The chapter is organized as follows. First, we provide some background on the growth of the hedge fund industry and discuss the gradual institutionalization of hedge funds. Second, we discuss the regulatory environment of banks and hedge funds as well as their fee structures. Next, we highlight the impact that the massive deleveraging of the financial system had on the hedge fund industry. We use two case studies to illustrate our arguments. Finally, we provide some insight into the future of the hedge fund industry and how the recent crisis has affected the demand and nature of alternative investments.

10.2 THE GROWTH OF THE HEDGE FUND INDUSTRY

The tremendous growth of the hedge fund industry over the last 10 years has brought the alternative investment industry very much into the mainstream. From an estimated $39 billion under management in 1990, the hedge fund industry as of the end of June 2008 totaled 10,000 funds globally with over $1.9 trillion in assets under management. The demand for alternative investments has not been limited to the traditional clientele of high net worth investors. In fact, the growth has largely been fueled by demand from pension funds, sovereign funds, and especially financial institutions. In fact, financial institutions had much to gain both from the strong performance of hedge funds as well as from substantial brokerage fees.

Notwithstanding the exceptional growth of the hedge fund industry, it still only makes up a small fraction of other markets. As of the end of June 2008, the global mutual fund industry consisted of over 68,000 mutual funds, with a total of $24 trillion in assets under management. Besides being much smaller than mutual funds, the size of the hedge fund industry is dwarfed by other institutional investors such as pension funds and insurance companies, which, in 2006, controlled about $22.7 and $17.4 trillion in assets, respectively. As of January 2007, the total hedge fund industry was less than one-third the size of the $5.8 trillion market for mortgage-backed securities.

Given the dynamic nature of many hedge fund strategies, they do however account for a disproportionate amount of the daily volume in certain financial securities. Market reports claim that approximately 20% of all trading volume on the New York Stock Exchange and 30%–35% on the London Stock Exchange are performed by hedge funds. They also play an increasingly dominant role in many debt markets, accounting for 45% of trading volume in emerging market bonds, about 50% in distressed debt, and 25% of high-yield bonds. Other markets, such as the convertible bond market cater almost exclusively to hedge funds. Feng (2004) estimates that 75% of actively traded convertible bonds are held by hedge funds. Hedge funds have also played a substantial role in the growth of the credit default swap (CDS) market. Greenwich Partners estimates that they account for 55% of the CDS trading volume.* All this trading activity is, of course, also of great value to the large brokerage houses that help finance and structure the trades, and indubitably pocket generous commissions.

---

* Since CDSs are designed to enable the manager to trade corporate credit long and short without the interest rate risk sensitivities, they offer a natural fit for hedge fund managers to actively manage credit exposure. CDSs are used by hedge funds mainly for hedging purposes (convertible arbitrage) and relative value arbitrage (long short corporate credit). CDSs are very convenient instruments for active managers. The only issue with CDS is that since it is traded over-the-counter (OTC) it relies on the strength of the brokers. The probable shift of CDS toward standardized exchange-traded contracts will address the issue.

10.3 THE REGULATORY ENVIRONMENT

The evolution of the regulatory environment was undoubtedly the true catalyst for the credit crisis. Disintermediation of the banking system through the securitization of mortgages and loans was the immediate cause; however, the seeds from which this crisis has grown were planted almost a decade ago.

Through the Gramm–Leach–Bliley Financial Modernization Act of 1998, and the repeal of Glass–Steagall in 1999, the new model for financial institutions in the United States was established. This model was largely based on a European style model of universal banking rather than on the U.S. New Deal's strict sectoral separation. As a result of this change, competition between commercial banks and investment banks for securities business increased dramatically, forcing the investment banks into ever more risky reliance on proprietary trading–speculating with their own capital and using leverage to increase returns. To add fuel to the fire, at a meeting on April 28, 2004, the Securities and Exchange Commission (SEC) unanimously voted to exempt the largest investment banks from a regulation that limited the amount of debt their brokerage units could accumulate. The exemption would unshackle billions of dollars previously held in reserve as a cushion against losses on their investments. These funds could then flow up to the parent company, enabling it to invest in the fast-growing but opaque world of mortgage-backed securities, credit derivatives, and other exotic instruments.

During this same period, as financial institutions were leveraging up, the average hedge fund leverage dropped significantly. Researchers at the Bank for International Settlements estimate that from 1998 to 2004, the average hedge fund leverage dropped from about 8 times assets to 3 times assets. This reduction in leverage can in part be explained by a shift in the dominant strategies from global macro to equity long-short. A 2007 study of hedge fund leverage by the Organization for Economic Co-operation and Development (OECD) estimated that average hedge fund leverage (including leverage from borrowed funds and implicit leverage from derivatives) was 3.9 to 1, with the bulk of leverage coming from derivatives.*

---

* There is often confusion between leverage and risk. A typical long/short market neutral hedge fund will hold more leverage (long plus short) but will expose the investor to less market risk than the typical mutual fund. In other words, when hedge funds buy protection against market moves, this increases their leverage but reduces their risk. In the case of banks, however, the increase of leverage observed over the last few years corresponded to an increase of risk.

The International Monetary Fund put forth an even more surprising estimate in October 2008, pinning the average global hedge fund leverage
from borrowed funds to a ratio of 1.4:1.

10.3.1 Hedge Funds Regulation

Although hedge funds are generally regarded as being unregulated,
there exists a substantial body of federal and state law that restricts the
activities of the funds and their managers and requires certain mandatory disclosures. Hedge funds are governed by the entity law of the state
or offshore jurisdiction in which they are organized along with the law
of contract governing their operating agreements. As investment advisers to the funds they manage, hedge fund managers are also governed by
federal investment adviser law. As issuers of securities and as purchasers
and sellers of the securities of other companies, hedge funds are likewise
governed by federal securities regulation. However, the funds operate
so as to be totally excluded from federal law applicable to investment
companies.

Nonetheless, hedge funds are fully subject to federal prohibitions on
fraud, market manipulation, and insider trading, and must make public
disclosures in connection with trading registered securities.

10.3.1.1 The Legal Structure

A hedge fund consists of three basic entities: investors, the fund itself, and
the investment adviser/management company. U.S.-based hedge funds are
typically structured as limited partnerships or limited liability companies.
A hedge fund–limited partnership is made up of two types of partners:
limited partners and the general partner.

Limited partners of a hedge fund are passive investors whose decision
making is limited to deciding when and how much capital to contribute or
withdraw, subject to capital redemption restrictions under the fund's operating agreement. The limited partners provide capital as the fund's investors
and are not liable for the fund's debts, although they are subject to losing all
of their investment capital and any profits not yet distributed. To avoid losing
their limited liability, limited partners do not participate in management
decisions.

The general partner of a hedge fund–limited partnership is the fund's
portfolio manager and investment adviser and is responsible for managing all aspects of the hedge fund business, including managing the fund's
investment portfolio. Limited partnership law gives the general partner

complete control over the activities of the partnership and the terms of the partnership agreement.

10.3.1.2 Investment Company and Investment Adviser Law

Because a hedge fund consists of an investment fund and an investment adviser, its activities fall within the scope of federal regulation under the Investment Company Act of 1940 and the Investment Advisers Act of 1940. All hedge fund managers are subject to provisions of the Advisers Act and a significant number are registered investment advisers.

The investment activities of hedge funds would deem them an "investment company" under the Company Act, however funds are excluded from the definition of investment company so long as they satisfy one of the following conditions: (1) do not exceed 100 investors and sell their securities only through a private sale and (2) only sell securities to "qualified purchasers" through a private sale. Qualified purchasers include natural persons owning at least $5 million in investments and certain companies with at least $100 million in securities investments. Because hedge funds are generally excluded from the Company Act, the funds are not subject to the restrictions on the investment activities of investment companies imposed by the Company Act and its regulations. This exclusion permits hedge funds to employ leverage, short sales, and derivatives without having to comply with the Act's restrictions with respect to those activities.

Hedge funds are fully subject to the prohibitions against fraud under the Securities Act of 1933, the Securities and Exchange Act of 1934, and the Investment Advisers Act of 1940. Since fraud includes making misleading statements or omissions, hedge funds typically make comprehensive disclosures to avoid later being found liable for omitting any important fact to investors. For instance, under the Advisers Act, a hedge fund can be found liable for lying to investors about investment strategies, experience and credentials, risks associated with the fund, and valuation of the fund's assets.

Furthermore, under the Exchange Act and its regulations, hedge funds are prohibited from trading upon material inside information, from engaging in abusive short-selling, and from manipulating the prices of securities and other financial instruments used by any other type of practice.

10.3.1.3 Disclosures Relating to Trading Registered Securities

Hedge funds must also comply with various requirements under the Exchange Act arising out of their investments in public companies. First, all hedge funds and their managers are required to disclose large shareholdings of public companies. Hedge funds must disclose beneficial ownership of greater than 5% in a class of voting shares of securities registered under the Exchange Act, and disclose whether the purpose of such ownership is to acquire or influence the issuer. In connection with preventing insider trading, hedge funds, upon acquiring a 10% ownership stake in any issuer's class of voting equity securities, must disclose such ownership, any other equity ownership in the company, and any subsequent changes in such ownership. In addition, to increase publicly available knowledge about institutional shareholdings, hedge funds owning more than $100 million in stock traded on a national exchange are required to disclose, on a quarterly basis, all of their equity holdings to the public, and, on a weekly basis, certain short-sale positions to the SEC.

10.4 MANAGEMENT COMPENSATION

In 2007, John Paulson, now famous for his short subprime trade, was criticized for registering the largest Wall Street salary in history (over $3 billion). It is important to remember, however, that some of his investors quadruple their investment that year. On the other side of spectrum, in 2008 bankers who had brought on the disaster rewarded themselves with $18.4 billion, the sixth largest bonus pool ever paid out. The same year, the Associated Press reported that an estimated $1.6 billion of bailout funds coughed up by U.S. taxpayers ended up in the pockets of top executives.

Much has been made of the hefty fees charged by hedge funds, however, until the recent crisis, there has been little discussion regarding the exorbitant bonuses paid out by modern financial institutions. Although we are by no means implying that hedge fund managers are not overpaid, the actual structure of their remuneration contract results in their incentives being considerably better aligned with those of the investors than is the case with top executives and asset managers in the major banks.

Under the terms of the applicable operating agreement, the hedge fund management company is compensated by a management fee, typically ranging from 1% to 2% of the underlying fund's net asset value (NAV), which may be calculated monthly or quarterly. The management fee covers expenses for operating and administering the fund such as for overhead, personnel salary, office leases, and physical capital costs. Management fees are typically used throughout the asset management industry including by publicly registered mutual funds.

A distinguishing and defining feature of hedge funds, however, is that their operating agreements have provisions compensating managers based upon the performance of the funds, typically calculated on an annual basis. Hedge fund performance-based fees typically range from 10% to 20% (but can reach up to 50%) of profits in excess of prior losses and net of management fees. These performance fees have often been criticized on two fronts: (1) they allow managers to take a share of profit but provide no mechanism for them to share losses and (2) the fees give managers an incentive to take excessive risk rather than targeting high long-term returns. In practice, however, hedge funds' performance fees are limited by two types of contractual provisions: a high watermark provision and a hurdle rate.

The high watermark provision limits the performance fee allocation to positive gains above the amount of the investor's capital contribution. A high watermark requires any losses from previous years to first be recouped, meaning that an investor must actually receive a net positive return on their investment before a manager is paid a performance fee. The "hurdle rate" is typically used in conjunction with a high watermark. A performance fee subject to a hurdle rate will not be allocated to the manager until the fund's annualized performance exceeds a benchmark rate, such as the yield on a Treasury security, a LIBOR rate, or a fixed percentage. This links performance fees to the ability of the manager to provide a higher return than an alternative, usually lower risk, investment.

Beyond the contractual provisions that protect hedge fund investors from paying unwarranted fees, the risk tolerance of the manager is often closely aligned to that of the investors as he also contributes to the investment pool. Managers often co-invest a significant portion of their own capital directly in the underlying funds that they manage. Agarwal et al. (2009) estimate that the average investment by managers accounts for 7.1% of fund assets with the median manager owning 2.4% of the fund. Hedge fund investors often desire co-investment by managers to align the manager's incentives with their own.

10.4.1 The Crisis and Deleveraging of Financial Markets

Over the past 15 years, hedge funds have generally delivered what they advertised—absolute returns with low correlation to the market and a volatility somewhere between fixed-income securities and the stock market. Obviously it is a bit naïve to lump together such a wide range of strategies, however, if we forgive ourselves that indiscretion, we can observe in Figure 10.1 that the Hedge Fund Research Indices (HFRI) Composite hedge fund index provided stable returns during the period 1993–2007. Between

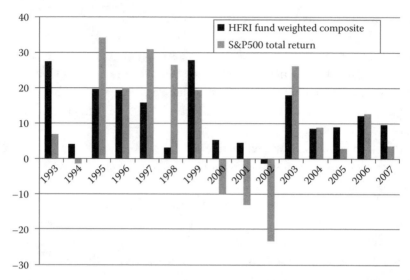

FIGURE 10.1   Annual performance of HFRI fund weighted composite and S&P 500 index.

1993 and 2007, the HFRI Composite index provided systematically positive returns with the exception of a 1% loss during the 2002 calendar year. Let us recall that although the 1990s were characterized by a strong bull market, the sample period was by no means an uneventful one. The sample spans the Asian financial crisis in 1997; the Russian financial crisis of 1998; the Long Term Capital Markets debacle of 1998; the bursting of the dot.com bubble in 2001; and the ensuing recession that lasted almost 2 years. During this period, the returns on the S&P500 were not only considerably more volatile but the mean return of 9.7% was considerably lower than the 12.2% generated by the average hedge fund investment. Even if we adjust for the known biases in hedge fund returns, the latter clearly offered substantial value (not to mention diversification benefits) to investors.

In 2008, the impressive streak of absolute returns came to a thunderous halt. Hedge funds were unable to shelter themselves from the widespread carnage in financial markets, and the loss on the HFRI weighted index was an abysmal 20% for the year. Although certain strategies such as Commodity Trading Advisors (CTAs), Macro Funds, and Short bias funds managed to eke out a profit during the credit crisis, the vast majority of funds suffered unprecedented losses. Ironically, however, 2008 also represents the best relative performance for the HFRI weighted index against the S&P500 index (+18%).

TABLE 10.1　Performance of a Sample of HFRI during 2008

	Merger Arbitrage (%)	Equity Market Neutral (%)	Short Bias (%)	Emerging Markets (%)	Equity Hedge (%)	Event-Driven (%)	Macro (%)	FI-CA (%)
January	−1.78	−1.04	4.76	−5.55	−4.47	−2.53	1.08	−1.40
February	0.88	1.45	2.47	2.71	1.31	0.41	4.22	−0.78
March	−0.86	−0.57	1.33	−4.32	−2.84	−1.41	−1.22	−4.16
April	1.40	0.16	−3.56	2.85	2.45	1.00	−0.03	1.31
May	0.93	1.12	−1.50	1.40	2.38	1.52	1.20	0.77
June	−1.43	1.37	6.37	−3.79	−2.44	−1.60	1.19	−2.28
July	−0.39	−1.13	−0.19	−3.35	−2.84	−1.43	−2.62	−1.63
August	0.32	−1.38	−1.04	−4.94	−2.17	−0.50	−1.18	−1.10
September	−2.90	−2.87	5.12	−10.38	−8.14	−6.01	−1.21	−11.81
October	−2.42	−0.53	9.41	−13.81	−9.23	−8.01	1.76	−16.37
November	−0.25	0.05	4.83	−3.91	−3.53	−3.51	0.76	−2.67
December	1.54	−2.84	−2.07	−0.30	0.28	−1.45	1.14	0.38
Total 2008	−4.94	−6.23	25.92	−43.38	−29.26	−23.52	5.10	−39.73

FI-CA, Fixed income-convertible arbitrage.

In Table 10.1, we provide a month-by-month overview of the performance of the HFRI subindices for some of the most dominant hedge fund strategies. Four strategies, specifically emerging markets, equity hedge, event driven, and fixed income/convertible arbitrage were by far the worst performers and their end of year returns were comparable to those of the major international equity indices. If we take a closer look at the returns, we note that the bulk of the losses transpired during the months of September and October, which also happened to be the worst months for equities. But the cause of the atrocious performance of hedge funds during those 2 months goes far beyond the sell-off in equity markets. Hedge fund strategies rely heavily on the proper and timely functioning of capital markets, and unfortunately during the second half of 2008, there were many irregularities in the global financial markets.

The real issue for hedge funds started only with the collapse of Lehman and the panic it triggered within the rank of financial institutions. This created a vacuum in the credit market, with banks fighting to survive and pulling back on every available line of credit. Banks were desperately trying to deleverage their own balance sheets to absorb the asset-backed commercial paper (ABCP)/special purpose vehicle (SIV)/credit crunch losses in order to maintain capital adequacy. Consequently, certain hedge fund strategies that required financing from the banks (i.e., fixed-income arbitrage) were hit hard as any form of loan made to hedge fund industry was severed resulting in forced liquidation of the underlying positions. The September crisis was also particularly brutal to small hedge funds because the brokerage community opted to reduce their exposure by cutting credit lines to funds that they felt would not generate much future business for the firm. This may be legitimate from a business standpoint but it had the effect of inflating the loss inflicted to some investors by forcing funds to liquidate positions in a difficult market.*

Beyond the liquidity and credit constraints, hedge funds were adversely affected by two significant market factors during the crisis. The first was the short-selling ban imposed by the SEC on certain securities, and the second, and most significant, was their reliance on the services of their increasingly distressed and predatory primary brokers. The massive redemptions resulting from the deleveraging of financial institutions amplified the effect of these irregularities.

10.4.2 Impact of the Short-Sale Ban

Strategies such as those implemented by convertible arbitrage managers require the ability to short the equity of the issuers of the convertible security. The effectiveness of the strategy is based on the manager's ability to buy cheap Gamma exposure via the conversion option and delta hedge the market risk by shorting the underlying equity. The short-sale ban that was put into effect by the SEC in September of 2008 left Convertible Arbitrage managers unable to hedge their exposures, and given the illiquidity in the corporate bond market, they were also unable to sell their bond holdings without incurring a serious markdown. The important drawdowns that resulted for this strategy in September and October (−11.81% and −16.37%, respectively) were more a result of the unforeseen regulatory intervention rather than a market risk embedded in the strategy. This is not to say that Convertible Arbitrage funds would not have suffered during the market collapse (widening credit spreads significantly reduced the value of their bond holdings), however, the amplitude of the shock would have been substantially reduced if they were able to maintain and adjust

---

* In most hedge funds legal documents (offering memorandum), there's a clause giving the board of directors of a fund the right to suspend redemptions. They will do so if they believe it to be in the best interest of investors. Once again, it's all about alignment of interest, in that case between investors of a same fund. Throughout the crisis, this very convenient but unpopular feature allowed many managers to avoid forced liquidation of already depreciated illiquid assets.

their hedge. A similar case can be made for most hedge fund strategies (Event Driven, Equity Hedge, Merger Arbitrage, Equity Market Neutral) that use systematic shorting of certain securities in order to control and limit risk exposures.

The short-sale ban was a very convenient way for the regulator to force buyers back into the market because in order to close a short position a manager had to buy back the shares. This artificial demand for stocks at the detriment of hedge fund investors was considered to be politically acceptable at that time. However, the benefit of the ban was short lived and only technical since market soon continued to correct downward based on fundamentals. This abrupt change in the law only further increased the perception that the regulator was not in control, triggering even more redemptions out of an already fragile hedge fund industry.

10.4.3 The Role of the Broker

Hedge funds rely extensively on their primary brokers in order to operate effectively. Tremendous pressure was placed on this relationship throughout the banking crisis putting to the test many structural issues that will have to be addressed in the future. Regulation alone will not be sufficient. It is fundamental that there be an alignment of interests and incentives between market participants since it is the only way to maintain the system equilibrium with negligible intervention. In the financial industry, the most difficult relationships to manage are the ones involving brokers. Throughout the crisis and in many other instances, there has been clear extraction of economic value by banks from other counterparties (including from other banks). As financial intermediaries, banks are insiders to all transactions; however they will execute and act on the basis of their own interest. Their private partnership structure gives them no incentive to manage long-term collective interest and we will try to demonstrate in the two following cases, the catastrophic impact that this had on the hedge fund industry.

Case #1: Amaranth—A Sign of Things to Come

Prior to considering an example of hedge fund dislocation within the context of 2008 crisis, it is important to note that there was already a clear indication of the potential for a clear conflict of interest between hedge funds and their main suppliers of services. We believe that the Amaranth situation should have been used as a trigger by managers and the government to put some distance between the investors and the service providers (brokers, custodians, etc.). Amaranth is a well-publicized case but very few people took the necessary time to go through the litigation documentation that emerged after the collapse. This now public information was not relayed in the press probably because it constitutes a challenge to the general perception that hedge funds managers are adverse to fair market practices.

The case of Amaranth is interesting in the context of the current crisis because it illustrates the issues associated with the restructuring of a fund in a distressed position. The objective here is not to analyze whether their oversized trade in natural gas future contracts was ill-advised or if the fact that their positions were leaked to competitors adversely affecting the performance of the trade. Rather, we will attempt to illustrate how the consequences of the losses on the natural gas trade could and should have been a lot less severe for investors.

On November 13, 2007, Amaranth filed a complaint against JP Morgan Chase. Although this complaint is probably part of a legal strategy by Amaranth in relation with other litigation cases, it is however very illustrative of the type of interaction that could be between the broker and the hedge fund.

In mid-September, due to some heavy losses in the natural gas market, Amaranth was unable to post the required margin to JP Morgan that was acting as the clearing agent for the fund. At that time, JP Morgan held collateral from Amaranth to protect against default on the natural gas trade. The risk for Amaranth and its investors was that JP Morgan would cease to do business with the fund and dispose of the collateral at a heavy discount, effectively wiping out the funds equity. According to court documents (see reference), the portfolio manager, Nick Maounis, contacted a senior partner at Goldman Sachs to organize the transfer of a significant portion of the Natural Gas trade to Goldman (75%) and the balance to Merrill Lynch. Over the weekend, the details of the transaction were finalized, resulting in a concession of $1.85 billion to be paid to Goldman Sachs. If executed, this trade would have reduced the loss to investors from −63% down to an estimated −30%. The next day, a conference call was held between Goldman Sachs, JP Morgan, and the NYMEX to execute the transaction. This was meant to be a mere formality. However, contrary to all expectations and without disclosing any reasons, JP Morgan decided to block the trade with Goldman and Merrill. On the day of the planned transaction, the market kept moving against the fund, resulting in an additional loss of several hundred million dollars. While JP Morgan was getting ready to take control of the portfolio, Ken Griffin of Citadel contacted Amaranth to offer "help" and to see if a deal was possible between Amaranth and Citadel. Citadel was at that time one of the largest hedge funds in the world. Intense discussions took place throughout the day and again the details of a deal were negotiated and agreed upon. Not only would Citadel get the same $1.85 billion concession Goldman was supposed to receive but Amaranth would also absorb two-thirds of the loss of the previous day resulting from the failed trade with Goldman. However, Citadel was later contacted by the co-chief executive officer of JP Morgan who declared Amaranth insolvent. JP Morgan's purpose here was not to prevent the restructuring of Amaranth, but rather to create the necessary conditions for JP Morgan to act as an intermediary to the trade between Amaranth and Citadel. In a matter of a few hours, this resulted in a payment of $725 million to JP Morgan that would have otherwise been entirely left out of the deal.

During subsequent conference in New York in November 2006, the same senior officer responsible for the trade at JP Morgan made the following statement: "We are not exposed from a credit perspective, materially, which allows us to respond quickly to opportunities when they come up, Amaranth was one obvious example of that[...] I imagine there will be others as we go through time where our ability to be on the inside, but not compromised, is extremely powerful." In layman's terms, brokers are licensed to kill.

A disconcerting feature of distress markets is that legal action creates mismatched timeframe. Technically a broker could legally seize the assets of a fund, in effect, forcing it to collapse. The damage to the fund is immediate and irreversible, whereas the outcome of litigation, even if successful, is long and onerous. It is very clear that it is that legal framework does not present a viable course to resolve this embedded conflict of interest. This issue must be resolved ex ante by mitigating possible conflicts of interest. Prior to 2008, very few managers appreciated the limitations of the legal agreements despite the fact that there had been several examples of wrongful disposal of assets by brokers in the 1998 crisis.

Case #2: NAV Triggers and Predatory Banking

This second case is more recent and based on a fund that collapsed in November 2008. A liquidator has already been named and much litigation is pending. This fund was a multibillion dollar multistrategy portfolio with a bias toward fixed income. Throughout 2008, the manager accumulated AAA CMBS paper at spreads that were already perceived to be at distressed levels. A significant portion of the allocation was comprised of municipals bonds, which offer lower liquidity outside of the IRA/401(k) season.

In the months prior to the collapse, the fund was holding $800 million in cash reserves (more than half of the equity of the fund at that time) against positions that were in most cases leveraged fixed-income trades. The degradation of market conditions and the increase in margin requirements by banks dwindled most of this cash reserve. The losses experienced in October 2008 placed the fund in default with regard to their International Security Dealers Association (ISDA) agreement via a mechanism known as the NAV trigger. This is a common feature of these contracts allowing the broker to terminate trades if the fund experiences substantial losses over a certain period of time. The brokers were quick to enact this clause, and in doing so, extract a very significant amount of wealth from the fund.

One of the brokers had a total claim of $100 million on a large CMBS total return swap with the fund. The loss occurred on Friday October 21 but the spreads tightened quickly thereafter. Under the ISDA contract provisions, the broker had the right to pick the exact time to estimate the value of the payment due but was not forced to trade in the market. So the broker sent in a "proof of claim" based on the Friday price taking official ownership of the assets owned as collateral. By the following Monday, the market loss had been recovered but the $100 million legal claim was still valid. There is no doubt that the claim will appear to be legitimate and the fact that the market rebounded thereafter is irrelevant since the bank assumed the risk. However, it still illustrate the fact that, in times of market dislocation, legitimate claims can have a very adverse effect. The market collapse created an unprecedented spike in prices that gave rise to a free option on the assets of a fund by the brokers since fair market values decouple from market prices. The industry needs to revisit the contractual framework used to settle trades during times of market dislocations.

In the case of the allocation to U.S. municipal bonds, the broker's claim was significantly larger, reaching $700 million. On the day prior to the fund defaulting on the position, the Municipal market completely blew out on relatively low volume. There was clear indication at that time that the market was driven by the same bank with which the fund defaulted. Furthermore, the price at which the claim was valued was highly disputable. In the absence of market liquidity, the broker had the ability to unilaterally fix the level of the claim at whatever price he deemed appropriate. In the case of the fund, the losses on the markdown appear to be three times market consensus.

Another point is that the bank acted has both custodian and prime broker to the fund. Through a New York court, they were able to obtain a garnishment order enabling them to freeze the custody account in the United States, and then move the assets from the original custody account in London back to New York where legislation was more favorable. The objective was to use their privileged status as custodians to assert a priority claim on that pool against other creditors (other brokerage firms). In effect, the bank in that case would be able to use the displacement of the fund to extract value not from investors but from nine other banks that were lenders to the fund. This illustrates the predatory nature of the banks, both toward the funds as well as toward each other. Also disconcerting is the fact that during the restructuring effort, the same financial institutions stopped executing wire instructions sent by the manager. This alone induced additional losses to the fund since it blocked all settlement attempts between the fund and some of its lenders.

This case highlights the potential conflict of interest when the same financial institution acts as banker (wires), broker (credit lines), and custodian of the assets. This case is by no means unique, and there have been numerous other examples of mismanagement and abuse during the crises. In one case, Sentinel, assets that were under custody for a fund were used illegitimately as collateral for a loan by the brokerage arm of the custodian. The second issue pertains to the potentially devastating effect of incorporating NAV triggers into ISDA agreements. Prior to the crises, nobody had realized that the exercise of these triggers could lead to forced liquidation of so many funds at the same time. At some point, this even represented a threat to certain very large funds such as Citadel and could have had monstrous consequences. These clauses were initially designed as an exit strategy for brokers in the case of an isolated fund failure, and by no means intended as a way to manage counterparty risk during a financial crisis.

10.5 CONCLUSION

In 2008, hedge funds failed to deliver on their promise of absolute returns, resulting in significant redemptions as investors shifted away from active money management. However, their disappointing performance stemmed largely from the fact that market infrastructure collapsed beneath them and not because the "hedge fund model" was flawed. In fact, until mid-2008, the credit crunch was perceived by many hedge fund managers as the incubation phase for a fantastic set of new opportunities. In the months leading up to the crisis, more than 100 credit opportunistic funds were launched, raising billions of dollars in assets, and waiting for the right time to invest.

The straw that broke the camel's back was the collapse of Lehman. It was the trigger for a total loss of investors' confidence in the financial system. Banks stopped trading with one another, focusing solely on their own survival. No one fathomed a few years ago that century-old Wall Street firms would cease to exist. The most critical element that was overlooked was how quickly the domino effect would kick in. Nothing was ring fenced. With hindsight we can clearly see that the system was built out of a large complex network of trades with unsustainable levels of leverage and that the brokerage community was at the heart of this construct.

One view that is beginning to emerge within the investment community is that the de facto nationalization of U.S. banks is a first step toward realignment of interests. Some observers expressed the view that banks were similar to utility companies and, as such, should be more heavily regulated. However, one needs to bear in mind that the health and the efficiency of the banking system constitutes a key competitive advantage to a country and overregulation is rarely desirable.

Investors and managers have become addicted to financial services provided by investment banks in one stop shop format. In many cases, brokerage firms were not only providing OTC trading, credit line and repo agreements but they also acted as administrator, custodian, real estate providers, IT consultants, and capital introduction specialists. Investors need strong truly independent service providers with clear alignment of interests. It means, for example, the development of more exchanges in place of the OTC markets, such as some of the initiatives put forth for the CDS market. It means less reliance on complex structures that are manufactured by entities that will ultimately not own them. A few regulatory changes are, of course, also needed and it could probably start with a reform of the different regulatory bodies in the United States since there has been clear case of conflict of interest between, for instance, the Commodity Futures Trading Commission (CFTC), the National Futures Association (NFA), and the SEC.*

Certain reforms will also be required in order to better regulate and protect the hedge fund industry. A first such step was made in January 2009 with the introduction in Congress of the "Hedge Fund Transparency

---

* In the case of Sentinel Management Group Inc., the CFTC protected the FCMs (future commission merchants) by using assets from segregated pools owned by SEC-regulated investors.

Act" requiring SEC registration of all private investment pools, the filing of an annual information form with the SEC, and cooperation with SEC inquiries and requests for information. Even though the hedge fund industry is largely misunderstood and far too reliant on the proper functioning and integrity of other financial intermediaries, as long as investors believe in the role of active management and the capability of managers to generate alpha, the hedge fund industry will survive and continue to play an important role in financial markets.

REFERENCES

Agarwal, V., Daniel, N. and Naik, N. (2009) Role of managerial incentives and discretion in hedge fund performance. *Journal of Finance*, forthcoming.

Amaranth L.L.C. and Amaranth Advisors L.L.C. against J.P. Morgan Chase & Co., J.P. Morgan (August 10, 2007). New York.

Amaranth. (November 13, 2007) Chase Bank, N.A. and J.P. Morgan Futures Inc. Available at http://online.wsj.com/public/resources/documents/amaranthjpm.pdf

Arner, D. (2008) *The Global Credit Crisis of 2008: Causes and Consequences.* Asian Institute of International Financial Law, University of Hong Kong.

Blundell-Wignall, A. (2007) An overview of hedge funds and structured products: Issues in leverage and risk. *Financial Markets and Trends*, 1(92): 1–20.

Feng J. (2004) *Hedge Fund Strategies Drive Market Direction in US and Euro Converts*, August 5th. Greenwich Associates: Greenwich, CT.

Getmansky, M. (2005) The life cycle of hedge funds: Fund flows, size, and performance. Working Paper, Isenberg School of Management, University of Massachusetts, Amherst, MA.

Hedge Fund Research. (2008) Global hedge fund industry report: Third quarter Chicago, IL.

Ineichen, A. and Silberstein, K. (2008) AIMA's roadmap to hedge funds. AIMA publications, Charlottesville, VA.

International Monetary Fund. (April, 2007) Global stability report. IMF, Washington, DC.

International Monetary Fund. (2008) Global financial stability report: Financial stress and deleveraging—Macro-financial implications and policy. Washington, DC.

Investment Company Institute. (2008) Worldwide mutual fund assets and flows, Second quarter, Washington, DC.

McGuire, P. (March, 2005) Time-varying exposures and leverage in hedge funds. *BIS Quarterly Review*, 59–72. Basel, Switzerland.

Shadab, H. (2009) Hedge funds and the financial market, testimony before the house committee on oversight and government reform. The Mercatus Center, George Mason University, Fairfax, VA.

US Securities & Exchange Commission, 2008, Testimony concerning the regulation of hedge funds. Washington, DC.

CHAPTER 11

Solving the Banking Crisis: A Private Capital Solution

François-Serge Lhabitant*

CONTENTS

The 2007 crisis spread internationally across the financial system and resulted in an unprecedented global banking systemic contagion. After their traditional hunt for culprits, governments, regulators, and politicians have explored the usual solutions, that is, providing some financial assistance, purchasing toxic assets, and changing the regulation to keep the banks afloat. All these solutions are known to

---

* The views expressed in this chapter are those of the author and may not reflect the opinion or position of any of the institutions he is affiliated with. We thank Caroline Farrelly for her editorial assistance.

be subject to moral hazard by potentially encouraging excessive risk taking and giving banks the incentive to engage in gambling for resurrection. In this chapter, we discuss an alternative solution, namely, their collaboration with hedge funds and private pools of capital in order to rescue the banking system.

11.1 WHEN KINGS BECAME BEGGARS

The seeds of the current banking crisis were sown in the mid-1990s, primarily at the point of intersection between the U.S. interest rate, real estate, and credit markets. The combination of a decade of ballooning U.S. house prices with record low interest rates and high employment led to an exponential takeoff in the demand for mortgages. Financial institutions initially responded very positively and allowed millions of U.S. families to achieve their ultimate dream—owning their own home. Initially, the related mortgages were typically granted by a neighborhood financial institution, which would hold them and service them throughout their lifetimes. Profits were small and primarily derived from the spread between the interest rate paid by the financial institution on its deposits and the higher rate charged on its mortgages. If customers defaulted on their loans, the financial institution was liable to its depositors for payment. It was therefore in its best interest to carefully screen its potential borrowers' ability to repay before providing them with any loan.

As long as property prices rose, there was always a smooth exit solution for both stressed homeowners and lenders and default rates remained extraordinarily low. Unfortunately, the low interest rate environment also triggered an insatiable appetite for investments with higher potential returns. Since traditional investments were unable to deliver higher yields, investors had to take on bigger risks to obtain better returns. Investing in subprime mortgages then emerged as one of the possible solutions. These loans were given to families or individuals deemed to be less creditworthy, and offered higher expected yields than the prime ones. This resulted in a dramatic increase in subprime home mortgage originations, from 8% in 2001 to 21% in 2005. It is now estimated that approximately 80% of these subprime loans were transferred in one form or another to investors through a process called securitization. Interestingly, according to Thomson Reuters (TRI), the two firms that were the most aggressive in mortgage investments were Bear Stearns and Lehman Brothers. The former issued $100 billion in U.S. mortgage-backed securities compared to an industry total of over $1 trillion in 2006, whilst the latter issued $96 billion out of an industry total of $922 billion in 2007.

In theory, securitization was a great financial innovation. It allowed originators to manage their balance sheets, lower their funding costs, and access additional financing sources. It provided investors with an opportunity to participate in new types of securities with higher promised returns and lower historical risks. At the systemic level, it increased liquidity and facilitated better risk sharing across market participants. In practice, however, securitization opened the door to moral hazard and mutated progressively into an uncontrollable monster. Through securitization, Wall Street firms would buy up risky illiquid assets such as mortgages, bundle them, and sell them to investors—an easy way to earn fees whilst getting rid of the nondesirable associated risks. Moreover, the development of the securitization pipeline led to an explosion of demand for underlying risky assets. Numerous mortgage originators began competing for market share by aggressively pursuing the "originate to distribute" business model. That is, they originated loans with no intention of owning and servicing them. These loans were typically purchased and warehoused by investment banks, with the goal of securitizing and later widely distributing them to investors for a fee, thereby transferring the assets off their balance sheets. As a result, nobody had a real incentive to perform traditional strict credit due diligence. Loose credit standards and dubious lending practices became the norm, since the final investors were too far away from the underlying borrowers to understand what was really happening.

When home prices started weakening, many homeowners also had to face the first batch of interest rate resets. They quickly realized they were unable to refinance at affordable conditions, but the value of their mortgage was higher than the value of their home. In nonrecourse states, an easy solution was to default on their mortgages, walk away from their homes, and hand back their keys. But these first signals of the up-and-coming crisis were largely understated. In the second quarter of 2007, for instance, the IMF Global Financial Stability Report stated that: "… weakness has been contained to certain portions of the subprime market (and to a lesser extent, the Alt-A market), and is not likely to pose a serious systemic threat. Stress tests conducted by investment banks show that, even under scenarios of nationwide house price declines that are historically unprecedented, most investors with exposure to subprime mortgages through securitized structures will not face losses.". On May 17, 2007, Ben Bernanke, Chairman of the Federal Reserve, declared: "We do not expect significant spillovers from the subprime market to the rest of the economy or the financial system." As we all know, despite these reassuring statements, the crisis continued to spread.

In August 2007, two Bear Stearns hedge funds that had invested in subprime-related securities collapsed. This was followed by several mortgage brokers and highly leveraged hedge funds going under. Money market funds broke the buck and had to suspend withdrawals. The disease then inexorably attacked financial institutions, government agencies, corporate borrowers, and even consumers in every major economy worldwide. Mortgage markets disappeared, credit markets seized up, equity markets began to free fall, asset managers experienced life-threatening redemptions, and numerous vulnerable mortgage originators filed for bankruptcy. Asset-backed securities (ABSs) and collateralized debt obligations (CDOs) were massively downgraded by credit rating agencies, triggering off a broad repricing of credit risk. Banks refused to lend to one another. More worryingly, they also refused to lend to businesses and individual customers, removing the money flow which normally lubricates the economy. When retail investors began to mistrust banks, they hid their cash and gold "under the mattress."

The crisis is certainly not yet over, but it is already unprecedented. As a reminder, here are a few examples of some of the unprecedented events observed over the past few months: Fannie Mae and Freddie Mac went into conservatorship, American International Group (AIG) had to be rescued and nationalized, Northern Rock was nationalized, Lehman Brothers went bankrupt, Merrill Lynch was sold to Bank of America, Washington Mutual, Indy Mac, and Wachovia failed, HBOS was taken over, and a 75 year era of independent investment banks ended abruptly when Goldman Sachs and Morgan Stanley were forced to convert into bank holding companies.

As the flight to safety drove U.S. Treasury securities to negative yields, central bankers around the world shifted toward a monetary-easing mindset. In a desperate attempt to restart the banking system engine, they poured record amounts of liquidity into their local financial markets. Governments and regulators intervened on several occasions to facilitate the takeover of banks in difficulties, extend bank deposit protection schemes, take banks into public ownership, and offer to purchase toxic assets with public money. The positive impact, if any, was not really visible. In the United States, for instance, despite more than $8 trillion of government loans, investments, and guarantees, the housing market has still not stabilized, unemployment is rising and net lending continues to

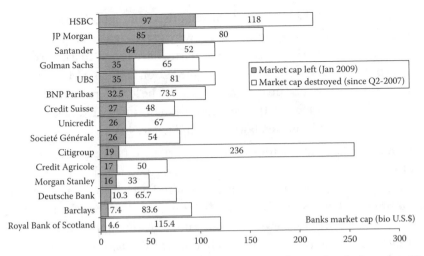

FIGURE 11.1 Evolution of the market capitalization of major banks between Q2 2007 and Q1 2009.

contract rapidly. The "shadow banking" system of securitization remains frozen. The total market capitalization of the world's banks has declined by approximately $1.6 trillion (see Figure 11.1) and the major banks are still teetering on the verge of insolvency. Last but not least, the collapse in value of all asset classes has crushed the net worth of every investor. According to a report from market research firm Spectrem Group, the number of American households with a net worth of $1 million or more fell from 9.2 million in 2007 to 6.7 million in 2008, and the number of affluent households (those with a net worth of $500,000 or more) declined from 15.7 million to 11.3 million.

11.2 WHO IS TO BLAME?

Whenever a financial crisis hits, crowd-pleasing words like "greed," "speculators," and other assorted bogeymen are widely used. Most politicians usually expend a considerable amount of energy searching for scapegoats—a good excuse not to learn from mistakes made. This time has been no different, except that the list of villains has grown tremendously in parallel with the intensity of the U.S. election debates... and the severity of the losses. As retirement dreams fade away, the media and public have become full of outrage at almost everyone with any sort of tie to the financial markets. Let us start by mentioning some of the names that are directly involved with the real estate/subprime crisis:

Greedy lenders. At the root of the current crisis were subprime mortgages, and at the root of the subprime problem were greedy lenders and their brokers. For years, these financial intermediaries have provided easy access to credit because they were primarily rewarded on the basis of short-term volumes. Since they had no direct exposure to the long-term consequences of possible future defaults, they abandoned their underwriting standards, which would have protected borrowers and lenders alike.

Credit rating agencies. Rating agencies contributed to the economic crisis by turning a blind eye to the real underlying risks attached to the securities they graded as "AAA." Their conflict of interest was obvious, as they were paid by securities, issuers only if deals completed. Thus they had a strong incentive to be overly generous in their ratings assignations. Furthermore, rating agencies failed to progressively downgrade these securities as the situation deteriorated. Ultimately, they aggravated the crisis by reacting with violent wholesale downgrades once the market had collapsed.

Home appraisers. Home appraisers were urged by their clients to inflate the value of their homes so that they could obtain more financing. Many of them conceded as this brought them more business. By ignoring intrinsic value, referrals kept coming in. A few honest appraisers petitioned the Congress to intervene in this widespread fraud, but nobody acted.

Congress. In the United States, some politicians have pointed fingers at advocates for affordable housing. In particular, Congress enacted the Community Reinvestment Act (CRA) to encourage banks to meet the credit needs of their local communities, including low- and moderate-income neighborhoods. As a result, irresponsible subprime borrowers were targeted as a lucrative source of income for lenders and were encouraged to borrow, despite it being known from the beginning that such borrowers would never be able to repay their loans.

The second group of frequently mentioned culprits includes financial engineers and their innovative products.

Financial engineers. Financial engineers have often been blamed for inventing financial derivatives and, in particular credit default swaps (CDSs), ABSs, CDOs as well as other financial products. More broadly through securitization, risk was allowed to be sliced up and distributed around the globe. This financial innovation was initially perceived as an efficient way to add some liquidity to a portfolio of illiquid credits and to reduce risk by spreading it across a larger number of market participants.

The reality is that these ever-fancier artificial securities and derivatives resulted in a significant loss of transparency in the global financial system* and weakened the incentives of financial intermediaries to monitor the behavior of borrowers. Many of them were marked to the model rather than the market. This was fine as long as everybody used the same mathematical models for valuations and markets were calm. However, in times of market turbulence, doubts arose concerning the mathematical models themselves and even the underlying assets from which value had been derived. As a result, when the first defaults appeared, sellers attempted to get out of the business altogether but there were no buyers. This killed the price discovery process turning these assets into toxic and illiquid waste on the banks' balance sheets.

Though the banking sector may have caused the financial crisis, regulators should also take some share of the blame.

The Federal Reserve. Many believe that Alan Greenspan, the former Federal Reserve Chairman, is in a way responsible for the crisis, or has at least contributed to it. In response to the post-9/11 recession and the collapse of the new economy bubble, Greenspan dropped federal fund rates to 1% to stimulate economic growth. Some will say "too low for too long." This set off an inflationary spiral in housing and a desperate hunt for yield by fixed-income investors. In addition, low interest rates encouraged banks to target subprime customers with variable rate mortgages and other sweeteners, which turned out to be poisonous a few years down the road.

Congress (again). After years of lobbying by banks, the Congress passed in 1999 a legislation loosening some of the New Deal-era bank rules, including the strict separation between retail banks, insurance companies, and investment banks, which had been established by the Glass–Steagall Act. This put the business model of banks under growing competitive pressure and led banks to take on increasing risk. Suddenly, several massive financial services, conglomerates were formed. Commercial banks migrated toward investment banks' turf and investment banks took on more leverage.

The SEC. Prior to 2004, broker/dealer net-capital rules limited firms to a maximum debt-to-net-capital ratio of 12 to 1. But in 2004, the SEC granted

---

* Wray (2007) quotes the example of large U.S. banks that sold CDOs backed by subprime mortgages to its customers. Many of these CDOs had a put clause allowing their holders to sell them back to the originator at the original value—something that had not been accounted for on the bank's balance sheet.

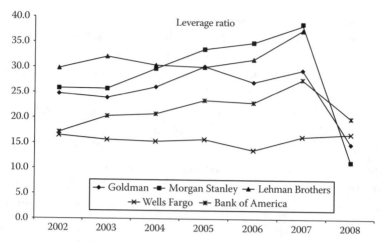

FIGURE 11.2 Evolution of the leverage ratio (adjusted assets divided by tangible shareholders' equity) of selected investment banks in the United States.

five firms (Goldman Sachs, Merrill Lynch, Lehman Brothers, Bear Stearns, and Morgan Stanley) an exemption, allowing them to exceed this leverage rule. They promptly levered up to 20, 30, and even 40 to 1 (see Figure 11.2). We all know the end of the story.

Basel I and II. The Basel I Capital Accord, completed under the supervision of the Bank of International Settlements and endorsed by the Group of Ten central bank governors, encouraged banks to (1) expand risky activities off balance sheet via securitization and (2) increase their exposure to certain types of assets deemed less risky, such as interest rate derivatives, mortgages, and other agency securities.* The new bank regulations under the Basel II accord encouraged the use of portfolio models and credit ratings to assess the riskiness of a portfolio. This strongly reduced the incentive for banks to challenge existing models and ratings, as long as they remained favorable for their capital requirements.

Finally, in Continental Europe, several leaders facing fractious electorates concerned about the stability of public finances, rising unemployment, and social protests are eager for an explanation absolving them of responsibility. They have pointed fingers at different scapegoats.

Executive bonuses. Following a few individual cases of obviously excessive compensation, particularly in the banking and insurance world, proposals

---

* As an illustration, under the Basel I framework, the risk associated with holding Fannie Mae and Freddie Mac debt was 20% of the risk of holding an otherwise identical debt from a high-quality industrial company.

have ranged from raising a massive tax on stock options to placing a cap on executive salaries and temporarily banning bonuses and options.

Offshore financial centers. Although their relationship with the current crisis was unclear, offshore financial centers have also been attacked. Why? According to OECD Secretary General Angel Gurria, the level of assets held offshore by individuals or companies, either to evade taxes or escape the political instability in their home countries, is somewhere between $5 and $7 trillion. Therefore offshore financial centers represent a vast source of potential tax revenue for onshore centers, which could be used to meet the demands of government spending.

Hedge funds. Once again, hedge funds have been blamed for the current meltdown. As an illustration, Franz Müntefering, the German vice-chancellor, repeated his notorious comparison between hedge fund investors and "locusts." In France, Nicolas Sarkozy urged for a new tax on "speculative" money flows and asked openly for the creation of a registry of hedge funds.

Needless to say, we strongly disagree with the latter choice of scapegoat. Attacking hedge funds is tempting, but simplistic and not fundamentally dissimilar to blaming public transport passengers for a bus crash. As discussed by Amenc (2008), their speculative, unregulated, and less transparent nature make them easy targets. In addition, the recently discovered Ponzi scheme run by Madoff has damaged investors' confidence and affected the credibility in hedge fund regulators.* Instead of continually whining about hedge funds and taking aim, politicians and their advisors should at least try to understand the most basic facts about them. As an illustration, let us consider the following three quotes from the French president Nicolas Sarkozy:

- "Who can tolerate a hedge fund buying a company with debts, firing 25 per cent of staff and then reimbursing them by selling it in pieces? Not me." Embarrassing, to say the least. This opinion is a caricatured example of mixing up hedge funds and private equity firms.

- "We want regulation of hedge funds." The idea that hedge funds fall outside the scope of regulation is out of date. First, there is domestic regulation on hedge funds and a lot of it actually comes from the EU. Second, it is an opportune time to remember that banks and prime

---

* Note that as discussed by Gregoriou and Lhabitant (2009), Madoff was in fact not a hedge fund and was a relatively easy case to spot by any serious due diligence process.

brokers control the markets, not hedge funds. Thirdly, we should not forget that it is the most heavily regulated institutions in the world, that is, banks and brokers that have failed during the recent crisis. Shouldn't we therefore focus on better regulating these institutions?

- "[We want]... hedge funds to be registered." It is a fact that most of the largest hedge funds are already SEC or FSA registered since some of their large institutional investors insist upon it. By contrast, most of the smaller funds that are not registered have not done so because they cannot afford the extra burden, hassle, and costs involved. So, unless registration aims at shutting down the smaller players, the usefulness of such a wish needs to be challenged. What is probably needed is hedge fund registration *and* increased disclosure on a regular, timely, and confidential basis to *more sophisticated and better staffed regulators*. But this is another debate.

A more fundamental criticism made to hedge funds concerns their ability to sell short and their use of leverage. It is thought this may have compounded the recent falls in bank share prices. Here again, we believe this criticism is largely unfounded. Several countries implemented short-selling bans in September 2008 with the aim of limiting negative market movements: the United States banned short-selling shares of nearly 1000 companies for almost a month, the United Kingdom banned short selling in shares of 34 commercial banks and organizations holding banking licenses, and Australia banned all short selling. The extent of these bans varied from one country to another; a recent The Alternative Investment Management Association (AIMA) research conducted by the Cass Business School, City University, London, evidenced that these bans had little impact. In fact, in several cases, they caused even more havoc in the underlying shares by restricting their liquidity, widening bid-ask spreads and further pushing down prices, in the absence of the traditional market price discovery mechanisms. The usual culprit in extreme market turbulence, short selling, may not have been a factor on this occasion.

Last but not least, some have argued that hedge funds are highly leveraged investment vehicles and have therefore the potential to destabilize markets. Once again, this claim is not supported by any evidence. First, aside from a few exceptions which all went bust in the first days of the crisis, hedge funds had never been as leveraged as banks. Secondly, all the most recent studies on hedge funds by the Financial Services Authority (FSA), the ECB, Merrill

Lynch, Bernstein Global Wealth Management, Bridgewater, Morgan Stanley Prime Brokerage, and the BIS evidenced average leverage levels (defined as assets divided by equity capital) to be around 2 times at the beginning of the crisis. These levels are now down to no leverage at all.

11.3   TOWARD A SOLUTION

So far, very few people have proposed a solution. If the goal is the preservation of a functional banking system, there are a series of routes that can be or have been somehow explored. Let us mention some of them.

A return to the old fashioned banking model. A simple alternative is to change the regulatory system to limit banks, activities to taking in deposits and lending them out in a safe and controllable way. To paraphrase Warren Buffett, banking needs to be simple enough that an idiot can run it because someday one will. While this sounds like a great solution, the reality is that it cannot be applied today, unless the balance sheet of the banking system is cleared first.

Massively inject public money into financial institutions. Many governments and sovereign wealth funds have purchased newly issued preferred shares of publicly traded bank holding companies. In most instances, these preferred shares can be converted into common shares at some point in time, thereby diluting the shares of outstanding common stock. Dividends, if they ever get paid, are juicy. Nevertheless, the operation remains attractive for banks as these preferred shares can count toward "core" tier 1 capital and thus significantly improve regulatory capital ratios. This solution has already been attempted on several occasions, with very large losses so far for the early buyers of preferred shares. It will therefore be difficult to convince them to go for a second round.

Create a "Bad Bank." Another simple way to clear toxic assets from the books of the banking system is to nationalize these assets at market price. That is, all banks would be forced to disclose their toxic assets, mark them down to current market prices, and sell them to a specially created "bad bank." The "bad bank" would then deal with these assets allowing the cleared banking system to operate normally. Unfortunately, so far, most policy makers seem to have rejected the "bad bank" approach with the exception of Switzerland. A likely reason is that most of the banking system would have gone bust if forced to write down assets at market prices. An alternative would be for the "bad bank" to turn into a really bad bank, buying assets at inflated prices.

Nationalize banks. Governments can inject equity in failed and systemically important banks, write down their shareholder equity to zero, impose haircuts or debt-to-equity conversions on unsecured creditors, recapitalize balance sheets with fresh equity, replace incumbent management, and restructure. They could run the resulting entity on strictly commercial terms with the goal of privatizing it at some point to extract maximum value for taxpayers' money. This would open the door to a dual banking system made of standalone banks free of government influence, and government influenced/controlled banks. The latter category is exposed to political risk whereas the former is exposed to dilution risk and dividend cuts. The key questions then become: whether nationalization is better than failure, whether governments can be more successful than private equity firms in rebuilding banks, as well as how to align the interests of rescued banks and taxpayers and how to prevent a massive competitive disadvantage for banks that managed their affairs more prudently and did not require government help. The difficulty is the magnitude of the problem, which may lead to a full-scale nationalization of most of the banking system.

An alternative solution, which has not yet been widely explored, is government collaboration with hedge funds, private equity funds, and more generally private pools of capital. Let us recall that historically, investment banks have been the very powerful force that shaped industrial markets. They helped finance the railroads of the nineteenth century, the heavy industries of the early twentieth century, and the technology innovators of recent decades. Private pools of capital could play a similar role for financial markets, as (1) they still have lots of available or committed capital; (2) they are not regulated and can, therefore, concentrate their positions in nontraded and noninvestment grade assets; and (3) they are only accessible to sophisticated investors and would, therefore, not put retail investors' money at risk. However, private pools of capital alone are not sufficient given the size of the problem and the asymmetry of information. Some form of action and encouragement by the government is necessary.

In the United States, the recently announced Term Asset-Backed Securities Lending Facility (TALF) program seems to start an exploration in this direction. Under TALF, the Federal Reserve Bank of New York will provide low-cost, nonrecourse funding to any eligible borrower with eligible collateral. Eligible borrowers include any type of investment fund (including hedge funds, private equity funds, mutual funds, etc.) that is organized in the United States and managed by an investment manager that has its principal place of business located in the United States. Hedge funds based outside the United States can still invest in the TALF but they must use a unit organized and managed in the United States. Eligible collateral includes all U.S. dollar-denominated nonsynthetic ABS with a credit rating in the highest long-term or short-term investment-grade rating category. To summarize, TALF will allow for a private pool of capital to borrow money so as to purchase some of the distressed assets from banks, which sit on their balance sheets. Historically, such credit facilities would have been granted by investment banks and prime brokers, but these players are now out of business. Under TALF, the Federal Reserve Bank of New York de facto acts as a giant centralized prime broker for hedge funds and private equity firms. Not surprisingly, the biggest private pools of capital have expressed their interest in the program. The major concerns and discussions seem to focus on the exact rules of the game, how these rules could change retroactively in the future, and the lack of transparency on some aspects of the program. Future taxation of capital gains, a topic that has not yet been openly discussed, is also a source of concern.

In Continental Europe, a TALF-like solution would be seen as a blasphemy and highly unlikely to be considered at any point, as one should never partner with the devils. Governments will not force oversized financial Behemoths to mark down their assets. European private pools of capital will therefore need to either invest in the United States or wait for fatal bankruptcies to occur in order to start looking at opportunities.

11.4 CONCLUSION

The most complex financial crisis since the 1930s has devastated the banking system and is now affecting the real world economy. Throughout the world, governments, regulators, and central bankers have taken steps to reform the financial system, make it "work better," and restore confidence. In this process, some seem to have abandoned market solutions in favor of populist activities like blaming speculators, banning short selling and leverage. Others still have faith in capitalism, despite its recent failures and are looking for private capital-driven solutions to the banking crisis. Provided they act responsibly, hedge funds could have a key role to play in the latter line of thought. Their legal form might evolve but their function is likely to survive.

REFERENCES

Amenc, N. (2008) Three early lessons from the subprime lending crisis: A French answer to president Sarkozy. EDHEC Discussion Paper, Lille, France.

Bernanke, B. (May 17, 2007) The subprime mortgage market. Chairman of the Board of Governors of the U.S. Federal Reserve System.

Gregoriou, G.N. and F.S. Lhabitant (2009) Madoff: A riot of red flags. The *Journal of Wealth Management*, 12(1): 89–98.

CHAPTER 12

Hedge Funds, Financial Leverage, and the 2008 Systematic Crisis: Are They Victims or Killers?

Ruggero Bertelli

CONTENTS

Financial leverage seems to have a huge responsibility in contributing to systemic crises. This chapter determines the maximum level of financial leverage that each of the hedge-fund strategies is capable of supporting ex-ante. If there are a large number of hedge funds that use leverage to a maximum level, the risk of systemic crisis arises and the absolute return objective may be in jeopardy.

12.1 INTRODUCTION

In 2008, a great deal of debt was used to magnify returns on investments, which eventually became dangerous for the stability of the global financial system. Capital requirements are imposed on banks, but the amount of leverage used by the nonregulated "shadow banking system" is a gray area, where financial intermediaries such as hedge funds helped to destabilize the world markets. How much leverage do hedge funds actually use? Obviously, it is difficult to answer this question, given the extreme differentiation among hedge-fund strategies or funds. The report of the President's Working Group on Financial Markets stated that LTCM leveraged their capital to as much as 28 times in 1997 and 1998. However, in 2004, the IMF stated that "since 1998, credit providers and hedge funds have developed a better understanding of leverage and, broadly speaking, hedge fund leverage is at relatively moderate levels today. At present, many equity hedge funds report leverage typically less than two times capital, and other styles and strategies are similarly reporting leverage at or below historical norms" (IMF, 2004, p. 50).

Table 12.1 summarizes some descriptive statistics for certain hedge-fund strategies, and a general overview of the leverage is examined in this chapter.

Obviously, the leverage used by hedge funds that had failed is completely different. For the period of June 2007–August 2008, the asset-weighted leverage of large fixed-income hedge-fund failures was 16 times (ratio of asset to equity capital) (IMF, 2008). Lowenstein (2001) affirmed that LTCM registered the "fantastic figure" of a 100–1 leverage. In general, it is not correct to conclude that hedge funds failed only because of the excess use of leverage (see Christory et al., 2007).

For example, the bank core capital ratio in the EURO area (i.e., the ratio of a bank's core equity capital to its total risk-weighted assets) in 2007 ranged from a minimum of 6.50% (risk-weighted assets are 15.38 times

TABLE 12.1   Data on the Degree of Leverage by Certain Hedge-Fund Strategies

	IMF 2004 Data (1997–2003 Period)			Schneeweis et al. (2005). (January 2000–March 2003 Data)			IMF (2008) (Typical Range)	
	Average	Max	Min	Average	Max	Min	Max	Min
Convertible arbitrage	3.0	7.0	1.0	3.0	5.0	1.0	9.0	5.0
Equity hedge	1.4	20.0	0.7	1.2	10.0	1.0	2.0	1.5
Event driven	1.4	10.0	1.0	1.2	3.0	1.0	na	na
Distressed securities	1.2	3.0	1.0	1.0	2.4	1.0	na	na
Merger arbitrage	1.6	10.0	1.0	1.3	2.8	1.0	2.0	1.5
Equity market neutral	1.8	3.0	1.0	2.0	3.5	1.0	na	na
Macro	2.4	5.0	1.0	na	na	na	4.0	4.0
Fixed income arbitrage	2.0	12.0	1.0	na	na	na	10.0	10.0

the equity capital) to a maximum of 10.70% (9.35 times) (see ECB, 2008; the data refers to "large and complex banking groups"). It is important to consider that the capital ratio refers to the risk-weighted assets and not the total asset, so that the ratio underestimates the actual leverage of banks. The ECB (2008, p. 45) reported evidence of "a rapid and widespread deleveraging in the hedge fund sector. The share of surveyed hedge funds reporting that they were leveraged less than 1× (i.e., that their gross investments did not exceed capital) reached record levels in September and October 2008." Is this de-leveraging process physiological, that is, consistent with the financial market conditions or is it a result of excess (not justified, morally hazardous) use of leverage by hedge funds? "In times of stress, achieving call option-like returns, some hedge fund managers may opt to reduce risk-taking and instead concentrate on capital preservation. This seems to be precisely what many hedge fund managers have been doing since the start of the recent turmoil, either voluntarily or due to cuts in financing extended by banks. The reduced availability of leverage, and the cautious attitudes of hedge fund managers that are justified at the fund level, are detrimental to the functioning of financial markets, because they imply asset sales and deprive markets of their most active participants" (ECB, 2008, p. 45). In these conditions, the combination of decreasing asset values, higher volatility, rising collateral haircuts, and investor redemption have resulted in an increasing frequency of hedge-fund failures (IMF, 2008).

The aim of this chapter is to demonstrate that the increasing hedge-fund default risk in recent months is due to the behavior of the lenders, who had to reduce their credit activity in an extremely short period of time, without the possibility of discriminating between the borrowers. "In fact, there is very little evidence to suggest that hedge funds caused the financial crisis or that they contributed to its severity in any significant way" (Brown et al., 2009, Chapter 6: Executive Summary). During the first half of 2008, the leverage employed by the majority of hedge funds was not excessive; instead, they were reducing the exposure when the turmoil in the banking sector intensified.

12.2 DE-LEVERAGING HEDGE-FUND STRATEGIES

12.2.1 Iterative Procedure

The objective of this procedure is to obtain a risk–return evaluation of the hedge-fund strategies that consider only the "business risk" (asset volatility) of the strategy. The effect of "financial risk" (i.e., the impact on risk and rewards of the use of leverage) used by the hedge-fund manager is discounted as a separate source of return and a risk management tool (Ineichen, 2007). This approach is significantly the same as in McGuire and Tsatsaronis (2008, p. 4), whereby the authors state that "….leverage can be thought of, broadly, as increasing the sensitivity of portfolio returns to underlying risk factors." The level of financial leverage actually used by the hedge-fund managers is not known and we have used a procedure to understand this level from the market data on interest rates and hedge-fund net asset values.

We start from this simple relation that connects the return on equity of the hedge fund (hence considering leverage) (re) and the return on the assets of the HF (ra):

$$re_t = ra_t + [ra_t - (i_t + sp_t)](\lambda_t - 1) \tag{12.1}$$

where
  i is the free-risk rate
  sp is the credit-risk spread
  λ is the leverage, that is, the total asset divided by equity capital
  t represents each time (month) of the period from 1 to N

We would like to determine the λ_t values given re_t, i_t, sp_t so that the Sharpe ratio of the hedge fund (she_{t-m}) is equal to the Sharpe ratio of the de-leveraged hedge fund (sha_{t-m}), where m is the number of the month that we include in the historic calculation of the Sharpe ratio.

$$she_{t-m} = \frac{rem_{t-m} - im_{t-m}}{\sigma e_{t-m}} \tag{12.2}$$

$$sha_{t-m} = \frac{ram_{t-m} - im_{t-m}}{\sigma a_{t-m}} \tag{12.3}$$

In Equations 12.2 and 12.3, the letter m after the symbol indicates the average of the returns of the period considered and σ refers to the standard deviation of returns on equity and assets. Equation 12.1 derives what is required to estimate the spread, by adding to the risk-free rate, to obtain the cost of debt. The hedge-fund industry operates with an interest rate close to the risk-free rate, because hedge funds employ implicit funding available in the derivative markets, as well as use initial margins and lines of credit from prime brokers who are mostly interested in brokerage fees, instead of the credit risk of operations.

We estimate an implicit maximum level of spread given the market conditions. When this implicit spread is higher, it implies that the hedge-fund managers can build leveraged positions that have a Sharpe ratio higher than the unleveraged position, if they can obtain credit at a cost below that implicit spread. In the opposite case (implicit spread close to zero), operators cannot build leveraged position without a reduction in the Sharpe ratio below the Sharpe ratio of the unleveraged position.

We have used the same Equation 12.1 and the same procedure described in a control case, comprising only a long non-leveraged position, and have searched for the level of spread (sp_t) that assures that (λ_t) equals to 1. The relevant control case chosen is the S&P 500 index, assuming that this index efficiently reflects the general situation of the financial markets.

To obtain the implicit sp_t from the S&P 500 and the λ_t for each hedge-fund strategy, we have used an iterative procedure, where the Sharpe ratios have been calculated over a 36 month period. Given these two variables and the monthly hedge-fund returns, it is possible to estimate the net monthly returns of the de-leveraged hedge-fund strategy (i.e., return on assets). The variable λ_t represents the minimum level of leverage that is necessary to obtain a Sharpe ratio of the hedge fund at least equal to the Sharpe ratio of the unleveraged strategy.

12.2.2 Data

The procedure described was applied to 11 hedge-fund strategies using the HFRI (Hedge Fund Research Index) indices from January 1990 to December 2008. The risk-free rate used was the LIBOR US Interbank 1 month offered rate. LIBOR rates are usually fixed on every business day by the British Bankers' Association. The LIBOR rate represents the rate at which each bank on the panel could borrow Eurodollars from other banks, for specific maturities. We preferred the Eurodollar Interbank rate to determine an international actual break-even bank rate.

12.2.3 De-Leveraging Procedure Results

The implicit spreads that stem from the S&P 500 data have registered a range from 0 to 37 basis points.

Figure 12.1 shows the inverse relation between the volatility of the market and the implicit spread calculated. It is clear that the high volatility during the 1999–2004 period makes it difficult to improve the risk-reward ratio of an unleveraged strategy using debt. During the low volatility environment of 2005–2007, it was ideal for the hedge funds to use leverage to increase the efficiency of the investment strategies. During the crisis of 2008, high volatility caused the implicit spread to decrease to zero: in this

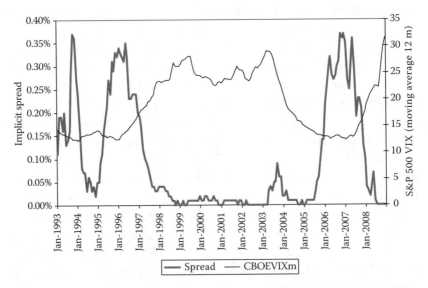

FIGURE 12.1 Implicit spread and the S&P 500 VIX. The figure shows the inverse relation between the volatility conditions of the markets and the maximum spread that allows constructing a leveraged position with at least the same Sharpe ratio.

case, it was not possible to maintain a leveraged position and improve the strategy risk-reward ratio.

For each strategy, as represented by the HFRI indices, we applied the de-leveraged procedure described in Section 12.2.3. The de-leveraged process is graphically depicted in Figure 12.2 for the HFRI Fund Weighted Hedge Fund Index (HFRIFWC).

Figure 12.3 shows the minimum level of leverage of each month that is necessary to obtain an ex-post 36 month Sharpe ratio, identical both in the case of the original strategy and the de-leveraged HFRI strategy. It is worth noting that the level of leverage is extremely variable, from 1 (no debt) to 40. When the strategy uses a level of leverage below this minimum, the ex-post efficiency (measured by the Sharpe ratio) of the strategy is lower than the ex-post efficiency of the same strategy without leverage. It is clear that when the minimum level is 1, even a relatively low use of debt can improve the risk-reward ratio.

Hedge funds used a great deal of leverage in the second half of 2007, until the first half of 2008; however, from the second half of 2005 to the first half of 2007, they were not required to use much leverage. The complete results for the 11 strategies are summarized in Table 12.2.

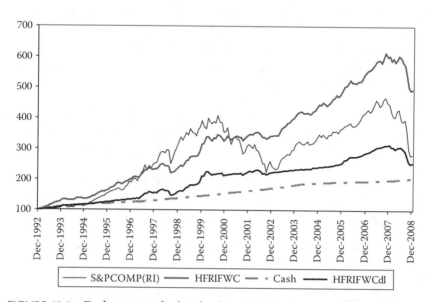

FIGURE 12.2　De-leveraging hedge-fund index. This figure compares the original HFRI INDEX and the transformed de-leveraged index (HFRIFWCdl), S&P 500, and CASH investments are shown along with benchmark.

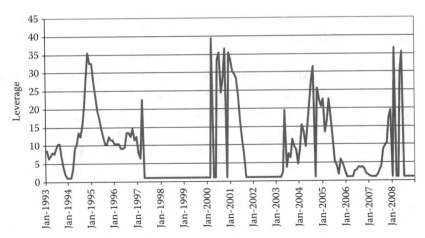

FIGURE 12.3 Leverage of the hedge-fund index. This figure shows the monthly level of leverage resulting from the de-leveraging procedure used. This level must be considered as the minimum level of leverage that is required to obtain an identical Sharpe ratio of the strategy and of the de-leveraged strategy.

12.2.4 De-Leveraging Procedure Remarks

Table 12.2 demonstrates that the performance of the hedge-fund industry during the last 15 years heavily depended on the use of leverage.

1. No de-leveraged strategy outperformed the S&P 500 Index in terms of absolute performance. However, 6 out of 11 leveraged strategies outperformed the market.

2. Using the Sharpe ratio as a measure of efficiency, six de-leveraged strategies were found to outperform the market, but 11 out of 11 leveraged strategies showed a better Sharpe ratio than the S&P 500.

3. Three relative-value strategies used leverage to obtain positive Sharpe ratios, where the de-leveraged strategies registered negative Sharpe ratios.

4. The average of the minimum leverage appears significantly high for all strategies, from three times to more than nine times the equity capital.

The hedge funds require leverage to affirm their role in the asset management industry. This consideration induces to explore the systematic risks of the use of leverage, especially when the financial intermediaries themselves are involved in a de-leveraging process.

TABLE 12.2   Results of the De-Leveraging Process for 11 Hedge-Fund Strategies

HFRI Indices (Jan. 1993–Dec. 2008)	re	ra	σε	σα	SHE	SHA	Leverage
HFRI fund weighted hedge fund	10.27%	6.00%	7.24%	5.99%	0.84	0.31	9.10
HFRI fund of funds composite	6.83%	4.71%	6.34%	5.54%	0.42	0.10	6.71
HFRI equity hedge (total)	11.57%	7.36%	9.28%	7.51%	0.80	0.43	8.61
HFRI equity hedge: Quant. directional	12.25%	5.50%	13.38%	11.55%	0.60	0.11	8.02
HFRI equity hedge: Equity market neutral	6.86%	5.88%	3.33%	2.90%	0.81	0.59	3.90
HFRI macro (total)	11.70%	5.53%	7.43%	4.50%	1.01	0.30	9.20
HFRI relative value: Fixed income	6.02%	**3.27%**	6.64%	6.08%	0.28	−0.15	7.37
HFRI event-driven (total)	11.02%	6.00%	6.89%	5.58%	0.99	0.33	8.37
HFRI relative value (total)	8.58%	5.99%	4.38%	3.64%	1.01	0.50	5.92
HFRI relative value: Fixed income-corporate	5.87%	**2.44%**	5.59%	4.63%	0.30	−0.37	7.07
HFRI relative value: Multistrategy	6.32%	**4.14%**	4.31%	3.96%	0.50	−0.01	3.02
Cash		4.17%		0.50%			
Cost of debt (risk-free rate + spread)		4.27%		0.51%			
S&P500		**7.57%**		**14.58%**		**0.23**	

Note: re, annual average return of the original strategy; ra, annual average return of the de-leveraged strategy; σε, annualized standard deviation of the original strategy; σα, annualized standard deviation of the de-leveraged strategy; SHE, ex-post Sharpe ratio for the original strategy; SHA, ex-post Sharpe ratio for the de-leveraged strategy; leverage, average of the monthly minimum leverage. In bold case, the returns below the free-risk investment (cash).

12.3   MEASURING THE FINANCIAL RISK OF HEDGE-FUND STRATEGIES

12.3.1   Probability of Default in a Merton Approach

The de-leveraging procedure provided us with the volatility (output) of the de-leveraged strategy (σa_{t-n}, where n is the number of months we used to calculate the volatility). Given the level of leverage for each month (λ_t) and starting from the net asset value of the hedge-fund index, we can easily calculate the market value of the asset for each month with these inputs.

We have applied inputs to the Merton model (Merton, 1974) to determine the default probability of the strategy, given the business risk (asset volatility) and the financial risk (degree of leverage). The meaning of this probability is understandable considering that the leverage determined is the minimum degree of leverage for achieving an ex-post efficiency (measured by the Sharpe ratio) equal to that of the unleveraged strategy. Hence, by using a Merton approach, we can obtain a time series of default probability (DP_t). To apply the Merton model, we have used a 1 year default probability on a 1 year maturity debt, as a proxy of the actual default probability that stems from a complex structure of debt that deals with a "mark-to-market."

Figure 12.4 is an example of the default probability associated with the HFRI, and can be interpreted as a general index of default risk of the hedge-fund industry. It is worth noting that this probability is a function of both the business risk and the financial risk. Hence, if the minimum leverage calculated is close to 1, then the default probability is close to 0, for any value of volatility. Accordingly, from July 2008, the default probability in Figure 12.4 is close to zero (following the 10.78% level of June 2008), which explains the "success" of the dramatic de-leveraging process that hedge funds have operated according to IMF (2008).

From the Standard and Poor's Global Fixed Income Research, the 5% 1 year default probability has to be considered as the threshold that divided the investment-grade and speculative-grade credit positions. Thus, we analyzed the DP_t time series for each strategy by searching for the maximum value of the default probability and recording the date in which the

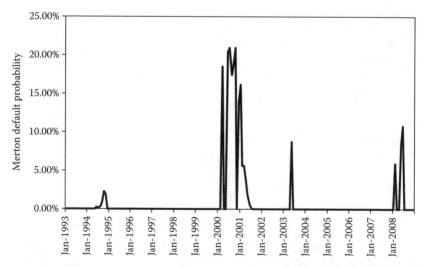

FIGURE 12.4   Merton default probability of the HFR hedge-fund index.

value has been reached. If the value exceeded the 5% threshold, then we considered this as a real danger for the solvency of the hedge funds.

During the period from January 2000 to August 2001, equity strategies, macro strategies, event-driven strategies, and the fund of hedge-fund strategy presented an extremely dangerous default probability (a speculative-grade risk) in the same period (Table 12.3). Surprisingly, the 2007–2008 period did not appear as a dangerous period according to the data of this table. The explanation comprises the combination of business risk (volatility) and financial risk that has been defined as the minimum degree of leverage. The 2000–2001 period was characterized by a relatively high level of the default probability, because, while the hedge funds needed leverage, the market volatility increased (see Figure 12.1). However, in the second half of 2008, hedge funds did not require leverage (according to our definition of minimum leverage), while the volatility was dramatically increasing.

This does not mean that they were not actually using leverage. If they actually used leverage (needed or not), then what were the risks for the hedge-fund industry and the banking system?

12.3.2 Defining the Maximum Leverage of Hedge-Fund Strategies

Using the Merton approach to examine the default probability, the objective is now to determine the maximum leverage of the strategies coherent with a given default probability level. It is clear that this level depends

TABLE 12.3   HFR Index Maximum Default Probability

HFRI Index	PROBDEF$_{max}$	Date
HFRI fund weighted hedge fund	**20.97%**	31/10/2000
HFRI fund of funds composite	**19.59%**	30/06/2000
HFRI EQUITY HEDGE (TOTAL)	**24.42%**	31/07/2000
HFRI EH: Quant. directional	**37.92%**	31/08/2001
HFRI EH: Equity market neutral	2.57%	30/11/1999
HFRI macro (total)	**14.49%**	31/01/2000
HFRI RV: Fixed Inc.	0.17%	31/01/1995
HFRI event-driven (total)	**5.28%**	31/07/2000
HFRI relative value (total)	1.09%	31/03/2008
HFRI RV: Fixed Inc.–corporate	1.74%	31/10/1994
HFRI RV: Multistrategy	0.00%	30/11/1994

Note: This table presents the maximum default probability for each strategy according to the Merton model during the period from January 1993 to December 2008. The last column of the table presents the date in which the value has been reached. The probabilities in bold represent the value above the 5% threshold, that is, the speculative grade default probability according to S&P data.

on the evolution of business risk of the strategy. The idea is to calculate a constant leverage value maintained from 1993 to 2008, so that the default probability reaches a definite threshold at least in 1 month. If the manager has leverage beyond this level, then the default risk is not acceptable and the lenders have to ask for de-leveraging intervention.

The maximum leverage is determined by historical simulation. At the start of the period (January 1993), we applied a given level of leverage, and then for each month, we calculated the default probability using the Merton model. For each month, the inputs are the given leverage (constant) and the asset value, from which it is possible to determine the amount of debt (defined in very general terms as the difference between the total assets and the net asset value of the fund). The asset value is obtained from the return on the asset time series gathered from the de-leveraging procedure. The same procedure allowed us to calculate the asset volatility time series. Finally, we used the risk-free rate time series as an input of the model. In the Merton approach, the default probability is the probability associated at the event exercise of the put option written on the asset value with strike price equal to the nominal value of debt at maturity.

Using an iterative process, we determined the leverage (i.e., maximum) that returns a given level of default probability in at least 1 month of the period examined. We chose the 5% threshold associated with the 1 year default probability of the speculative-grade corporate bonds (according to the S&P data that we used in Table 12.3). Once this level of leverage is found, we obtained a default probability time series associated with this maximum level of financial risk. For the HFRIFWC, the result is shown in Figure 12.5. The so-called maximum leverage is at 6.73 times level, thus obtaining the time series of default probability.

Figure 12.5 is different from Figure 12.4, because in the latter figure, the default probability was calculated considering the "minimum" leverage, that is, the level of leverage under which the strategy registered an ex-post Sharpe ratio equal to that of the de-leveraged strategy. However, the former utilized the "maximum" leverage, that is, the leverage that could have been actually used without a perception of a concrete default risk by the lenders (measured by the threshold above which the rating should be set to "speculative").

Our example (which refers to a general index) demonstrates that in April 1999, lenders should have requested a reduction of the financial leverage used by hedge funds, given that the 5% probability of default had been reached.

Figure 12.5 shows that from June 2003 to August 2008, lenders did not believe that it was risky to provide funding to hedge funds. Meanwhile,

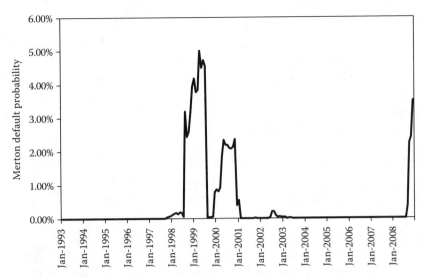

FIGURE 12.5 Merton default probability of hedge-fund index with leverage = 6.73 times. The figure shows the default probability of the hedge-fund index at the given 6.73 level of leverage. This leverage is considered "maximum" because this stated leverage in conjunction with the asset volatility of the strategy determined a 5% default probability in at least 1 month of the period (in this case, April 1999, when the volatility of the markets peaked—see Figure 12.1).

Figure 12.3 shows the level of minimum leverage (i.e., of financial risk). From June 2003 to September 2005, the minimum leverage reached extremely high values of 20–30 times. However, the business risk of the strategy did not imply a concrete risk of default. In the first part of 2008, the minimum leverage returned to high levels (beyond 35 times) and lenders felt the risk of default associated with the increasing business risk of the strategy. Table 12.4 shows the results of our "maximum" leverage exercise for each strategy.

For 8 strategies out of 11, the 5% default probability was reached in December 2008. This implies that lenders had to ask for a de-leveraging process to reduce the concrete default risk that they were experiencing.

12.4 CONCLUDING REMARKS ABOUT THE 2008 DE-LEVERAGING PROCESS AND THE DEFAULT RISK OF HEDGE FUNDS

Table 12.4 demonstrates that 2008 was characterized by a historically significant level of default probability for all hedge-fund strategies. Changes in the financing conditions and risk factors in 2008 affected the performance of the fixed-income strategies more than other strategies. There

TABLE 12.4   Maximum Leverage for Hedge-Fund Strategies

HFRI Index	LEVAMAX	PROBDEF	Date
HFRI fund weighted hedge fund	6.73	5.00%	30/04/99
HFRI fund of funds composite	7.15	5.00%	31/12/08
HFRI equity hedge (total)	5.14	5.00%	31/12/08
HFRI EH: Quant. directional	3.32	5.00%	30/11/00
HFRI EH: Equity market neutral	13.05	5.00%	31/12/08
HFRI macro (total)	10.69	5.00%	31/10/97
HFRI RV:Fixed Inc.	3.61	5.00%	31/12/08
HFRI event-driven (total)	6.81	5.00%	31/12/08
HFRI relative value (total)	6.79	5.00%	31/12/08
HFRI RV:Fixed Inc.–corporate	5.87	5.00%	31/12/08
HFRI RV:Multistrategy	6.49	5.00%	31/12/08

Note: This table shows the maximum level of leverage for each hedge-fund strategy, given the 5% default probability used as a threshold. The maximum leverage is the level that determines a 5% default probability in at least 1 month.

was also a differentiation in the leverage registered among the strategies and the hedge-fund industry in general (IMF, 2008). Figure 12.6 examines four hedge-fund strategies during the period of 2006–2008. The figure shows that the minimum leverage reached the highest value at the end of 2007. In 2008, the minimum leverage dropped to 1 for all hedge-fund strategies with the exception of relative-value hedge funds.

Using our approach (during the crisis), it was not necessary for hedge funds to maintain a high degree of leverage to obtain good risk-reward ratios. Superior returns in the hedge-fund industry depended heavily on the use of leverage (see Table 12.2). Owing to excessive leverage, hedge funds witnessed poor returns in 2008. In addition, their default probabilities started to increase from September 2008, reflecting the increase in the business risk of hedge-fund strategies and not the use of debt. Our estimation of the default probability stems from a given leverage (that we called maximum level) and the asset volatility.

The correlation among the default probabilities during 2008 demonstrates that diversification among strategies did not work. While the average value of correlation was 0.9586 in 2008 (this number is the simple average of correlations in Table 12.5), it was 0.2468 in 2007 (Table 12.6). We calculated the same average in 2006 and 2005: 0.1563 and 0.2269, respectively (correlation matrices are not showed for brevity). The above correlations indicate that the default probability of all hedge-fund strategies have

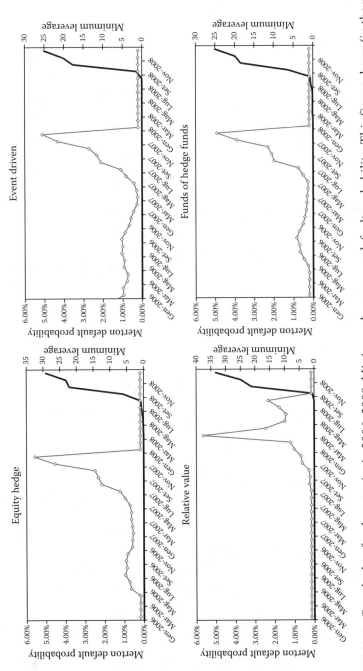

FIGURE 12.6　Four hedge-fund strategies in 2006–2008: Minimum leverage vs. default probability. This figure shows (in the right scale) the minimum leverage of equity hedge, relative-value, event-driven, and funds of hedge-fund strategies. The default probability (left scale) is represented by the solid line. In all the four cases, the leverage reached a maximum value at the end of 2007 or in the first month of 2008, while the default probability started to increase in the second half of 2008.

TABLE 12.5 Correlation Matrix Among the 2008 Default Probabilities

	HFRIF WC	HFRIF FP	HFRIE QH	HFRIN HD	HFRIE MN	HFRIM AC	HFRIC OA	HFRIE VD	HFRIR VR	HFRIFI H	HFRIFI T
HFRIFWC	1										
HFRIFFP	0.991	1									
HFRIEQH	0.994	0.999	1								
HFRINHD	0.995	0.974	0.978	1							
HFRIEMN	0.881	0.837	0.835	0.919	1						
HFRIMAC	0.919	0.875	0.876	0.950	0.990	1					
HFRICOA	0.992	0.988	0.993	0.978	0.820	0.871	1				
HFRIEVD	0.996	0.989	0.993	0.984	0.842	0.890	0.999	1			
HFRIRVR	0.996	0.983	0.988	0.991	0.866	0.912	0.995	0.998	1		
HFRIFIH	0.994	0.984	0.989	0.985	0.846	0.896	0.997	0.999	0.999	1	
HFRIFIT	0.998	0.987	0.991	0.990	0.861	0.907	0.996	0.999	1.000	0.999	1

Note: This table shows the correlations in the following order: HFRI weighted hedge fund (HFRIFWC); fund of funds composite (HFRIFFC); HFRI equity hedge (total) (HFRIFFP); HFRI equity hedge: quant. directional (HFRINHD); HFRI equity hedge: Equity market neutral (HFRIEMN); HFRI Macro (total) (HFRIMAC); (HFRIEQH); HFRI relative value: Fixed income convertible arbitrage (HFRICOA); HFRI event driven (total) (HFRIEVD); HFRI relative value (total) (HFRIRVR); HFRI relative value: Fixed income-corporate (HFRIFIH); and HFRI relative value: Multistrategy (HFRIFIT).

TABLE 12.6 Correlation Matrix Among the 2007 Default Probabilities

	HFRIF WC	HFRIF FP	HFRIE QH	HFRIN HD	HFRIE MN	HFRIM AC	HFRIC OA	HFRIE VD	HFRIR VR	HFRIFI H	HFRIFI T
HFRIFWC	1										
HFRIFFP	0.953	1									
HFRIEQH	0.893	0.716	1								
HFRINHD	-0.176	-0.187	-0.129	1							
HFRIEMN	0.282	0.406	0.056	-0.207	1						
HFRIMAC	-0.127	-0.136	-0.093	0.360	-0.150	1					
HFRICOA	-0.136	-0.145	-0.100	0.923	-0.160	-0.027	1				
HFRIEVD	-0.127	-0.136	-0.093	0.897	-0.150	-0.091	0.998	1			
HFRIRVR	-0.216	-0.230	-0.158	0.138	-0.254	0.466	-0.044	-0.074	1		
HFRIFIH	-0.134	-0.143	-0.099	0.919	-0.158	-0.037	1.000	0.999	-0.049	1	
HFRIFIT	-0.145	-0.155	-0.106	0.949	-0.171	0.046	0.997	0.991	-0.010	0.997	1

Note: This table shows the correlations in the following order: HFRI weighted hedge fund (HFRIFWC); fund of funds composite (HFRIFFP); HFRI equity hedge (total) (HFRIEQH); HFRI equity hedge: quant. directional (HFRINHD); HFRI equity hedge: Equity market neutral (HFRIEMN); HFRI Macro (total) (HFRIMAC); HFRI relative value: Fixed income convertible arbitrage (HFRICOA); HFRI event driven (total) (HFRIEVD); HFRI relative value (total) (HFRIRVR); HFRI relative value: Fixed income-corporate (HFRIFIH); and HFRI relative value: Multistrategy (HFRIFIT).

started to increase since September 2008, and the finding can be interpreted in two different ways. The first and more simplistic one is based on the sudden increase in the volatility of the markets, whereby all hedge-fund strategies suffered from this dramatic increase in volatility, which caused an increase in their "business risk," that is, the volatility of their de-leveraged investment strategy. Increasing default probability explains why "margin financing from prime brokers has been cut and haircuts and fees on repo financing have increased" (IMF, 2008, p. 41). In other words, the business risk of the strategy became so high that banks were forced to ask hedge funds for an immediate de-leveraging process. The second explanation is more intriguing, whereby lenders asked hedge funds for immediate de-leveraging because of their specific liquidity problems (see Adrian and Shin, 2008; ECB, 2008), independently from the concrete default risk of hedge funds. These unjustified requests explain the sudden increase in correlation of a hedge fund's default probability.

In conclusion, it is true that the hedge funds used and actually need leverage to support their competitive advantage in the asset management industry. This peculiarity did not cause and did not amplify the 2008 financial crisis. The dramatic—as well as correlated among the different strategies—increase in their default probability is mostly explained by the requests of the lenders for an immediate de-leveraging in one of the most difficult market environment since 1929.

REFERENCES

Adrian, T. and Shin, H. S. (2008) Liquidity and leverage. Working Paper, Federal Reserve Bank of New York Staff Reports, no. 328, May revised January 2009, New York.

Brown, S. et al. (2009) Hedge funds in the aftermath of the financial crisis. In: Acharya, V. V. and Richardson, M. (Eds.), *Restoring Financial Stability*, NYU Stern White Papers, Wiley: New York.

Christory, C., Daul, S., and Giraud, J. R. (2007) Quantification of hedge fund default risk. Working Paper, EDHEC Risk and Asset Management Research Centre, Nice Cedex 3, France.

ECB (2008) Financial stability review, December. European Central Bank, Frankfurt am Main, Germany.

IMF (2004) Global financial stability report, September. International Monetary Fund, Washington DC.

IMF (2008) Global financial stability report, October. International Monetary Fund Washington DC.

Ineichen, A. M. (2007) *Asymmetric Returns*. John Wiley & Sons: Hoboken, NJ.

Lowenstein, R. (2001) *When Genius Failed*. Random House: New York.

McGuire, P. and Tsatsaronis, K. (2008) Estimating hedge fund leverage. Working Paper, Bank of International Settlement, no. 260, September, Basel, Switzerland.

Merton, R. C. (1974) On the pricing of corporate debt: The risk structure of interest rates. *Journal of Finance*, 29(2): 449–470.

Schneeweis, T., Martin, G. A., Kazemi, H. B., and Karavas, V. N. (2005) The impact of leverage on hedge fund risk and return. *The Journal of Alternative Investments*, 7(4): 10–21.

CHAPTER 13

Evaluation of Evidence for Banking Equity Market Volatility in the Emerging Economy of China

Jack Penm and R.D. Terrell

CONTENTS

This chapter applies a new time-series approach—alternating ARCH modeling—to a specific bank equity market typical of emerging economies. Any financial and banking crisis within emerging markets may well lead to rapid and widespread contagion to other financial markets and banking sectors throughout the world. The outcomes of our research indicate that our alternating structure-ARCH (AS-ARCH) specification offers a better statistical fit than other time-series approaches. The findings confirm the causal relationship between bank stock return volatility and Shanghai's business cycle.

13.1 INTRODUCTION

Modern finance theory and empirical evidence both suggest that financial volatility might embed information about future economic activity. However, previous studies focus on testing the relationship between equity volatility and the business cycle within developed markets, with significantly less attention being given to modeling of equity volatility and the business cycle within emerging markets. Within the global economy as a whole, emerging market economies are becoming increasingly important. Any financial and banking crisis within emerging markets may well lead to rapid and widespread contagion to other financial markets and banking sectors throughout the world. Thus, a clear understanding of the dynamics of the macroeconomic and financial variables within emerging markets will be significant and valuable for the developed markets.

With a total of about $1.15 trillion market capitalization in 2008, China's banking stock group becomes the most dominant industry sector in Shanghai, which ranks in first place within the emerging stock markets. China's banking stock group in Shanghai is chosen as the market for close examination in this study. This study examines the dynamic relationship between bank equity volatility, system shifts, and the business cycle in Shanghai's bank stock market. The data are obtained from the China Economic Databases (CED), which comprise monthly bank stock returns of the Shanghai Stock Exchange. The causality between bank stock return volatility and Shanghai's business cycle is strongly evident.

Due to the inherent characteristic of greater fluctuations within emerging stock markets, standard GARCH models are able to determine the evolving nature of volatility; however, they fail to capture structural shifts in data that are caused by low probability events, such as the 1997–1998

Asian financial crisis. The alternating structure ARCH (AS-ARCH) model suggested by Edwards and Susmel (2001) allows the conditional variance to experience any jumps between discrete states or systems. This is consistent with the popular conception of underlying states within an economy, such as high- and low-volatility states in the stock market. Moreover, the leverage effect can also be incorporated into AS-ARCH to indicate the nature of the asymmetric volatility–return relationship.

This study examines the dynamic relationship between bank equity volatility, system shifts, and the business cycle in Shanghai's stock market. Our findings confirm that the AS-ARCH specification, compared to the use of other specifications, offers a better statistical fit. The causality between stock return volatility and Shanghai's business cycle is strongly evident.

13.2 MORTGAGE BANK TURMOIL, INCLUDING THE UNITED STATES AND THE UNITED KINGDOM SUBPRIME LENDING CRISES

In the current bank mortgage market, the prime mortgage market and the subprime mortgage market are two major segments. Borrower costs associated with prime lending are predominantly determined by the capacity to make a down payment with a satisfactory credit history. However borrower costs associated with subprime lending are essentially not based on such requirements. The subprime mortgage market lends money to people who do not meet the credit or documentation standards for prime mortgages. Compared to conventional prime lending, subprime lending can most easily be characterized as high risk lending. Since subprime borrowers often have credit problems or low incomes, there is a greater chance that they will be unable to pay back their debts, making subprime lending inherently risky for lenders. Consequently, banks and other lenders must charge higher interest rates on subprime mortgages in order to compensate for these risks, including default and delinquency. Over the past decade, these rates have been about 2% higher than prime interest rates, making subprime lending potentially very lucrative.

Subprime mortgages lie at the centre of recent U.S. banking turmoil in housing and credit markets. The subprime industry rapidly expanded in the mid- and early-1990s, when the housing sectors in major industrialized countries were experiencing a seemingly never-ending housing boom. Historically, banks and other lenders considered the lending risks to be too substantial to issue large amounts of subprime lending. This housing boom changed this consideration, leading banks and other lenders to neglect the hidden lending risks behind those subprime loans. Banks and other lenders initiated subprime lending in considerable amounts and subprime mortgages in significant numbers. In order to attract more borrowers, banks and other lenders created several new subprime lending approaches. Many subprime borrowers got loans that allowed them to make no down payment or to initially pay interest only. If the seemingly never-ending housing boom could remain, subprime lending borrowers with lower credit rating and lower income would be able to refinance their homes or simply sell them when they had difficulties keeping up with their payments. If the housing market slipped into a slump, and interest rates were rising, those strained subprime lending borrowers could hardly refinance or even sell their homes to clear their debts.

Unfortunately, beginning in early 2007, the U.S. housing market entered a slump. As indicated in the "subprime lending" article of Wikipedia, the subprime mortgage market went into a meltdown. A sharp increase in the interest rates of the subprime market resulted in a large number of defaults and foreclosures. As the large number of late payments and foreclosures in the subprime market rose, banks and other lenders were forced to cease their lending operations. The impact of this subprime meltdown had a large impact on virtually all major international banks, including Merrill Lynch, Deutsche Bank AG, Bear Stearns, and Citigroup. The inability of these major banks and the related companies to cope with the pressures caused prices in the mortgage-backed securities market to collapse. Amongst others, Merrill Lynch and Citigroup reported huge losses, and bad company news caused the Dow Jones industrial average to fall by hundreds of points. Some larger banks were then reluctant to provide mortgage funding to their borrowers or even to each other. Tightening credit led to a shrinkage in spending and investment, and a slow overall growth of the U.S. housing market and total economy.

Subsequently the controversy surrounding subprime lending expanded, and spilled over to wider financial markets, and then inevitably had an extreme impact on overseas markets as well. In the United Kingdom, some researchers observed that the United Kingdom's subprime crisis could become worse than that of the United States. Property prices were considerably overvalued, lending criteria had been greatly loosened, and subsequently arrears began to escalate. The August 3, 2007 press release of the Council of Mortgage Lenders suggested that lenders foreclosed on more than 10,000 properties over the period January 2007–June 2007 in the United Kingdom. The August 28, 2007 news release of the U.K.

Personal Loan Store indicated that according to the 2007 data, the ratio of personal income to household debt was 0.617 in the United Kingdom compared with 0.704 in the United States and 0.735 in Japan. The price of British home had been driven mainly by the United Kingdom's small geographic size, high levels of immigration, and very low levels of house building. Therefore the United Kingdom began to face a subprime crisis on a similar scale to the United States.

13.3 MODELS OF EQUITY VOLATILITY DYNAMICS

Most researchers agree that equity volatility is explained partly by financial leverage, indicating that equity volatility tends to rise in response to "bad news," and fall in response to "good news." This is widely referred to as the "leverage effect," which causes higher volatility to be associated with lower average returns. A negative stock return causes the value of the equity to go down, meaning that the leverage of the firm increases, as does the future volatility of the equity; conversely, positive stock returns reduce both firm leverage and future equity volatility. Therefore, leverage does indeed cause systematic and asymmetric changes in equity volatility.

Nonlinear models have been introduced in order to deal with the above problems such as the TARCH, the TGARCH, and the EGARCH models (Penm and Terrell, 2003). These models allow good news and bad news to have different impacts on volatility, whereas the linear GARCH model does not. The parameters that describe such extensions to the GARCH process appear to be highly statistically significant; however, they are not stable over time. Indeed, the forecasts of the GARCH, the TARCH, and the EGARCH models have been shown to have limited ability to account for financial time series with volatility that undergoes occasional system shifts. These system shifts may be due to certain abrupt events, such as a stock market crash.

Edwards and Susmel (2001) propose that stock returns with the dynamics of switching variance are characterized by different ARCH processes with shifts to account for structural changes. Colavecchio and Funke (2009) suggest that once ARCH parameters are allowed to change between systems, according to an unobserved Markov alternating–state variable, the estimated ARCH effects will be much less persistent than those of a standard ARCH model. In other words, Markov-alternating parameters enable volatility to experience discrete shifts or changes in the persistence parameters, which implies that structural breaks in variance can account for the high persistence of the estimated conditional variance. Such a specification can significantly improve the performance of market volatility forecasting by accounting for volatility persistence in an appropriate manner.

13.4 DATA AND METHODOLOGY

13.4.1 Data

The data are obtained from the CED, which comprise monthly stock returns of the Shanghai Stock Exchange and the corresponding economic indicators provided by CED. A total of 360 monthly observations are provided covering the period from June 1975 through May 2005, a period which covered many business cycles in Shanghai. Further, we use the period from June 2005 through May 2008 for the forecasting of out-of-sample performances. The indices of corresponding economic indicators are measurements that move with the aggregate economy, and which vary directly and simultaneously with the business cycle, thereby providing information on the current state of the economy; these statistics will always tend to move up and down with the expansions and contractions of the business cycle.

13.4.2 Estimation Methods

Due to the frequent structural shifts of the economy in Shanghai, the AS-ARCH model, hypothesizing the existence of a single latent variable, is chosen to model the selected return series. This latent variable is decided by an unobserved Markov chain and determines the scale of equity volatility. Forecast comparisons are undertaken between the AS-ARCH model and other volatility models, including the ARCH, the GARCH, the TARCH, the TGARCH, and the EGARCH models, in order to determine whether AS-ARCH provides a best fit. The nature of the Granger causality between the corresponding economic index and the AS-ARCH for equity volatility is then tested.

13.4.3 The AS-ARCH Model

In order to capture structural shifts, an AS-ARCH model (K, q) is proposed as follows.

Let x_t denote the return observation and Λ_{t-1} denote the information set until $t-1$ with the process u_t being the residual from the first-order autoregression:

$$x_t = \varphi_0 + \varphi_1 x_{t-1} + u_t, \quad u_t \big| \Lambda_{t-1} \sim N(0, h_t)$$

The system-alternating model for the above conditional mean can be written as follows:

$$x_t = \mu_{S_t} + \tilde{x}_t$$

$$\tilde{x}_t = \varphi_1 \tilde{x}_{t-1} + \varphi_2 \tilde{x}_{t-2} + \cdots + \varphi_q \tilde{x}_{t-q} + \varepsilon_t$$

We assume that $\tilde{x}_t$ follows a zero-mean autoregression (q) with the abrupt shifts in the average level of returns being captured by μ_{S_t}; the residual term is set at $u_t = \sqrt{g_{s_t}} \times \tilde{u}_t$, where g_{s_t} are the scale parameters capturing the system change. S_t is an unobserved latent variable with possible outcomes of $1, 2, 3, \ldots, k$, and represents the volatility phase of observations. The error term u_t is multiplied by the scale factor $\sqrt{g_1}$ when the process is in the system $s_t = 1$, multiplied by $\sqrt{g_2}$ when $s_t = 2$, and so on. Suppose that we normalize $g_1 = 1$, then g_2 can be interpreted as the ratio of the average variance of stock returns at state 2, S_2 as compared to that at state 1, S_1.

The following equations confirm that stock returns are characterized by autoregression and that the error term u_t follows a qth-order ARCH process, where ξ stands for the leverage effect.

$$\tilde{u}_t = v_t \sqrt{h_t}, \quad v_t \sim N(0,1)$$

$$h_t = a_0 + a_1 \tilde{u}_{t-1}^2 + a_2 \tilde{u}_{t-2}^2 + \cdots + a_q \tilde{u}_{t-q}^2 + \xi \cdot d_{t-1} \cdot \tilde{u}_{t-1}^2$$

$$\begin{cases} d_{t-1} = 1 \\ d_{t-1} = 0 \end{cases} \text{ if } u_{t-1} > 0, \text{ then } d_{t-1} = 0; \ u_{t-1} \leq 0, \text{ then } d_{t-1} = 1$$

In the presence of leverage effect ($\xi \neq 0$), we will refer to this as an AS-ARCH (K, q) specification. The estimations of the AS-ARCH model should then determine the parameters of the ARCH terms along with the transition probabilities between the different systems. This model therefore requires a formulation of the probability law causing volatility to change between systems. The transition probabilities between the systems follow a hidden Markov chain, with K indicating that the transition probabilities p_{ij} are constant and independent of x_t for all t. The estimation of the AS-ARCH model is undertaken using the maximum likelihood method, in which it is possible to maximize all of the population parameters mentioned above, with the constraints that

$$g_1 = 1, \sum_{j=1}^{K} p_{ij} = 1 \text{ for } i = 1, 2, \ldots, K \text{ and } 0 \le p_{ij} \le 1 \text{ for } i, j = 1, 2, \ldots, K.$$

Given knowledge of the population parameters, it is straightforward to characterize the filter and smoothed probabilities. The respective filter and smoothed probabilities are subsequently specified by $p(s_t, s_{t-1}, \ldots, s_{t-q} | x_t, x_{t-1}, \ldots)$ and $p(s_t | x_T, x_{T-1}, \ldots)$; where T is the sample size. The filter probability represents the conditional probability that the system at date t is s_t, while at $t-1$ it was s_{t-1}. The smoothed probability represents inferences over the actual system at date t based upon the data available until T. For example, for a two-system model, the smoothed probabilities at date t are represented by a 2×1 vector including the estimated probabilities for the two systems involved. This vector represents *expost* inferences made on the system of the studied variable at date t, based upon the entire sample of observations.

13.5 EMPIRICAL RESULTS

13.5.1 Model Fitting and Forecasting

We fit the monthly return data using all volatility model specifications described in Section 13.2. The findings are reported in Table 13.1. The choice of the order of the models, for $K = 4, 3, 2, 1$, and $q = 4, 3, 2, 1$, is based upon the significance of the parameters and the AIC criterion. Further, two different loss functions, other than the log likelihood, are used to undertake a comparison on the basis of forecasting performance. The statistics, as indicated in Table 13.1, are then viewed as the model selection criteria. Table 13.1 shows the out-of-sample forecasting performances of all models involved. These results indicate that the AS-ARCH approach has the best out-of-sample outcomes.

TABLE 13.1   Comparison of 1 Month Forecasts for Different Models

Model	Log Likelihood	MSE	|LE|	Out-of-Sample (MSE × 0.001)
AS-ARCH (3, 2)	343.833	0.000520	1.6800	0.363
ARCH (2)	332.207	0.000588	1.8105	0.365
GARCH (1, 1)	343.625	0.000615	1.7022	0.378
TGARCH (1, 1)	343.611	0.000610	1.7035	0.371
TARCH (3)	343.328	0.000581	1.7002	0.377
EGARCH (1, 1)	342.117	0.000652	1.6908	0.372

$$\text{MSE} = T^{-1} \sum_{t=1}^{T} (\hat{u}_t^2 - \sigma_t^2)^2, |\text{LE}| = T^{-1} \sum_{t=1}^{T} \left| \ln(\hat{u}_t^2) - \ln(\sigma_t^2) \right|$$

The estimated parameters based upon the model specification of AS-ARCH (3,2) with p-values in parentheses are as follows:

$$x_t = \underset{(0.002)}{0.0075} + \underset{(0.002)}{0.1283}\, x_{t-1} + u_t$$

$$h_t = \underset{(0.002)}{-0.0281} + \underset{(0.000)}{0.0428}\, \tilde{u}_{t-1}^2 + \underset{(0.001)}{0.0435}\, \tilde{u}_{t-2}^2 + \underset{(0.001)}{8.071}\, \tilde{u}_{t-1}^2 \cdot d_{t-1}$$

$$g_1 = 1; \quad \hat{g}_2 = \underset{(0.001)}{2.4825}; \quad \hat{g}_3 = \underset{(0.001)}{12.2781}.$$

$$\hat{P} = \begin{bmatrix} 0.939 & 0 & 0.018 \\ 0.017 & 0.958 & 0.063 \\ 0 & 0.028 & 0.915 \end{bmatrix}$$

There are significant distinctions between the scale system parameters in the above estimation, with $\hat{g}_3$ being much greater than $\hat{g}_2$, and g_1 being less than $\hat{g}_2$. Hence, the results suggest that a three-state AS-ARCH model is appropriate. The stock return volatility in Shanghai can be characterized as having three distinct states, namely low-, medium-, and high-volatility state. The outcomes also reveal that the variance in the high-volatility state ($s_t = 3$) is, on average, around 12 times as high as that in the low-volatility state, while variance in the medium-volatility state ($s_t = 2$) is two and half times as high as that in the low-volatility state.

The estimated transition probabilities of the three states are also shown to be persistent, but to different degrees. We would expect, for example, to see state 1 lasting for $(1 - \hat{P}_{11})^{-1}$, which is 66.7 months, whereas states 2 and 3 should last for 35.7 months and 12 months, respectively. Furthermore, state 1 is never preceded by state 2 ($\hat{P}_{21}=0$) nor is it ever followed by state 3 ($\hat{P}_{13}=0$). Finally, the coefficient of the leverage effect parameter is statistically significant with an expected positive sign demonstrating the existence of volatility asymmetry in Shanghai's stock market. That is, a stock price increase or decrease has a greater effect on subsequent volatility.

13.5.2 Causality Test Results

The augmented Dickey–Fuller test on data involved indicates that both the corresponding economic index and the year-on-year growth rate of the corresponding economic index are stationary.

The test results indicating the existence of Granger causality are summarized in Table 13.2, which reveals that the monthly change in the corresponding economic index clearly leads to low-volatility filter probabilities. In Table 13.2 the insignificant test results are omitted; FP1, FP2, and FP3 denote the filter probability for low-, medium-, and high-volatility state, respectively. SP1, SP2, and SP3 denote the smoothed probability for low-, medium-, and high-volatility state. EI denotes the corresponding economic index; EY denotes the year-on-year growth rate of the corresponding economic index. Numbers in parentheses are p-values; *** and ** denote significance levels of 0.01 and 0.05, respectively.

The filter probabilities for medium- and high-volatility states significantly precede the year-on-year growth rate of the corresponding economic index at lag 4. Consequently, this outcome indicates that the filter probabilities can predict the fluctuation in the business cycle.

On the other hand, the results of smoothed probabilities provide strong support for the causal relationship between medium/high volatilities and the corresponding economic index. The smoothed probability for a medium volatility precedes changes in the corresponding economic indicator across all lags, with the exception of lag 1, and the results for high-volatility states are also significantly ahead of the index. Likewise, the results on the annual change in the corresponding indicators also point to a lead–lag causal relationship of similar fashion.

There is still one element of evidence of bidirectional causality in the smoothed probability for the medium-volatility state at lags 2 and 3. Conversely, there are few unidirectional causalities (except lag 1) between the smoothed probabilities for the low-volatility state and the business cycle indicator across all lags, at a significance level of 0.05. Hence, such

TABLE 13.2 Test Results Indicating the Existence of Significant Granger Causality between Probability for Volatility and the Business Cycle

Lag	EI → FP1	EI → FPY	FP2 → EY	FP3 → EY	EY → SP1	SP2 → EY	SP3 → EY	EY → SP3
(1)	14.05***				15.131***			
	(0.0001)				(0.0001)			
(2)	8.991***	11.23***				3.075**		4.253**
	(0.0000)	(0.001)				(0.043)		(0.000)
(3)	6.351***	7.495***				4.221***	4.563***	
	(0.0002)	(0.00006)				(0.006)	(0.001)	
(4)	4.727***	5.873***	2.215***	1.621***		3.415***	3.535***	
	(0.0008)	(0.001)	(0.0001)	(0.0001)		(0.007)	(0.001)	

evidence strongly reinforces the value of the information provided by bank equity volatility, particularly in an anomalous or highly volatile manner, in signaling the fluctuations in Shanghai's business cycle.

13.6 CONCLUSIONS

China's equity market has grown by leaps and bounds, and foreign investment has played a key role in this expansion. At the beginning of 2008, the number of companies listed on China's domestic A-share market in Shanghai and Shenzhen increased about 20-fold that of the 1993 level, with total market capitalization increasing an incredible 128 times. The boom was accompanied by a number of deregulation measures, yet the opening of the Chinese financial markets is still far from complete. Over the past year, the Chinese equity market has experienced a significant correction, partly due to the global banking and stock market turmoil and partly due to aggressive tightening of central bank policies to address surging inflation. One measure that the Chinese government has enacted in recent months to address short-term market pressures has been to continue to raise the amount that foreign institutional investors can place in equity markets, thus attracting increased interest in Chinese investment markets.

This chapter has investigated the behavior of Shanghai's stock return volatility over recent decades, in particularly the most recent years, through a comparison of the use of an AS-ARCH model with other time-series models. The findings have demonstrated that an AS-ARCH specification offers a better statistical fit to the dataset, incorporating the leverage effect. In the case of significant volatility in an emerging stock market, the empirical evidence using the data before May 2008 strongly suggests that as a result of frequent system shifts within emerging economies, the AS-ARCH model fits more appropriately. A high-volatility system is, to some degree, associated with structural shift events; furthermore, we provide strong confirmation of the pattern of Granger causality between stock return volatility and the corresponding indicator of the business cycle. This suggests that the smoothed probabilities for medium- and high-volatility systems lead both to the monthly and to the annual change in the corresponding economic index, although there is much less evidence of a bidirectional relationship. Our outcome, using the data before May 2008, is the evidence of China's recent transformation from a closed to a relatively open economy with more open capital markets. This process is likely to continue. This is positive for the long-term development of the Chinese stock markets and economy.

REFERENCES

Edwards, S. and Susmel, R. (2001) Volatility dependence and contagion in emerging equity markets. *Journal of Development Economics*, 66(2): 505–532.

Colavecchio, R. and Funke, M. (2009) Volatility dependence across Asia-Pacific onshore and offshore currency forwards markets. *Journal of Asian Economics*, 20(2): 174–196.

Penm, J. and Terrell, R.D. (2003) *Collaborative Research in Quantitative Finance and Economics*. Evergreen Publishing: ACT, Australia.

II

Global, European, and Emerging Markets' Perspectives

Global Perspective on the Banking Crisis and Recovery: An Analysis of Domestic vs. Foreign Banks

Mahmud Hossain, Pankaj K. Jain, and Sandra Mortal

CONTENTS

Shareholder wealth losses in the 2007–2008 economic crisis are the most severe for foreign banks with substantial U.S. operations and U.S. banks operating internationally. These are followed by U.S. banks with purely domestic operations. Interestingly, banks with any international operations, even if outside the United States, suffer substantial losses. Foreign banks with mainly domestic operations suffer the least damage but are not untouched by the effects of this global liquidity crisis. We also conduct a country-wise analysis and a multivariate regression to understand the determinants of bank stock losses. Finally, the effectiveness of subsequent government interventions and bailouts is analyzed.

14.1 INTRODUCTION

The global financial crisis following U.S. financial turmoil erupted in September 2008. This crisis has been driven by defaults on U.S. housing loans to customers and the subsequent collapse of liquidity in interbank funding markets. The turmoil became much more far-reaching after the failures of Lehman Brothers and Washington Mutual, and government takeovers of Fannie Mae, Freddie Mac, and American International Group (AIG). Even though the financial world suffered from several financial crises earlier, they were not nearly as severe as the one being witnessed now. Because of the globalization of financial markets, almost the entire world economy is now suffering from the consequences of the U.S. financial crisis.

The objective of this study is threefold. First, we compare the negative impact of the financial crisis on the stock prices of U.S. banks vs. overseas banks. Second, we examine the crisis events and policy interventions that carried the most impact on equity valuations of financial stocks. Third, we examine various firm-specific attributes associated with stock price decline of U.S. banks and foreign banks.

Our comprehensive global sample includes 2467 banks and their stock returns from 2007–2009. Using domicile information in conjunction with geographic distribution of their operations, we form five bank portfolios. We find losses are most severe for foreign banks with substantial U.S. operations, and U.S. banks operating internationally. These are followed by U.S. banks with purely domestic operations. Interestingly foreign banks with international operations outside United States are severely affected as well due to the highly integrated nature of financial markets. Foreign banks with mainly home operations suffer the least damage but are not untouched by the effects of this global liquidity crisis.

Next, we assess the positive or negative effect of each crisis event or policy intervention. We analyze several news sources such as BBC news, CNN Money, and Washington Post, to build our timeline of banking crisis and recovery. The crisis events that stand out in severely damaging the equity valuations of financials are Chapter 11 bankruptcy filing by Lehman Brothers, Securities and Exchange Commission (SEC) short selling ban, rejection of bailout legislation by house of representatives, Paulson's announcement that troubled asset relief program (TARP) funds will not be used to buy illiquid mortgage securities, NBER formal recession declaration, and crisis announcements from large individual investment banks and other finance companies. The policy interventions that carried the most weight on stock markets include global expansion of swap lines by central banks, purchase of bank preferred stock by U.S. Treasury, Fed rate cut to 0%, and finally the provision of guarantees, liquidity, and capital by Fed and Federal Deposit Insurance Corporation (FDIC) to large individual banks such as Citibank and Wachovia.

We present a country-by-country analysis of losses suffered by bank shareholders. U.S.-based banks lost an average of 30% and median loss is more severe at 42%. Irish and British banks stand out with significantly higher losses. Banks from emerging countries like Bangladesh and Peru are largely untouched by the mortgage crisis because they mainly serve their home markets.

Next, we focus on bank leverage as the gargantuan leverage of large banks has often been blamed for the precipitous losses which have quickly wiped out their equity capital. The high leverage portfolio loses almost 50% in value whereas the low leverage portfolio losses are 33%. The risk of leverage becomes evident in periods of crisis like these.

At the next stage, we perform multiple regression analysis to investigate whether some firm-specific characteristics—firm size, debt, and growth opportunities—have a systematic association with the magnitude of share value lost by the above mentioned five types of banks. We infer from our empirical findings that firm size—as proxied by market capitalization—has a positive link with the magnitude of the losses; however we observe that growth opportunities—as proxied by market to book ratio—and debt level cannot explain the level of losses after controlling for other factors. Based on our regression analysis, we also conclude that U.S. banks operating domestically as well as internationally and overseas banks operating in United States sustained maximum loss. Absence of U.S. operations somewhat contained the extent of damage. Nevertheless, non-U.S. international operations resulted in higher losses for foreign banks relative to other foreign banks that serve mainly their home markets.

The remainder of this chapter is organized as follows. The next section contains a literature review of how financial crisis has affected stock markets in the past. Section 14.3 mentions the data sources and classifications. Section 14.4 provides empirical methods and explains our findings. Section 14.5 concludes this chapter.

14.2 LITERATURE REVIEW

The impact of financial crisis and bailouts on stock valuations is of key interest to policymakers, investors, and the key players in the financial markets. Several researches address various facets of financial crisis.

Richardson (2007) analyzes the possible reasons for banking system failure during the time of Great Depression of 1930s and identifies withdrawals of deposits, illiquidity of assets, and Federal Reserve's reluctance to act as underlying causes for the banking system collapse. However, Staikouras (2004) contends high inflation, low reserves, overvalued currencies, openness of economy, and regulatory environment are the determining factors of banking crisis. Soral et al. (2006) show that systematic fraud associated with weak enforcement of conventional regulatory principles is a potential source of banking crisis. Rochet (2003) argues even though financial deregulation and globalization mainly initiated the banking crises in the past 25 years, it is political interference that amplified these crises. He further asserts bank supervisory boards across the world face a fundamental commitment problem; independence and accountability of the bank supervisors must be ensured for a successful reform of bank supervision.

Kamath (1980) examines the factors determining the premiums and discounts on commercial bank common stocks. His empirical findings indicate profitability is the most influential variable associated with the stock price fluctuations of banking companies. His study also exhibits Beta coefficient—surrogating systematic risk—and the volume of stock trading has a favorable impact on banking company stocks. He further documents lower payout ratios are also favorably viewed by stock market.

The financial and economic crisis in 2007–2008 is unique and far more extensive than anything seen before. In the past, severe and systemic financial crises affecting the entire country, governments, or regions have been seen in many emerging markets but they were not as globally pervasive as the current crisis. Mexico suffered from a severe currency and debt crisis in 1994, Southeast Asia in 1997–1998, Russia in 1998, Brazil in 1999, and Argentina in 2001. Several studies analyze various aspects of these financial crises. David-Friday and Gordon (2005) examine whether the association between Mexican firms' stock prices and their book values, earnings, and cash flows changed during the 1994 currency crisis. They observe that valuation coefficient in book value did not change significantly during the crisis period, even though its incremental explanatory power increased; however, the valuation coefficient on and incremental explanatory power of earnings declined during this crisis period. Applying an event study approach—using daily stock market data—Wilson et al. (2000, p. 1) document investors did not perceive devaluation itself as negative. Further, they find that investors did not anticipate the "...peso devaluation, the declining reserve levels of Mexico's central bank, and increasing sovereign default risk of Mexico. Their empirical findings suggest that equity investors did not positively respond to remedial action taken by governmental authorities, such as the Clinton bailout plans."

The financial turbulence in Southwest Asia is claimed to have been started by the Thai devaluation in 1997, followed by Korea, Indonesia, Malaysia and the Philippines. Jiangli et al. (2008) investigate whether lending relationships benefit firms by making credit more available during financial crisis and find mixed results for various countries involved. Mitton (2002) examines whether weak corporate governance and accounting disclosure was associated with South Asian financial crisis during 1997–1998. Using sample firms from Indonesia, Korea, Malaysia, the Philippines and Thailand, he documents a positive association between stock price performance, disclosure quality, and level of outside ownership concentration. Baek et al. (2004) also observe that during the 1997 Korean financial crisis, firms with higher ownership concentration and better disclosure quality experienced a smaller reduction in their equity value. The Russian financial crisis in 1998–1999 was caused by a combination of several factors: currency crisis, debt crisis, and banking crisis. The ruble was substantially devalued, the aggregate capital of the banking system declined to almost negative, and the payment system came completely to a halt. The Russian government declared a moratorium on private financial debt and default on all ruble-denominated public debt to deal with this financial crisis (Selim, 2005).

Maniam et al. (2004) identify several factors that led Argentina to financial crisis: lack of government's ability to implement fiscal policies, rising government debt, and unfavorable external factors. They argue government should have devalued the peso after achieving price stability and growth.

Using 47 banking crises episodes in 35 industrial and emerging market economies between the 1970s and 2003, Apanard (2009) examines the association between banking regulation and supervision, and the acuteness of banking crises—measured in terms of the magnitude of output loss. He finds comprehensive deposit insurance and restrictions on bank activities influence the severity of banking crisis. However, he does not observe any association between bank supervision and severity of bank crisis.

While the above-mentioned studies examine the potential causes of financial meltdowns and how stock prices of commercial banks were affected in different times, to our knowledge this chapter is the first to address how bank shareholders' wealth impairment has been affected by the recent financial crisis—which is global and much more pervasive in nature compared to any financial crises the world experienced earlier, Moreover, this study enriches the extant literature on financial crises by empirically testing the differential effects—in terms of magnitude and timing—of United States-originated financial disaster on stock prices of different types of global banks (differentiated based on their country of origin and domestic and/or foreign operations).

14.3 DATA SOURCES AND CLASSIFICATIONS

We begin with the universe of 3775 listings in the banking sector world-wide tracked in Datastream International. Of these, 1308 are multiple listings representing banks that cross-list their stocks on more than one exchange within or outside their home country. This leaves us with 2467 unique banks to analyze. We obtain monthly and daily stock returns for each firm from 2007 to 2009. Datastream also provides the country where each bank is incorporated as well as all the exchanges where the bank's stock is cross-listed. For single-listed banks, we obtain additional information from bank regulators to ascertain whether or not the bank has international operations. We use this domicile information in conjunction with geographic distribution of their operations to form the five portfolios. The FDIC provides the details of both domestic and foreign branch locations of covered U.S. banks. The Federal Reserve Board (Fed) provides on its Web site the structure data for the U.S. offices of foreign banking organizations. The details of these portfolios and the number of banks within each portfolio are shown in Table 14.1.

14.4 EMPIRICAL METHODS AND RESULTS

We begin by plotting monthly bank stock returns from January 2007 to December 2008 in Figure 14.1. The damaging effects of the crisis are pervasive. No portfolio of bank stocks is spared from the shareholder wealth

TABLE 14.1   Sample Distribution and Portfolio Formation

Label	Description	Number of Banks
UU	United States with largely local operations	907
UI	U.S. banks with substantial international operations	441
FH	Foreign banks operating largely within their home country	778
FU	Foreign banks operating in the United States	109
FI	Foreign banks operating internationally other than the United States	232
	Number of unique banks in the sample	2467

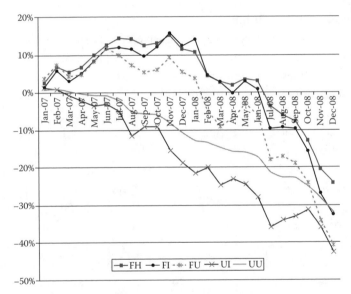

FIGURE 14.1   Worldwide cumulative losses in bank stocks during the financial crisis of 2007–2008.

destruction irrespective of the country of origin or area of operations. The portfolio of large U.S. banks operating on international scale suffered the most, losing 43% of its value. Foreign banks operating in United States followed closely with 41% loss. Although the markets were relatively quick in identifying the potential losses of U.S. banks, the realization that foreign banks operating in United States would sink as well came late to the stock investors. U.S. banks with largely domestic operations lost 32%. Smaller position due to smaller size or conservative philosophy, which is reflected in avoiding international risks, could be potential explanations of why losses were somewhat contained for such banks. Surprisingly, foreign banks with international operations outside the United States took a 32% hit as well, perhaps due to the spillover effects of the crisis or the global integration of the banking transactions. Foreign banks operating mainly with their home countries are expected to be least affected by the mortgage crisis originating in the United States. Nevertheless, the portfolio of such foreign banks was not completely aloof and lost 24% of its value in our sample period.

Monthly returns are cumulated from January 2007 for U.S. banks operating domestically (UU), U.S. banks with substantial international operations (UI), foreign banks with U.S. operations (FU), foreign banks operating internationally other than in United States (FI), and foreign banks operating mainly within their home country (FH).

Next, we assess the positive or negative effect of each crisis event or policy intervention in Table 14.2. We analyze several news sources such as BBC news, CNN Money, and Washington Post, to build our timeline of banking crisis and recovery. The table lists crisis and bailout events for 2007–2009. The first column classifies each event into positive, negative, or ambiguous. The classification is based on the theoretical prediction or desired effect of any policy intervention. The second column contains the actual return, as described below. The third column has the exact event date and the last column contains the description of the event.

We use the Datastream daily U.S.-Financials Index to measure the impact of these events. Event day returns are computed by comparing the index on event date t and its level 1 day before, that is, $(I_t - I_{t-1})/Ii_{t-1}$. The average daily return in the financials index throughout the sample period of January 2007–January 2009 is −0.13%. Standard deviation of returns during this period is 2.91%. We use this information to assess the statistical significance of returns associated with each event. If the actual returns are more than one standard deviation away from the mean, then we label the return with one asterisk. If the actual returns are more than two standard deviations away from the mean, then we label the return with two asterisks.

The table helps us identify the crisis events that stand out in severely damaging the equity valuations. The news announcements associated with strongest negative returns and investor pessimism are Chapter 11 bankruptcy filing by Lehman Brothers, SEC short selling ban, rejection of the bailout legislation by the house of representatives, Paulson's announcement that TARP funds will not be used to buy illiquid mortgage securities,

TABLE 14.2 Analysis of Crisis Events and Policy Interventions

Pr	Return	Date	Event
N	−3.28%*	2/27/2007	Freddie Mac announces that it will no longer buy the most risky subprime mortgages and related securities.
N	−0.45%	4/2/2007	New Century Financial Corporation files for Chapter 11.
N	0.33%	6/1/2007	Standard & Poor's and Moody's Investor Service downgrade over 100 bonds backed by second-lien subprime mortgages.
N	−1.68%	6/7/2007	Bear Stearns informs investors that it is suspending redemptions from its high-grade structured credit strategies enhanced leverage fund.
A	−0.06%	6/28/2007	The Federal Open Market Committee (FOMC) votes to maintain its target for the federal funds rate at 5.25%.
N	0.38%	7/11/2007	Standard & Poor's places 612 securities backed by subprime mortgages on a credit watch.
N	−2.75%	7/24/2007	Countrywide Financial Corporation warns of "difficult conditions."
N	−1.54%	7/31/2007	Bear Stearns liquidates two hedge funds that invested in mortgage-backed securities.
N	4.17%*	8/6/2007	American Home Mortgage Investment Corporation files for Chapter 11.
A	0.85%	8/7/2007	The FOMC votes to maintain its target for the federal funds rate at 5.25%.
N	−3.34%*	8/9/2007	BNP Paribas, France's largest bank, halts redemptions on three investment funds.
P	−0.18%	8/10/2007	The Fed announces that it "will provide reserves as necessary…to promote trading in the federal funds market at rates close to the FOMC's target rate of 5.25%."
N	3.18%*	8/16/2007	Fitch Ratings downgrades Countrywide Financial Corporation to BBB+, and Countrywide borrows the entire $11.5 billion available in its credit lines with other banks.
P	3.62%*	8/17/2007	The Fed votes to reduce the primary credit rate 50 basis points to 5.75% and also increases the maximum primary credit borrowing term to 30 days, renewable by the borrower.

(continued)

TABLE 14.2 (continued)   Analysis of Crisis Events and Policy Interventions

Pr	Return	Date	Event
P	0.11%	9/14/2007	The chancellor of the exchequer authorizes the Bank of England to provide liquidity support for Northern Rock.
P	4.04%*	9/18/2007	The Fed votes to reduce the primary credit rate 50 basis points to 5.25%.
P	−0.74%	10/10/2007	U.S. Treasury Secretary Paulson announces the HOPE NOW initiative.
A	−1.70%	10/15/2007	Citigroup, Bank of America, and JPMorgan Chase announce plans for an $80 billion Master Liquidity Enhancement Conduit to.
P	1.02%	10/31/2007	The Fed votes to reduce the primary credit rate 25 basis points to 5.00%.
N	−4.11%*	11/1/2007	Financial market pressures intensify, reflected in diminished liquidity in interbank funding markets.
P	−4.38%*	12/11/2007	The Fed votes to reduce the primary credit rate 25 basis points to 4.75%.
P	−1.01%	12/12/2007	The Fed announces the creation of a term auction facility (TAF) in which fixed amounts of term funds will be auctioned to depository institutions against a wide variety of collateral.
P	1.99%	12/21/2007	The Fed announces that TAF auctions will be conducted every 2 weeks.
P	1.99%	12/21/2007	Citigroup, JPMorgan Chase, and Bank of America abandon plans for the Master Liquidity Enhancement Conduit.
P	−0.50%	1/11/2008	Bank of America announces that it will purchase Countrywide Financial in an all-stock transaction.
N	−1.54%	1/18/2008	Fitch Ratings downgrades Ambac Financial Group's insurance financial strength rating to AA.
P	2.24%	1/22/2008	The Fed votes to reduce the primary credit rate 75 basis points to 4%.
P	−1.14%	1/30/2008	The Fed votes to reduce the primary credit rate 50 basis points to 3.5%.
P	0.94%	2/13/2008	President Bush signs the Economic Stimulus Act of 2008.
A	0.01%	2/17/2008	Northern Rock is taken into state ownership by the Treasury of the United Kingdom.
N	−0.47%	3/5/2008	Carlyle Capital Corporation receives a default notice after failing to meet margin calls on its mortgage bond fund.

TABLE 14.2 (continued) Analysis of Crisis Events and Policy Interventions

Pr	Return	Date	Event
P	0.19%	3/7/2008	The Fed announces $50 billion TAF auctions on March 10 and March 24.
P	6.42%**	3/11/2008	The Fed announces the creation of the term securities lending facility (TSLF), which will lend up to $200 billion of Treasury.
P	−3.49%*	3/14/2008	The Fed approves the financing arrangement announced by JPMorgan Chase and Bear Stearns It also announces they are "monitoring market developments closely and will continue to provide liquidity as necessary."
P	#N/A	3/16/2008	The Fed establishes the primary dealer credit facility (PDCF), extending credit to primary dealers.
P	7.21%**	3/18/2008	The Fed votes to reduce the primary credit rate 75 basis points to 2.50%.
P	0.77%	3/24/2008	The Federal Reserve Bank of New York announces that it will provide term financing to facilitate JPMorgan Chase & Co.'s acquisition of The Bear Stearns Companies Inc.
P	−0.70%	4/30/2008	The Fed votes to reduce the primary credit rate 25 basis points to 2.25%.
P	0.17%	5/2/2008	The FOMC expands the list of eligible collateral for Schedule 2 TSLF auctions to include AAA/Aaa-rated asset-backed securities.
P	1.81%	6/5/2008	The Fed announces approval of the notice of Bank of America to acquire Countrywide Financial Corporation.
N	1.81%	6/5/2008	
A	0.34%	6/25/2008	The FOMC votes to maintain its target for the federal funds rate at 2.00%.
N	−2.09%	7/11/2008	The Office of Thrift Supervision closes IndyMac Bank, F.S.B.
P	−4.25%*	7/13/2008	The Fed authorizes the Federal Reserve Bank of New York to lend to Fannie Mae and Freddie Mac, if such lending necessary.
P	−4.25%*	7/13/2008	The U.S. Treasury Department announces a temporary increase in the credit lines of Fannie Mae and Freddie Mac.

(continued)

TABLE 14.2 (continued)   Analysis of Crisis Events and Policy Interventions

Pr	Return	Date	Event
A	−2.61%	7/15/2008	SEC issues an emergency order temporarily prohibiting naked short selling in the securities of Fannie Mae, Freddie Mac, and primary dealers.
P	1.79%	7/30/2008	President Bush signs into law the Housing and Economic Recovery Act of 2008.
P	1.79%	7/30/2008	The Fed extends the TSLF and PDCF through January 30, 2009, introduces auctions of options on $50 billion of draws on the TSLF, and introduces 84 day TAF loans.
A	4.39%*	8/5/2008	The FOMC votes to maintain its target for the federal funds rate at 2.00%.
N	−3.09%*	8/17/2008	FOMC releases a statement about the current financial market turmoil, and notes that the "downside risks to growth have increased appreciably."
N	4.02%*	9/7/2008	The Federal Housing Finance Agency (FHFA) places Fannie Mae and Freddie Mac in government conservatorship.
P	−8.46%*	9/14/2008	The Fed expands the list of eligible collaterals for the PDCF to include any collateral that can be pledged in the tri-party repo system of the two major clearing banks.
P	−8.46%**	9/15/2008	Bank of America announces its intent to purchase Merrill Lynch & Co. for $50 billion.
N	−8.46%**	9/15/2008	Lehman Brothers Holdings Incorporated files for Chapter 11.
P	5.47%*	9/16/2008	The Fed authorizes the Federal Reserve Bank of New York to lend up to $85 billion to the AIG.
A	5.47%*	9/16/2008	The FOMC votes to maintain its target for the federal funds rate at 2.00%.
N	5.47%*	9/16/2008	The net asset value of shares in the Reserve Primary Money Fund falls below $1.
P	−7.47%**	9/17/2008	The U.S. Treasury Department announces a supplementary financing program consisting of a series of Treasury bill issues that will provide cash for use in Federal Reserve initiatives.
A	−7.47%**	9/17/2008	The SEC announces a temporary emergency ban on short selling in the stocks of all companies in the financial sector.

TABLE 14.2 (continued)   Analysis of Crisis Events and Policy Interventions

Pr	Return	Date	Event
P	9.94%**	9/18/2008	The FOMC expands existing swap lines by $180 billion and authorizes new swap lines with the Bank of Japan, Bank of England, and Bank of Canada.
P	10.45%**	9/19/2008	The Fed announces the creation of the asset-backed commercial paper money market mutual fund liquidity facility (AMLF) to extend nonrecourse loans.
P	10.45%**	9/19/2008	The U.S. Treasury Department announces a temporary guaranty program that will make available up to $50 billion to promote investments.
P	−8.39%*	9/20/2008	The U.S. Treasury Department submits draft legislation to Congress for authority to purchase troubled assets.
P	−8.39%	9/21/2008	The Fed approves applications of investment banking companies Goldman Sachs and Morgan Stanley to become bank holding companies.
P	−0.89%	9/24/2008	The FOMC establishes new swap lines with the Reserve Bank of Australia and the Sveriges Riksbank for up to $10 billion each and with the Danmarks Nationalbank and the Norges Bank for up to $5 billion each.
N	1.97%	9/25/2008	The Office of Thrift Supervision closes Washington Mutual Bank. JPMorgan Chase acquires the banking operations of Washington Mutual Bank.
P	2.31%	9/26/2008	The FOMC increases existing swap lines with the ECB by $10 billion and the Swiss National Bank by $3 billion.
P	−12.88%**	9/29/2008	The FOMC authorizes a $330 billion expansion of swap lines with Bank of Canada, Bank of England, Bank of Japan, Danmarks Nationalbank, ECB, Norges Bank, Reserve Bank of Australia, Sveriges Riksbank, and Swiss National Bank.
P	−12.88%**	9/29/2008	The U.S. Treasury Department opens its temporary guarantee program for money market funds.
P	−12.88%**	9/29/2008	The FDIC announces that Citigroup will purchase the banking operations of Wachovia Corporation.

(*continued*)

TABLE 14.2 (continued)   Analysis of Crisis Events and Policy Interventions

Pr	Return	Date	Event
N	−12.88%**	9/29/2008	The U.S. House of Representatives rejects legislation submitted by the Treasury Department requesting authority to purchase troubled assets.
P	−3.20%*	10/3/2008	Wells Fargo announces a competing proposal to purchase Wachovia Corporation.
P	−3.20%*	10/3/2008	Congress passes and President Bush signs into law the Emergency Economic Stabilization Act which establishes the $700 billion TARP.
P	−4.09%*	10/6/2008	The Fed announces that the fed will pay interest on depository institutions' required and excess reserve balances.
P	−9.78%**	10/7/2008	The Fed announces the creation of the commercial paper funding facility (CPFF), which will provide a liquidity backstop to U.S. issuers of commercial paper.
P	−9.78%**	10/7/2008	The FDIC announces an increase in deposit insurance coverage to $250,000 per depositor.
P	−3.19%*	10/8/2008	The Fed authorizes the Federal Reserve Bank of New York to borrow up to $37.8 billion in investment-grade, fixed-income securities from AIG in return for cash collateral.
P	−3.19%*	10/8/2008	The Fed votes to reduce the primary credit rate 50 basis points to 1.75%.
P	9.53%	10/12/2008	The Fed approves an application by Wells Fargo & Co. to acquire Wachovia Corporation.
P	9.53%**	10/13/2008	The FOMC increases existing swap lines with foreign central banks: The Bank of England, European Central Bank, and Swiss National Bank.
P	4.07%*	10/14/2008	The FOMC increases its swap line with the Bank of Japan.
P	4.07%*	10/14/2008	The U.S. Treasury will make available $250 billion of capital to U.S. financial institutions.
P	4.07%*	10/14/2008	The FDIC creates a new temporary liquidity guarantee program to guarantee the senior debt of all FDIC-insured institutions, holding companies, and deposits in non-interest-bearing deposit transactions.

TABLE 14.2 (continued)   Analysis of Crisis Events and Policy Interventions

Pr	Return	Date	Event
P	−1.88%	10/21/2008	The Fed announces creation of the money market investor funding facility (MMIFF).
P	−6.25%**	10/22/2008	The Fed announces that it will alter the formula used to determine the interest rate paid to depository institutions on excess reserve balances.
P	−3.45%*	10/24/2008	PNC Financial Services Group Inc. purchases National City Corporation, creating the fifth largest U.S. bank.
P	11.04%**	10/28/2008	The U.S. Treasury Department purchases a total of $125 billion in preferred stock in nine U.S. banks.
P	11.04%**	10/28/2008	The FOMC and Reserve Bank of New Zealand establish a $15 billion swap line.
P	−1.71%	10/29/2008	The Fed reduces the primary credit rate 50 basis points to 1.25%.
P	−1.71%	10/29/2008	The FOMC also establishes swap lines with the Banco Central do Brasil, Banco de Mexico, Bank of Korea, and the Monetary Authority of Singapore.
P	−1.71%	10/29/2008	The International Monetary Fund (IMF) announces the creation of a short-term liquidity facility for market-access countries.
P	−7.55%**	11/5/2008	The Fed announces that it will alter the formula used to determine the interest rate paid to depository institutions on required and excess reserve balances.
P	−4.25%*	11/10/2008	The Fed approves the applications of American Express and American Express Travel Related Services to become bank holding companies.
P	−4.25%*	11/10/2008	The Treasury Department agrees to purchase $40 billion of AIG preferred shares under the TARP program
P	−2.12%	11/11/2008	The U.S. Treasury Department announces a new streamlined loan modification program with cooperation from the Federal Housing Finance Agency (FHFA), Department of Housing and Urban Development, and the HOPE NOW alliance.
N	−6.03%**	11/12/2008	U.S. Treasury Secretary Paulson formally announces that the Treasury has decided not to use TARP funds to purchase illiquid mortgage-related assets.

(continued)

TABLE 14.2 (continued)   Analysis of Crisis Events and Policy Interventions

Pr	Return	Date	Event
P	−4.89%*	11/14/2008	The U.S. Treasury Department purchases a total of $33.5 billion in preferred stock in 21 U.S. banks under the capital purchase program.
A	−5.30%	11/15/2008	Three large U.S. life insurance companies seek TARP funding: Lincoln National, Hartford Financial Services Group, and Genworth Financial.
A	−0.44%	11/18/2008	Executives of Ford, General Motors, and Chrysler testify before Congress, requesting access to the TARP for federal loans.
P	−9.35%**	11/20/2008	Fannie Mae and Freddie Mac announce that they will suspend mortgage foreclosures until January 2009.
A	5.06%*	11/21/2008	The U.S. Treasury Department announces that it will help liquidate The Reserve Fund's U.S. Government Fund.
P	5.06%*	11/21/2008	The U.S. Treasury Department purchases a total of $3 billion in preferred stock in 23 U.S. banks.
P	14.38%	11/23/2008	The U.S. Treasury Department, Fed, and FDIC jointly announce an agreement with Citigroup to provide a package of guarantees, liquidity access, and capital.
P	3.14%*	11/25/2008	The Fed announces the creation of the term asset-backed securities lending facility (TALF), under which the Federal Reserve Bank of New York will lend up to $200 billion on a nonrecourse basis to holders of AAA-rated asset-backed securities.
P	3.14%*	11/25/2008	The Fed announces a new program to purchase direct obligations of housing-related, government-sponsored enterprises (GSEs)—Fannie Mae, Freddie Mac, and Federal Home Loan Banks—and MBS backed by the GSEs.
P	4.73%*	11/26/2008	The Fed approves the notice of Bank of America Corporation to acquire Merrill Lynch and Company.
P	7.14%**	12/2/2008	The Fed announces that it will extend three liquidity facilities, PDCF, the Asset-AMLF, and TSLF through April 30, 2009.
P	4.84%*	12/3/2008	The SEC approves measures to increase transparency and accountability at credit rating agencies.

TABLE 14.2 (continued)   Analysis of Crisis Events and Policy Interventions

Pr	Return	Date	Event
P	7.71%**	12/5/2008	The U.S. Treasury Department purchases a total of $4 billion in preferred stock in 35 U.S. banks.
P	−0.28%	12/10/2008	The FDIC reiterates the guarantee of federal deposit insurance in the event of a bank failure.
N	−7.69%**	12/11/2008	The Business Cycle Dating Committee of the National Bureau of Economic Research announces that the economy is going through a recession.
P	2.16%	12/12/2008	The U.S. Treasury Department purchases a total of $6.25 billion in preferred stock in 28 U.S. banks.
P	−3.51%*	12/15/2008	The Fed approves the application of PNC Financial Services to acquire National City Corporation.
P	10.03%**	12/16/2008	The FOMC votes to establish a target range for the effective federal funds rate of 0%–0.25%. The Fed votes to reduce the primary credit rate 75 basis points to 0.50%.
P	0.61%	12/19/2008	The U.S. Treasury Department authorizes loans of up to $13.4 billion for General Motors and $4.0 billion for Chrysler from the TARP.
P	0.61%	12/19/2008	The Fed announces revised terms and conditions of the TALF.
P	0.61%	12/19/2008	The U.S. Treasury Department purchases a total of $27.9 billion in preferred stock in 49 U.S. banks under the capital purchase program.
P	−3.32%*	12/22/2008	The Fed approves the application of CIT Group Inc., an $81 billion financing company, to become a bank holding company.
P	−1.33%	12/23/2008	The U.S. Treasury Department purchases a total of $15.1 billion in preferred stock from 43 U.S. banks under the capital purchase program.
P	1.10%	12/24/2008	The Fed approves the applications of GMAC LLC and IB Finance Holding Company, LLC (IBFHC) to become bank holding companies.

(continued)

TABLE 14.2 (continued)   Analysis of Crisis Events and Policy Interventions

Pr	Return	Date	Event
P	−1.21%	12/29/2008	The U.S. Treasury Department announces that it will purchase $5 billion in senior preferred equity from GMAC as part of its program to assist the domestic automotive industry and also agrees to lend up to $1 billion to GM, so GM can help fund GMAC.
P	3.44%*	12/30/2008	The Fed announces that it expects to begin to purchase mortgage-backed securities backed by Fannie Mae, Freddie Mac, and Ginnie.
A	3.44%*	12/30/2008	The U.S. SEC releases a report that recommends against the suspension of fair value accounting standards.
P	3.10%*	12/31/2008	The U.S. Treasury Department purchases a total of $1.91 billion in preferred stock from seven U.S. banks under the capital purchase program.
P	−1.60%	1/5/2009	The Federal Reserve Bank of New York begins purchasing fixed-rate mortgage-backed securities guaranteed by Fannie Mae, Freddie Mac, and Ginnie Mae.
P	−4.74%*	1/7/2009	The Fed announces to (1) expand the set of institutions eligible to participate in the MMIFF and (2) reduce the minimum yield on assets eligible to be sold to the MMIFF.

Note: Events are collected from BBC news, CNN Money, and Washington Post. Positive (P), negative (N), or ambiguous (A) return prediction (Pr) in the first column is based on the theoretical prediction or desired effect of any policy intervention. Datastream daily U.S.-Financials Index is used to compute event day returns by comparing the index on event date t and its level one day before, i.e., $(I_t - I_{t-1})/I_{t-1}$. The average daily return in the financials index throughout the sample period of January 2007 to January 2009 is −0.13%. Standard deviation of returns during this period is 2.91%. If the actual returns are more than one (two) standard deviation away from the mean, they are labeled with * (**).

NBER formal recession declaration, and crisis announcements from large individual investment banks and other finance companies.

The effectiveness of various bailout programs varied significantly. Government actions associated with strongest positive market returns are global expansion of swap lines by central banks, purchase of bank preferred stock by U.S. Treasury, Fed rate cut to 0%, and finally the provision of guarantees, liquidity, and capital by Fed and FDIC to large individual banks such as Citibank and Wachovia. Not all treasury proposals were met with optimism and the permission to convert investment banks into bank holding companies were not viewed positively by bank stock investors.

We present a country-by-country analysis of losses suffered by bank shareholders in Figure 14.2. Losses are computed from total return index (*ri*) for each bank from January 01, 2007 to January 01, 2009 as follows:

$$\text{Loss} = \left(ri_{09} - ri_{07}\right)/ri_{07} \qquad (14.1)$$

The above calculation requires that a bank exists and remains in our sample throughout the sample period. Thus, we have a reduced sample size of 2016 banks that meet this criterion. U.S.-based banks lost an average of 30%. Irish and British banks stand out with significantly higher losses. Banks from emerging countries like Bangladesh and Peru are largely untouched by the mortgage crisis because they mainly serve their home markets. Bank returns over the sample period of January 01, 2007–January 01, 2009, for 2016 banks for which returns index data is available in Datastream.

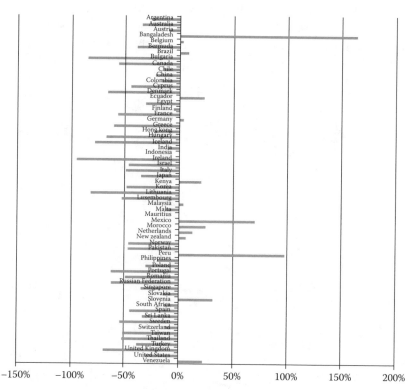

FIGURE 14.2 Countrywide bank stock losses.

Next, we focus on bank leverage as the gargantuan leverage of large banks has often been blamed for the precipitous losses which have quickly wiped out their equity capital. Once again we cumulate monthly returns from January 2007 to December 2008 in Figure 14.3. Here the two portfolios are banks with high and low leverage computed by dividing the total debt at the beginning of 2007 by total assets on the same date. Sample is divided into two portfolios representing banks with high leverage and low leverage, respectively, using the median leverage as the cutoff point. Given the international scope of our study, the data for debt level is missing for banks for several emerging markets. Thus, our leverage analysis is based on 2237 banks compared to our full sample of 2467 banks.

The high leverage portfolio loses almost 50% in value whereas the low leverage portfolio losses are 33%. The risk of leverage becomes evident in periods of crisis like these.

Monthly returns are cumulated from January 2007 for banks with high and low leverage computed by dividing the total debt at the beginning of 2007 by total assets on the same date. Sample is divided into two equal parts using the median leverage as the cutoff point.

As evident from the analysis thus far, there is significant cross-sectional variation in the magnitude of losses suffered by the shareholders of various

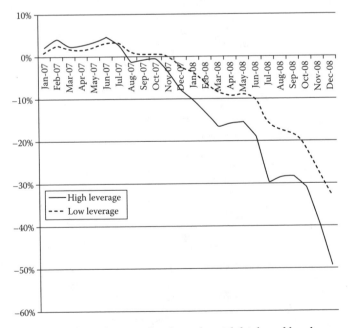

FIGURE 14.3 Cumulative losses in bank stocks with high and low leverage.

banks. To analyze the various determinants of the magnitude of losses, we estimate a multivariate regression equation in Table 14.3. The dependent variable loss during the crisis period is as defined in Equation 14.1. The following regression model is estimated linking bank losses to its size, leverage, valuation multiple, and dummy variables for bank portfolios defined earlier:

$$Loss = \alpha_0 + \beta_1 Market\ Cap. + \beta_2 Debt + \beta_3 Market\text{-to-book} + \beta_4 FU + \beta_4 FI + \beta_4 UU + \beta_4 UI + \varepsilon \tag{14.2}$$

The dummy variable for one of the bank portfolios needs to be dropped to ensure that the model has a full rank. We dropped the dummy for foreign banks operating mainly in their home countries (FH). Thus, the losses for that category are absorbed in the intercept term. The other dummy variables respectively indicate the additional losses for the remaining portfolios, after controlling for other bank characteristics. In an alternative regression model, we replace debt with the leverage ratio computed by dividing total debt by total assets. The results are almost identical. The leverage variable takes a negative and statistically significant coefficient

TABLE 14.3   Stock Loss Regressions

	Parameter Estimate	$Pr > |t|$
Intercept	−0.2542	<.0001
Market capitalization (billions)	7.1698	0.0766
Debt level (billions)	−0.0045	0.2029
Market-to-book ratio	0.0001	0.1731
Foreign banks operating in the United States (FU)	−0.2459	<.0001
Foreign banks operating internationally outside U.S. (FI)	−0.0855	0.0511
U.S. banks with mainly domestic operations (UU)	−0.2004	<.0001
U.S. bank with significant international operations (UI)	−0.2103	<.0001
R-squared	5.38%	
Adj R-squared	4.82%	

Note: Dependent variable is the percentage loss computed from total return index for each bank on January 01, 2007 and January 01, 2009: Loss = $(ri_{09} - ri_{07})/ri_{07}$. The following regression model is then estmated linking bank losses to its size, leverage, valuation multiple, and dummy variables for bank portfolios defined earlier:

$$Loss = \alpha 0 + \beta 1 Market\ Cap. + \beta 2 Debt + \beta 3 Market\text{-to-book} + \beta_4 FU + \beta_4 FI + \beta_4 UU + \beta_4 UI + \varepsilon.$$

and the direction and significance of other variables are the same as those reported in Table 14.3.

Bigger banks lost less as implied by the positive coefficient on the market capitalization variable. Consistent with the leverage results reported in Figure 14.3, the regression coefficient for bank debt is negative but after controlling for all other variables it is not statistically significant at conventional levels. The coefficient on market to book ratio is statistically insignificant implying that price multiples before the crisis are not a major determinant of shareholder wealth destruction. U.S. operations caused the most damage to the value of a bank's stock as is apparent from large negative coefficients for FU, UU, and UI. Absence of U.S. operations somewhat contained the extent of the damage. Nevertheless, international operations resulted in higher losses in FI portfolio relative to FH portfolio.

14.5 CONCLUSION

This research tests the stock price losses incurred during the time of recent global financial crisis for five sets of banks: U.S. banks operating domestically, U.S. firms operating domestically while having substantial international operations, foreign banks operating mainly in their home country, foreign banks with significant U.S. operations, and foreign banks operating in their home country as well as internationally other than the United States. We observe that the financial crisis has the most negative stock price impact on U.S. banks operating internationally and foreign banks with U.S. operations, followed by U.S. banks with domestic operations. Because of globalization, foreign banks with any international operation lost substantially. Even though, the impact of financial turmoil is lowest on foreign banks operating within their home market, the loss sustained by these banks is not negligible.

Next, we assess the positive or negative effect of each crisis event or policy intervention. We analyze several news sources such as BBC news, CNN Money, and Washington Post to build our timeline of banking crisis and recovery. The crisis events that stand out in severely damaging the equity valuations of financials are Chapter 11 bankruptcy filing by Lehman Brothers, SEC short selling ban, rejection of bailout legislation by house of representatives, Paulson's announcement that TARP funds will not be used to buy illiquid mortgage securities, NBER formal recession declaration, and crisis announcements from large individual investment banks and other finance companies. The government bailout programs that helped bank stock recovery include global expansion of swap lines by central banks, purchase of bank preferred stock by U.S. Treasury, Fed rate cut to 0%, and finally the provision of guarantees, liquidity, and capital by Fed and FDIC to large individual banks such as Citibank and Wachovia.

We then examine whether some firm-specific characteristics can explain the magnitude of the losses. While we observe a positive association between firm size and returns, debt level and growth opportunities cannot explain cross-sectional variation in losses across firms in the regression framework although leverage ratio is systematically positively correlated with losses in univariate analysis. We also infer, based on our multivariate regression analysis, that all banks operating in United States—that is, U.S. banks with only U.S. operations, U.S. banks with domestic and foreign operations, and foreign banks operating in United States—experienced the most losses. Nevertheless foreign banks with operations outside the United States also sustained significant losses. Overall, these results have important policy implications for bank regulators, depositors, and bank stock investors.

REFERENCES

Apanard, A. P. (2009) Banking regulation and the output cost of banking crises. *Journal of International Financial Markets, Institutions and Money* 19(2): 240–257.

Baek, J., Kang, J., and Park, K. (2004) Corporate governance and firm value: Evidence from Korean financial crisis. *Journal of Financial Economics* 71(1): 265–313.

David-Friday, P. and Gordon, E. A. (2005) Relative valuation roles of equity book value, net income and cash flows during a macroeconomic shock: The case of Mexico and the 1994 currency crisis. *Journal of International Accounting Research* 4(1): 1–21.

Jiangli, W., Unal, H., and Yom, C. (2008) Relationship lending, accounting disclosure, and credit availability during the Asian financial crisis. *Journal of Money, Credit and Banking* 40(1): 25–55.

Kamath, R. (1980) Determinants of premiums and discounts on commercial bank common stocks. *Review of Business and Economic Research* 16(1): 54–72.

Maniam, B., Leavell, H., and Patel, V. (2004) Financial crisis in emerging markets: Case of Argentina. *Journal of American Academy of Business* 4(1/2): 434–438.

Mitton, T. (2002) A cross-firm analysis of the impact of corporate governance on the East Asian financial crisis. *Journal of Financial Economics* 64: 215–241.

Richardson, G. (2007) The check is in the mail: Correspondent clearing and the collapse of the banking system 1930 to 1933. *Journal of Economic History* 67(3): 643–671.

Rochet, J. (2003) Why are there so many banking crises? *CESIFO Economic Studies* 49(2): 141–156.

Selim, T. H. (2005) A comparative essay on the causes of recent financial crises. *Business Review* 3(2): 303–309.

Soral, H. M., Iscan, T. B., and Hebb, G. (2006) Fraud, banking crisis and regulatory enforcement: Evidence from micro-level transaction data. *European Journal of Law and Economics* 21(2): 179–197.

Staikouras, S. K. (2004) A chronicle of banking and currency crisis. *Applied Economic Letters* 11: 873–878.

Wilson, B., Saunders, A., and Caprio, Jr. G. (2000) Financial fragility and Mexico's 1994 peso crisis: An event-window analysis of market-valuation effects. *Journal of Money, Credit, and Banking* 32(3): 450–473.

CHAPTER 15

Overcoming Institutional Myopia and Bankers' Self-Dealing Behavior: Coping with the Impact of the Global Financial Crisis on European Securitization Markets

Ulrich Hommel and Julia Reichert

CONTENTS

Deficient governance systems of banking firms are one of the causes of the recent financial crisis. Institutional myopia and lax constraints for self-dealing by bankers have led to the buildup of untenable risk positions in the banking industry. This chapter looks at the challenges ahead from a European perspective. In particular, it will look at different regulatory alternatives and how they may contribute to resolving the issues underpinning the current woes of the banking industry.

15.1 INTRODUCTION

An examination of the history of financial markets would have provided regulators, investors, and the banking community with ample and timely signals of a crisis in the making. Credit volume growth was outpacing economic growth by a significant margin and such periods have always led to an accumulation of bad credit risks and the ultimate bursting of a credit bubble. The Golden 1920s followed by the Great Depression serve as a case in point. While all such supergrowth cycles have been rationalized by experts on the basis of supposedly special economic circumstances at the time, subsequent events have proven them wrong in every instance.

The rapid growth of securitization markets coupled with the investors' indiscriminate willingness to acquire complex and nontransparent assets has been one of the most important triggers of the ongoing financial crisis. Securitization markets have been also the source of infection for the

European banking system. In an effort to add a European perspective to this edited volume, we, therefore, discuss how the European securitization business became increasingly linked to the U.S. subprime market and how structural market flaws made European institutions particularly susceptible to the crisis. After developing a comprehensive understanding of the unique features of the "European" banking crisis, we turn to a discussion of potential regulatory responses in order to shield securitization markets from similar events in the future.

The remainder of this chapter is organized as follows. Section 15.2 summarizes recent developments of European securitization markets and highlights key differences to the U.S. counterpart. In Section 15.3, we explain why European financial institutions actually ended up suffering the most from the financial crisis. Section 15.4 sheds more light on the ultimate sources of the crisis—institutional myopia and individual self-dealing. Finally, alternative regulatory approaches to coping with these issues are analyzed in Section 15.5.

15.2 DEVELOPMENT OF THE EUROPEAN SECURITIZATION MARKET

Securitization has gained significantly in importance throughout Europe starting in the late 1990s. Issuance volume grew from €78 billion in 2000 to €418 billion in 2006 (ESF, 2008). Despite a more than fivefold increase, the European market has, however, remained small compared to the global securitization volume of approximately $3000 billion for the same year (ASF et al., 2008).

The European securitization market is still fairly heterogeneous, in particular with respect to the asset class mix and the size of issuance. Observed differences are mainly due to a regulation-induced market fragmentation along national boundaries creating a unique competitive environment in every European country.* With €615.5 billion in outstanding volume as of Q4/2008, the U.K. market is by far the largest individual market within Europe, followed by Spain, the Netherlands, Italy, and Germany (Figure 15.1) (ESF, 2008). The European securitization market in total has been dominated by residential mortgage-backed securities (RMBS), which comprise more than 60% of the outstanding volume in Europe as of Q4/2008. On a country basis, their importance varies with RMBS making up 23% of

---

* For further details on regulatory differences of European markets see, for example, Kothari (2006, pp. 115–147).

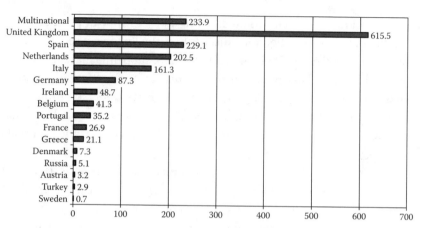

FIGURE 15.1 Outstanding securitization volume by country of collateral, Q4/2008 (€ billion).

the outstanding volume in Germany, 71% in Spain, 74% in the United Kingdom, and 90% in the Netherlands (Figure 15.2) (ESF, 2008).

Despite these differences within Europe, there are some key commonalities that distinguish the European securitization market from the U.S. counterpart. In this context, the U.K. market constitutes an exception as it resembles the U.S. market in many structural and regulatory aspects and thus distinguishes itself from rest of Europe. The U.K. mortgage market—as its U.S. counterpart—has for example a considerable number of specialist lenders providing finance to niche borrowers (e.g., subprime borrowers). As a consequence, the U.K. securitization market has a significant nonconforming (subprime) and investment property (buy-to-let) segment. This is atypical for Continental Europe, as mortgage securitizations are mostly of prime quality (Adams, 2005). Another similarity between the U.K. and the U.S. market lies in the importance of mortgage-backed market growth as the driver for overall market growth. The British real estate market has seen an enormous boom in recent years, combined with high loan-to-value (LTV) ratios and a marked increase of private indebtedness. This led to an increase in gross lending volumes of approximately 200% from £120 billion in 2000 to £363 billion in 2007 (Bank of England, 2009), which in turn fueled securitization volumes even further. In contrast, in most Continental European countries (with the exception of Spain) house prices have increased more moderately or have even stagnated as in the case of Germany (Dechent, 2008). Furthermore, an important difference between the U.S./U.K. markets and the Continental European markets lies in the differing business

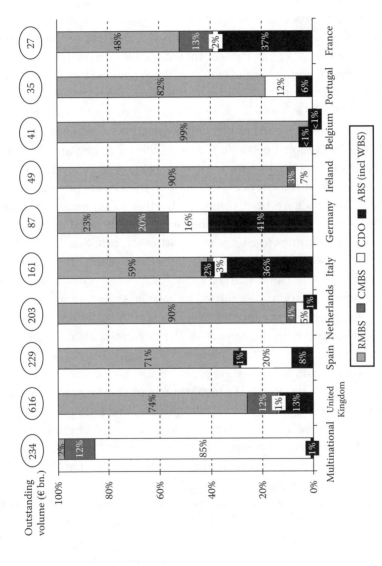

FIGURE 15.2   Asset class mix by country of collateral, selected European markets, Q4/2008.

models and degrees of specialization. Many U.S. and some U.K. mortgage firms have been operating under the originate-to-distribute business model, that is, they have originated loans only to distribute or securitize them within days.* In contrast, it has been much more common for Continental European banks to securitize existing assets that they had held on their balance sheet for some time as a means of capital relief and refinancing. Regulators have also frequently undertaken specific steps to mitigate the risk that loans earmarked for subsequent securitization would be subjected to less stringent credit risk standards than those retained by the bank. As, for instance, stipulated by a circular of the German banking supervision agency, BAFIN, the loans included in a securitization deal need to be chosen at random (on the basis of predefined selection characteristics) from the existing institutional portfolio (BAFIN, 1997). Furthermore, the loan officer handling a certain loan file must be unaware of whether the loan has already been securitized away or is still on the bank's balance sheet. Similar business standards have been common in other Continental European countries as well (but often noncodified), which has curtailed institutional and individual incentives to engage in opportunistic behavior.

In summary, some key differences between the (Continental) European and the U.S. securitization market apply: European securitizations are mostly of prime quality, are dominated by balance-sheet transactions, and display a lower degree of specialization. These differences are of importance when considering the impact of the current crisis on European securitization markets. In fact, many of the market-immanent structural weaknesses and securitization practices that have contributed to the high delinquency rates in the U.S. subprime market have been much less prevalent in (Continental) Europe. Delinquencies and rating downgrades have nevertheless been rising—but at much more modest rates than in the United States. Since the outbreak of the crisis in Q3/2007, the ratio of upgrades to downgrades has been approximately 1:34 for U.S. issues, whereas it remained at a level of about 1:12 in Europe.† Consequently, the share of U.S. securities with a Moody's Aaa rating fell to 63.0% in Q4/2008 (81.8% in Q1/2008) (Figure 15.3), whereas the share of securities rated Ba or below rose from 4.1% in Q1/2008 to 14.0% in Q4/2008 (Figure 15.4 and Figure 15.5). European ratings remained more favorable with 81.1% of the

---

* See Duffie (2007) for a discussion of the effects of credit risk transfers.

† From Q3/2007 to Q4/2008 Moody's performed 217 upgrades and 2,645 downgrades of European issues (ratio 1:12.2) compared to 1,740 upgrades and 58,682 downgrades (ratio 1: 33.7) for U.S. issues.

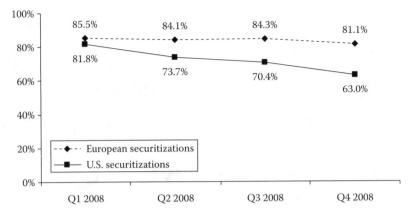

FIGURE 15.3   Share of outstanding Moody's ratings Aaa, 2008.

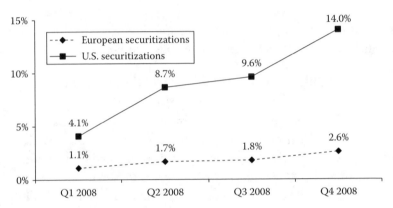

FIGURE 15.4   Share of outstanding Moody's ratings Ba and below, 2008.

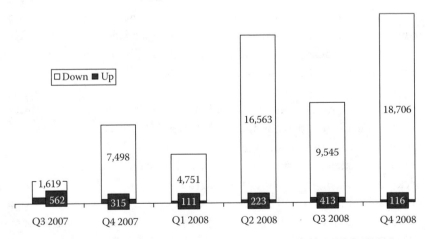

FIGURE 15.5   Moody's rating actions, U.S. securitizations, Q3/2007–Q4/2008.

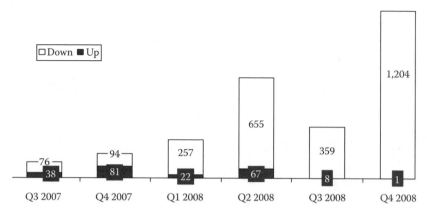

FIGURE 15.6 Moody's rating actions, European securitizations, Q3/2007–Q4/2008.

outstanding securities being rated Aaa and only 2.6% having a rating of Ba or below as of Q4/2008 (Figure 15.6) (ESF, 2008).

15.3 HOW DID EUROPEAN BANKS END UP BEING AMONGST THOSE SUFFERING MOST?

In a first wave, European banks were hit by the financial crisis mainly because of their role as ABS investors and sponsors of asset-backed commercial paper (ABCP) programs. Many European banks had invested heavily in the U.S. subprime market. As delinquency rates started to rapidly increase in 2007, market prices for RMBS tranches tumbled and banks were forced to write off significant portions of their RMBS investments. While ABCP programs have originally been designed to securitize constant income streams such as leasing receivables, they have in recent years been increasingly used to refinance long-term assets (such as mortgages) with (cheaper) short-term debt. Of the top 20 ABCP program administrators as of March 2007, 14 were of European origin, 1 was Asian, and 6 were U.S. institutions (Moody's, 2007) (Figure 15.7).

ABCP programs employ so-called structured-investment vehicles (SIVs), typically set up as conduits. While the SIVs are bankruptcy-remote, the program sponsor usually provides a liquidity facility to the SIV. As credit markets have started to dry up, rolling over short-term financing became increasingly difficult and program sponsors were called upon to bridge liquidity gaps and in some cases were even forced to consolidate the assets that had originally been strictly off-balance-sheet. As a consequence, European banks were hit hard with deteriorating equity positions

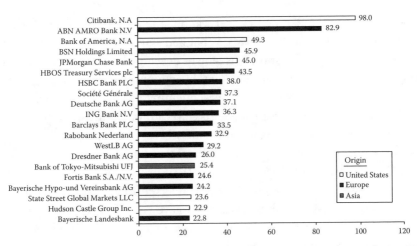

FIGURE 15.7   Top 20 ABCP program administrators, outstanding volume Q1 2007 ($ billion).

and additional refinancing requirements. Some midsized banks (e.g., the German IKB) had accumulated significant risk positions relative to their balance sheets and were, therefore, the first ones asking for a government bailout as the only means of preventing insolvency.

European banks were subsequently hit by a second wave of losses when the financial crisis started spreading to the real economy. With more and more workers being forced out of employment, mortgage delinquency rates have been rising throughout Europe. Economies that have suffered the most from a housing price bubble have been hit particularly hard with rapid drops in house prices leading to rising loss ratios for mortgages in default. The impact of the first wave on Continental European markets has, however, been considerably stronger, has affected financial institutions more swiftly, and rather unexpectedly. The remainder of this chapter will provide a further discussion of the causes of the crisis as well as the regulatory implications. Since many of the structural weaknesses on the originator side play less of a role in (Continental) Europe, our discussion mainly focuses on the investors' perspective.

15.4 INSTITUTIONAL MYOPIA AND SELF-DEALING BEHAVIOR

At the heart of the financial crisis lies a problem of institutional myopia as well as self-dealing behavior of market participants. Many of the active players in the securitization market acted in a rather short-sighted manner and without much concern about the long-term effects for the entire market or even themselves. What had started as unchecked cream-skimming behavior by some has slowly developed into widespread and opportunistic profit-taking behavior by many. The entire market was eventually suffering from a prisoners' dilemma problem with most institutions no longer applying proper risk precautions. Turning a blind eye toward excessive risk exposures ultimately became an acceptable practice in the entire industry. In the end, all market participants were exploiting the limitations of Basle I/II rather than protecting their institutions against them. This development is best referred to as institutional myopia. Myopic decision-making structures gave rise to previously unheard of self-dealing opportunities with bonuses typically paid out on the basis of non-risk-adjusted performance.

15.4.1 Information Asymmetries and Incentive Problems

The general conditions under which myopic and opportunistic behavior manifested itself were characterized by information asymmetries.* A broad base of the academic literature studies the effects of information asymmetries on security design (see, for example, Riddiough, 1997; DeMarzo and Duffie, 1999; DeMarzo, 2005). More recently, Ashcraft and Schuermann (2008) identified seven informational frictions throughout the securitization process, which have contributed to the formation of the crisis. These frictions involved all sorts of players from originators, arranging investment banks to investors. Originating banks disguised the true quality of securitized loans, in some cases even collaborated with the borrower to make false interpretations of the loan application and gradually reduced lending standards.† Originators did not particularly care either, as they did not keep any default risk on their own books (Crouhy et al. 2008). Demyanyk and van Hemert (2008) show that credit quality deteriorated during the years leading up the crisis and that securitizing investment banks must have been, to some extent, aware of it. Investment banks had little incentive to perform a proper due diligence on the assets acquired from mortgage banks knowing

---

* Empirical evidence for the existence of information asymmetries is, for example, provided by Bernardo and Cornell (1997). They find significant differences in the ex ante valuation of Collateralized Mortgage Obligations (CMOs) by investors, which they attribute to private information regarding valuation.
† For an analysis on lending standards see Dell'Ariccia and Marquez (2006), Dell'Ariccia et al. (2008), or Keys et al. (2008).

that investors could or would not verify their efforts. Finally, investors acquired structured finance instruments whose true risk profile they could not possibly understand.

Earlier research suggests that market pressures to maintain institutional reputation should have prevented banks from exploiting information asymmetries (Duffie and Gârleanu, 2001). Ambrose et al. (2005) found that loans securitized between 1995 and 1997 have experienced lower ex post defaults than those retained in the institutional portfolio, which is consistent with this hypothesis. Investors have strongly relied on bank reputation and thereby misjudged its disciplining role in the context of changing market structures and business models. It can be concluded with 20/20 hindsight that they had in fact carelessly delegated a core investment function to financial intermediaries that did not really deserve such far-reaching trust. Lax reporting was insufficiently penalized and the rising complexity of securitized portfolios (due to the packaging and repackaging of risks) failed to trigger sufficient precautionary efforts by investors as part of their general due diligence. It, however, would have been also the responsibility of bank regulators and financial market oversight agencies to implement proper checks and balances and to tighten regulatory standards in the face of rising systemic risk.

15.4.2 Deficient Bank Governance and Risk Management Systems

Proper risk governance would have prevented financial institutions from investing in securities, which were only understood by a handful of people. Accepting financial risks is obviously one of the core functions of banking institutions—even if it is just done temporarily as part of risk transformation. Full transparency of the aggregate risk position is an essential prerequisite for effective risk governance which must above all ensure the adequate provision of equity capital to back up potential losses. While the majority of banks were not in a position to correctly judge the full scope of their risk exposures, they nevertheless kept on investing in structured financial products and no internal controls prevented them from doing so. As more and more of these positions developed into "bad risks," banks were increasingly caught in a so-called liquidity trap. Write-offs led to the deterioration of bank capital positions and rating downgrades triggered additional capital requirements. As market participants began to draw on existing liquidity facilities, a contagion effect started to spread through the entire banking system. The interbank lending market eventually came to a complete halt as liquidity constraints tightened up and the perceived uncertainty over the willingness and ability of governments to bail out failing banks exploded following the bankruptcy of Lehman Brothers.

The banks' risk appetite had also been fueled by flawed compensation systems. Bonus payments were typically based on short-term (annual) performance indicators and were either immediately vested or could easily be transformed into an equivalent cash payment using supplementary OTC transactions (i.e., forward sale of employee stock or employee options with lockup restrictions). The supposedly "high earning" investment bankers had also been quite apt to negotiate favorable bonus arrangements with unlimited upsides and a bonus floor substantially exceeding corresponding guarantees found in corporate and retail banking. The lack of any penalties for switching employers further encouraged myopic behavior by bankers.

Bankers certainly had no reason to be cautious since another and better-paying job was typically just around the corner and since the current job was only as good as the time until the next market downturn. They must have also been fully aware that any banking crisis would ultimately be played out according to the principle "privatize gains, socialize losses." European legal systems do not assign any criminal or civil penalties for proven foolishness and incompetence. Hence, it is not surprising that quite a number of senior European bank managers have been more occupied with what they should not know in order to protect their lifetime pension guarantees rather than develop a full grasp of the problem for the institution (Caprio et al., 2008).

Some European banks have so far needed more government bailout support than others. This is certainly a reflection of the respective bank's business model, the quality of its risk management system, the integrity of its compensation structures, and the depth of its equity capitalization. It is, however, safe to say that banks which have so far not requested public support must have had a reasonably functioning risk management system in place at the beginning of the crisis.

15.4.3 Lack of External Control

The emergence of institutional myopia and the rising prevalence of self-dealing behavior were supported by lax national regulations and the lack of regulatory oversight. Regulating financial institutions is certainly a balancing act with too much and too little regulation imposing harm on financial institutions and the real economy—especially in a regulatory environment with nationally fragmented structures and the tacit acceptance of a race toward the least restrictive common denominator. Caprio et al. (2008) called the spiral toward less and less regulatory oversight an "endless and unfair game of action and response."

As is widely acknowledged today, regulators simply lacked the quantitative capabilities as well as adequate financial resources to engage in comprehensive risk monitoring as a prerequisite to enforcing financial discipline. This was particularly true for securitized investments and it seemed that regulators deliberately closed their eyes and thus further exacerbated the problem. By allowing banks to shift their most toxic exposures off-balance-sheet and by thereby exempting those investments from fundamental capital adequacy and transparency standards, they had actually laid the groundwork for Europe's financial crisis. One should also not forget in this context that politicians of many European countries (in particular Germany) have played a dual role before and during the crisis. They have been instrumental in pushing the liberalization of financial markets forward as members of the national legislature or as members of government and, at the same time, they were signing off on the off-balance-sheet activities of state-owned or semipublic banks as members of supervisory boards or as external board members. The recent woes of the German Landesbanken (which are jointly owned by the state and the regional savings banks) serve as a case in point—gaining first-hand knowledge of how the business works had not improved regulatory outcomes and intimate knowledge of regulatory objectives had apparently no bearing on monitoring outcomes.

15.5 REGULATORY IMPLICATIONS

Already since the early days of the crisis, regulatory reforms of a more fundamental nature were demanded by politicians, regulators, and banking associations alike. The proposals roughly fall into one of two categories. The first set of reforms focuses on the securitization process and seeks to resolve the information and incentive problems inherent in these types of transactions. The second set of reforms targets more generally the flaws in bank governance and regulatory oversight.

15.5.1 Securitization Process

15.5.1.1 Alignment of Interests

The European Commission has targeted the differing interests between securitizers and investors with its proposed amendment of the Capital Requirements Directive. According to this proposal, "it should be required from investors to make sure that originators and sponsors retain a material share (not less than 5 per cent) of the risks" (European Commission, 2008). The regulatory intention is clear: risk sharing should align the interests of the securitizing bank with those of investors and thereby reduce the incentives for opportunistic profit-taking. The EU Commission's move has, however, been heavily criticized by the banking and securitization community, above all because of the projected increase in lending costs (BDB, 2008) and the operational difficulties of implementing the new standard (TSI, 2008). It was further argued that risk retention fails to address some of the more important regulatory issues (BBA et al., 2008; BDB, 2008).

National fragmentation of financial market regulation has in this context also been used as a counter argument which de facto favors the adoption of the least restrictive common denominator. In particular, it has been argued that a European solo attempt poses the danger of putting European banks at a competitive disadvantage relative to non-European rivals (TSI, 2008). The main problem of European institutions has been imperfect and incomplete risk recognition in their role as investors and the risk retention provision aims at protecting European banks against the consequences of their own lack of judgment. The unilateral implementation of such a rule will, however, necessarily punish banks with sound risk management systems as well as prudent investment strategies and may, therefore, reduce the overall competitiveness of the European banking industry.

15.5.1.2 Increasing Transparency
An alternative approach to resolving the conflict between securitizer and investor is to increase the transparency of securitization transactions. Investors have typically possessed little information on the assets underlying a transaction and ongoing reporting has been rather lackluster. Hence, investors have been overly dependent on the judgment of rating agency as well as the securitizers' supposedly good reputation (ASF et al., 2008). Surely, investors will be more cautious looking forward and banks will be more forthcoming in supplying information in an effort to rebuild investor confidence. The information to be provided mandatorily to the investor community—in addition to the offering memorandum—should definitely include detailed credit quality information and distributions, cash flow projections, and stress scenarios. Furthermore, the documentation for new issues should include information on the loan originators and the transaction sponsors and whether these parties intend to retain any part of the transaction risk (e.g., the first-loss piece or FLP).

The EU Commission has argued that "investors should make their decisions only after conducting a thorough due diligence, for which they need adequate information about the securitisations" and that transparency requirements should be expanded accordingly (European Commission, 2008). Investors should be enabled to perform their own due diligence and make their own value and risk assessment. In addition, investors should receive monthly performance reports including payment analyses and delinquency reports for each underlying pool. Both, transaction data and performance information, should be made publicly available. In this context, it is noteworthy that the EU Commission places the primary responsibility on the investors rather than the originators. The Global Joint Initiative to Restore Confidence in the Securitization Markets, an initiative of major international securitization organizations, has also identified "improved disclosure of information on underlying assets for RMBS" as one of its key priorities in trying to restore confidence in the securitization markets (ASF et al., 2008).

Some politicians and academics have naively argued that the rapid buildup of a publicly accessible risk register is a prerequisite to overcoming the financial crisis. If done so in the very near term, such a move may, however, seriously backfire. While the lack of risk transparency has certainly contributed to the current credit market freeze, it has also allowed banks to silently de-leverage their portfolios and to trim down their risk positions. If the regulators were to turn on the spotlight and thereby reveal the true magnitude of "toxic assets" on bank balance sheets, it would probably lead to the instantaneous recognition that most banking institutions are essentially bankrupt. Instead of simply providing banks with funding guarantees, governments would then be forced to implement large-scale equity recapitalization programs leading to real fiscal outflows and the de facto nationalization of the entire banking industry.

15.5.1.3 Standardization

Even with all the required information available, forcing each investor to perform a stand-alone due diligence is clearly not a resource-minimizing approach. After all, the complexity of products drives costs and investors would not even consider such assets if they have to carry out an investment analysis from scratch every time. Standardizing the transaction design would, therefore, ease market entry barriers on the demand side and would also add additional discipline to the market by enhancing the comparability of different issues and the performance of different securitizers.

The German Pfandbrief is a perfect example for successful standardization (but even this instrument has not been completely unscathed by the crisis).* As a consequence, the Pfandbrief market has suffered much less from widening of credit spreads compared to asset-backed securities or other covered bond markets. It would surely be difficult to apply similar standards to other securitization types, especially since flexibility in transaction design is one of the key advantages of securitization. Establishing some form of a standard, however, appears inevitable in the light of the current crisis.

15.5.1.4 Capabilities of Investors

The assessment of ABS investment risk imposes considerable demands on the quantitative literacy of investors. As we know today, not all financial institutions have possessed the necessary sophistication to fulfill these requirements. The Counterparty Risk Management Policy Group (CRMPG) has, therefore, proposed to restrict the range of financial institutions permitted to invest in securitized products by implementing a formal licensing procedure. Structured financial instruments should only be sold to "sophisticated investors" with the proven capability to analyze the risk/return trade-off for such instruments, to price these assets, to quantify risk exposures and to run stress tests "within a matter of hours" (CRMPG, 2008).

15.5.2 Bank Governance and Oversight

15.5.2.1 Risk Management

The financial crisis has revealed significant shortcomings in risk monitoring and risk management for all types of banking institutions. Senior management often lacked knowledge and control of the institutional "risk book" and organizational risk management functions did not possess the required capabilities to handle stress scenarios. On top, the relevance of liquidity risk and its interlinkage with capital adequacy requirements was systematically ignored. The most comprehensive review of risk management practices has probably been carried out by CPRMG (2008) and has led to a number of important policy recommendations. In particular, they

---

* Banks issuing a Pfandbrief must be licensed to do so by BaFin, which supervises the Pfandbrief market. With a mortgage Pfandbrief, only up to 60% of the collateral value must be included in the cover fund backing the Pfandbrief. The collateral value must be derived prudentially by an appraiser. In the event of an issuer's insolvency, the claims of the Pfandbrief creditors are privileged by a preferential right.

propose that all information regarding risk taking, monitoring, and management is to be consolidated centrally in an independent risk function which needs to possess all the necessary capabilities to properly assess exposures. Top management should be informed regularly on all financial risk exposures and should also carry the responsibility for setting institutional risk tolerance levels.

15.5.2.2 Long-Term Incentives

Compensation and incentive systems play a central role in shaping the risk-taking behavior of bank managers. In the past, there has clearly existed a symbiotic relationship between institutional myopia of financial institutions and the short-term focus of institutional bonus systems. With large compensation packages at stake, some managers have been tempted to take disproportionate risks, especially if the risks were only expected to become visible after several years. Senior managers had certainly no reason to worry about creating a performance minefield with their actions if their risk of ever stepping into it again was negligible. By now, there appears to be an agreement among national regulators and legislatures across Europe that—looking forward—bonus arrangements must be risk-adjusted and must also be capped, that golden parachutes guarantees for senior bank executives must be restricted, and that board remuneration committees must be able to act more independently.

One approach is currently pioneered by UBS AG. On November 17, 2008, it announced the implementation of a new bonus/malus system for senior managers and main risk takers (UBS AG, 2008). Starting from 2009, cash and share bonuses will be kept in ring-fenced accounts for up to 5 years and bonuses accumulated in good years will be charged against "maluses" of bad years. A "malus" will be awarded for financial losses at group or divisional level, asset write-downs, personal misconduct, breaches of risk rules, and for missing performance targets.

15.5.2.3 Oversight

Financial supervision has failed to spot and prevent the financial crisis. The shortcomings of the supervisory systems of some of the most developed countries in the world are striking and fundamental reforms appear to be inevitable. To ensure more effective supervision, authority and capabilities of supervising bodies must be strengthened and unified across national boundaries. Regulators must gain access to reliable and comprehensive information. Their reach must include off-balance-sheet positions and so far unregulated nonbanks (e.g., mortgage firms, hedge funds), which must be bound by the same regulatory standards as banks. More importantly, the incentive structures of regulatory agencies must be revised to provide stronger incentives for regulators to keep up with new financial developments and, if necessary, invest more in the buildup of analytical capabilities. The fact that regulators have been to some extent aware of the rising systemic risk in the subprime market, but have nevertheless failed to take effective action further points in this direction. Similar to the banks' board members exerting too little effort on internal control, regulators underinvested in oversight—partly out of fear of missing out on additional economic growth opportunities.

Effective bank supervision requires functional organizational structures and the unambiguous assignment of responsibilities. The role of supervising the U.S. banking sector has historically been shared by several agencies with the actions of one agency not necessarily influencing the behavior of others.* In Europe, financial service supervision has been somewhat more centralized with fewer agencies sharing supervisory functions, but only on a national level. Despite the establishment of a single currency area and a single central bank (for the majority of countries), bank supervision has remained a national undertaking—largely for political reasons. The existence of a single European institution for banking supervision would have clearly enforced more consistent European standards for the rescue or support of ailing banks and would have prevented the emergence of national discrepancies. Looking forward, the establishment of a single supervisory body should facilitate the implementation of comprehensive standards and would certainly put a stop to the race toward the least restrictive common denominator—at least within Europe. While politicians across Europe have argued in favor of stronger cooperation on a European level, this has largely been lip service so far. Instead of more intense cooperation within the EU, we have rather seen unilateral attempts of individual member countries to reform their supervisory structures.

15.6 CONCLUSIONS

While banking institutions have already come a long way to de-leverage their portfolios, governments are still searching for a workable approach

---

* Agencies involved in bank supervision included the Federal Reserve, the Office of the Comptroller of the Currency (OCC) (for national banks), the Office of Thrift Supervision (for federal savings associations), the SEC (for investment banks), and the Federal Deposit Insurance Corporation (FDIC) (responsible for deposit insurance at most banks).

to stabilize the banking sector and to rejuvenate credit markets. Especially, European politicians appear to have become involuntary converts to Keynesian economics who are handing out loans, loan guarantees, and equity capital on an unprecedented scale and without a clear metric to measure expected effectiveness. European legislatures have also not made significant inroads to deal with securitization-specific crisis triggers or with the more general deficiencies in bank governance and oversight. A lot of saber rattling can, however, be heard in political hallways, which has, however, so far failed to ignite a constructive debate on the creation of a new international governance framework for financial markets—mainly due to the diversity of national interests. One must wonder whether we really have learned our lessons from history after all.

REFERENCES

Adams, P. (2005) An introduction to the European securitization markets. *Journal of Structured Finance*, 11(3): 33–39.

Ambrose, B. M., LaCour-Little, M., and Sanders, A. B. (2005) Does regulatory capital arbitrage, reputation, or asymmetric information drive securitization? *Journal of Financial Services Research*, 28(1–3): 113–133.

American Securitization Forum (ASF), SIFMA, Australian Securitisation Forum and European Securitisation Forum (2008) Restoring confidence in the securitization markets. Global joint initiative to restore confidence in the securitization markets. New York.

Ashcraft, A. B. and Schuermann, T. (2008) Understanding the securitization of subprime mortgage credit. Working Paper, Federal Reserve Bank of New York, New York.

Bank of England (2009) Monthly lending secured on dwellings, gross lending not seasonally adjusted (A5.3). Monetary and financial statistics tables, available at http://www.bankofengland.co.uk.

Bernardo, A. E. and Cornell, B. (1997) The valuation of complex derivatives by major investment firms. *Journal of Finance*, 52(2): 785–798.

British Bankers Association (BBA), London Investment Banking Association, International Swaps and Derivatives Association, Inc. and European Securitisation Forum (2008) Joint associations' response to European commission consultation on CRD potential changes. Securitisation, London, U.K.

BAFIN (1997) Sale of customer receivables in connection with asset-backed securities transactions by German Credit Institutions. Circular 04/97, Bundesanstalt für Finanzdienstleistungsaufsicht, Frankfurt, Germany.

BDB (2008) Association comments on the EU's review of the capital requirements directive (CRD). Press Release, Bundesverband deutscher Banken, Frankfurt, Germany.

Caprio, Jr. G., Demirgüç-Kunt, A., and Kane, E. J. (2008) The 2007 meltdown in structured securitization. Searching for lessons, not scapegoats. Working Paper, The World Bank, Washington, DC.

CRMPG (2008) Containing systemic risk: The road to reform. The report of the CRMPG III. Available at http://www.crmpolicygroup.org.

Crouhy, M. G., Jarrow, R. A., and Turnbull, S. M. (2008) The subprime credit crisis of 07. *Journal of Derivatives*, 16(1): 81–110.

Dechent, J. (2008) Häuserpreisindex—Projektfortschritt und Erste Ergebnisse für Bestehende Wohngebäude. *Wirtschaft und Statistik*, 2008(1): 69–81.

Dell'Ariccia, G. and Marquez, R. (2006) Lending booms and lending standards. *Journal of Finance*, 61(5): 2511–2546.

Dell'Ariccia, G., Igan, D., and Laeven, L. (2008) Credit booms and lending standards: Evidence from the U.S. subprime mortgage market. Working Paper, International Monetary Fund, Washington, DC.

DeMarzo, P. M. (2005) The pooling and tranching of securities: A model of informed intermediation. *The Review of Financial Studies*, 18(1): 1–35.

DeMarzo, P. M. and Duffie, D. (1999) A liquidity-based model of security design. *Econometrica*, 67(1): 65–99.

Demyanyk, Y. and van Hemert, O. (2008) Understanding the subprime mortgage crisis. *The Review of Financial Studies*, forthcoming.

Duffie, D. (2007) Innovations in credit risk transfer: Implications for financial stability. Working Paper, Stansford University, California.

Duffie, D. and Gârleanu, N. (2001) Risk and valuation of collateralized debt obligations. *Financial Analysts Journal*, 57(1): 41–59.

European Commission (2008) Proposal for a directive of the European parliament and of the council amending directives 2006/48/EC and 2006/49/EC as regards banks affiliated to central institutions, certain own funds items, large exposures, supervisory arrangements, and crisis management. Brussels, Belgium.

ESF (2008) ESF securitisation data report Q4 2008. Data Report, European Securitisation Forum, London, U.K.

Keys, B. J., Mukherjee, T., Seru, A., and Vig, V. (2008) Did securitization lead to lax screening? Evidence from subprime loans. Working Paper, University of Michigan, Ann Arbor, MI.

Kothari, V. (2006) *Securitization. The Financial Instrument of the Future*. John Wiley & Sons (Asia) Pte Ltd, Singapore.

Moody's (2007) ABCP 2nd quarter 2007 program index. Program Index, Moody's Investors Service, New York.

Riddiough, T. J. (1997) Optimal design and governance of asset-backed securities. *Journal of Financial Intermediation*, 6(2): 121–152.

TSI (2008) TSI nimmt Stellung zu McCreevys Vorschlag Kredithandel und Risikotransfer einzuschränken. Press Release, True Sale International, Frankfurt, Germany.

UBS AG (2008) Compensation report. UBS's new compensation model. Available at http://www.ubs.com.

CHAPTER 16

The Millennium's Credit Crunch and Lender of Last Resort: A Review of the Literature

Vicente Jakas*

CONTENTS

---

* Any opinions expressed in this chapter are those of the author and do not constitute policy of the Deutsche Bank.

This chapter reviews the literature and discusses the concept of lender of last resort (LOLR) with reference to the problem of systemic risk within the current crises perspective. The origin and criticisms of the modern theory on systemic risk as well as the origin of the current credit crunch are briefly reviewed. Similarly, the policy implications arising from the concept of the LOLR, the moral hazard, the costs of bearing emergency assistance, and the associated systemic risk are all discussed. Central banks are expected to assume a coordinating and supervisory role, and act as crises managers with the authority to oblige banks to monitor their risks and penalize bank management where appropriate.

16.1 INTRODUCTION

The fragility of the financial system as a consequence of systemic risk has been a matter of concern for a long time (see Thornton (1802), MacKay (1841), Bagehot (1873), and MacLeod (1883)). Systemic risk was thought to be caused by the irrational and subsequently herding behavior of investors who, all of a sudden, might decide to withdraw their liquid assets from an institution. The figure of an LOLR was then suggested by these authors as a way of reducing the probability of a financial collapse. However, since then, they were disturbed by the contradicting effects an LOLR would have upon the stability of the financial system. In this regard, the literature has been divided into that of supporters and opponents of the LOLR. The latter would prefer arrangements such as deposit insurance contracts and/or the provision of own capital requirements, whereas those in favor of the LOLR appear to be confronted with the so-called problem of eligibility, which consists in choosing the features or criteria an institution under financial distress should fulfill in order to be eligible for the LOLR rescue.

Since the latest reviews of this topic by Freixas et al. (1990), Kaufman (1994), and Dowd (1992), significant political changes have taken place. These go from increasing macroeconomic policy coordination within states (due to an ever-increasing financial interaction across economic blocks) to the full integration of the European Monetary Union and major reforms of international institutions such as the World Bank and the International Monetary Fund (IMF). Naturally, in a fairly large number of papers concerning the impact of such changes upon systemic risk, the LOLR and its international version have appeared. The latest events surrounding the current credit crunch have revived discussions about the LOLR function. As a result, the information seems to be scattered enough to justify revisiting the literature on the LOLR produced in the last decades. Motivated by this, the author decided to review and bring back to discussion some old and other rather new developments in the LOLR theory.

To this end, this chapter has been organized as follows: Section 16.2 is devoted to systemic risk and its origin. Particular attention is paid to define systemic risk and how systemic risk finds its roots in the behavior of economic agents, actors, or market participants, as well as the reasons for bank failures. Section 16.3 introduces the main discussions concerning the problem that the mere existence of an LOLR may constitute the source of the moral hazard. Therefore, this section also discusses alternatives to central banks intervention as well as the apparent conflict between price stability and the LOLR function with reference to the European Central Bank (ECB) interventions during the credit crunch. Another aspect discussed in this section refers to cases when Governments act as an LOLR; here the author makes mention of the U.S. and European Union (EU) experience, as well as its short- and long-term consequences. Section 16.4 outlines the major theoretical underpinnings surrounding the apparent consensus on the fact that if there should be an LOLR, only solvent individual institutions should be eligible for an LOLR rescue. Section 16.5 gives some examples of the eligibility criteria and tries to outline current issues and challenges that could arise from an eventuality of the kind. Section 16.6 also concerns the current challenges for an international version of the LOLR. The literature has tried to establish criteria and determined the role of such an institution. Some authors show that it is rather impossible a version of its kind, others appear to identify it as a new role for the IMF. Comments and remarks are produced in Section 16.7.

16.2 SYSTEMIC RISK AND ITS ORIGINS

A financial crisis "…is a sudden actual or potential breakdown of an important part of the credit system. Financial crises and panics have been part of the financial system for centuries" (Kindleberger, 1978, p. 49; MacKay, 1841, p. 32). They are associated with a loss of confidence in the standing of some financial institutions or assets. In the modern financial system, the chain of credit is based on tightly interlinked expectations of the ability of many different debtors to meet payments. In this context, a crisis can spread rapidly creating a contagion through the financial system and, if unchecked, can significantly affect the behavior of the economy as a whole. In economic theory, panics can be modeled as cases of multiple equilibrium and possibly dependent on the herding behavior (Fisher, 1999).

According to Kaufman and Scott (2003, p. 372), systemic risk refers to "the risk or probability of breakdown in an entire system, as opposed to breakdowns in individual parts or components, and is evidenced by co-movements (correlation) among most or all the parts. [… Hence] evidenced by high correlation and clustering of bank failures." Such a definition of systemic risk suggests a dramatic and unexpected structural change in equilibrium over an entire system as opposed to an isolated single institution. This could involve a *clustering* of bank failures that can lead to the collapse of the entire financial infrastructure. In other words, systemic risk refers to the sum of the single institutional failures as a result of correlated reactions to the same event or a set of information available at a given moment in time that contributes altogether to the entire system's breakdown.

16.2.1 The Behavior of Economic Agents, Actors, or Market Participants

Often, the literature has referred to *contagious systemic risk*, introducing a distinction between rational and irrational contagious systemic risk. Rational systemic risk refers to the contagious process triggered by information-based events (Allan and Gale, 2000). According to these authors, economic agents act rationally according to a set of information and, as a consequence, there are high probabilities of reaching the same conclusion and exhibiting the so-called herding behavior (Kaminsky and Reinhart, 1998). On the contrary, irrational behavior refers to economic agents acting without an apparent reason, either not based on, or uncorrelated to a set of information and events.

According to Diamond and Dybvig (1983), economic agents behave randomly, but if they conclude that other economic agents will withdraw at a given moment, they will try to withdraw immediately, thus triggering a run on the financial intermediary. Similarly, Wallace (1988) and Champ et al. (1996) assume that random behavior is not limited to economic agents but can be seen in a wider cross section of the population. Fama

(1985) asserted that market participants know that the bank transforms liquid liabilities into illiquid assets. On the other hand, Azariadis (1980), and Cass and Shell (1983, p. 200) postulated that the randomness in the behavior of individual agents would imply a "commonly observed random variable in the economy." This would suggest that the agents' behavior is the result of a more structured analysis of information rather than simply randomness (see also Allan and Gale, 2000). Such an opinion is further echoed by Dowd (1992), and Morris and Shin (1999). These authors refer to the fact that the information that is available to the agents and the reactions ex post are evidences suggesting no randomness in the agents' behavior. Most of the critics on randomness refer to the fact that the literature has been based on peculiar mutual funds that bear little resemblance on the real world.

16.2.2 Reasons for Bank Failures

It is widely accepted that banks can be a major source of systemic risk. Paradoxically, they are also the frontline protection against system failures. This is because of the different functions banks perform in the financial system. Banks participate in the creation of money by way of deposit liabilities. They manage or take part in the payments system by providing a sound and stable mechanism to allow payments. Through the creation of indirect financial securities, they act as pivotal financial intermediaries between lenders and borrowers. Banks may also be regarded as agents of information who contribute to the supply of information. Economic actors may choose to limit the availability of public information, but are nonetheless willing to share it with a bank in order to obtain the requisite finance. Finally, banks are maturity transformers, which means that they take liquid deposits and invest part of the proceeds in the form of illiquid assets. By doing so, banks pool risk and enhance economic welfare.

Banking firms thus exist because the capital market is not perfect and cannot supply the full range of instruments. Therefore, banking firms fill this void by providing financial services and are able, in most cases, to supply specific instruments, tailored on request.

Banks fail because of insolvency or illiquidity, as seen during the current credit crunch. A general panic results in mass withdrawals by depositors or investors who believe the bank does not have adequate reserves to meet the demand. Although the impact of general poor economic fundamentals cannot be entirely controlled by banks, they nonetheless have the resources at their disposal, which can be utilized to try and avoid financial distress.

Even though depositors are naturally concerned about the profitability of the bank, if they believe that banks do not have sufficient liquidity as a result of toxic assets, a panic run will occur. An alternative explanation of what precisely triggers bank runs is provided by Jacklin and Bhattacharya (1988). According to Jacklin and Bhattacharya, interim information about the bank's investment in the risky long-lived assets is what induces depositors to early withdrawals. A demand that the bank cannot support when resources are limited, leads to information-based bank runs.

It is often argued that the higher the early return, the higher the incentive for depositors to early withdrawals. It thus follows that if early withdrawals are anticipated, banks should make adequate reserve provisions to meet the demand. But maintaining reserves reduces the earning capacity of banks. The point here is that banks should strive to find the right balance between earnings in the long run, and maintaining adequate liquidity to meet the demand in the short run.

The issue here is how financial institutions deal with the information they gather on borrowers and lenders. Inspired by Bryant (1980), Diamond and Dybvig (1983) put forward a microeconomic model that encapsulates two functions of the banking sector. By examining the issues of maturity matching between assets and liabilities, and the provision of insurance to bank customers (depositors) against the liquidity risk, these authors made some important observations about bank crises. They demonstrated that bank deposit contracts can be optimal and, at the same time, give rise to banking panics. Specifically, the Diamond and Dybvig model accounts for (1) the existence of banks as a risk-sharing agreement between depositors against unexpected liquidity needs, (2) bank runs viewed as an act of collective irrationality by rational depositors, and (3) the introduction of deposit insurance as an efficient mechanism to prevent bank runs. Diamond and Dybvig also looked at how to cope with the panics as an explanation for bank runs, the consequences of open economies, and different solutions for preventing runs. Similarly, they proposed that, in order to resolve the problem of liquidity shock, banks should design optimal insurance contracts, screen loans, and credit profiles with modern technology.

16.3 LENDER OF LAST RESORT AND MORAL HAZARD

In case of a domestic bank run, the first institutional crisis manager is the central bank. The Fed's management of several arrangements, such as Bear Stearns, Wachovia, Merrill Lynch, AIG, and Washington Mutual, in the United States, was somehow observed during this crisis. And some of its failures, such as Lehman Brothers, were also observed. In the case of the ECB in Europe, this kind of intervention was not observed. Now going back to the theory, the literature deals with the extent to which central banks should intervene to cope with systemic risks and to ensure a desirable level of financial stability. Since the Bullionist Henry Thornton first introduced the concept of the LOLR back in 1802, the intervention of central banks has been linked to the idea of a *supra-institution* that provides liquidity to the institutions under liquidity constraints, if and only if they were solvent. In 1873, Walter Bagehot, also a bullionist, established that banks seeking liquidity after runs should have a higher rate to avoid the moral hazard.

The moral hazard is a concept introduced by MacLeod in 1883, which denotes the careless attitude investors may have once they know that the central bank will come and help their investing institutions in case of liquidity shortcomings.

16.3.1 Alternatives to Central Banks Interventions

In modern financial systems, the debate over the role of central banks persists, and the idea that their intervention can lead to more indiscipline has been supported by many authors. For example, according to Freixas et al. (1998a), central banks interventions encourage banks to take more risks with the understanding that they will be bailed out. If the money market is well developed, and the economy is liquid and well functioning, why should the central bank intervene?

Freixas prefers deposit insurance to intervention. Deposit insurance is a provision from the banking authorities that can achieve the same level of safety nets. Similarly, Santomero (1997) observed that the existence of flat insurance premiums independent of the level of risk undertaken by the agents suggests that the expected returns of the agents are likely to be a function of the expected profits generated from the insurance premium, rather than from their portfolio risk. Therefore, financial intermediaries are not being penalized in a way that is proportional to the level of risk of their investments.

Another alternative would be the suspension of convertibility of deposits. This is a common practice in developing or emerging market economies, which is used when widespread financial crises occur. It involves individual participants as well as institutions suddenly withdrawing and hence demanding payment on their deposits. In this way, wealth moves out of the financial system into other forms of investments. When this happens, however, systemic risk, and monetary and fiscal policies failures are often mistaken (see Kaufman and Scott, 2003).

Finally, the use of shareholders' funds as a safety net has also been suggested as an alternative mechanism to avoid liquidity shortcomings in the financial intermediary industry. Basel Accord on capital requirements promotes the use of risk assessments to allocate capital as a function of risk. Therefore, institutions operating at higher levels of risk will require a higher capital than those incurring lower levels of risk. The EU has introduced the Capital Adequacy Directive III (known as CAD III), which is almost a version of the Basel Accord introduced as part of the EU Legislation, which is still pending to be approved by most member states. CAD III and Basel II define three main types of risks: credit, market, and operational risks. Criticisms to the Basel II Accord, however, have been put forward.*

16.3.2 The European Central Bank: Apparent Conflict between Price Stability and the LOLR Function

It has been argued recently in the financial press about the possible conflict between the central bank's objective of price stability and that of the LOLR. In the short term, this can be ruled out in the case of the ECB, as long as the ECB is not mandated to perform the role of an LOLR. In the strict sense of the word, there is no reason to believe that the ECB will act as such. However, it might appear that this argument could be flawed depending on the horizon used for this analysis. As the crises unraveled, the ECB reduced interest rates, increased its balance dramatically, and accepted qualitatively lower collaterals from banks for the weekly tender. The only reason why this behavior has not pushed prices upward is that forecasts for economic growth have been revised downward since mid-2007; a smaller-than-expected output gap points to less inflation pressure and lower interest rates. Even though interest rates have massively fallen since November 2008, two months after Lehman's collapse, banks continue to be reluctant to lend, as risk premiums remained high, and financial intermediaries continue to reduce their balance sheet, to improve their profitability ratios as well as to avoid incurring losses due to expensive funding costs.

16.3.3 The Government as an LOLR: U.S. and EU Experience, Short- and Long-Term Consequences

European governments as well as the United States have engaged in economic programs aimed at pumping liquidity into banks. In some circumstances, this has resulted in the government acquiring equity and/or the

---

* For further analysis refer to Lind (2005).

so-called toxic assets as well as giving state guarantees to the affected banks. In the short term, this could give a boost of confidence to the markets giving access to funds, especially to those institutions affected by the crises. However, if the intervention does not take place rapidly and effectively, little gains are to be seen in the short run. Inadequate and ineffective government intervention might result in longer periods of state intervention. Long periods of government intervention result in counterproductive effects, as nonsponsored banks will suffer from the competitive disadvantage of not having a government sponsor. This is because funding will become more expensive for them in comparative terms. Another aspect is how the government interventions are expected to be financed, and what are the macroeconomic consequences of subsequent increases in government deficits and, hence, increases on the long-term interest rates. Not to mention the moral hazard effects that they could have in the long run, as institutions will not have incentives to control their risks and assign resources accordingly. The aim of government intervention should be limited to achieve stability for the banking system in the short-run. Government intervention can cause a moral hazard increase in systemic risk as a consequence of weaker financial systems with poorer risk control mechanisms and inefficient allocation of resources. The increase in regulation increases operating costs. For those who have an access to government sponsorship, these higher operating costs are financed by lower funding costs, as the government sponsorship gives them a cheaper access to the financial markets. However, those without government sponsorship become less profitable. Under these conditions, the optimum solution is to become government sponsored.

16.4 ILLIQUID BUT SOLVENT: LOLR FOR INDIVIDUAL INSTITUTIONS

A number of authors have arrived at the conclusion that the LOLR can be used to address large-scale crises. For example, Goodfriend and King (1988) defined the LOLR as being part of the monetary policy aimed at smoothing interest rates. An opposite view is that the LOLR should be used only for supporting individual institutions and not the market as a whole, as postulated by Goodhart (1999).

The literature suggests that using the concept of the LOLR to support and manage financial market crises constitutes a source of the moral hazard, which can also exacerbate systemic risk (Kaufman and Scott, 2003). This is so, because market participants may be eager to take more risks with the assumption that they will be bailed out. The LOLR should only be designed to provide liquidity if it is in response to an *adverse shock*, which raises the demand for liquidity that cannot be obtained from normal sources. This implies, in the first place, the existence of a widespread shock that is putting a single institution, or a particular market segment, under unexpected liquidity constraints; and secondly, that there are difficulties in accessing liquidity from interbank and/or international financial resources. Using the LOLR to restore wide financial market equilibrium has other hidden dangers. For example, if the liquidity shortage is triggered by an abrupt reversal in the international capital flow, the large-scale provision of the LOLR facility necessary to ensure liquidity of the entire system can lead to inflationary expectations and large exchange-rate depreciations as postulated by Sachs (1999). This in turn can produce a vicious cycle that may result in, or deepen, the liquidity and solvency problems of banks and enterprises. Nonetheless, Ishii and Habermeier (2002) suggested that, when crises occur, good institutional arrangements in these areas will allow markets and institutions to improve more quickly and at low financial costs.

By contrast, Freixas and Rochet (1997), following the line of argument of Kaufman (1991), argued that such an implicit rescue would discourage the management, shareholders, and depositors from achieving control over their portfolio risk, as well as from monitoring the activities of the banks where they have deposits. This may be the reason why one of the justifications of the LOLR stands from the point of view of the eligibility of the LOLR facility. There is consensus in the literature that even though a bank might be illiquid at some point in time, this does not imply insolvency. For institutions in such situations, Guttentag and Herring (1983) proposed that they could be provided with liquidity for a given period of time, without incurring significant losses in the process, i.e., until the helped institution is able to transform the required illiquid assets into liquid. This is so because in modern finance, even if the interbank markets are efficient, an illiquid but solvent bank may still not be able to find liquidity (Rochet and Vives, 2003). This seems to justify the LOLR for those institutions that are illiquid but solvent. Bagehot (1873) arrived at the same conclusion more than a century ago, as he asserted that only illiquid but solvent institutions should be eligible for the LOLR rescue.

16.5 ELIGIBILITY CRITERIA

When eligibility becomes a criteria, which determines whether a bank may qualify for the LOLR, central banks lay down the selection terms by collecting data on every single institution that has sizable impact on the financial system. This will not only imply the gathering of every single financial statement but also the information about the quality of the institutions' investments. Central banks must analyze the significant movements on the market participants' wealth in order to spot trends and threats. Finally, central banks determine the liquidity capacity as a function of the returns' volatilities, which, in other words, are the risks that institutions undertake and their ability to hedge them adequately.

Eligibility criteria may vary slightly from one central bank to another. The Bank of Canada (BOC), for example, is the ultimate source of liquid funds to the Canadian financial system. As such, it routinely provides liquidity to facilitate payments settlement and responds in various ways to exceptional or emergency situations. The BOC has three distinct roles as an LOLR. In the first place, the Bank facilitates the settlement of payments systems by routinely extending overnight credits to participants in the large value transfer system (LVTS) through the standing liquidity facility (SLF). Secondly, the BOC covers temporary end-of-day shortfalls in settlement balances that can arise in the daily settlement of payments. Thirdly, the BOC can provide emergency lending assistance (ELA) to solvent financial institutions that require more substantial and prolonged credits. The ELA is intended to overcome a market failure associated with financial institutions that have a significant share of their liabilities as "deposits" (fixed-value promises to pay, redeemable at a very short notice) and whose assets are generally highly illiquid. However, the Bank of Canada Act requires that such lending must be secured by collateral pledged by the borrowing institution. It is the policy of the Bank to lend only to institutions that are judged to be solvent in order to mitigate the moral hazard that can arise from such potential intervention, and to avoid damaging the interests of unsecured creditors. In conditions of severe and unusual stresses on the financial system, the Bank has the authority to provide liquidity. The terms of the BOC's provision of credit through the SLF are a routine activity, which is given under the following terms (BOC, 2004).

A more complex example of the LOLR function is seen in Europe. An interesting illustration is the evidence reported by Barth, Caprio, and Levine (2001) on the variation across the EU countries in supervisory institutions and practices. Their conclusion is that supervisory arrangements within the EU are as diverse as in the rest of the world. Boot (2006) explained that the difficulties arise from the fact that the ECB has primary stability responsibilities when it comes to the payment system. But the

ECB does not have an explicit task of preserving the stability of the financial system in general. This works though, through the so-called European System of Central Banks, as this responsibility is left to the national central banks. These national central banks also have the LOLR role, and not the ECB. Therefore, this formal description is of importance, because the practical allocation of tasks in the Euro-system could deviate considerably, particularly because of the Euro-area-wide consequences of the manifestation of systemic risks. It is clear that any serious role of the ECB in LOLR operations (and crisis management) should go hand in hand with some burden-sharing arrangements to cover potential losses in those operations. However, the credibility and authority at the ECB level will thus give a powerful boost to information sharing, and this could distinctly improve the efficiency and effectiveness of the LOLR operations.

In the past, it was thought that some financial institutions are "too big to fail." But the increasing transparency of the eligibility significantly reduces the moral hazard and other "constructive ambiguity" to use an expression from Corrigan (1990). This increasing transparency has also reduced the need to shroud the LOLR process in secrecy, as some authors recommend (Enoch, Stella, and Khamis, 1997). Ideally, most central banks would like to address the liquidity crisis in financial institutions without committing funds. They can act as coordinating agents. Freixas et al. (1998a) explained the significant role that central banks may undertake as coordinating agents. First, the central bank assumes the role of a "crisis manager." From this position, it is possible for the central bank to coordinate the agents' actions, and, therefore, prevent a speculative gridlock. "By guaranteeing the credit lines of all banks, the central bank eliminates any incentive for early liquidation. Also as a coordinating agent, the central bank can organize the bypass of the defaulting bank and provide liquidity to those that depend on the defaulting bank. Finally, because the existence of an interbank market may loosen market discipline, there is a plausible need for monitoring and supervising, with the regulatory agency having the right to close down a bank in spite of the absence of any liquidity crisis at that bank" (Freixas et al., 1998a, p. 184).

16.6 IS THERE A PLACE FOR AN INTERNATIONAL LOLR?

Sachs (1999, p. 182) defines that a liquidity crisis would be the "circumstance where a borrower cannot obtain short-term funds despite the fact that the rate of return on the short-term borrowing would exceed the market cost of capital." For Sachs, liquidity crises could be as a result of panics, debt overhang, and public sector collapse. Within this context, there are four roles for an international LOLR (ILOLR). An ILOLR should first of all forestall panic simply by being there. Second, it should engage in lending to a financial institution or system when panic is already occurring. Third, it should lend into a debt overhang, as in many bankruptcy legislation. And finally, its fourth role is to lend into a situation when public sector institutions collapse, to help a state to consolidate its political power.

Sachs' (1999, p. 184) view is that "...controls on short-term capital inflows to banking sectors of emerging markets are necessary and that prudential limits on capital inflows plus flexible exchange rates would have prevented many of these crises from ever happening." Sachs (pp. 184–185) criticises BIS regulations, which, first, he finds "...are too asset oriented and not adequately liability focused in the first place. Hence, they do not focus on the risk of liquidity crises. And second, they cause a sharp bias toward short-term lending, because of the way that we do the risk weighting in the capital adequacy standards, giving such low risk to interbank short-term loans. These standards have made the whole system more vulnerable to financial panic."

Capie (1998) explored the possibility of the existence of an ILOLR as well as the extent to which an international institution can perform the role of an ILOLR. It has often been assumed that the IMF could play this role. The IMF's ability to play this role has been tested during the financial crises in Asia and Latin America, which spread to other continents (see studies by Stanley Fischer (1999), Frankel (1999), and Cline (2005)). A strand of argument asserts that the function of such a lender is to provide the market with liquidity in times of need, and not to rescue individual institutions (Goodfriend and King, 1988; Goodhart, 1999).

However, other researchers, like Capie, noted that the IMF was unable to obtain enough resources to tackle the Mexico 1995 and Indonesia 1998 crises. In these cases, the relatively large size of the funds needed would have required that national central banks cede the right to issue their currencies to the ILOLR. Capie expressed strong reservations about the feasibility of an ILOLR in the form that is proposed today. He argued that "a lender-of-last-resort is what it is by virtue of the fact that it alone provides the ultimate means of payment. There is no international money and so there can be no international lender-of-last-resort." In other words, large international rescues can only incur more moral hazard. In addition, in the same way that an LOLR would not want to have special arrangements for any individual bank, which could later be running into liquidity problems, an international institution, which is focused on the country in difficulties, would not want to have special arrangements with the country of concern. It can be inferred from Capie's argument that such an arrangement would violate one of the fundamental rules of the LOLR, which is not to provide liquidity to individual countries but to the market as a whole.

"A different view was presented by Corsetti et al. (2003, p. 441), they showed "...how an international financial institution could help to prevent liquidity runs via coordination of agents' expectations, by raising the number of investors willing to lend to the country for any given level of the economic fundamentals". They asserted that "the influence" of such an institution is increasing in the size of its interventions and that the precision of its information: more liquidity support and better information make agents more willing to roll over their debt and reduces the probability of a crisis." This contradicts Capie because it is "...different from the conventional view that stresses debtor moral hazard. It is argued that official lending will be an incentive for governments to implement policies, despite the costs 'by worsening the expected return on these policies, destructive liquidity runs may well discourage governments from undertaking them, unless they can count on contingent liquidity assistance.'"

However, recent literature seems to follow Capie's assertions. This is the case with Schwartz (1999), who argued that the IMF, as well as other international institutions, cannot effectively perform the role of an ILOLR. He observed that the support provided by the IMF and the World Bank to countries in financial crisis has had a relatively limited impact on some recipients, subsequently incapable of preventing future similar economic problems or simply sinking into further crises. Jeanne and Wiplosz (2001, p. 5) asserted that for an ILOLR to be able to prevent banking crises, it would require "...two important changes in the global financial architecture. The first one would be a global issuing of an international currency, and the second would have to be operated by an 'international banking fund' closely involved in the supervision of domestic banking systems." Some more ways of reducing the moral hazard derived from the ILOLR have been outlined by Mishkin (2000. p. 12), who suggested eight principles for this purpose: "(1) restore confidence to the financial system; (2) provide liquidity to restart the financial system; (3) provide liquidity as fast as possible; (4) restore balance sheets; (5) punish owners of insolvent institutions; (6) encourage adequate prudential supervision; (7) engage in LOLR operations only for countries that are serious about implementing necessary reforms; and (8) engage in lender-of-last resort operations infrequently and only for short periods of time."

A study from Keleher (1999. p. 1) at the Joint Economic Committee Study on the ILOLR concerning the IMF and the Federal Reserve concluded that "...under existing institutional arrangements, the IMF cannot serve as a genuine LOLR. Specifically, the IMF cannot create reserves, cannot make quick decisions, and does not act in a transparent manner in order to qualify for an ILOLR. The Federal Reserve, however, does meet the essential requirements of an ILOLR. It can quickly create international reserves and money, although it has not openly embraced ILOLR responsibilities. The Federal Reserve can easily implement this function by employing several readily available market price indicators and global price measures." However, a question remains whether the other nations would accept the authority of the U.S. Federal Reserve, since national central banks are often seen as an expression of sovereignty of national states.

Maybe, more than an ILOLR what is needed is an international supervisory body accepted by all countries around the globe. This is simply because institutions with government sponsorship will become more attractive to investors despite the increase in regulations. As mentioned earlier, those institutions that have an access to government sponsorship will see that the higher operating costs resulting from more regulations are compensated by lower funding costs, as the government sponsorship gives them cheaper access to the financial markets. However, those institutions with government sponsorship and with a lack of regulations (such as Basel II, etc.) will become even more profitable than those that do comply. Under these conditions, the optimum solution is not only to become government sponsored but also to comply with as little regulations as possible as long as the sponsoring government tolerates it.

16.7 CONCLUSIONS AND FINAL REMARKS

This chapter presents a review of the current literature on the relation between the LOLR and systemic risk. It also analyzes the policy implications, the moral hazard, and the costs of emergency assistance, when financial institutions and countries find themselves in major crises. Systemic risk provides a sound justification for the existence of an LOLR. There are moral hazards especially if systemic risk subsequently increases. The literature suggests that the LOLR rescue may be more effective for individual institutions, as opposed to the market as a whole, especially for institutions under liquidity constraints but solvent. An international version of the LOLR does not seem possible due to the lack of an international currency, the lack of macroeconomic policy coordination, and the level of the moral hazard involved. The central bank is expected to assume a coordinating and supervisory role, as well as the role of a crises manager with the authority to oblige banks to monitor their risks and penalize bank management when appropriate.

BIBLIOGRAPHY

Akerlof, G. A. (1970) The market for lemons: Quantitative uncertainty and the market mechanism. *Quarterly Journal of Economics*, 84(3): 488–500.

Aharony, J. and Swary, I. (1992) Contagion effects of bank failures: Evidence from capital markets. *Journal of Business* (July), 56(3): 305–322.

Aharony, J. and Swary, I. (1992) Local versus nation-wide contagion effects of bank failures: The case of the south west. Working Paper, Israel Institute of Business Research, Tel Aviv, Israel.

Aharony, J. and Swary, I. (1996) Additional evidence on the information-based contagion effects of bank failures. *Journal of Banking and Finance*, 20(1): 57–69.

Allan, F. (1991) The market for information and origin of financial intermediation. *Journal of Financial Intermediation*, 1(1): 3–30.

Allan, F. and Gale, D. (2000) Financial contagion. *Journal of Political Economy*, 108(1): 1–33.

Allan, F. and Gale, D. (1997) Innovation in financial services, relationships and risk sharing. Working Paper, pp. 97–126, University of Pennsylvania, Philadelphia, PA.

Angelini, P. and Giannini, C. (1994) On the economics of interbank payment systems. *Economic Notes*, 23(2): 191–215.

Angelini, P. G., Meresca, G., and Russo, D. (1996) Systemic risk in the netting system. *Journal of Banking and Finance*, 20(1): 853–868.

Avery, R. B., Belton, T. M., and Goldberg, M. A. (1988) Market discipline in regulating bank risk: New evidence from the capital markets. *Journal of Money, Credit, and Banking*, 20(1): 597–610.

Azariadis, C. (1980) Self-fulfilling prophecies. *Journal of Economic Theory*, 25(1): 380–396.

Bagehot, W. (1873) *Lombard Street: A Description of the Money Market*, Henry S. King, London, U.K.

Barth, J., Caprio, G., and Levine, R. (2001) Banking systems around the globe: Do regulations and ownership affect performance and stability? In *Prudential Supervision: What Works and What Doesn't*, Miskin, F.S., (Ed) pp. 31–96, University of Chicago Press, Chicago.

Benston, G. and Smith, C. W. (1976) A transaction cost approach to the theory of financial intermediation. *Journal of Finance*, 31(2): 215–231.

Bernanke, B. S. (1990) Clearing and settlement during the crash. *Review of Financial Studies*, 3(1): 133–151.

Bernanke, B. and Gertler, M. (1995) Inside the black box: The credit channel of monetary policy transmission. *Journal of Economic Perspectives*, 9(1): 27–48.

Bhattacharya, S. and Thakor, A. V. (1993) Contemporary banking theory. *Journal of Financial Intermediation*, 9(1): 7–25.

Bhattacharya, S. and Fulghieri, P. (1994) Uncertain liquidity and interbank contracting. *Economics Letters*, 44(1): 287–294.

Bhattacharya, S. and Gale, D. (1987) Preference shocks, liquidity and Central Bank policy. In *New Approaches to Monetary Economics*, Barnett, W. and Singleton, K. (Ed.), Cambridge University Press, Cambridge, NY.

Bordo, M. D. (1990) The lender of last resort: Alternative views and historical experience. Federal Reserve Bank of Richmond Economic Review, Jan/Feb, pp. 18–29.

Birmimer, A. F. (1989), Distinguished lecture on economics in government: Central banking and systemic risks in capital markets. *Journal of Economic Perspectives*, 3(1): 3–16.

Bryant, J. (1980) A model of reserves, bank runs and deposit insurance. *Journal of Banking and Finance*, 43(1): 749–761.

Boot, A. W. A. (2006) Supervisory arrangements, LOLR and crisis management in a single European banking market. Online document from the University of Amsterdam and CEPR. Visit http://www.accf.nl/pages/members/Riksbank-April21-2006-paper-AWABOOT.pdf.

Calomiris, C. and Gorton, G. (1991) The origins of banking panics: Models, facts and bank regulation, In *Financial Markets and Financial Crises*, Hubbard, G. R. (Ed.), University of Chicago Press, Chicago, IL.

Capie, F., Goodhart, C. A. E., Fisher, S., and Schnadt, N. (1994) *The Future of Central Banking: The Tercentenary Symposium of the Bank of England*, Cambridge University Press, Cambridge, U.K.

Capie, F. (1998) Can there be an international lender-of-last-resort? *International Finance*, 1(2): 311–325.

Caprio, G., Dooley, M. P., Leipziger, D. M., and Walsh, C. E. (1996). The lender of last resort function under a currency board: The case of Argentina. *Open Economies Review*, 7(1): 625–650. Available at SSRN: http://ssrn.com/abstract=620520.

Cass, D. and Shell, K. (1983) Do sunspots matter? *Journal of Political Economy*, 91(2): 193–227.

Champ, B. et al. (1996) Currency elasticity and banking panics: Theory and evidence. *Canadian Journal of Economics*, 29(1): 828–864.

Char, V. (1989). Banking without deposit insurance or bank panics: Lessons from a model of the U.S. National Banking System. *Federal Reserve Bank of Minneapolis Quarterly Review*, 13(1): 3–19.

Chari, V. and Jagannathan, R. (1988). Banking panics, information, and rational expectations equilibrium. *Journal of Finance*, 43(1): 749–760.

Cline, W. R. (2005) The case for a lender-of-last-resort role for the IMF, Center for Global Development and Institute for International Economics September 23, 2005. Paper prepared for the Conference on IMF Reform, Institute for International Economics, Washington, DC, September 23. Available at http://www.iie.com/publications/papers/cline0905imf.pdf

Corrigan, E. G. (1990) Statement before U.S. Senate Committee on Banking, Housing and Urban Affairs, United States Senate, Washington, DC.

Corsetti, G., Guimaraes, B., and Roubini, N. (2003) International lending of last resort and moral hazard: A model of IMF's catalytic finance. *Journal of Monetary Economics*, 2006, v53(April 3), 441–471. NBER Working Paper No. 10125.

Diamond, D. and Dybvig, P. H. (1983) Bank runs, deposit insurance, and liquidity. *The Journal of Political Economics*, 9(3): 401–415.

Diamond, D. (1984) Financial intermediation and delegated monitoring. *Review of Economic Studies*, 51(3): 393–414.

Diamond, D. (1991), Monitoring and reputation: The choice between bank loans and directly placed debt. *Journal of Political Economy*, 99(4): 689–721.

Diamond, D. (1996), Financial intermediation as delegated monitoring: A simple example. Federal Reserve Bank of Richmond, *Economic Quarterly* (82/83,), pp. 51–66.

Diamond, D. (1997), Liquidity banks and markets. *Journal of Political Economy*, 105(5): 928–956.

Dowd, K. (1992), Models of banking instability: A partial review of the literature. *Journal of Economic Surveys*, 6(2): 107–132.

ECB, 2003, Memorandum of understanding on high-level principles of cooperation, Press release, March 10, Frankfurt, Germany.

ECB, 2005, Memorandum of understanding on cooperation between the banking supervisors, central banks and finance ministries of the European Union in financial crises situations, Press release, May 18, Frankfurt, Germany.

EFC, 2001, Report on financial crisis management, (Brouwer report), Document from the Economic and Financial Committee, EFC/ECFIN/251/01, Brussels, Belgium.

EFC, 2002, Financial regulation, supervision and stability, Document from the Economic and Financial Committee, EF76/ECOFIN 324, 10 October, Brussels, Belgium.

Enoch, C., Stella, P., and Khamis, M. (1997) Transparency and ambiguity in central bank safety net operations. IMF Working Paper No. 97/138.

Fama, E. (1985) What's different about banks? *Journal of Monetary Economics*, 15(1): 29–39.

Fisher, S. (1999) On the need for an international lender of last resort, Paper presented at the Conference in the American Economic Association and the American Finance Association New York. On IMF site: http://www.imf.org/external/np/speeches/1999/010399.htm

Franke, G. and Krahnen, J. P. (2005) Default risk sharing between banks and markets: The contribution of collateralized debt obligations, University of Konstanz, Germany and NBER Working Paper No. 11741. Nov. 2005, Cambridge, MA.

Frankel, J. A. (1999) International lender of last resort. Rethinking the International Monetary System Conference held at Wequasett Inn, Chatham MA, Federal Reserve Bank of Boston, MA. See also: http://ksghome.harvard.edu/~jfrankel/weqfrbosllr.PDF.

Freixas, X., Paragi, B., and Rochet, J.-C. (1998a) Systemic risk, interbank relations and liquidity provision by the central bank. Available at http://www.imes. boj.or.jp/cbrc/cbrc-08.pdf

Freixas, X., Giannini, C., Hoggarth, G., and Sousa, F. (1990) Lender of last resort: A review of the literature, Financial Stability Review, November, Bank of England, pp. 151–167.

Freixas, X. and Rochet, J.-C. (1997) *Microeconomics of Banking*. MIT Press, Boston, MA.

Goodfriend, M. and King, R. G. (1988) Financial deregulation, monetary policy and central banking. *Federal Reserve Bank of Richmond Economic Review*, 74(3): 35–40.

Goodhart, C. A. E. (1999) Myths about the lender of last resort. *International Finance*, 2(3): 54–60.

Goodhart, C. A. E. (2000) In *Financial Crises, Contagion, and the Lender of Last Resort*. Oxford University Press, Oxford, U.K.

Goodhart, C. A. E. and Illing, G. (2000) In *Which Lender of Last Resort for Europe?*, Central Banking Publications, London, U.K.

Greenbaum, S. I., Kanatas, G., and Venezia, I. (1989) Equilibrium loan pricing under the bank client relationship. *Journal of Banking and Finance*, 13(1): 221–235.

Greenbaum, S. I. and Thakor, A. V. (1987) Bank funding modes: Securitisation vs deposits. *Journal of Banking and Finance*, 11(1): 379–2401.

Grossman, S. J. and Stiglitz, J. (1980) On the impossibility of informationally efficient markets. *American Economic Review*, 70(3): 393–408.

Grossmann, R. (1993) The macroeconomic consequences of bank failures under the national banking system. *Explorations in Economic History*, 30(1): 294–320.

Guttentag, J. and Herring, R. (1983) The lender of last resort function in an international context. Princeton Essays in International Finance, Princeton, NJ.

Hasan, I. and Dwyer, G. (1994) Bank runs in the free banking period. *Journal of Money, Credit and Banking*, 26(2): 271–288.

Herrala, R. (2001) An assessment of alternative lender of last resort schemes. Bank of Finland Discussion Paper No. 1/2001. Available at SSRN: http://ssrn.com/ abstract=318141

Holmström, B. and Tirole, J. (1997), Financial intermediation, loanable funds, and the real sector. *Quarterly Journal of Economics*, 112(3): 663–691.

Humphrey, D. B. (1986), Payments finality and risk of settlement failure. In *Technology and the Regulation of Financial Markets: Securities, Futures, and Banking*, Saunders, A. and White, L. J. (Eds.), pp. 97–120. Lexington Books, Lexington, MA.

Huang, H. and Goodhart, C. A. E. (1999) A model of the lender of last resort. IMF Working Paper No. 99/39, Washington, DC. Available at SSRN: http://ssrn. com/abstract=880566

Ishii, S. and Habermeier, K. (2002) Capital account liberalisation and financial sector stability. IMF Occasional Paper 211, Washington, DC.

Huang, H. (2000) A simple model of an international lender of last resort. IMF Working Paper No. 00/75. Available at SSRN: http://ssrn.com/abstract=879585

Jacklin, C. J. (1987) Demand deposits, trading restrictions, and risk sharing, in *Contractual Arrangements for Intertemporal Trade*, Edward C. P. and Neil W. (Eds.), University of Minnesota Press, Minneapolis, MN.

Jacklin, C. J. and Bhattacharya, S. (1988). Distinguishing panics and information-based bank runs: Welfare and policy implications. *Journal of Political Economy*, 96(3): 568–592.

Jeanne, O. and Wiplosz, C. (2001) The international lender of last resort: How large is large enough? IMF Working Paper, Washington, DC and NBER Working Paper No. 8381. Issued in July 2001.

Kahn, C. and Roberds, W. (1996a) On the role of bank coalitions in the provisions of liquidity. Working Paper, June, Federal Reserve Bank of Atlanta, Atlanta, GA.

Kahn, C. and Roberds, W. (1996b) Payment system settlement and bank incentives. Working Paper, September, Federal Reserve Bank of Atlanta, Atlanta, GA.

Kaminsky, G. and Reinhart, C. (1998) On crises, contagion, and confusion. Working Paper, December. George Washington University, Washington, DC.

Kaufman, G. (1991) Lender of last resort: A contemporary perspective. *Journal of Financial Services Research*, 5(2): 95–110.

Kaufman, G. (1994) Bank contagion: A review of the theory and evidence. *Journal of Financial Services Research*, 8(2): 123–150.

Kaufman, G. and Scott, K. (2003) What is systemic risk and do bank regulators retard or contribute to it? *The Independent Review*, 7(3): 371–391.

Keleher, R. (1999) An international lender of last resort, The IMF, and the Federal Reserve. Joint Economic Committee Study for the United States Congress, Washington, DC.

Kindleberger, C. P. (1978) *Manias, Panics, and Crashes: A History of Financial Crises*. John Wiley and Sons, Hoboken, NJ.

Laeven, L. and Levine, R. (2005) Is there a diversification discount in financial conglomerates?, NBER Working Paper No. 11499, Cambridge, MA.

Leland, H. E. and Pyle, D. H. (1977) Informational asymmetries, financial structure and financial intermediation. *The Journal of Finance*, 32(2): 371–387.

Lind, G. (2005) Basel II—The new framework for bank capital, Sveriges Risksbank Economic Review 2/2005, pp. 32–36.

McAndrews, J. J. and Wasilyew, G. (1995) Simulations of failure in a payment system. Working Paper, no. 95-19, Philadelphia, PA. Federal Reserve Bank of Atlanta, June.

MacLeod, H. D. (1883) *The Theory and Practice of Banking*. Longmans Green and Company, London, U.K.

MacLeod, H. D. (1889) *The Theory of Credit*. Longmans Green and Company London, U.K.

MacKay, C. (1841) *Extraordinary Popular Delusions and the Madness of Crowds*. Barnes & Noble, New York.

Mishkin, F. (1995) Comments on systemic risk, in *Banking Financial Markets and Systemic Risk: Research in Financial Services, Private and Public Policy*, JAI Press, Greenwich, CT.

Mishkin, F. (1999) Moral hazard and reform of the government safety net. Paper prepared for FRB Chicago conference *Lessons from Recent Global Financial Crises*, Chicago, IL. September 30–October 2.

Mishkin, F. (2000) The international lender of last resort: What are the issues? National Bureau of Economic Research at: http://www0.gsb.columbia.edu/faculty/fmishkin/PDFpapers/00KIEL.pdf.

Morris, S. and Shin, H. (1999) Coordination risk and the price of debt, Cowles Foundation Discussion Papers, 1241R, Cowles Foundation, Yale University, New Haven, CT.

Niskanen, M. (2002) Lender of last resort and the moral hazard problem. Bank of Finland Working Paper No. 17/2002. Available at SSRN: http://ssrn.com/abstract=332321.

Pagratis, S. (2005) Prudential liquidity regulation and the insurance aspect of lender of last resort. Bank of England Working Paper Series. Available at SSRN: http://ssrn.com/abstract=557231.

Radelet, S. and Sachs, J. (1998) The East Asian financial crisis: Diagnosis, remedies, prospects. *Brookings Papers on Economic Activity*, 1998(1): 1–90.

Repullo, R. (2005) Liquidity, risk-taking and the lender of last resort. CEPR Discussion Paper No. 4967 Available at SSRN: http://ssrn.com/abstract=771491.

Rochet, J.-C. and Tirole, J. (1996a) Interbank lending and systemic risk. *Journal of Money Credit and Banking*, 28(4): 733–762.

Rochet, J.-C. and Vives, X. (2003) Coordination failures and the lender of last resort: Was Bagehot right after all? *Journal of the European Economic Association*, 2(6): 1116–1147.

Sachs, J. (1999) The international lender of last resort: What are the alternatives? Rethinking the International Monetary System. *Conference Proceedings Series No. 43*. Federal Reserve Bank of Boston, Boston, MA.

Schinasi, G. J. and Teixeira, P. G. (2006) The lender of last resort in the European single financial market. IMF Working Paper No. 06/127, Washington, DC. Available at SSRN: http://ssrn.com/abstract=910692.

Scholtens, B. and Van Wensveen, D. (2003) The theory of financial intermediation: An essay on what it does (not) explain. SUERF The European Money and Finance Forum, Vienna, Austria.

Santomero, A. (1997) Deposit Insurance: Do we need it and why? Working Paper, The Wharton Financial Centre.

Santomero, A. and Hoffman, P. (1998) Problem bank resolution: Evaluating the options. The Wharton School Financial Institutions Centre Discussion Paper 98-05, Philadelphia, PA.

Schoenmaker, D. (1996a), Contagion risk in banking. L.S.E. Financial Markets Group Discussion Paper, no. 239, London School of Economics, London, U.K.

Schwartz, A. J. (1999) Is there a need for an international lender of last resort? National Bureau of Economic Research. Available at http://www.somc.rochester.edu/mar99/schwartz399.pdf.

Thornton, H. (1802) (1939) *An Enquiry into the Nature and Effects of the Paper Credit of Great Britain*. (edited with an Introduction by F. A. von Hayek) George Allen and Uniwin, London, U.K.

Wallace, N. (1988). Another attempt to explain an illiquid banking system: The Diamond and Dybvig model with sequential service taken seriously. *Federal Reserve Bank of Minneapolis Quarterly Review*, 14(1): 11–23.

CHAPTER 17

Emerging Stock Markets and the Current Financial Crisis: Emergence of a New Puzzle?

Mohamed El Hedi Arouri, Fredj Jawadi, and Duc Khuong Nguyen

CONTENTS

This chapter aims to investigate the intensity and the strength of the impacts of the current international financial crisis on emerging stock markets. We first discuss the extent of international capital market linkages and major mechanisms of crisis shock transmission from the United States to emerging market economies since the beginning of the subprime distress. We then conduct an empirical study of the short- and long-run relationships between the United States and four emerging markets (Argentina, Mexico, South Korea, and Thailand) using a multivariate cointegration model. Our findings show significant, but asymmetric effects of the current crisis on selected emerging stock markets due particularly to the regional differences. More interestingly, export-dependent Asian markets seem to be more affected by the crisis than commodity-price-dependent Latin American markets. Finally, the results point to the emergence of a new puzzle that the efforts by emerging markets to reduce their dependences on the U.S. economy through developing internal demands does not permit to spare them from the current crisis.

17.1 INTRODUCTION

Emerging stock markets have been extensively studied in financial economics literature as they constitute a new, but quite challenging asset class for international portfolio investments. That is, high expected returns supported by high economic growth rates are coupled with their high volatility and low correlations with the rest of the world. In the early 1970s, both academic researchers and portfolio managers mainly asked the question of whether the inclusion of emerging market assets permits to reduce the risk of an international diversified portfolio (see, e.g., Grubel, 1968; Solnik, 1974). With the ongoing process of economic reforms and market openings that have taken place in emerging countries since the beginning of 1980s, the essential question is none other than to understand the effects of these changes on diversification benefits and financial stability. These above issues are all the more so important and challenging that global investors could have always in mind the disastrous consequences of various crisis events that shake the universe of emerging economies during the last three decades. Nevertheless, emerging markets still appear to attract foreign investors even though they become more economically and financially integrated with the world system (see, e.g., Bekaert and Harvey,

1995; Gerard et al., 2003; Carrieri et al., 2007). A more mature asset class should be then an explanation.

The advent of the 2007 subprime crisis in the United States that currently leads to a global economic recession may, however, set up a new deal for emerging markets and international investments. Indeed, when the U.S. housing markets started to ignite hedge funds failures and to damage the U.S. financial and banking sectors, many specialists demonstrated that emerging markets will be scarcely affected for several reasons. First, almost emerging markets have a high growth rate and a strong trade and fiscal balances. They accumulate, since the year of 2002, considerable current account surplus compared to the structural deficit for the United States. A prime example is China with a surplus representing more than 9.4% of the GDP in 2006. Second, in line with the current account surplus, the actual foreign reserves of emerging markets have exceeded 3 trillion compared to only 1 trillion in 2000 and represent bout 72% of the world reserves. This position of strength should naturally allow emerging markets to reduce their financial dependences on developed countries (i.e., one of the profound causes of the 1997 Asian financial crisis), to improve the liquidity levels and to stabilize financial markets. Finally, from a purely economic point of view, emerging market economies are being more independent of cyclical movements in the U.S. economy due to their efforts in stimulating internal demands of manufactured products. As a result, they will not suffer from the U.S. recession, or at least not much. But recently, more economists and financiers share the view that the economic slowdown in the United States would generate heavy influences on emerging countries as their expected economic growth declines sharply, and their stock markets experienced free falls in response to the financial crisis shocks, albeit smaller than those of developed countries.

The reasoning that emerging markets are not decoupled from the current financial crisis can be further explained by their major economic challenges. On the one hand, the recent return of high inflation due to sudden increases in fuel and food prices is likely to provoke dramatic changes in emerging market economies that are deeply dependent on oil imports. In addition to the dilemma of inflation and economic growth, some emerging countries feel the impact of declines in commodity prices following the global recession (i.e., the case of Latin American emerging countries) whereas the others suffer from the sharp decreases in their main exports (i.e., the case of most Asian emerging countries). On the other hand, less dependence on international capital flows does not mean that the impact of the U.S. crisis is not large on emerging markets since the latter experienced higher financial integration with the world during recent periods. Furthermore, strong comovements in financial sectors may also lead to comovement of the output level.

According to the above discussions, it is clear that the effect of the current crisis on emerging stock markets needs to be carefully analyzed as it is a matter of interests for both investors and policymakers. Accordingly, in this chapter we ask whether the current financial crisis affects emerging stock markets, and to a certain extent, how large are its impacts on the considered markets? To achieve this objective, we conduct an empirical study of the short- and long-run relationships between the United States and four of the most important emerging markets (Argentina, Mexico, South Korea, and Thailand) using a multivariate cointegration approach. The model is advantageous in the sense that it enables us to gauge the financial interdependences between sample markets, and to discuss the magnitude of shock transmission around the current financial crisis based essentially on their integration degree.

Using an up-to-date dataset, our findings show significant, but asymmetric effects of the current crisis on selected emerging stock markets due particularly to the regional differences. More precisely, export-dependent Asian markets seem to be more affected by the crisis than commodity-price-dependent Latin American markets. The results also point to the emergence of a new puzzle that the efforts by emerging markets to reduce their dependences on the U.S. economy through developing internal demands does not permit to spare them from the current crisis.

The remainder of the chapter is organized as follows. In Section 17.2, we briefly discuss the extent of international capital market linkages and major mechanisms of crisis shock transmission from the United States to emerging market economies since the beginning of the subprime distress. Section 17.3 presents the method used to assess the degree of market integration between individual stock markets of the sample. The evolution of the U.S. stock market and industrial production index represents the original source of the crisis shocks. Unlike previous studies that examine the issue of stock market linkages only on the short-term basis, our focus is on both short- and long-run dynamic relationships. Empirical results are also reported and discussed in this section. Section 17.4 provides some concluding remarks.

17.2 MARKET INTERDEPENDENCES AND MECHANISMS OF CRISIS SHOCK TRANSMISSIONS

As far as the question of studying the impact of a shock in one country on other country is concerned, empirical research usually takes a close look on two main channels of shock transmissions: the trade and financial linkages among different national economies (see, for instance, Caramazza et al., 2004). The latter can be effectively employed to show how crisis and economic recession events in the U.S. economy can affect other market places and countries.

Of the two channels, it is common that trade linkage or integration plays a crucial role as it reflects the degree to which finished products and services are exchanged between economies. For instance, Frankel and Rose (1998) show that business cycles become correlated positively as trade integration progresses. In other words, more trade integration leads to notably favor the quick transmission of shocks across countries. Despite the increasing role of European Union in international trade activities and the outperformance of emerging economies, the United States still appears to be the biggest importer in the world, and as a result, most emerging economies, especially export-oriented Asian economies, should not be completely decoupled from the United States. Then, it is reasonable to think that the global economic downturn affects the growth performance of emerging market economies through the existing linkages. One may note that the impact is definitively felt by several Asian emerging markets. For example, China's exports fell by 17.5% in February, 2009 from a year earlier while South Korea's exports declined by 32.8% in January 2009. Notice also that the net exports account for one-third in South Korea and almost half in Thailand.

The cross-market financial linkages constitute another channel through which shocks in the United States can be transmitted to emerging markets. This channel is more immediate because of the facility of cross-border capital flows greatly stimulated by the ongoing process of market openings in emerging countries. Thus, a decline in asset prices in the United States due to the financial crisis or economic slowdown would lead to lower and more volatile asset prices in emerging markets with high linkages. As emerging markets are now more open to foreign funds and they experience higher comovements with world stock markets in recent periods (see, e.g., Chen et al., 2002; Johnson and Soenen, 2003; Barari, 2004; Fujii, 2005), there is reason to expect some rapid and large responses to the original shocks.

In what follows, we concentrate on the analysis of finance links between emerging and U.S. stock markets to explore the possibility of crisis transmission. The rationale for doing so is based on the fact that finance links could be even a more useful way to detect contagion effects rather than a simple channel of shock transmission in times of crisis.

17.3 ECONOMETRIC IMPLEMENTATION AND EMPIRICAL RESULTS

As indicated previously, this section describes our econometric model to investigate the possible linkages between sample markets. After discussing the statistical properties of the data used, we proceed to test the hypothesis of integration between the emerging markets and the U.S. stock market using univariate and multivariate cointegration techniques. Note that the rejection of the integration hypothesis is informative of the fact that the U.S. and emerging stock markets are *a priori* segmented, whereas validating a multivariate vector autoregressive model (VAR) model suggests further evidence of significant impacts of the said crisis on emerging stock markets.

17.3.1 Preliminary Results

Our dataset consists of monthly stock market indices from two Latin American emerging countries (Argentina and Mexico), two Asian emerging countries (South Korea and Thailand), the U.S. stock market index, and the U.S. Industrial Production Index (CSA) over the period from December 1987–January 2009. Stock market indices are obtained from Morgan Stanley Capital International (MSCI), whereas the U.S. Industrial Production Index is extracted from the Federal Reserve Board database. All data are expressed in U.S. dollars in order to provide homogenous data and to avoid currency risk. To develop a global view of the stock price dynamics, we plot the five market indices together in Figure 17.1.

Figure 17.1 suggests firstly that all stock prices are not *a priori* stationary. Secondly, both Asian (respectively Latin American) stock prices seem to evolve together indicating some evidence of regional integration. Thirdly, the evolution of emerging stock prices toward the U.S. index did seem to be neither linear nor symmetrical. Indeed, before 1994, the emerging market indices appear to relatively follow the U.S. index due to the effects of synchronization, liberalization, and deregulation of the emerging financial systems, and the dependence of their economies and infrastructures on the U.S. economy, but after the Tequila effect, emerging

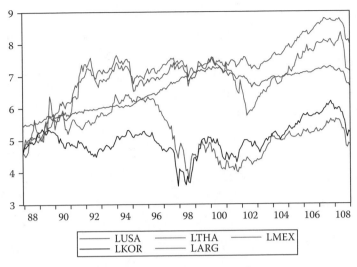

FIGURE 17.1 Logarithmic representation of the stock price movements.

markets are being more independent notably thanks to the stimulation of their internal demands. This implies lower dependence anchoring by the local emerging market fundamentals. However, this did not last for a long time because of the current international financial crisis, all emerging markets have followed the U.S. markets and they frighteningly plunged into an economic recession and their economic reality suggests the emergence of a new puzzle. The latter can be apprehended while investigating the linkages between the U.S. and emerging stock markets by making use of the cointegration tools.

We then check the integration order of stock prices using the Augmented Dickey–Fuller (ADF) and the Philips–Perron (PP) tests. Our findings suggest that all stock indices as well as the U.S. industrial production index are not stationary in the level but that are stationary in the first difference, suggesting that studied series are integrated of the first order, noted I(1).*

In order to have a preliminary insight about the impacts of the current international crisis on emerging stock markets, we compute the correlation coefficients among sample markets over two periods: December 1987–June 2007 (Table 17.1) and December 1987–January 2009 (Table 17.2). The first one excludes the financial crisis whereas over the second period, we include the effects of financial crisis. From Tables 17.1 and 17.2, correlation

---

* These results are available upon request to the authors.

TABLE 17.1   Correlation Matrix (December 1987–June 2007)

	RUSA	**RTHA**	**RMEX**	**RKOR**	**RARG**
RUSA	1.00	0.43	0.50	0.36	0.27
RTHA		1.00	0.37	0.50	0.20
RMEX			1.00	0.29	0.43
RKOR				1.00	0.10
RARG					1.00

TABLE 17.2   Correlation Matrix (December 1987–January 2009)

	RUSA	**RTHA**	**RMEX**	**RKOR**	**RARG**
RUSA	1.00	0.46	0.54	0.41	0.32
RTHA		1.00	0.41	0.53	0.25
RMEX			1.00	0.33	0.46
RKOR				1.00	0.15
RARG					1.00

matrices put up several important findings. Firstly, significant correlations of emerging markets and notably the Latin American ones with the U.S. market indicate important linkages between them. More interestingly, all these linkages are increasing and are more apparent over the whole sample period when including the current international crisis. This shows that the current crisis has affected emerging markets even though policymakers of these countries have tried to stimulate internal demands and reduce their dependence toward the United States since 1994. Secondly, given the bilateral interdependences that increased over the whole period, it would be worth noting that each of our sample markets potentially exhibits some degree of market integration with the others and that the financial crisis has increased their comovements.

The analysis of the descriptive statistics over these two periods also supports these findings. Indeed, from Tables 17.3 and 17.4, a significant decrease of stock returns for all studied indices is suggested. Moreover, stock markets seem to be more volatile over the whole sample period with a more important volatility for emerging stock markets. Symmetry and normality hypothesis are rejected for all stock indices, suggesting *a priori* complex and asymmetrical patterns of stock price dynamics.

However, it is important to note that these statistics only yield a static analysis of the financial crisis effects. In order to investigate this issue in a dynamic framework, we employ cointegration tools that allow

TABLE 17.3 Descriptive Statistics (December 1987–June 2007)

	RUSA	RTHA	RMEX	RKOR	RARG
Mean	0.01	0.003	0.01	0.005	0.01
Median	0.01	0.01	0.02	0.001	0.01
Maximum	0.10	0.35	0.25	0.53	0.66
Minimum	−0.15	−0.41	−0.41	−0.37	−0.48
Standard deviation	0.03	0.11	0.09	0.10	0.14
Skewness	−0.57	−0.41	−0.96	0.29	0.57
Kurtosis	4.08	5.00	6.07	6.23	6.73
Jarque–Bera	24.6	46.0	128.0	105.3	148.7
Probability	0.00	0.00	0.00	0.00	0.00
Observations	234	234	234	234	234

TABLE 17.4 Descriptive Statistics (December 1987–January 2009)

	RUSA	RTHA	RMEX	RKOR	RARG
Mean	0.004	0.001	0.01	0.002	0.01
Median	0.01	0.01	0.02	−0.003	0.01
Maximum	0.11	0.35	0.25	0.53	0.66
Minimum	−0.18	−0.41	−0.41	−0.37	−0.53
Standard deviation	0.04	0.11	0.09	0.10	0.14
Skewness	−0.84	−0.48	−1.01	0.22	0.36
Kurtosis	4.95	5.09	6.22	5.91	6.85
Jarque–Bera	70.6	56.0	153.1	91.7	162.5
Probability	0.00	0.00	0.00	0.00	0.00
Observations	253	253	253	253	253

studying at the same time the financial crisis impacts and financial linkages between the markets under consideration in both the short- and long-run horizons.

17.3.2 Cointegration Tests

The cointegration theory that has been developed by Granger (1981), Engle and Granger (1987), among others, stipulates that economic variables may undergo some short-term disruptions while sharing similar properties. Further, they can establish stable relations and converge in the long term toward the equilibrium. Formally, let X_t and Y_t be two variables that are I(1). X_t and Y_t are said to be cointegrated if it is possible to find a stationary linear combination z_t between these two variables, so that the cointegration relationship is given as follows:

$$z_t = X_t - a_0 - a_1 Y_t \tag{17.1}$$

where z_t designates the error term of the cointegration relationship.

In practice, we first estimate this long-term relationship by regressing the emerging stock index on the U.S. stock market index. We then test the linear cointegration hypothesis by applying the ADF tests to z_t. Summarized findings are reported in Tables 17.5 and 17.6 for both periods under consideration.

From the ADF statistics presented in Tables 17.5 and 17.6 and the critical values of Engle and Yoo (1987), we reject the hypothesis of linear cointegration for all emerging market indices for both periods. However, the rejection is less obvious over the whole period (1987–2009) since all ADF statistics have increased, implying also that the financial crisis has activated the linkages and comovements between the U.S. and emerging stock markets.

TABLE 17.5   Linear Cointegration Test (December 1987–June 2007)

Price Series	Constant	LUSA	$\bar{R}^2$	ADF (p, Model)
Mexico	−0.57 (−1.47)	1.17* (19.82)	0.62	−1.69 (1, a)
Argentina	0.97* (2.23)	0.89* (13.4)	0.43	−2.39 (0, a)
Thailand	9.18* (21.7)	−0.61* (−9.44)	0.28	−2.27 (0, a)
South Korea	4.84* (14.76)	0.02 (0.31)	0.02	−1.21 (0, a)

Note: The values between brackets are the t-statistics. (*) and (a) designate respectively the significance at 5% and a model without constant and linear trend. The order p is the number of lags retained while applying the cointegration test.

TABLE 17.6   Linear Cointegration Test (December 1987–January 2009)

Price Series	Constant	LUSA	$\bar{R}^2$	ADF (p, Model)
Mexico	−1.16* (−2.98)	1.27* (21.52)	0.65	−1.87 (0, a)
Argentina	0.55 (1.32)	0.97* (15.15)	0.48	−2.53 (0, a)
Thailand	8.7* (21.1)	−0.53* (−8.48)	0.22	−2.40 (0, a)
South Korea	4.14 (11.9)	0.913 (2.51)	0.02	−1.79 (0, a)

Note: The values between brackets are the t-statistics. (*) and (a) designate respectively the significance at 5% and a model without constant and linear trend. The order p is the number of lags retained while applying the cointegration test.

In order to ensure the robustness of the findings and provide an in-depth analysis of the issue, we applied the Johansen's (1988) trace test which enables us to simultaneously test for the cointegration hypothesis and the number of cointegrated relationships between markets. The main results obtained over the two periods are reported in Tables 17.7 and 17.8.

Overall, we reject the hypothesis of linear cointegration according to Johansen tests, and according to cointegration tests, emerging stock markets are not cointegrated with the U.S. market. Consequently, we turn our attention to short-term dynamics of these markets and attempt to study the dynamic causality relationships between emerging and U.S. markets over the considered periods without having to take the error-correction terms into account.

17.3.3 Granger Causality Tests

These tests may enable us to investigate the effects of one market's shocks (i.e., the financial crisis) on the other markets in the sample. To answer this question, we proceeded as follows. We first performed the Granger causality

TABLE 17.7   Johansen Trace Test (December 1987–June 2007)

Hypothesized No. of CE(s)	Eigenvalue	Trace Statistic	5% Critical Value	P-Value
None	0.10	57.03	69.81	0.33
At most 1	0.05	31.25	47.85	0.65
At most 2	0.04	17.73	29.79	0.58
At most 3	0.02	7.29	15.49	0.54
At most 4	0.003	0.91	3.84	0.33

Note: We don't reject the null hypothesis of non-cointegration because the trace statistic (57.03) is under the critical value (69.81).

TABLE 17.8   Johansen Trace Test (December 1987–January 2009)

Hypothesized No. of CE(s)	Eigenvalue	Trace Statistic	5% Critical Value	P-Value
None	0.10	66.21	69.81	0.09
At most 1	0.05	37.79	47.85	0.31
At most 2	0.04	22.67	29.79	0.26
At most 3	0.03	11.79	15.49	0.16
At most 4	0.021	5.32	3.84	0.02

Note: We don't reject the null hypothesis of non-cointegration because the trace statistic (66.21) is under the critical value (69.81).

test to examine whether stock returns in emerging stock markets are caused by return innovations in the U.S. market. We then estimated a VAR to investigate the dynamic adjustments of returns in four emerging markets to the U.S. stock market in the short term within a four-market VAR system. Note that not only we introduced the U.S. stock market index as an explanatory variable, but also, we introduce in addition the U.S. industrial production index to capture the effects of the U.S. recession on emerging stock markets.

Firstly, as a brief reminder of Granger's causality test, let X and Y be two variables. According to Granger (1969), the variable X causes the variable Y if X values provide statistically significant information about the future values of Y. Thus, the Granger causality test examines the null hypothesis of noncausality against its alternative of causality, and is based on a likelihood ratio test. Secondly, we apply this test to investigate the causality hypothesis between the four emerging markets and the U.S. stock market over the two sample periods retained. Our findings, reported in Tables 17.9 and 17.10 indicate that we do not reject the null hypothesis of absence of causality over the period 1987–2007 (the hypothesis is accepted for South Korea only at a 10% statistical level), suggesting that emerging markets are not significantly dependent on the U.S. market. However, over the second period, the null hypothesis is strongly and significantly rejected for Asian emerging stock markets and only at 11% and 15% for Argentina and Mexico, respectively. This implies further evidence of significant synchronization between the U.S. and emerging stock markets. It also suggests some evidence of new puzzle for emerging markets that even though they made considerable efforts to disconnect from the U.S. market, their tendency to follow exceed the effects of the mean reversion of their local fundamentals.

Finally, we represent these linkages between the U.S. and emerging stock markets while estimating the VAR model that was introduced by Sims (1980). Indeed, it permits us to examine both the lead–lag effects and the causality effects within the five-market system, while its moving

TABLE 17.9 Granger Causality Test (December 1987–June 2007)*

Series	*P*-Value
Mexico	0.64
Argentina	0.35
South Korea	0.10
Thailand	0.17

* Tables 17.9 and 17.10 present only the results of Granger causality test regarding the United States.

TABLE 17.10 Granger Causality Test (December 1987–January 2009)

Series	*P*-Value
Mexico	0.15
Argentina	0.11
South Korea	0.04
Thailand	0.08

average representation with orthogonalized residuals enables the impulse response of a specific market to shocks caused by another market in the system to be computed. The main advantage of this modeling is to simultaneously reproduce the adjustment dynamics of a vector of return variables (Mexico, Argentina, South Korea, Thailand, and the United States), enabling us to apprehend the behavior of several variables and study their responses following a shock which may affect any one of them. Note that unrestricted VAR is estimated by the least squares method while a restricted VAR is estimated by the maximum likelihood method. Empirical results are reported in Tables 17.11 and 17.12.

In practice, the lag number is determined using the information criteria but also a Likelihood ratio test testing a VAR($p+1$) against VAR(p). For our application, we retain $p = 2$.* Our estimation results of a VAR(2) over both sample periods show several important findings. Firstly, the dependence toward the American market is statistically significant but it

TABLE 17.11   VAR Estimation (December 1987–June 2007)

	RARG	**RMEX**	**RKOR**	**RTHA**
RARG(−1)	0.02 [0.29]	0.01 [0.24]	−0.04 [−0.80]	−0.02 [−0.32]
RARG(−2)	−0.12 [−1.75]	−0.01 [−0.49]	−0.05 [−1.13]	−0.05 [−1.05]
RMEX(−1)	0.02 [0.23]	0.09 [1.42]	0.12 [1.51]	0.09 [1.13]
RMEX(−2)	0.23 [1.95]	0.01 [0.27]	−0.06 [−0.81]	−0.01 [−0.08]
RKOR(−1)	0.04 [0.40]	−0.02 [−0.35]	0.01 [0.08]	0.05 [0.67]
RKOR(−2)	−0.03 [−0.37]	−0.05 [−0.92]	−0.03 [−0.49]	−0.06 [−0.86]
RTHA(−1)	0.05 [0.55]	0.01 [0.28]	0.06 [0.94]	0.03 [0.42]
RTHA(−2)	0.01 [0.12]	0.14 [2.57]	0.17 [2.48]	0.17 [2.39]
C	0.01 [0.83]	0.01 [1.90]	−0.00 [−0.04]	−0.00 [−0.57]
RUSA	1.091 [4.45]	1.13 [8.35]	0.97 [5.72]	1.23 [7.03]
RPI	−2.08 [−1.25]	−2.31 [−2.52]	−0.74 [−0.65]	−1.52 [−1.28]
R-squared	*0.10*	*0.30*	*0.17*	*0.22*
Adj. *R*-squared	0.06	0.27	0.14	0.19
F-statistic	2.67	9.68	4.81	6.58
Log likelihood	126.4	264.3	211.8	203.7
Akaike AIC	−0.99	−2.18	−1.73	−1.66
Schwarz SC	−0.83	−2.02	−1.56	−1.49

Note: RMEX, RARG, RUS, RTHA, RKOR, and RPI are respectively Mexican, Argentinean, United States, Thailand, Korean stock returns, and U.S. industrial production. [.] designates the *t*-ratio.

---

* The details of the determination of lag numbers are available upon request.

TABLE 17.12   VAR Estimation (December 1987–January 2009)

	RARG	RMEX	RKOR	RTHA
RARG(−1)	0.03 [0.45]	0.01 [0.43]	−0.02 [−0.54]	−0.02 [−0.47]
RARG(−2)	−0.12 [−1.80]	−0.02 [−0.71]	−0.06 [−1.44]	−0.06 [−1.32]
RMEX(−1)	0.03 [0.32]	0.08 [1.33]	0.13 [1.65]	0.10 [1.25]
RMEX(−2)	0.23 [1.97]	0.02 [0.44]	−0.05 [−0.74]	−0.004 [−0.05]
RKOR(−1)	0.057 [0.57]	−0.02 [−0.43]	−0.01 [−0.11]	0.04 [0.57]
RKOR(−2)	−0.01 [−0.14]	−0.04 [−0.78]	−0.02 [−0.36]	−0.05 [−0.78]
RTHA(−1)	0.04 [0.50]	0.02 [0.40]	0.07 [1.17]	0.02 [0.42]
RTHA(−2)	0.010 [0.11]	0.12 [2.39]	0.15 [2.36]	0.15 [2.27]
C	0.00 [0.42]	0.01 [1.47]	−0.00 [−0.29]	−0.00 [−0.53]
RUSA	1.23 [5.73]	1.21 [10.25]	1.06 [7.19]	1.27 [8.25]
RPI	−1.20 [−1.90]	−1.70 [−2.3]	−0.76 [−1.83]	−1.75 [−1.84]
R-squared	*0.14*	*0.34*	*0.22*	*0.25*
Adj. *R*-squared	0.10	0.31	0.18	0.22
F-statistic	3.92	12.54	6.77	8.31
Log likelihood	139.0	289.7	232.7	223.1
Akaike AIC	−1.02	−2.22	−1.76	−1.69
Schwarz SC	−0.86	−2.06	−1.61	−1.53

Note: RMEX, RARG, RUS, RTHA, RKOR, and RPI are respectively Mexican, Argentinean, United States, Thailand, Korean stock returns, and U.S. industrial production. [.] designates the *t*-ratio.

is notably stronger over the second period because of the financial crisis effects. Secondly, we highlight that the U.S. industrial production affects negatively and significantly the emerging stock markets, essentially after the crisis. This result is very interesting and suggests the above new puzzle for which it seems that over the second period, emerging markets are more driven by the U.S. economy than by the past. Thirdly, according to determination coefficients and information criteria, VAR models are more relevant over the second period. Finally, the U.S. stock market effects seem to be more significant than the regional effects suggesting a new puzzle relative to "American News." This also confirms our analysis above as well as the Granger causality test, and suggests that each shock affecting the New York stock market is transmitted to the other emerging markets at least in the short term.

Finally, in order to illustrate more explicitly this new puzzle, the dependence structure between these markets, and the relationship with the U.S. financial system, we estimate the impulsion response function and try to explain the consequences of a shock affecting the U.S. returns on those of the emerging stock markets under consideration over the two sample periods. Precisely, we orthogonalize the shock using Cholesky decomposition, define the shock as a standard deviation and study the effect of this shock over a 10 month period.*

Figure 17.2 shows, for Argentina, that the effect of a U.S. shock is immediate, and its consequences subside after 3 months. For Mexico, the effect is less important over the first period. However, the shock has negative effect for Asian markets and is more remarkable for South Korea. It is amortizing and disappears after 4–5 months. In Figure 17.3, we have the same effects but several interesting findings are found. Firstly, for the United States, the disappearance of a shock is longer. This fact typically reflects the actual anxiety characterizing the U.S. markets perhaps because of the current crisis. Secondly, the dependence of Mexico toward the U.S. financial system is more significant after the financial crisis. Finally, for all emerging markets, while comparing impulsion response functions before and after the crisis, we suggest further evidence of persistence in the disappearance

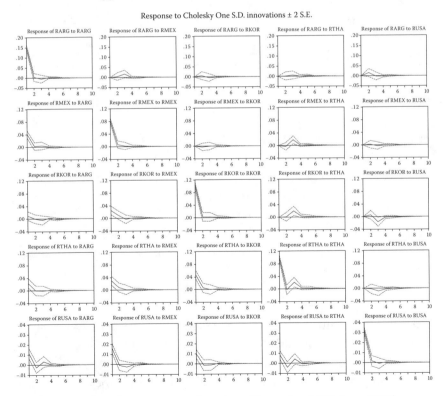

Response to Cholesky One S.D. innovations ± 2 S.E.

FIGURE 17.2 Impulsion response functions (December 1987–June 2007).

---

* See Hamilton (1994) for more details.

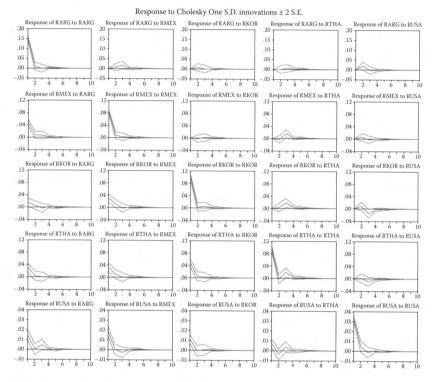

FIGURE 17.3   Impulsion response functions (December 1987–January 2009).

of stock market shocks and we highlight that the effects of these shocks are more significant over the second period.

Overall, these findings suggest some evidence of short-term linkages between the U.S. market and the emerging stock markets, and highlight the dependence of their financial systems on the American one notably over the second period. They reflect thus the effects of the present financial crisis and the transmission of the subprime crisis to emerging Asian and Latin American stock markets.

17.4 CONCLUSION

This chapter examines the extent to which emerging markets are affected by the recent financial crisis, originated from the United States and most of the developed countries. We base our analysis on a multivariate cointegration approach and study in particular the issue of stock market integration between four major emerging markets (Argentina, Mexico, South Korea, and Thailand) and the U.S. stock markets in both the short run and long run. As financial linkage is one of the principal channels through which crisis events can be transmitted across markets. An acceptance of the integration hypothesis during the crisis would suggest the existence of spillover effects from the United States to emerging markets of interest. Notice that the empirical method used also permits to capture the intensity and the strength of the crisis effects.

Over a period of two decades, we identify strong evidence of short-term financial dependences of four emerging markets on the U.S. markets, especially in the follow-up of the subprime crisis and economic recession symptoms. We also find that the responses of all emerging markets to the United States tend to persist more after the crisis. Our results, however, does not work in favor of long-term relationships between the individual markets we study, which typically suggest that on a long-run basis it is not possible to predict changes in emerging markets from the movements in the United States and inversely.

Altogether, the obtained results point to the existence of a new puzzle for emerging markets economies. That is, they are not decoupling from the U.S. shocks, and to the broader extent from those of the developed world despite the fact that they are less financially dependent on the latter in terms of both capital flows and external demands for exports.

REFERENCES

Barari, M. (2004) Equity market integration in Latin America: A time-varying integration score analysis. *International Review of Financial Analysis*, 13(5): 649–668.

Bekaert, G. and Harvey, C. (1995) Time-varying world market integration. *Journal of Finance*, 50(2): 403–444.

Caramazza, F., Ricci, L. A., and Salgado, R. (2004) International contagion in currency crises. *Journal of International Money and Finance*, 23(1): 51–70.

Carrieri, F., Errunza, V., and Hogan, K. (2007) Characterizing world market integration through time. *Journal of Financial and Quantitative Analysis*, 42(4): 511–540.

Chen, G., Firth, M., and Rui, O. M. (2002) Stock market linkages: Evidence from Latin America. *Journal of Banking and Finance*, 26(6): 1113–1141.

Engle, R. F. and Granger, C. W. J. (1987) Cointegration and error correction: Representation, estimation and testing. *Econometrica*, 55(2): 251–276.

Engle, R. F. and Yoo, S. (1987) Forecasting and testing in cointegrated systems. *Journal of Econometrics*, 35(1): 143–159.

Frankel, J. and Rose, A. (1998) The endogeneity of the optimum currency area criteria. *Economic Journal*, 108(449): 1009–1025.

Fujii, E. (2005) Intra and inter-regional causal linkages of emerging stock markets: Evidence from Asia and Latin America in and out of crises. *Journal of International Financial Markets, Institutions & Money*, 15(4): 315–342.

Gerard, B., Thanyalakpark, K., and Batten, J. (2003) Are the East Asian markets integrated? Evidence from the ICAPM. *Journal of Economics and Business*, 55(5–6): 585–607.

Granger, C. W. J. (1969) Investigating causal relations by econometric models and cross-spectral methods. *Econometrica*, 37(3): 424–438.

Granger, C. W. J. (1981) Cointegrating variables and error correcting models. Working Paper, University of California, San Diego, CA.

Grubel, H. (1968) Internationally diversified portfolios: Welfare gains and capital flows. *American Economic Review*, 58(5): 1299–1314.

Johansen, S. (1988) Statistical analysis of cointegration vectors. *Journal of Economic Dynamics and Control*, 12(2–3): 231–254.

Johnson, B. and Soenen, L. (2003) Economic integration and stock market comovement in the Americas. *Journal of Multinational Financial Management*, 13(1): 85–100.

Sims, C. A. (1980) Macroeconomics and reality. *Econometrica*, 48(1): 1–48.

Solnik, B. (1974) Why not diversify internationally rather than domestically? *Financial Analysts Journal*, 30(4): 48–54.

CHAPTER 18

The Financial Crisis and Loan Impairment Provisioning in Asian Banks

Tyrone M. Carlin, Nigel Finch, and Guy W. Ford

CONTENTS

This chapter examines the response of a sample of Asian banks to the recognition of loan loss provision in the face of a gathering economic storm. Drawing on empirical data from 2006 through 2008, this chapter focuses on the level of loan loss provisioning undertaken by the banks, with a view to generating insights into the effectiveness of the approach to loan impairment and provisioning prescribed by IAS 39—*Financial Instruments: Measurement and Recognition*. Given that the focus of impairment decision-making, under IAS 39, is historically oriented rather than future oriented, we argue that this may result in the diminution in the decision usefulness of the content of bank financial statements in the face of imminent, though not yet manifested, economic distress. Despite mounting evidence that substantial portions of the globe's financial and economic fabric lay in a state of severe distress, our analysis of the financial disclosures of the sample of Asian banks shows a picture at odds with this larger reality. We argue that this response is shaped by the requirements of the newly introduced accounting standard and that a broadening of the legitimate sources of evidence upon which loan impairment recognition decisions may be based, pursuant to IAS 39, should be a matter of priority.

18.1 INTRODUCTION

Perhaps the greatest impression forged into the minds of policy makers and regulators in the wake of the global financial crisis that began to make its presence felt from 2007 is related to the pace of deterioration in conditions. In July 2007, soon after the term "subprime" had begun to take on a sinister connotation in minds around the world, the U.S. Federal Reserve Chairman, Bernanke, forecast that losses relating to these loans might amount to $100 billion. Not long would pass before the International Monetary Fund (IMF) would be pronouncing that the losses stemming from the burgeoning banking and credit markets' crisis would not be measured in billions, but in trillions.

In April 2008, the IMF ventured the view that losses would exceed $1 trillion (IMF, 2008a). By October of that year, this had been revised to a total loss estimate of $1.4 trillion (IMF, 2008c). The best estimate, as in January 2009, stood at $2.2 trillion (IMF, 2009), notwithstanding the fact of massive and coordinated fiscal and monetary stimulus responses by central banks and governments around the world as they attempted to prevent a financial market meltdown contaminating the globe's real economy.

Though the implications of the crisis are global in their reach and significance, the financial complexes of the United States and Europe have been at the vortex of the storm. It is these institutions that had the largest direct exposures to subprime and similar loan classes and to complex-structured financial products (Duffie, 2008). In turn, it was the U.S. and European financial institutions that suffered the greatest proportion of direct loan impairments and security write-downs stemming from the crisis, with

Asian banks figuring in the direct loss equation only to a relatively minor extent (IMF, 2008b).

However, the shocks to liquidity, the cost of credit, and the impediment to economic growth induced by the financial crisis have implications for financial institutions throughout Asia, even though, to a large extent, Asian institutions have not been in the vanguard of the first wave of loss exposure. Consequently, drawing on empirical data from 2006 through 2008, this chapter focuses on the position of a sample of Asian banks and their responses to the gathering economic storm.

In achieving this objective, the chapter is structured as follows. Section 18.2 discusses relevant elements of the IAS 39 approach to loan impairment and some of the challenges and difficulties inherent with the application of this approach. Section 18.3 sets out an overview of the data sample drawn upon and the testing methods applied to the empirical evidence drawn upon, for the purposes of the study. Section 18.4 contains an overview and discussion of the results, while Section 18.5 contains some conclusions.

18.2 LOAN IMPAIRMENT

Provisioning for loan losses plays a key role in determining the makeup and thus the transparency and representational faithfulness of banks' balance sheets. Decisions made with respect to loan loss provisioning also have the capacity to wield substantial influence over the level of volatility and cyclicality evident in reported bank earnings. Because the loan impairment provision decision is one of such potentially high significance, it has been associated with considerable controversy.

For example, in jurisdictions where regulations permit the exercise of substantial discretion in the generation of loan loss impairment provisions and charges, there have been many thoroughly documented cases of systematic under-provisioning in the face of adverse financial climates, with the result of substantial degradation in the decision usefulness of information sets contained in bank financial statements (Genay, 1998).

The literature also documents how discretion has been exercised to smooth capital requirements and earnings (Beatty et al., 1995). This has been documented across a wide variety of jurisdictions and across extended time frames (Wahlen, 1994; Kim and Kross, 1998; Ahmed et al., 1999). Further, there is evidence to the effect that in many cases, provisioning levels have failed to change either at the same rate as, or remain contemporaneous with changes in the size of bank loan portfolios (Jimenez and Saurina, 2005).

From a regulatory perspective, discretion in loan impairment provisioning may provide a greater capacity to build up substantial buffers against deterioration in credit quality prior to the emergence of objectively verifiable data evidencing the existence of impairment in individual loans or portfolios of loans (Borio and Lowe, 2001). This forward-looking, uncertainty-tolerant approach to loan impairment provisioning lies in stark contrast to the dominant flavor of contemporary accounting rules on the subject, which emphasize the primacy of objective and verifiable evidence over future-oriented conjecture (Glover et al., 2005).

In effect, according to a typical prudential regulatory management approach, impairment provisions are best characterized as anchored within an expected losses model, while the contemporary accounting rulemaking approach to loan impairment provisioning is anchored within an incurred loss tradition.

This incurred loss approach is clearly reflected in the provisions of IAS 39, which governs loan impairment accounting. Paragraph 58 of the standard directs entities reporting in accordance with the standard to assess, at each reporting date, whether there is any objective evidence consistent with the proposition that an individual loan or portfolio of loans is impaired.

Objective evidence of impairment is said to spring from the occurrence of some loss event or events,* which, individually or together, impact on the value of the estimated future cash flows attributable to a loan or portfolio of loans. Examples of events giving rise to evidence of impairment include evidence of financial distress on the part of an obligor, delinquency or default in relation to interest or principal payments, and the entry into or high probability of entry into bankruptcy or financial reorganization.†

IAS 39, Paragraph 59 (f) (ii), also countenances the view that impairment may be established where national or local economic conditions that correlate with defaults, including increases in unemployment rates, decreases in property prices, or adverse changes in business conditions identifiable as touching on borrowers from a geographic region where loans have been advanced, are visible.

While this may appear to be an invitation to include a forward-looking dimension into the loan impairment assessment decision, the better view

---

* IAS 39, paragraph 59.
† See IAS 39, paragraph 59. The list discussed above is not exhaustive, and IAS 39 provides further examples.

is that the focus of impairment decision-making in IAS 39 is historically oriented rather than future oriented. Paragraph 59 of the standard emphasizes this form of focus, stipulating that "losses expected as a result of future events, no matter how likely, are not recognised."

The insistence on the availability of objective evidence of impairment threaded through IAS 39 raises a paradox of potential significance. On the one hand, discretion in the framing and timing of provisions and write-downs has been the source of a notorious capacity for financial statement manipulation. Consequently, the removal of the capacity to create provisions or write down the value of assets by reason of impairment, without a requirement that such decisions be grounded in some objectively verifiable evidentiary base, should improve the transparency and reliability of financial statement data (Briloff, 1972; Mulford and Comiskey, 2002; Tweedie, 2007).

On the other hand, a strict insistence on a historically grounded approach to the loan-impairment recognition decision may result in the exclusion of potentially value-relevant information from financial statements. The financial crisis, which commenced in 2007, rapidly resulted in a self-reinforcing cycle of deleveraging (Brunnermeier, 2008). This form of de-leveraging is typically associated with severe restrictions on credit availability and systematic decreases in asset prices and material economic contraction (Minsky, 1993). In turn, these phenomena have been demonstrated to be strongly associated with higher default rates and lower recovery rates subsequent to default (Altman and Paternack, 2006). Further, there is strong evidence of a substantial synchronization between key global economies on dimensions including output, investment, credit availability and to a lesser extent, consumption (Claessens et al., 2008).

Thus, while the epicenters of economic stress resulting from the financial crisis lay in the United States and Europe, a material roll on consequences for Asian markets and economies were highly likely, from the outset (Lim, 2008). Yet IAS 39 is constructed on the premise that notwithstanding the likelihood of losses stemming from future events, recognition of value diminution is not possible until the loss-trigger events themselves have occurred, and verifiable evidence of these is available.

Although IAS 39 countenances the possibility of having regard for macroeconomic conditions that have been shown to correlate with increased loan losses and defaults (e.g., increase in unemployment rates or decline in property prices), the standard stipulates that any macroeconomic referents drawn upon as a basis for justifying loan-impairment decisions should be closely related to the location of the loan portfolios—the subject of potential default. The practical consequence of this stipulation is that macroeconomic indicators relating to jurisdictions other than those in which bank loan portfolios are held may not be taken into account in the determination of loan impairment, even in circumstances where the variables in question are likely to represent leading indicators of future conditions in the jurisdiction in which the loan portfolios are domiciled.

One consequence of this apparent deficiency may be the inducement of an upward bias in reported earnings in periods immediately prior to the commencement of economic adjustments, and a more savage adjustment of earnings and balance sheets during the period of economic adjustment (Cavallo and Majnoni, 2001). Of course, whether this conjectured information-content deficiency might arise in the types of circumstances contemplated above is an empirical question.

Given the likelihood of macroeconomic contagion effects (IMF, 2009), the position of Asian banks in the immediate wake of the U.S. and European financial crisis represents a useful setting in which to test the proposition, that the IAS 39 focuses on existing local or national verifiable evidence relating to loan impairment, might lead to a reduction in the decision usefulness of the content of bank financial statements. The approach taken, and data drawn upon for the purpose of developing insights into this proposition are discussed in the next section.

18.3 DATA AND METHODS

The conjecture investigated in this chapter is that the approach taken to the issue of loan impairment in IAS 39 may result in the diminution in the decision usefulness of the content of bank financial statements in the face of imminent, though not yet manifested, economic distress. Evidence useful for the purposes of testing the validity of this conjecture was gathered from a sample of Asian* bank financial statements spanning 2006 through 2008. Given the robust performance of the global and Asian macroeconomy during 2006, that year provides a baseline against which, data from later periods during which financial and economic distress was becoming increasingly evident, may be compared (IMF, 2006, 2007, 2008d).

The data drawn upon for the purposes of the analysis was based on a sample consisting of 10 Asia-based banks and was constructed by selecting

---

* Australia is included in Asia for the purposes of this chapter.

two large banks each, from Australia, China, Hong Kong, Malaysia, and Singapore. These jurisdictions represent countries that have each adopted International Accounting Standards (IAS) over the focal period for the analysis (2006–2008 inclusive). Further, each of these countries plays a significant economic role in the Asia-Pacific region (RBA, 2008).

To increase the comparability and consistency of data drawn upon for the purposes of the analysis, only banks with an equivalent "AA" or "A" institutional credit rating were included in the final research sample. With the exception of Malaysian domiciled banks, Standard & Poor's institutional credit ratings were used to assess the banks' financial credentials. For the Malaysian banks in the sample, institutional credit ratings were obtained from Rating Agency Malaysia (RAM), Malaysia's leading rating agency.* To ensure an even spread, five banks were selected with an equivalent "AA" rating and five banks were selected with an equivalent "A" rating.

Financial data relating to the operations of the ten banks for the period 2006, 2007, and 2008 was obtained. This included measures of the value of outstanding loan portfolios, total bank equity, and loan impairment charges and earnings. To control the variation in taxation rates across the jurisdictions studied, all earnings' measures employed for the purposes of the analysis were stated on a before-tax basis.

Several of the banks included in the sample use December 31, year ends. For these organizations, the most recent 2008 financial disclosure was in the form of an unaudited 6 month or 9 month financial report. Where this was the case, all non-balance sheet data was adjusted on a straight line pro-rata basis as a means of estimating full-year equivalent data.

The identity of each bank included in this sample, its jurisdiction, institutional credit rating, financial year month-end, and the total assets as in 2008 (expressed in equivalent U.S. dollars) is shown in Table 18.1.

Analysis was undertaken on the whole of sample- and credit-rating-based subsample basis. Arithmetic and weighted-average measures of key variables of interest were calculated for each of the 3 years—2006, 2007, and 2008. For the purpose of the calculation of weighted mean measures, loan to equity ratios, return on equity and provision expense to equity were weighted on the basis of the bank total equity expressed as a USD

---

* In 2005, Standard & Poor's and Rating Agency Malaysia entered into an affiliation agreement whereby the two firms have coordinated their analytical and business development activities in Malaysia which included Standard & Poor's sharing its global rating expertise and analytical criteria with RAM.

TABLE 18.1   Details of Research Sample

Bank Name	Country	Credit Rating	Month End	Assets ($ Million)
Commonwealth Banking Group	Australia	AA	June	314,338
ANZ Banking Group	Australia	AA	September	303,669
China Construction Bank Corporation	China	A	December	1,063,831
Industrial & Commercial Bank of China	China	A	December	1,365,355
The Hong Kong and Shanghai Banking Corporation	Hong Kong	AA	December	2,546,678
Hang Seng Bank	Hong Kong	AA	December	96,493
Malayan Banking Berhad	Malaysia	A	June	74,194
Public Bank Berhad	Malaysia	A	December	52,586
OCBC Group	Singapore	A	December	119,919
DBS Group	Singapore	AA	December	169,981
n = 10			Total Assets	6,107,043

equivalent. Provision expense to loans was weighted on the basis of total outstanding loans expressed as a USD equivalent, while provision expense to profit was weighted on the basis of net profit before tax, expressed as a USD equivalent. The results and a discussion thereof are set out in the next section.

18.4 RESULTS AND DISCUSSION

Over the period reviewed, there was some evidence of variation in the financial performance (as measured by return on equity). The whole of sample equity-weighted ROE increased from 19.2% in 2006 to 22.5% in 2008. Decomposition of the data into credit-rating based subsamples revealed that "A" rated banks reported substantial increases in ROE over the period reviewed (up from an equity-weighted mean of 17.7% in 2006 to 29% in 2008), while on the same measure, "AA" rated banks exhibited a decline in performance from 20.3% in 2006 to 16.8% in 2008.[*]

---

[*] This decline was dominated by higher charge offs, by one bank in particular, HSBC. It is likely that these were in turn driven by that organization's comparatively high exposure to U.S. markets, and in particular, to U.S. mortgage markets.

Irrespective of the pattern divergence evident between the return on equity experience of the "A" rated and "AA" rated banks, the data portrays the operating climate of the organizations included in the research sample as, essentially benign. This is especially evident when viewing the data through the whole of sample equity weighted lens, which depicts a consistent but gradual increase in bank ROE across the period reviewed. The ROE results data are set out in Table 18.2.

The data also reveals that from a leverage perspective, there was comparatively little change in the composition of the sample banks' balance sheets over the period studied, with the loan-to-equity ratio hovering at close to eight times, between 2006 and 2008 (Table 18.3). Again, nothing in this data suggests the presence of balance sheet shocks or other substantial dislocations to either the makeup of the balance sheet or the underlying businesses of the organizations scrutinized.

Over the period studied, the whole of sample loan-impairment expenses exhibited growth when measured as a proportion of outstanding loans, equity and before tax profits (for summaries of these results, see Tables 18.4 through 18.6, respectively). However, the sample analysis revealed the existence of ostensibly anomalous patterns. On a weighted-average basis, "A" rated banks recognized lower loan impairment costs than "AA" rated banks in each year reviewed.

Further, the level of loan impairment charges recorded by "A" rated banks fell in each successive year from 2006 through 2008, while "AA" rated banks exhibited the opposite pattern. The discrepancy between weighted and unweighted impairment measures suggests that, larger banks undertook higher impairment cost charge-offs than smaller banks. If balance sheet size is a useful proxy for wider international loan-portfolio mix exposure, one potential interpretation of the data is that by 2007 and 2008, banks included in the research sample were increasing their recognition of loan impairment in relation to ex-Asian loan exposures, but not to the same extent in relation to intra-Asian loan portfolios.

Taking a whole of sample perspective, it is difficult to reconcile the impairment cost data produced within Asian bank financial statements with the cataclysmic events in the global banking and finance system across 2007 and 2008. On an equity-weighted basis, loan impairment costs, as a proportion of outstanding loans, increased by only 25 basis points between 2006 and 2008.

Over the same period of time, a very substantial proportion of the equity of the U.S and European commercial banking complex vaporized,

TABLE 18.2 ROE (before Tax)

	FY 2006			FY 2007			FY 2008		
	"A" Banks (n=5)	"AA" Banks (n=5)	Total Sample (n=10)	"A" Banks (n=5)	"AA" Banks (n=5)	Total Sample (n=10)	"A" Banks (n=5)	"AA" Banks (n=5)	Total Sample (n=10)
Min.	15.3%	14.3%	14.3%	15.1%	13.1%	13.1%	10.8%	11.8%	10.8%
Max.	25.1%	43.6%	43.6%	30.1%	57.3%	57.3%	31.6%	37.9%	37.9%
Spread	9.8%	29.3%	29.3%	15.0%	44.2%	44.2%	20.8%	26.1%	27.0%
Median	19.9%	21.4%	20.6%	22.7%	26.6%	23.3%	21.2%	17.0%	19.1%
Wtd Avg.	17.7%	20.3%	19.2%	22.0%	20.3%	21.1%	29.0%	16.8%	22.5%
Avg.	20.2%	24.9%	22.6%	22.6%	28.3%	25.5%	21.3%	21.2%	21.2%

TABLE 18.3 Loan-to-Equity Ratio

	FY 2006			FY 2007			FY 2008		
	"A" Banks (n = 5)	"AA" Banks (n = 5)	Total Sample (n = 10)	"A" Banks (n = 5)	"AA" Banks (n = 5)	Total Sample (n = 10)	"A" Banks (n = 5)	"AA" Banks (n = 5)	Total Sample (n = 10)
Min.	4.09	4.05	4.05	4.24	4.59	4.24	4.33	5.21	4.33
Max.	8.60	12.86	12.86	9.95	13.10	13.10	11.72	13.82	13.82
Spread	4.51	8.81	8.81	5.72	8.51	8.87	7.4	8.61	9.50
Median	7.80	7.55	7.68	7.34	7.25	7.31	7.93	7.83	7.88
Wtd Avg.	7.63	8.24	7.98	7.21	8.22	7.77	7.52	8.50	8.04
Avg.	7.29	8.63	7.96	7.27	8.93	8.10	7.98	9.11	8.54

TABLE 18.4 Provision Expense to Loans

	FY 2006			FY 2007			FY 2008		
	"A" Banks (n=5)	"AA" Banks (n=5)	Total Sample (n=10)	"A" Banks (n=5)	"AA" Banks (n=5)	Total Sample (n=10)	"A" Banks (n=5)	"AA" Banks (n=5)	Total Sample (n=10)
Min.	0.00%	0.11%	0.00%	0.05%	0.14%	0.05%	0.12%	0.11%	0.11%
Max.	0.85%	1.22%	1.22%	0.84%	1.76%	1.76%	0.64%	1.92%	1.92%
Spread	0.85%	1.11%	1.22%	0.78%	1.62%	1.71%	0.52%	1.81%	1.81%
Median	0.67%	0.16%	0.36%	0.53%	0.23%	0.47%	0.49%	0.54%	0.51%
Wtd Avg.	0.73%	0.81%	0.78%	0.70%	1.15%	0.97%	0.58%	1.38%	1.03%
Avg.	0.55%	0.35%	0.45%	0.49%	0.58%	0.54%	0.43%	0.68%	0.56%

TABLE 18.5   Provision Expense to Equity

	FY 2006			FY 2007			FY 2008		
	"A" Banks (*n* = 5)	"AA" Banks (*n* = 5)	Total Sample (*n* = 10)	"A" Banks (*n* = 5)	"AA" Banks (*n* = 5)	Total Sample (*n* = 10)	"A" Banks (*n* = 5)	"AA" Banks (*n* = 5)	Total Sample (*n* = 10)
Min.	0.0%	0.6%	0.0%	0.2%	1.5%	0.2%	1.4%	0.7%	0.7%
Max.	6.0%	8.4%	8.4%	5.7%	11.3%	11.3%	4.9%	13.1%	13.1%
Spread	6.0%	7.8%	8.4%	5.5%	9.8%	11.1%	3.5%	12.4%	12.4%
Median	5.0%	1.4%	3.3%	3.9%	2.5%	3.2%	4.0%	3.4%	3.7%
Wtd Avg.	5.3%	6.1%	5.8%	4.7%	8.4%	6.8%	4.2%	10.3%	7.5%
Avg.	4.2%	2.7%	3.4%	3.6%	3.9%	3.8%	3.2%	5.3%	4.2%

TABLE 18.6 Provision Expense to Net Profit before Tax

| | FY 2006 | | | FY 2007 | | | FY 2008 | | |
	"A" Banks (n = 5)	"AA" Banks (n = 5)	Total Sample (n = 10)	"A" Banks (n = 5)	"AA" Banks (n = 5)	Total Sample (n = 10)	"A" Banks (n = 5)	"AA" Banks (n = 5)	Total Sample (n = 10)
Min.	0.1%	1.8%	0.10%	1.40%	2.60%	1.4%	10.3%	1.8%	1.8%
Max.	29.4%	32.4%	32.4%	22.3%	41.6%	41.6%	16.4%	49.5%	49.5%
Spread	29.3%	30.6%	32.3%	20.9%	39.0%	40.2%	6.1%	47.8%	47.8%
Median	18.1%	6.4%	11.7%	14.6%	8.8%	13.3%	12.2%	19.3%	13.4%
Wtd Avg.	22.8%	22.9%	22.9%	18.0%	28.4%	23.6%	13.0%	37.5%	22.7%
Avg.	17.2%	10.4%	13.8%	13.4%	15.2%	14.3%	12.9%	22.7%	17.8%

TABLE 18.7 Sensitivity of Profit to a 10 BPS Increase in Provisions

| | FY 2006 | | | FY 2007 | | | FY 2008 | | |
	"A" Banks (n = 5)	"AA" Banks (n = 5)	Total Sample (n = 10)	"A" Banks (n = 5)	"AA" Banks (n = 5)	Total Sample (n = 10)	"A" Banks (n = 5)	"AA" Banks (n = 5)	Total Sample (n = 10)
Max.	-4.90%	-5.29%	-5.29%	-3.43%	-4.92%	-4.92%	-10.81%	-7.42%	-10.81%
Min.	-2.40%	-1.70%	-1.70%	-2.81%	-1.19%	-1.19%	-2.44%	-1.60%	-1.60%
Spread	2.51%	3.59%	3.59%	0.62%	3.74%	3.74%	8.37%	5.82%	9.21%
Avg.	-3.66%	-3.73%	-3.69%	-3.19%	-3.70%	-3.44%	-4.64%	-4.87%	-4.75%

with massive injections of public funds and de facto nationalizations being key components of the suite of response measures put in place in a bid to combat the crisis. None of the flavor of this is evident in the content of the financial statements of the Asian banks studied. This appears consistent with the conjectures in relation to the impact of IAS 39 discussed above.

Over and above considerations relating to returns, leverage and impairment cost experience, a further noteworthy feature of the data appears to be the increasing sensitivity to loan-impairment charges exhibited by the sample banks over time. Whereas, the average profit sensitivity to a 10 basis point increase in loan impairment costs stood at 3.69% in 2006, this had grown to 4.75% by 2008 (see Table 18.7).

This suggests the capacity for more rapid deteriorations in reported earnings and in balance sheet conditions in the event that the Asian economic cycle experiences a material downturn in response to the 2007 and 2008 banking crisis. Were this to occur, the historically focused approach to impairment recognition, stipulated by the IAS, would drive a schism between the information content of financial statements produced by Asian banks, in a period contemporaneous with economic distress in leading ex-Asia economies, and in some subsequent period when the effect of that distress had left a more material imprint on the fortunes of Asian economies themselves.

18.5 CONCLUSION

By 2007, and certainly by 2008, it was clear that substantial portions of the globe's financial and economic fabric lay in a state of severe distress. Yet, an examination of the financial disclosures by a series of large Asian domiciled banks shows a picture at odds with this larger reality. As the global crisis worsened through 2007 and 2008, the response of Asian banks was muted. No doubt, an element of the explanation of the benign signal implicit in the Asian bank financial disclosures lies in their lower direct exposures to troubled asset classes and complex structured products than was the case with their U.S. and European counterparts (Mosharian, 2009).

However, it also seems strongly arguable that the impairment recognition procedures, stipulated by IAS 39, represent an element of any explanation for the patterns evident in the data. This should then be a matter of concern. If one of the objectives of the IFRS regime is to allow reporting entities (in this case, banks) to produce financial disclosures which are of greater assistance to users, by way of being constructed on a foundation of more decision-useful information, it may be that this objective is being poorly served by the current approach to evidence set out in IAS 39.

Unless the pan-Asian economic complex has substantially decoupled from the U.S. and European economies (and there is very little evidence consistent with this proposition), the study of Asian bank financial disclosures reported in this chapter strongly suggests that a broadening of the legitimate sources of evidence upon which loan impairment recognition decisions may be based pursuant to IAS 39 should be a matter of priority. A failure to reconsider this matter should be of substantial concern in those jurisdictions which have yet to adopt the IFRS but which are actively considering such a transition over the medium term.

REFERENCES

Ahmed, A., Takeda, C., and Thomas, S. (1999) Bank loan loss provisions: A re-examination of capital management, earnings management and signalling effects. *Journal of Accounting and Economics*, 28 (3): 1–25.

Altman, E. and Pasternack, B., (2006) Defaults and returns in the high yield bond market: The year 2005 in review and market outlook. *Journal of Applied Research in Accounting and Finance*, 1 (1): 3–29.

Beatty, V., Chamberlain, S., and Magliolo, J. (1995) Managing financial reports of commercial banks: The influence of taxes, regulatory capital and earnings. *Journal of Accounting Research*, 33 (2): 231–261.

Borio, C. and Lowe, P. (2001) To provision or not to provision. *BIS Quarterly Review*, September: 36–48, Basel, Switzerland.

Briloff, A. (1972) *Unaccountable Accounting—Games Accountants Play*, Harper & Row, New York.

Brunnermeier, M. (2008) Deciphering the liquidity and credit crunch 2007–08. Working Paper, NBER, Cambridge, MA.

Cavallo, M. and Majnoni, G. (2001) Do banks provision for bad loans in good times? Empirical evidence and policy implications. Working Paper, New York University, New York.

Claessens, S., Kose, M., and Terrones, M. (2008) What happens during recessions, crunches and busts? Working Paper, International Monetary Fund, Washington, DC.

Duffie, D. (2008) Innovations in credit risk transfer: Implications for financial stability. Working Paper, BIS, Basel, Switzerland.

Genay, H. (1998) Assessing the condition of Japanese banks: How informative are accounting earnings? *Economic Perspectives*, Federal Reserve Bank of Chicago, IL, pp. 12–34.

Glover, J., Ijiri, Y., Levine, C., and Liang, P. (2005) Separating facts from forecasts in financial statements. *Accounting Horizons*, 19(4): 267–282.

International Monetary Fund. (2006) *World Economic Outlook*, September, Washington, DC.

International Monetary Fund. (2007) *World Economic Outlook*, October, Washington, DC.

International Monetary Fund. (2008a) *Global Financial Stability Report*, April, Washington, DC.

International Monetary Fund. (2008b) *Global Financial Stability Report—Market Update*, July 28, Washington, DC.

International Monetary Fund. (2008c) *Global Financial Stability Report*, October, Washington, DC.

International Monetary Fund. (2008d) *World Economic Outlook*, October, Washington, DC.

International Monetary Fund. (2009) *Global Financial Stability Report—Market Update*, January 28, Washington, DC.

Jimenez, G. and Saurina, J. (2005) Credit cycles, credit risk and prudential regulation, Banco de Espana, pp. 1–38, Madrid.

Kim, M. and Kross, W. (1998) The impact of the 1989 change in capital standards on loan-loss provisions and loan write-offs. *Journal of Accounting and Economics*, 25(2): 69–99.

Lim, M. (2008) Old wine in new bottles: Subprime mortgage crisis—Causes and consequences. *Journal of Applied Research in Accounting and Finance*, 3(2): 3–13.

Minsky, H. (1993) Finance and stability: The limits of capitalism. Working Paper, Jerome Levy Economics Institute, Annandale-on-Hudson, New York.

Mosharian, F. (2009) Strength in numbers. *CFO*, February, p. 35.

Mulford, C. and Comiskey, E. (2002) *The Financial Numbers Game—Detecting Creative Accounting Practices*, John Wiley & Sons, New York.

Reserve Bank of Australia (RBA). (2008) *Financial Stability Review*, 24 September, Melbourne.

Tweedie, D. (2007) Can global standards be principle based? *Journal of Applied Research in Accounting and Finance*, 2(1): 3–8.

Wahlen, J. (1994) The nature of information in commercial bank loan loss disclosures. *The Accounting Review*, 69(3): 455–478.

CHAPTER 19

Currency and Maturity Mismatches in Latin America*

Marco Sorge and Chendi Zhang

CONTENTS

* The authors thank Tito Cordella for his valuable comments. We are grateful to Hui Jiang for her excellent research assistance.

Currency and maturity mismatches have been found to be associated with the current global financial crisis and many of the episodes of financial fragility recorded in the last decade. This chapter discusses (a) cross-country evidence on the extent, determinants, and possible adverse consequences of currency and maturity mismatches, with a focus on Latin American countries and (b) the roles of capital markets in mitigating such mismatches, thus reducing the systemic vulnerability of developing countries. We observed that the overall extent of mismatches for corporate debt in Latin American countries appears to have decreased in recent years. A growing array of financial instruments has been developed to hedge the risks of mismatches, though with mixed results.

19.1 INTRODUCTION

Financial turmoil over the past decade has stimulated research on various sources of vulnerability of economies around the world (see, e.g., Kaminsky and Reinhart, 1999). In particular, both maturity and currency mismatches have been found to be associated with many of the episodes of financial fragility recorded in the past decade.

Many developing countries have accumulated a considerable amount of short-term external debt not matched by foreign assets of similar maturities. This "maturity mismatch" has often been criticized as a source of financial instability: in case of a capital-account reversal, emerging economies are unable to renew their debts, and face a liquidity problem. Consequently, excessive reliance on short-term external debts leaves the emerging market countries vulnerable to exogenous shocks and may lead to self-fulfilling financial crises (see Chang and Velasco, 1999). To address these issues, many countries have embarked on policies promoting the development of long-term loan or bond markets with mixed results.

Like the first global financial crisis in the twenty-first century, the current financial and economic crisis has illustrated the adverse impact of reliance on short-term borrowing on the real economy, even in developed countries. For example, Northern Rock, a large British bank, relied significantly on short-term borrowing from the money market to finance its long-term mortgage lending business. Following the U.S. subprime mortgage crisis, liquidity problems at the bank have recently triggered the first bank run in the United Kingdom in more than a century. The bank was eventually nationalized in February 2008. Another example of the adverse consequences of short-term borrowing is the current banking and economic crisis in Iceland. In autumn 2008, three large Icelandic banks, i.e., Glitnir, Kaupthing, and Landsbanki, were nationalized. These banks were among the casualties of the global crisis, as their short-term funding from abroad evaporated due to the extreme fear in markets following the collapse of Lehman Brothers.

In addition to maturity mismatches, several emerging economies have financed a significant portion of their local-currency investments with foreign currency (FC)-denominated debt, incurring a "currency mismatch." In such cases, a significant real depreciation of the domestic currency can lead to a negative balance-sheet effect and a sudden deterioration in the repayment capacity of the public sector, with a potentially important contraction of economic activity. A number of studies have shown that currency mismatches have been a major factor triggering recent financial crises and raising the crisis resolution costs (see Allen et al., 2002).

This chapter addresses the following questions. First, we have presented empirical evidence on the extent of currency and maturity mismatches for Latin American countries using recent data from 1993 to 2007. Second, we have summarized the main factors identified by the empirical literature as determinants of mismatches, and have shed light on the links between mismatches and financial fragility, both at the sovereign and corporate levels. Third, we have discussed the roles of bond markets, financial derivatives, and capital markets in general, in mitigating currency and maturity mismatches in developing countries. The development of financial markets extends the space of insurable risks within emerging economies and enables them to self insure against exogenous shocks. For this purpose, a number of financial instruments (such as indexed bonds and currency derivatives) have been proposed to hedge against currency and macroeconomic risks.

Our main findings can be summarized as follows. First, the overall extent of mismatches for corporate debt in Latin American countries appears to have decreased in recent years, especially for the post-2001 period, when Argentina simultaneously recovered from its economic crisis. With regard to sovereign debt, FC debt and short-term (ST) debt in Latin America are observed to move in opposite directions since 2001. This seems to indicate that there is a structural break in the risk-coping behavior of lenders at about 2001, which may be due to a number of factors including the financial crisis in Argentina as well as the financial development and institutional reforms taking place in the region.

Second, the existing studies on the determinants of currency and maturity mismatches mainly fall into two broad groups. One view is that crisis-prone debt structures, such as currency and maturity mismatches, can be a symptom rather than the root cause of potential financial crisis, and, therefore, should focus on building institutional and policy credibility. In contrast, another approach emphasizes how the underdevelopment of domestic financial systems can restrict the hedging possibilities of investors, thus leading to mismatches.

Third, among the financial market instruments potentially available to the emerging market economies to hedge the risk of mismatches, both debt, indexed to nominal or real variables, and currency derivatives appear highly imperfect tools for country insurance. Debt indexed to nominal variables may exacerbate financial fragility, while GDP-linked securities induce moral hazard and possible manipulation of national accounts. Although currency-derivative markets in Latin America experienced explosive growth over the last few years, they are also subject to maturity mismatches. In fact, firms use ST derivatives to hedge long-term FC debt. The inability to rollover hedges during crises due to low market liquidity may convert the synthetic local-currency debt back to unhedged FC debt and expose the economy to significant systemic risk. A large amount of derivative transactions, especially Over the Counter (OTC) traded derivatives, may increase the risk of contagion during crises due to a rapid expansion of counterparty credit risk. In addition, the literature (e.g., Davies et al., 2003) has also warned against moral hazard and other potentially destabilizing effects of both insurance and risk-transfer instruments.

The rest of this chapter is organized as follows: Section 19.2 presents the cross-country evidence on the extent of mismatches and their potential adverse consequences. Section 19.3 reviews the literature on their possible causes of mismatches. Section 19.4 discusses a number of financial instruments potentially available to emerging economies to mitigate these mismatches, and reviews a few country experiences. Lastly, Section 19.5 presents the conclusion.

19.2 CROSS-COUNTRY EVIDENCE ON CURRENCY AND MATURITY MISMATCHES

19.2.1 Potential Adverse Consequences of Mismatches

The link between mismatches and financial fragility is well established in the empirical literature on financial crises. As Allen et al., (2002) emphasized in the balance-sheet approach to financial crises, ST debt may make the governments vulnerable to liquidity risk and debt rollover crises, while a significant amount of FC debt may lead to a sharp increase in government debt, and thus systemic vulnerability, when there is a severe depreciation of the local currency (Céspedes, 2004).

Several studies have shown that the likelihood of international financial crises increases with the level of ST (Rodrik and Velasco, 1999) and FC debt (Calvo et al., 2004) in a country. In addition, currency mismatches may be associated with more rigid exchange-rate regimes, a phenomenon labeled as "fear of floating" by Calvo and Reinhart (2002). Perry and Serven (2003) argued that pervasive mismatches in borrowers' portfolios were one of the key sources of vulnerability leading to Argentina's crisis during 1999–2002.

There is also a growing literature focusing on the adverse consequences of currency mismatches using firm-level data. In a study on the Asian crisis, Harvey and Roper (1999) found that Asian corporations "bet" on currency stability by greatly increasing the leverage in FC at a time of declining profitability. The risk exposure induced by currency mismatches significantly worsened the impact of the Asian crisis on the corporate sector. However, many firm-level studies have observed mixed effects of holding FC debt and ST debt on firm performance and investments (see Allayannis et al., 2003; Bleakley and Cowan, 2004). Although for some countries, there is evidence that firms holding more FC debt perform worse at times of devaluation, in other countries, the differential effect is insignificant or even positive.*

---

* The ambiguous results are possibly due to econometric issues. In particular, if firms in emerging markets internalize the risk of currency mismatches, the empirical findings on the balance-sheet effect of currency mismatches may be biased toward zero. First, firms may match the currency composition of their liabilities with that of their assets and income (Bleakley and Cowan, 2002), or use currency derivatives effectively hedging the exchange rate risk. Second, firms may endogenously decide to carry higher amount of FC debt if they are less vulnerable to exchange-rate fluctuations, e.g., if they are in the tradable sector (Cowan et al., 2005).

19.2.2 The Extent of Mismatches across Countries and over Time

Figure 19.1 presents the currency and maturity composition for corporate debt in Latin America over the 1993–2008 period. We observed that companies in Latin American countries employ a deceasing proportion of FC debt and ST debt over time, especially for the more recent post-2001 period. Indeed, since 2002, Argentina started to recover from its recent economic crisis during 1999–2002. As argued by Schmukler and Vespoeroni (2006), firms' access to international capital markets may also increase the corporate debt maturity. Interestingly, in the pre-2001 period, almost all Latin American countries experienced an increasing reliance on FC debt.

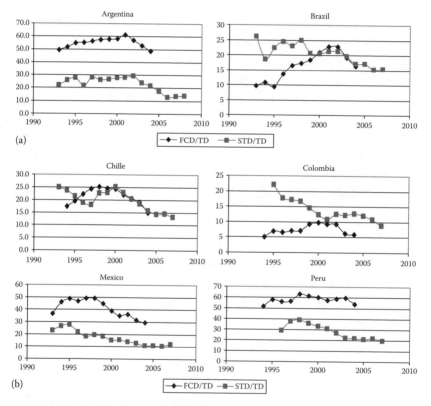

FIGURE 19.1 The development of corporate debt composition in Latin America. This figure plots the average foreign-currency debt (FCD) and short-term debt (STD), and the current proportion of long-term debt as a percentage of total liabilities (TD) for firms in six Latin American countries for the 1993–2008 period. The figures are based on the authors' calculations. (From Worldscope and the FSDI database at World Bank.)

In the developing world, Latin America is the region with the largest share of total (domestic plus external) sovereign debt denominated in FC. This may be associated with potentially higher currency mismatches. Table 19.1 shows that nearly all the *external* sovereign debt in Latin American countries is in FC. Even in their *domestic* markets, most Latin American countries issue large amounts of FC debt. For instance, 64% of the domestic sovereign debts in Argentina are denominated in FC.* The share of ST debt, i.e., debt with a maturity of less than 1 year, is notably lower (13%) in Latin America, which implies that the extent of maturity mismatches may be less than that of the currency dimension. However, this is mainly due to the fact that almost all external debt is long-term debt.[†] In domestic sovereign-debt

TABLE 19.1 Currency and Maturity Composition of Sovereign Debt in Latin America

	FCD/TD			STD/TD		
	Total	**External**	**Domestic**	**Total**	**External**	**Domestic**
Argentina	0.88	0.99	0.64	0.09		
Brazil	0.40	1.00	0.13	0.27	0.01	0.40
Bolivia	0.98	1.00	0.83			
Chile	0.19	1.00	0.06		0.06	
Colombia	0.61	1.00	0.20	0.01	0.02	0.00
Mexico	0.53	1.00	0.04	0.19	0.05	0.38
Peru	0.89	1.00	0.25	0.03	0.00	0.10
Uruguay	0.91	1.00	0.77	0.15	0.09	0.27
Venezuela	0.77	1.00	0.19		0.04	
Average	0.65	1.00	0.26	0.13	0.04	0.26

Source: Cowan, Levy-Yeyati, Panizza, and Sturzenegger (2006).
Note: This table reports the average FCD and STD as a fraction of total domestic and external sovereign debt (including bonds) for Latin American countries over the period of 1980–2005. The starting year is varied across the countries. The figures are based on the authors' calculations.

---

* After the hyperinflation and deposit-confiscation experience in the late 1980s, Argentina adopted a hard peg of the peso to the U.S. dollar, following the introduction of the Convertibility Law in 1991, in spite of its limited trade with the United States and inflexibility in price and wage adjustments to asymmetric shocks. The Convertibility led to the re-creation of a domestic banking system based on dollar-denominated deposits and loans (Perry and Serven, 2003).

[†] Emerging economies sometimes issue "putable" long-term bonds, which are bonds embedded with put options to allow lenders to recall the debt principal before maturity (Dodd, 2000). These seemingly long-term debts essentially function like ST debt. IMF (1999) estimated that there were $32 billion putable debts in the emerging economies in 1999, out of which $8 billion was from Brazil.

markets, several Latin American countries rely on ST debt: e.g., in Brazil and Mexico, ST debt represents about 40% of their domestic debt.

Panel A of Figure 19.2 plots the share of FC sovereign debt and ST sovereign debt over time for major Latin American economies. Since 2001,

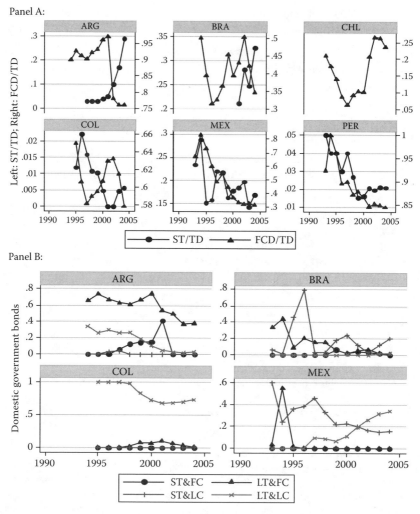

FIGURE 19.2 The development of sovereign debt composition in Latin America. Panel A plots the FCD and STD as a fraction of total domestic and external sovereign debt (including bonds) for six Latin American countries for the 1993–2004 period. Panel B plots the currency (foreign vs. local currency) and maturity (short term vs. long term) composition of domestic sovereign bonds for six Latin American countries for the same period. The figures are based on the authors' calculations. (From Cowan, Levy-Yeyati, Panizza, and Sturzenegger, 2006.)

FC debt and ST debt has been moving in opposite directions: FC debt is observed to drop significantly in almost all Latin American countries, whereas ST sovereign debt is found to rise at the same time in Argentina, Brazil, Colombia, and Peru. For instance, in Argentina, sovereign debt denominated in FC dropped from 96% in 2001 to 75% in 2004 of the total sovereign debt, while ST debt increased from 3% to 30%. This finding is consistent with De la Torre and Schmukler's (2004) hypothesis that currency and maturity mismatches are two alternative risk-coping mechanisms. On the other hand, we also observed that in the 1990s, both FC debt and ST debt decreased in Mexico and Peru. This seems to indicate that there is a structural break in the risk-coping behavior of lenders during 2001, which may be due to a number of factors including the financial crisis in Argentina as well as the financial development and institutional reforms taking place in the region.

19.3 CAUSES OF CURRENCY AND MATURITY MISMATCHES

Existing studies on the causes of currency and maturity mismatches can be classified into two broad groups. One view is that currency and maturity mismatches are optimal responses to systemic risks in emerging economies, such as lack of policy credibility and weaknesses in domestic institutions. Mismatches cannot be reduced without addressing the underlying deeper problems. Another view emphasizes the underdevelopment of domestic financial systems as causes of the mismatches. The current structure of financial instruments in emerging markets can be improved through the development of financial markets, including an adequate financial regulation and domestic investor base. A frequently cited example is the development of the private pension fund industry in Chile, along with the supporting regulatory framework. In the following section, we will discuss the empirical evidence on the determinants of mismatches along these two lines.

19.3.1 Policy and Institutional Factors

The existence of currency and maturity mismatches in developing countries may be attributed to their inability to borrow abroad in their own currencies and/or at long maturities, a phenomenon called "original sin" by Eichengreen et al., (2003). Their paper argues that domestic policies and institutions have little influence on original sin and on currency mismatches relative to factors largely beyond the control of the individual country, such as network externalities, transaction costs, and imperfections in global capital markets. On the other hand, Goldstein and Turner

(2004) defended the view that domestic financial markets, policies, and institutions are of central importance. For instance, larger currency and maturity mismatches have been associated with loose monetary policy, inadequate regulation, and poor institutions.

In particular, volatile and unsustainable policies increase the systemic risk in the economy, and lead creditors to use FC debt and ST debt to cope with these risks. In addition, lack of policy credibility also plays a significant role in the period leading up to crises, as governments tend to shift the composition of their debt toward FC debt and ST debt (e.g., Mexico in 1994 and Brazil in 1998). Detragiache and Spilimbergo (2002) showed that the positive association between ST debt and the likelihood of future debt crises is no longer significant after controlling for the fact that ST debt may be influenced by factors such as low credibility. This is consistent with the view that mismatches are a symptom, rather than a root cause, of a potential crisis.

The related macro literature on financial dollarization (Ize and Levy-Yeyati, 2003) also show that the credibility of macroeconomic policy (i.e., the relative variances of inflation and real exchange rates) and the quality of legal and political institutions are the key determinants of the cross-country variation in dollarization. Furthermore, empirical studies on corporate debt maturity observed that debt maturities are longer if there are better institutions such as the rule of law or democracy (Demirguc-Kunt and Maksimovic, 1999), or higher quality of credit information as a result of higher coverage of public and especially private credit registries, as well as better accounting standards (Sorge and Zhang, 2006).

Taken together, crisis-prone debt structures such as currency and maturity mismatches can be a symptom rather than the root cause of potential financial crisis. Initiatives aiming at eliminating the mismatches without addressing deeper problems such as policy volatility or institutional weakness may lead to risk reallocation rather than risk reduction (De la Torre and Schmukler, 2004). For instance, restrictions on ST debt may induce international investors to shift toward alternative risk-coping mechanisms, such as FC debt.

19.3.2 Underdeveloped Financial Markets

Several studies have emphasized how mismatches may arise due to the limited hedging possibilities offered by underdeveloped financial markets. For example, De la Torre and Schmukler (2004, p. 359) argued that "because of financial underdevelopment, which itself is not independent of the high systemic risks, explicit hedges (such as derivative products) are not normally available in emerging markets to achieve full defeasance of the mentioned systemic risks. Therefore, financial contracts that involve emerging economy assets have to resort to what we call risk-coping mechanisms (i.e., ST debt and FC debt)."

A number of papers have argued that the size and development of the financial system are among the key determinants of the ability of emerging economies to issue local-currency bonds in international capital markets (Eichengreen et al., 2003), while policies and institutions seem to play only an indirect role (Hausmann and Panizza, 2002). In addition, accessing international capital markets would allow firms in the emerging economies to extend their debt maturity (Schmukler and Vesperoni, 2006). Furthermore, a developed domestic financial market makes it easier for a country to issue debt in local currency, thus reducing currency mismatches.* Claessens et al. (2003) show that countries with a larger domestic financial system have a smaller amount of FC bonds as a percentage of total private and public bonds. In addition, Raddatz (2006) showed that financial development contributes to the reduction of macroeconomic volatility due to liquidity provision.

Moreover, creditors in developed countries can hedge currency risks of their portfolio using derivative contracts. In contrast, well-developed derivative markets are missing in most emerging economies, which further reduce the willingness of foreign investors to hold long-term local-currency (LTLC) debt in these countries.

Besides the scarcity of financial instruments and the low liquidity, emerging markets also suffer from inadequate financial legislation and a lack of a critical mass of institutional investors. In particular, pension funds are important institutional investors for the development of domestic LTLC debt markets. Funded pension systems are likely to induce significant domestic savings available for investment in domestic local-currency government debt. In addition, the long-term investment horizon of pension funds makes them natural investors for long-term bonds. Chile has been a pioneer in Latin America with the largest private pension fund system holding assets representing 28% of its GDP. Chile's FC debt as a percentage of total debt is also the lowest among the Latin American countries (see Figures 19.1 and 19.2).

---

* Some countries have resorted to capital controls to further the development of local currency bond markets. For instance, the presence of capital controls has been found to be associated with higher share of local-currency borrowing in the domestic credit market (Hausmann and Panizza, 2002).

19.4 FINANCIAL MARKET INSTRUMENTS TO MITIGATE MISMATCHES

In this section, we will discuss about the financial market instruments that may mitigate currency and maturity mismatches. In particular, we will discuss about the three types of financial instruments: long-term domestic bonds, indexed debt, and currency derivatives.

19.4.1 Domestic Government Debt

The development of LTLC bond markets has been at the heart of policy debate to mitigate mismatches. Burger and Warnock (2006) found that countries with stable inflation records and strong creditor protection have more developed local-currency bond markets and less foreign-currency bonds.

Although implementing institutional reforms and acquiring sufficient credibility may take several years or even decades, a number of emerging market countries developed LTLC bond markets almost immediately after stabilizing from high inflation. Panel B of Figure 19.2 presents the currency and maturity composition of domestic government bond markets for four Latin American countries. Over the past decade, Mexico has developed a relatively large LTLC bond market: it rose from zero in 1995 after its financial crisis, to a market representing 35% of the total domestic government bond market in 2004. In the meantime, the share of short-term local-currency bonds has dropped from 40% to 15%. In addition, Columbia also has a large LTLC bond market, capturing about 70% of the domestic bond market in 2004.

The development of long-term emerging market currency bond markets may help mitigate currency and maturity mismatches in developing countries. Thanks to their ability to raise funds at favorable rates, international financial institutions such as the World Bank have been among the first parties to issue bonds denominated in the currencies of emerging markets, tapping new investor bases interested in holding assets denominated in developing country currencies, but bearing no default risk.

19.4.2 Indexed Debt

Indexed debt, essentially a debt contract with an embedded forward contract contingent on variables such as inflation or GDP, has been at the forefront of the development of domestic government debt markets in a number of countries. Emerging economies that are unable to issue non-indexed long-term debt can become less reliant on ST debt by issuing inflation-indexed LTLC debt. In Latin America, in some countries (e.g., Chile, Argentina, and Columbia), the majority of domestic government debt is indexed to inflation, whereas other countries (e.g., Venezuela, Brazil, and Mexico) borrow a large amount of interest-rate indexed debt.

While building reputation in their policies and institutions, emerging countries need financial instruments that offer a form of insurance against external shocks. Several proposals and a few experiments have been made to issue debt indexed to real variables. Indexation to real variables involves higher interest payments when economic performance is relatively strong and lower payments when economic performance is weak. For example, given that the growth slowdown helps to predict debt crises (Detragiache and Spilimbergo, 2002), countries could issue GDP-indexed bonds paying out lower interests when GDP growth is weak, as a form of country insurance. Thus, real indexation would tend to stabilize the debt/GDP ratio, and reduce the likelihood of debt crises (Borensztein and Mauro, 2004). Emerging economies may also issue LTLC debt with a real indexation clause.

19.4.3 Currency Derivatives

A market for currency derivatives can help foreign investors to hedge systemic risks of their international portfolios, which may reduce the use of FC debt and ST debt as risk-coping mechanisms. Furthermore, currency derivatives also help domestic firms to manage currency mismatches between firms' assets and liabilities.

Figure 19.3 presents the development of exchange-traded currency derivatives, including currency futures and options, in Latin America, Africa, and Middle East. The currency derivative markets in these regions experienced explosive growth over the last 5 years: the turnover of currency futures reached $330 billion in 2006, while the currency option market had a lower turnover of $170 billion, but higher, outstanding amounts of about $30 billion. Furthermore, Figure 19.4 plots the daily turnover of OTC currency derivatives in 2001, 2004, and 2007 for six Latin American countries. Nearly all counties show a strong growth in currency derivatives turnover, except Brazil. In Latin America, Mexico has the largest currency derivatives market.

A number of firm-level studies have investigated the use of derivatives to hedge foreign-exchange risk. Papers based on both the US data and other developed countries (see, e.g., Gezcy et al., 1997) have found that firms that are more likely to use derivatives are typically larger firms with

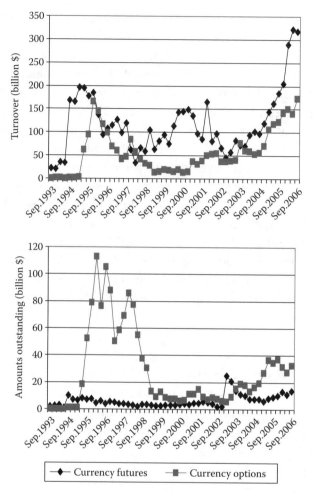

FIGURE 19.3 Exchange-traded currency derivatives in Latin America, Africa, and Middle East. This figure plots the annual turnover and outstanding amounts for exchange-traded currency derivatives (including futures and options) in Latin America, Africa, and Middle East for the 1993–2006 period. The figures are based on the authors' calculations. (From BIS statistics.)

higher leverage or better investment opportunities. In addition, derivatives and foreign income are substitutes for currency hedging. With regard to the use of currency derivatives in Latin America, Aguiar (2002) observed that Mexican firms only partially hedge the currency composition of their liabilities. In particular, Aguiar (2002) also showed that Mexican firms are not fully hedged and the currency depreciation during the Tequila crisis led

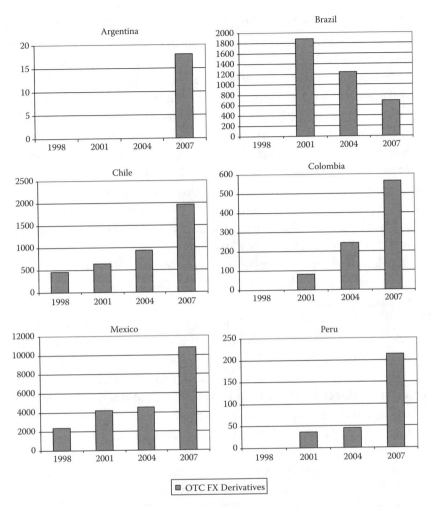

FIGURE 19.4 OTC foreign exchange derivatives in Latin America. The figure plots the daily average turnover (in million $) of, and OTC foreign-exchange derivatives (including forwards, swaps, and options) for six Latin American countries for the 1998–2007 period. The figures are based on the authors' calculations. (From BIS Triennial Central Bank survey of foreign exchange and derivatives market activity.)

to a reduction in the firms' net worth and a drop in investment. In addition, Cowan et al., (2005) found that the currency derivatives play a role in insulating firm-level investment from exchange-rate shocks in Chile. Furthermore, pension funds and exporters, who provide sizable foreign-currency paying positions in derivative contracts, are the main providers of foreign exchange hedging to corporate end-users in Chile (Chan-Lau, 2005).

The currency hedging practices of East Asian firms during the Asian financial crisis were examined by Allayannis et al. (2003). The most interesting finding is that the use of "synthetic local-currency debt" (i.e., FC debt hedged by currency derivatives) is associated with the biggest drop in firms' market value during the crisis, possibly due to the illiquidity of currency derivative markets. The authors argued that maturity mismatches also exists in derivative markets where firms use ST derivatives to hedge longer-term FC debt. This is akin to the maturity mismatches found in debt markets. The inability to rollover hedges during the crisis converted the synthetic local-currency debt back to unhedged FC debt. In Latin America, the most actively traded derivative contracts are also of short maturity (Fernandez, 2003). In particular, 99% of the currency derivative contracts in Chile have maturity less than 1 year (Chan-Lau, 2005).

The development of derivative markets may also help to reduce vulnerability to external shocks at the macro level. During the financial crisis in the 1990s, Chile defended the exchange rate of its currency, peso. This was in contrast to Australia that did not base its policy response to external shocks on a defense of the currency. Australia has a highly developed and liquid-currency derivative market, which allows banks to hedge their net FC liabilities. The currency derivatives market allows banks to separate credit risk from currency risk, and consequently, to intermediate international capital inflows. It also provides a mechanism for external insurance against currency depreciation, and thus reduces the need for FC debt.

In contrast, derivatives markets are not nearly as developed in Chile. Fernandez (2003) argued that regulations may constrain the development of the derivatives markets in Chile. A positive finding is that the volatility of the underlying assets dropped substantially in Chile following the introduction of options (Conrad, 1989). Jadresic and Selaive (2005) showed that the development of currency derivative markets in Chile did not increase the volatility of the spot markets, but reduced the aggregate exposure to currency risk.

However, a shortcoming of derivatives as an instrument for country insurance is that they may increase the risk of contagion during crises. First, the presence of a large amount of derivative transactions, especially OTC-traded derivatives, may create a rapid expansion of counterparty credit risk during financial crisis (Dodd, 2000). Second, in case of illiquid derivative markets during financial crises, market makers are forced to hedge their exposure to currency derivatives by short-selling local currency on the spot market. As a result, the domestic currency weakens further and the crisis deepens (Chan-Lau, 2005). Third, a liquid derivatives market facilitates the transmission of external shocks and financial crises across country borders (Moguillansky, 2002).

19.5 CONCLUSION

This chapter has reviewed the empirical literature on systemic vulnerabilities associated with currency and maturity mismatches in Latin America. We have presented cross-country evidence on the extent, determinants, and possible adverse consequences of currency and maturity mismatches, and have reviewed a wide spectrum of financial instruments potentially available to emerging economies to mitigate the vulnerabilities related to such mismatches. The main conclusions from the paper can be summarized as follows.

First, recently, firms in Latin American have employed a decreasing proportion of FC or ST corporate debt, especially for the post-2001 period. The amount of FC sovereign debt relative to total sovereign debt has also dropped significantly in almost all Latin American countries, since 2001. However, ST sovereign debt has risen at the same time in Argentina, Brazil, Colombia, and Peru. The pattern was different before 2001, when we observed that both FC sovereign debt and ST sovereign debts decreased, for example, in Mexico and Peru. There appears to be a structural break in the risk-coping behavior of the lenders during 2001, which may be due to a number of factors including the economic crisis in Argentina during 1999–2002, as well as the financial development and institutional reforms taking place in the region.

Second, while the link between currency and maturity mismatches and financial fragility is well established in the empirical literature of financial crises, researchers still appear divided with regard to the determinants of such imbalances. One view focuses on policy and institution building, while an alternative approach emphasizes the importance of the development of financial markets to increase hedging possibilities.

Third, emerging economies that are unable to issue non-indexed long-term debt can become less reliant on ST debt by issuing inflation-indexed LTLC bonds. In Latin America, in some countries (e.g., Chile, Argentina, and Colombia), the majority of domestic government debt is indexed to inflation, whereas other countries (e.g., Venezuela, Brazil, and Mexico) borrow a large amount of interest-rate indexed debt. The choice between inflation-indexed bonds and interest-rate indexed bonds possibly depends on the relative risks between inflation and interest-rate changes as well as the transparency and effectiveness of monetary policy.

Fourth, although currency derivative markets in Latin America experienced explosive growth over the last few years, there seems to remain a large unhedged foreign-exchange exposure. Furthermore, similar to the debt market, maturity mismatches also exist in derivative markets, where firms use ST derivatives to hedge longer-term FC debt. For example, nearly all the currency derivative contracts in Chile have maturity less than 1 year. The inability to rollover hedges during crises due to low market liquidity may convert the synthetic local-currency debt back to unhedged FC debt. An additional shortcoming of the derivatives as an instrument for country insurance is that they may increase the risk of contagion during crises. The presence of a large amount of derivative transactions, especially OTC-traded derivatives, may create a rapid expansion of counterparty credit risk during financial crisis.

This paper has raised issues for future research in various directions. One possible direction is to more thoroughly investigate the impact of the development of financial markets on currency and maturity mismatches, and especially any tradeoff between these mismatches. This could also explain why the relationship between the mismatches changes over time, which possibly depends on the financial development and institutional reforms taking place in the region.

Another direction of future research is to examine the determinants of the development of derivative markets around the world. In spite of the fact that derivative transactions experienced explosive growth around the world and showed significant implications for systemic vulnerabilities, research on financial development has been focused almost exclusively on the development of banking, equity, or bond markets. Little is known about the institutional and macroeconomic factors that may facilitate the growth of derivative markets, especially in emerging economies.

REFERENCES

Aguiar, M. (2002) Investment, devaluation, and foreign currency exposure: The case of Mexico, Working Paper, University of Chicago GSB, Chicago, IL.

Allayannis, G., Brown, G., and Klapper, L. (2003) Capital structure and financial risk: Evidence from foreign debt use in East Asia, *Journal of Finance* 58(6): 2667–2709.

Allen, M., Rosenberg, C., Keller, C., Setser, B., and Roubini, N. (2002) A balance sheet approach to financial crisis, Working Paper 02/210, IMF, Washington, DC.

Bleakley, H. and Cowan, K. (2002) Corporate dollar debt and depreciations: Much ado about nothing? Working Paper 02-5, Federal Reserve Bank of Boston, Boston, MA.

Bleakley, H. and Cowan, K. (2004) Maturity mismatch and financial crises: Evidence from emerging market corporations, Working Paper 2004-16, UCSD, San Diego, California.

Borensztein, E. and Mauro, P. (2004) The case for GDP-indexed bonds, *Economic Policy* 19(38): 165–216.

Burger, J. and Warnock, F. (2006) Local currency bond markets, *IMF Staff Papers*, 53(1): 133–146.

Calvo, A., Izquierdo, A., and Mejía, L. (2004) On the empirics of sudden stops: The relevance of balance-sheet effects, Working Paper 509, IADB, Washington, DC.

Calvo, G. and Reinhart, C. (2002) Fear of floating, *Quarterly Journal of Economics* 117(2): 379–408.

Céspedes, L. (2004) Financial frictions and real devaluations, Working Paper, Central Bank of Chile, Chile.

Chang, R. and Velasco, A. (1999) Liquidity crises in emerging markets: Theory and policy, Working Paper 7272, NBER, Cambridge, MA.

Chan-Lau, J. (2005) Hedging foreign exchange risk in Chile: Markets and instruments, Working Paper 05/37, IMF, Washington, DC.

Claessens, S., Klingebiel, D., and Schmukler, S. (2003) Government bonds in domestic and foreign-currency: The role of macroeconomic and institutional factors, Working Paper, World Bank, Washington, DC.

Conrad, J. (1989) The price effect of option introduction, *Journal of Finance* 44(2): 487–498.

Cowan, K., Hansen, L., and Herrera, E. (2005) Currency mismatches, balance sheet effects and hedging in Chilean non-financial corporations, Working Paper 521, IADB, Washington, DC.

Davies, N., Podpiera, R., and Das, U. (2003) Insurance and issues in financial soundness, Working Papers 03/138, IMF, Washington, DC.

De la Torre, A. and Schmukler, S. (2004) Coping with risks through mismatches: Domestic and international financial contracts for emerging economies, *International Finance* 7(3): 349–390.

Dermirguc-Kunt, A. and Maksimovic, V. (1999) Institutions, financial markets, and firm debt maturity, *Journal of Financial Economics* 54(3): 295–336.

Detragiache, E. and Spilimbergo, A. (2002) Empirical models of short-term debt and crises: do they test the creditor run hypothesis? Working Paper 01/2, IMF, Washington, DC.

Dodd, R. (2000) The role of derivatives in the East Asian financial crisis, Working Paper 20, CEPA, New York, NY.

Eichengreen, B., Hausmann, R., and Panizza, U. (2003) Currency mismatches, debt intolerance and original sin, Working Paper 10036, NBER, Cambridge, MA.

Fernandez, V. (2003) What determines market development? Lessons from Latin American derivatives markets with an emphasis on Chile, *Journal of Financial Intermediation* 12(3): 390–421.

Gezcy, C., Minton, B., and Schrand, C. (1997) Why firms use currency derivatives, *Journal of Finance* 52(4): 1323–1354.

Goldstein, M. and Turner, P. (2004) *Controlling Currency Mismatches in Emerging Markets*. Institute for International Economics Press: Washington, DC.

Harvey, C. and Roper, A. (1999) The Asian bet. In: Harwood, A., Litan, R., R. Pomerleano (Eds.), *The Crisis in Emerging Financial Markets*. Brookings Institution Press: Washington, DC.

Hausmann, R. and Panizza, U. (2002) The mystery of original sin: The Case of the missing apple. In: Eichengreen, B. and Hausmann, R. (Ed.), *Debt Denomination and Financial Instability in Emerging Market Economies*. University of Chicago Press: Chicago, IL.

IMF (1999) Involving the private sector in forestalling and resolving financial crises, Working Paper, IMF, Washington, DC.

Ize, A. and Levy Yeyati, E. (2003) Financial dollarization, *Journal of International Economics* 59(2): 323–347.

Jadresic, E. and Selaive, J. (2005) Is the FX derivatives market effective and efficient in reducing currency risk? Working Paper 325, Central Bank of Chile, Chile.

Kaminsky, G. and Reinhart, C. (1999) The Twin crises: Causes of banking and balance-of-payments problems, *American Economic Review* 89(3): 473–500.

Moguillansky, G. (2002) Non-financial corporate risk management and exchange rate volatility in Latin America, Discussion Paper 30, WIDER, Helsinki, Finland.

Perry, G. and Serven, L. (2003) The anatomy of a multiple crisis: Why was Argentina special and what can we learn from it? Working Paper 3081, World Bank, Washington, DC.

Raddatz, C. (2006) Liquidity needs and vulnerability to financial underdevelopment, *Journal of Financial Economics* 80(3): 677–722.

Rodrik, D. and Velasco, A. (1999) Short-term capital flows, Working Paper 7364, NBER, Cambridge, MA.

Schmukler, S. and Vesperoni, E. (2006) Financial globalization and debt maturity in emerging economies, *Journal of Development Economics*, 79(1): 183–207.

Sorge, M. and Zhang, C. (2006) Credit information quality and corporate debt maturity: Theory and evidence, Working Paper 4239, World Bank, Washington, DC.

CHAPTER 20

Dangers and Opportunities for the Russian Banking Sector: 2007–2008

Dean Fantazzini, Alexander Kudrov, and Andrew Zlotnik

CONTENTS

The Russian banking system has been in its worst crisis since 1998, a fact made particularly evident by the collapse in the share prices for every financial service company, together with the fall of Russian stock markets. However, unlike 1998, the banking system has been in a better position thanks to the previous macroeconomics boom, which lasted for almost 10 years. The highly fragmented structure of the banking sector still relies heavily on the state banks, but the contribution by foreign banks as well as local private banks has increased steadily in the last years. We have discussed the major improvements and weaknesses that currently characterize the Russian banking system, together with the systemic risk that still plays a major role in such a market.

20.1 INTRODUCTION

The Russian banking system has been in its worst crisis since 1998: On the one hand, this is a consequence of global financial and economic crisis, and on the other hand, there are specific country factors. First, Russian economy depends on a relatively small number of industries. Second, Russian firms have a large amount of foreign debt. Furthermore, when oil prices decrease, this leads to a decline in the ruble against the dollar and the euro. For example, in Figure 20.1, the most important Russian macro indicators that influence the local banking system are presented: it is evident that the steep fall in reserves and money mass starting from September 2008 to October 2008, together with the Russian stock index and the Brent oil price, was accompanied by a quick depreciation of the ruble against the dollar and the euro.

Another problem is represented by the amount of risky investments. In the last 10 years, risk managers in Russia were not really independent and unprejudiced: As risk managers know what the top management wants (usually, they want money), this has determined a large increase in risky investments that has generated high profits, but also high risk. Thus, the

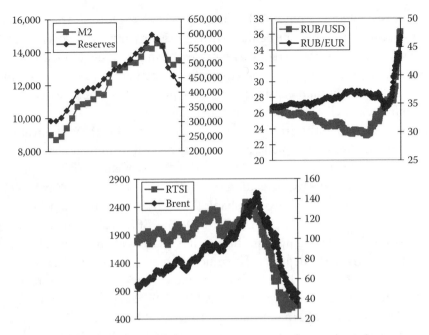

FIGURE 20.1   Russian macro indicators: M2, amount of reserves, exchange rates, RTS index, and oil price. (From Central Bank of Russia, RTS, Finam.)

entire economy became more risky. The main reason for this problem can be identified in the performance measurement of business management: bonuses depend on profits and high profits imply high risks. The history of the last quarter of a century clearly shows that a number of banks were merged or acquired after a financial crisis to avoid default, due to their poor strategies that focused only on a short-term horizon.

The main institutional problem is that *risk management measures risk,* but *does not manage it.* In this regard, this clearly implies that Basel II conception should be revised and that stress testing should lead to real protection.

20.2 DESCRIPTION OF THE RUSSIAN BANKING SECTOR

20.2.1 Current Situation

The current total number of banks in Russia has not changed significantly from the past years before the crisis. According to a report by the Central Bank of Russia (CBR), during the year 2008, the amount of banks decreased by 3.1%, from 1092 banks to 1058 banks in Russia. These banks were acquired by the state banks or by top private banks with government protection, and their market share was less than 5%. The main reasons of these deals were the impossibility of refinancing (owing to the seizure in the global credit markets since July), nonoptimal structure of assets and liabilities, as well as low-quality risk management.

The Russian authorities have tried to deal with this threat to protect the population, banks, and large firms: The deposit insurance agency now guarantees the deposits up to 20,000 dollars* in each bank, while the local Parliament passed a law extending the time for the value added tax (VAT) payments. Besides, Vnesheconombank (VEB) received $50 billion in refinancing facilities (5 years at LIBOR + 5). The CBR has also taken some measures for increasing liquidity: It has increased the limit of free-budget funds on the deposits of commercial banks, increased the number of banks up to 28 for which the budget funds are available, and decreased the rate of legal reserve requirements by 0.5%. Furthermore, the CBR and the Ministry of Finance have provided $38 billion in subordinated long-debt financing (maturity at 2019 at 8%): $20 billion for Sberbank, $8 billion for VTB, $1 billion for Rosselkhozbank, and $9 billion for other top banks.

It is important to highlight that the first 200 Russian banks have accumulated 95% of assets, while the first 50 banks have concentrated 80% of assets. On January 1, 2009, the total capital of the registered operating credit institutions amounted to $25.2 billion, more than 20.4% higher than the level of that capital on January 1, 2008 (see Figure 20.2).

As we can see from Figure 20.2, Sberbank and VTB are the key drivers of Russian banking system. These banks hold 50% ($118 billion) and 5% ($12 billion) of deposits, respectively, while the other top 20 banks hold only a market share of 18%. The total share of other banks in the top 50 is 10%, whereas the share of the remaining 850 banks (which have special license) is around 17%.

Sberbank is the largest bank in Russia and Eastern Europe. In many Russian regions, Sberbank is basically the only bank able to provide local administrations and citizens with banking services and important financial support for local investments and social programs (see also http://en.wikipedia.org/wiki/Sberbank). Like many banks around the world,

---

* In this chapter, we have assumed 35 ruble for 1 dollar for current and future events, while we have used 25–1 Rub/$ for the period of time before December, 2008.

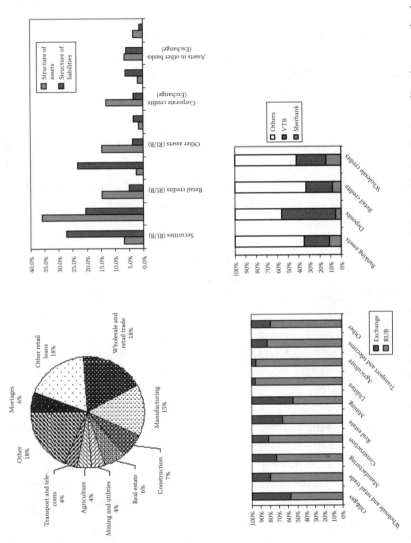

FIGURE 20.2   Currencies (ruble or foreign) in which the credits are denominated, and distribution of banking credits by sectors. (From Central Bank of Russia, Troika Dialog, Renaissance Capital.)

Sberbank has taken a beating amid the global financial crisis, with its shares plunging on Russian stock markets. On November 17, 2008, the Kommersant daily reported (quoting a plan approved by the bank's board) that Sberbank wants to shed up to 70,000 jobs by 2014, a quarter of its workforce.

The government has promised to provide $230 billion to the banking sector, where Sberbank and VTB banks could benefit up to $80 and $40 billion, respectively. At the end of November 2008, these banks received about half of these funds (Sberbank—$36 billion and VTB—$22 billion). The weighted average cost of state funds for Sberbank is currently 8.9%, while for VTB, it is 9.4%. Comparing these values with the cost of credit for commercial firms, which is equal to 18% (i.e., the maximum possible percent "recommended" by the government), we can observe that these two banks can potentially generate high margins. However, the problem is that considerable share of these credits will be written of. Sberbank announced that it will support utilities, retail agriculture, defense, and small business, while VTB will focus on oil and gas, chemistry, constructing, retail, and agricultural and food production.

20.2.2 General Trends in the Russian Banking Sector

The first important trend is the "outflow of retail and wholesale deposits" in 4Q 2008. Retail deposits accounted for almost 25% of banks' finance, and in general, the situation on the outflow of funds cannot be described as disastrous. The outflow of funds from retail deposits in state-owned banks was about 5%–6% in October 2008. With regard to private banks, the outflow was much more serious, having lost, on an average, 10%–12% of retail deposits in October 2008. According to some local data, Sberbank lost about 3.2% of its deposits in October, which accounted for about $4 billion. Corporate deposits represent about 40% of bank liabilities. In August 2008, the corporate deposits showed a steady growth, which could be attributed to the cash flows from foreign borrowing on contracts signed in July 2008. However, the likely increase in the number of corporate defaults could cause several problems to the banks. Nevertheless, in September and October 2008, clear trends were not observed in this regard: while there was a significant outflow of corporate deposits (around 20%) in some banks, there was an influx of funds in others. For example, there is no substantial outflow of funds from corporate deposits in Sberbank, but rather there is a flow of funds in settlement funds. This is consistent with the expectation that corporations are now sending their funds to those financial institutions that are more stable. Consequently, the probability of a fall of the financial sustainability for many smaller banks has increased, along with the probability of a fall of the ruble as a result of large-scale conversion of deposits denominated in rubles into dollars and euros. Leading Russian investment banks, Troika Dialog and Renaissance Capital, have forecasted a continuation of capital outflow in 2009.

The second dangerous fact is that "the wholesale funding (debt market) does not work." The only source of liquidity for the Russian banks is now represented by the resources of the CBR and the Ministry of Finance. Banks will not start recovering large-scale lending, until the debt market has stabilized. It is expected that the CBR will keep on playing a key role in maintaining the liquidity of the banking system: $26 billion were invested in the banking system in October 2008, significantly more than the value of the deposits outflows. Unfortunately, most of these funds have focused on the FOREX market. The list of banks that have received financial support by the CBR is much broader than that initially approved by the authorities and based on ratings: a large amount of financial institutions have received funding even though they do not have a rating (around $0.3–$2 billion in October). In most cases, this support has been carried out against the backdrop of the critical situation of banks, and in some cases, this support was motivated by political reasons. In November 2008, the CBR has amended the regulations to support not only banks with rating assigned by agencies such as S&P, Fitch and Moody's, but also banks with ratings assigned by national agencies recognized by CBR. The consequences of this situation on banks will depend mainly on their abilities and dimensions. As a result, for example, top banks are forecasted to remain quite stable, because they are supported by the state and the CBR. Till date, the risk of large-scale bankruptcies in the banking sector has remained low and is fully manageable.

The third new trend is the "new lending structure." Most private banks have reduced their lending activity and their lending portfolio has declined by 5%–7% on an average in October. Comparing the dynamics of the banks' loan portfolios with the amounts of funds received thanks to the state support, we can observe that in November 2008, the lending to the real sector was not the main priority for the top 30 Russian banks. Out of the $33 billion received in November 2008 from the CBR, the largest banks have used only one-third of this amount for lending activities, whereas they have used the remaining part to finance their own debts, mainly debts to the Ministry of Finance. In November 2008, the largest Russian banks returned their debts to the Ministry of Finance. However, in October 2008, the two largest state-controlled banks (VTB and Sberbank) significantly increased lending: Both by $5 billion. In general, the aggregate effect on the credit system can be described as neutral or slightly negative. However, the majority of new loans were taken by large borrowers, while small businesses are experiencing significant problems in refinancing their debts.

Another result of the difficult situation for lending is the deteriorating quality of assets. According to the aggregate balance sheet for 30 Russian banks issued by the CBR, the amount of outstanding debt in the major banks loaned to other credit organizations has increased seven times and reached $0.3 billion, whereas the amount loaned to nonfinancial organizations has increased by 25% and amounted to $7 billion. Much of these increases represent bad debts and they fall on the 30 largest Russian credit organizations. While such figures are the first signs of a general deterioration in the quality of credit portfolios, nevertheless, its importance for the banking system is not crucial. For example, the percentage of credit-related crimes increased only by 15%–20%.

The last (but not the least) new trend is represented by the fact that Russian banks increased their assets in the interbank foreign exchange market by $15–$20 billion in October 2008 alone. Until now, the CBR has taken no effective action to stop or mitigate the effects of these speculations. On the contrary, the devaluation of the ruble against a basket of currencies contributed to more aggressive speculative activities. Interestingly, in September 2008, the most aggressive players against ruble were the foreign banks, but in October 2008, they were Russian banks that used considerable facilities to increase their foreign exchange funds. The CBR plans to impose sanctions on those banks that are using funds from the state support to buy foreign currency.

20.3 ECONOMETRIC ANALYSES OF BANKING SECTORS

20.3.1 New Approach for Default Forecasting: Zero-Price Probability

The standard KMV-Merton model as well as other structural and statistical approaches for default forecasting assumes that the accountancy data represent the true picture of the company financial situation. With regard to the KMV-Merton model, the accountancy data used is the book face value of the firm's total liabilities. The usual practice considers the sum of short-term liabilities along with one-half of the long-term liabilities for total liabilities. This assumption, which is made by Moody's KMV for North American firms, ensures that the firms liabilities are not overstated (see also Vassalou and Xing, 2004; Hao, 2006). Even though Vassalou and Xing (2004) stated that using different percentages for long-term liabilities is not deemed to alter the main qualitative results, such an approach may not be robust to "window dressing" policies made to improve the financial score of a company or, in the worst case, to financial frauds. Besides, KMV itself admits that "in practice the market leverage moves around far too much to provide reasonable results" and particular iterative methods have to be used, instead (see Crosbie and Bohn, 2001).

The recent default of the food giant Parmalat in the 2003 and Enron in 2002 clearly showed how the debts reported in the certified balanced sheet can represent only a part of the true debt figures. In general, the debt values reported in the certified balance sheets are underestimated for two reasons: (1) to window dress the financial health of the company, in the best case; and (2) to hide financial fraud, in the worst case (see, e.g., Parmalat, Enron, etc.). Ketz (2003) discussed a wide variety of techniques to hide debts and financial risks. This explains why the Merton's default probabilities and firm values are usually underestimated with respect to other methods. Besides, the log-normal is not an appropriate distribution for price dynamics, as it underestimates the tail of the distribution. Furthermore, heteroskedasticity is not considered at all in the KMV-Merton model. Increasing volatility and leptokurtosis can be interpreted as a signal of informed trading (see Biais et al. (2005) and Hasbrouck (2007), for recent surveys about market microstructure studies).

The previous short description clearly highlights that using accounting data to infer the firm's default probability can be misleading and may result in a very poor estimate. To avoid such problems, we have used a recent approach proposed by Fantazzini (2008; 2009), which uses the null price as a default barrier to separate an operative firm from a defaulted one, and to estimate its default probability without resorting to accountancy data.

Let us consider the following two financial identities based on the "true" accountancy data at time T:

$$\begin{cases} E_T = A_T - B_T \\ E'_T = A_T = (A_T - B_T) + B_T = E_T + B_T \end{cases}$$

and consider the financial meanings and signs of E_T and E'_T according to the situation faced by the firm.

TABLE 20.1   Financial Meaning and Signs of E_T and E'_T

	$E_T = A_T - B_T$	$E'_T = A_T$
Operative	Equity belonging to shareholders (+)	Asset value (+)
Defaulted	Loss given default for debtholders (−)	Equity belonging to debtholders (+)

Table 20.1 shows that the quantity E_T is negative when the firm defaults, as it represents the loss given default for debtholders, while it is positive when the firm is operative, representing the equity belonging to share-holders, instead. A negative value for E_T is a direct consequence of the limited liability now in place in all modern western legislations. Besides, losses can be theoretically infinite like profits: for example, the effects of September 11, 2001 attacks on the airline companies, or the impact of the mad cow and bird flu diseases on agriculture companies. Thus, we can also resort to probability density functions with negative domain.

The main consequence of the previous discussion is that we can esti-mate the distant to default simply by using E_T, instead of d_2 as in Merton's framework, and the default probability by $\Pr[E_T \leq 0]$, as the firm defaults when E_T is zero or negative. Furthermore, given that $E_T = S \times P_T$, where P_T is the quoted stock price at time T and S is the number of shares, the default probability of a firm can be retrieved by estimating $\Pr[P_T \leq 0]$, that is, by using the zero-price probability (ZPP). While the quoted price P_T is a truncated variable that cannot be negative, the quantity E_T has no lower bound, as it has a different financial meaning on whether the firms is operative or defaulted: in the former case E_T is computed daily in (elec-tronic) financial markets, whereas in the latter case, the loss given default is computed in bankruptcy courts.

Stock prices are usually nonstationary I(1) variables and it is common to model their dynamics by considering the log-returns, so that the prices are guaranteed to be positive. However, we are interested in find-ing $\Pr[P_T \leq 0]$, as we have just shown that the null price can be used as a default barrier. A straightforward way to determine this is to consider a conditional model for the differences in prices levels $X_t = P_t - P_{t-1}$, instead of differences in log-prices.

An analytical close-form solution for $\Pr[P_T \leq 0]$ is available for a few special and unrealistic cases, such as normally distributed prices with homoskedastic variance. When this is not the case, simulation methods are required. If we are at time t and want to estimate the default probabil-ity at time $t + T$, we can use the following general algorithm:

Proposition 1 (ZPP estimation algorithm):

Step 1: Consider a generic conditional model for the differences of prices levels $X_t = P_t - P_{t-1}$, without the log-transformation, given the filtration at time t, F_t:

$$\begin{cases} X_t = E[X_t \mid F_t] + \varepsilon_t \\ \varepsilon_t = H_t^{1/2} \eta_t, \quad \eta_t \sim i.i.d.(0,1) \end{cases} \tag{20.1}$$

where $H_t^{1/2}$ is the conditional standard deviation.

Step 2: Simulate a high number N of price trajectories up to time $t + T$, using the estimated time series model (3.1) at Step 1.

Step 3: The default probability is simply the ratio n/N, where n is the number of times out of N when the price touched or crossed the barrier along the simulated trajectory.

This method entails a number of important benefits: we only need the stock prices and we do not need neither any firm's volatility σ_A nor the debt face value, like in Merton-style models. We can consider more realistic distributions than the log-normal and can estimate the default probability for any given time horizon $t + T$. Besides, the structure in Table 20.1 can be easily generalized to a general sector: Instead of having the equity for a single firm, we can have the equity belonging to all shareholders of a specific sector, for example, the financial sector. Therefore, by using a sector index instead of a single stock, the ZPP can also be used as an early warning system for systemic default of a general sector.

20.3.2 An Extreme Value Theory Approach to Value at Risk Estimation

It is well known that historical returns are usually modeled with the so-called fat-tailed distributions. In particular, the distributions of stock returns are fat-tailed (see, e.g., Jansen and De Vries, (1991)). In this case, to assess the extremal index, the Hill's estimator is applied.

Let us denote the Value at Risk (VaR) at the probability level p for the distribution function $F(x)$ of negative returns by x_p. Then, we have that:

$$F(x_p) = 1 - p.$$

To estimate the VaR at the 5% or 1% probability levels, we propose the following procedure, which provides an accurate approximation in case of small samples:

Proposition 2 (Robust estimation algorithm for VaR):

- For every consecutive 250 days (during the considered period of time), we compute the set of Hill's estimators ($\gamma(k)$) for the extremal index of the distribution function of negative returns: Suppose we have, within the considered period, m negative returns $X_1,\ldots,X_m$ and let $X_{(1)} \leq X_{(2)} \leq \ldots \leq X_{(m)}$ be their ordered statistics; then, the set of Hill's estimators ($\gamma(k)$) is defined as follows:

$$\gamma(k) = \frac{1}{k}\sum_{i=1}^{k}(\log X_{(m-i+1)} - \log X_{(m-k)}), \quad 1 \leq k \leq m-1$$

- Consider the following model for the sequence of Hill's estimators ($\gamma(k)$):

$$\gamma(k) = \gamma + \beta_1 k + \varepsilon_k, \quad k = 1,\ldots,\kappa \tag{20.2}$$

where $E(\gamma(k)) = \gamma + \beta_1 k$, $Var(\varepsilon_k) = \sigma^2/k$ and γ is the true value of the extremal index for the distribution of negative returns. We can then estimate γ by using the method of weighted least squares with a weighting $\kappa \times \kappa$ matrix W (that has ($\sqrt{1},\ldots,\sqrt{\kappa}$) on the main diagonal and zeroes elsewhere).

- To estimate the excess level or VaR at the probability level p ($0 < p < 1$) for the next day, we can use the following estimator:

$$\hat{x}_p = \frac{\left(\dfrac{r}{pn}\right)^{\hat{\gamma}} - 1}{1 - 2^{-\hat{\gamma}}}(X_{(n-r)} - X_{(n-2r)}) + X_{(n-r)} \tag{20.3}$$

where
   n is the number of negative returns
   $r = [\kappa/2]$ ([.]-integer part)
   $\hat{\gamma}$ is the estimator for the *extremal index* γ

$X_{(n-r)}$, $X_{(n-2r)}$ are $(n-r)$- and $(n-2r)$-ordered statistics of the absolute valued positive returns sequence $X_1,\ldots,X_n$, respectively.

The estimator of extremal index γ, used on the first step of the estimation algorithm for VaR, was proposed by Huisman et al. (2001), where it is recommended to take $\kappa = m/2$. Instead of selecting an optimal threshold for the

Hill's estimator of the extremal index, this approach allows to compute an optimal unbiased estimate of γ on the basis of the Hill's estimators set (with the thresholds $k = 1,\ldots,\kappa$). In the second step, we have used the consistent estimator of the excess level x_p proposed by Dekkers and De Haan (1989).

20.3.3 Empirical Analysis

Given the situation described in Section 20.2, we believe that it will be interesting to employ the previously described ZPP and EVT approaches to analyze the development of investor-risk perceptions in the last two years, 2007–2008. First, we deal with financial sector indexes to examine the systematic default risk in the banking sector. In addition to the Russian financial sector, we have also considered the American, English, and Italian financial sectors. While the former two have been badly hit by the subprime crisis, the Italian financial sector has suffered less because of its relatively small exposure to credit-structured products and the small amount of private debt. Particularly, we have considered the Russian RTS Financial Index, the American Dow Jones Financial Index, the English FTSE Banking Index, and the Italian MIBTEL Financial Index.

20.3.3.1 Financial Sector Indexes

We have considered an AR(1)-Threshold-GARCH(1,1) model for the differences in prices levels $X_t = P_t - P_{t-1}$, together with a Student's t distribution, to take the leverage effect as well as leptokurtosis in the data into account (see Glosten et al., 1993 for more details):

$$\begin{cases} X_t = \mu + \phi_1 X_{t-1} + \varepsilon_t \\ \varepsilon_t = \sqrt{h_t}\,\eta_t, \quad \eta_t \sim i.i.d.(0,1) \\ h_t = \omega + \alpha\varepsilon_{t-1}^2 + \gamma\varepsilon_{t-1}^2 D_{t-1} + \beta h_{t-1} \end{cases} \tag{20.4}$$

where $D_{t-1} = 1$ if $\varepsilon_{t-1} < 0$. We chose such a specification given its past success in modeling financial variables. As for the number N of simulated price trajectories to estimate the ZPP, we had set $N = 5000$. Due to space limits, we did not consider here the goodness-of-fit test procedures and bootstrapped confidence intervals for the estimated ZPPs (see Fantazzini, 2008; 2009 for more details).

To estimate the default probabilities, we considered a 1 year and 5 year ahead horizon. Following Giacomini and Komunjer (2005) and González-Rivera et al. (2004), we used a rolling estimation scheme, because it may be more robust to a possible parameter variation. Particularly, we re-estimated the parameters at each time t using an estimation sample containing 1000 most recent observations, that is, the observations from $t-1000+1$ to t. We estimated the default probabilities for the examined financial sectors in the years 2007–2008 for a total of 500 trading days up to December 31, 2008 (trading days are slightly different among the four stock exchanges, due to local festivities). Figures 20.3 and 20.4 show the default probabilities computed by using the ZPP, with a range between 0% and 100%.

The Russian financial index clearly shows a higher degree of riskiness than the other markets that was quite high already at the beginning of 2007 and peaked in October 2008. However, this higher risk is mostly due

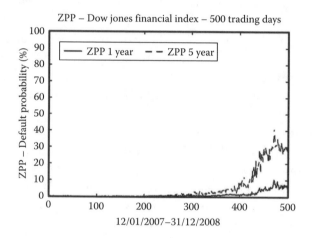

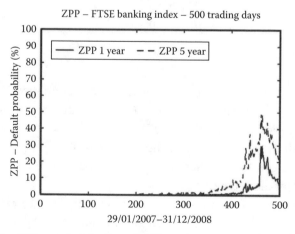

FIGURE 20.3 Estimated default probability: American and English financial sectors indexes.

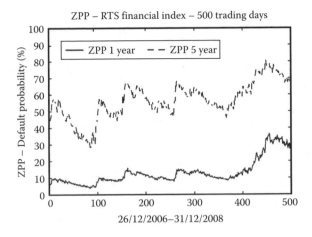

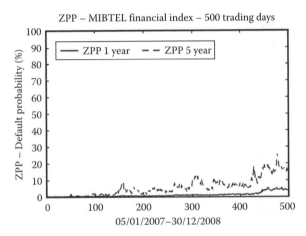

FIGURE 20.4 Estimated default probability: Russian and Italian financial sectors indexes.

to a higher country risk (Russia has a rating of BBB+), than the competing countries (United States and United Kingdom have AAA, while Italy A+). Besides, the increases in the default probabilities for the American and English banking sectors in 2008 are very large and reflect the difficulties that they have experienced so far. Interestingly, the Italian financial sector currently shows the smallest default probability (although it was still higher than the American and English ones till July 2008), thus confirming the smaller impact that the subprime crisis has had on Italian banks.

The previous conclusions are confirmed by the VaR estimates in Figures 20.5 and 20.6, which confirm the dramatic increase in the risk of American and English financial sectors, when compared with the Italian and Russian sectors.

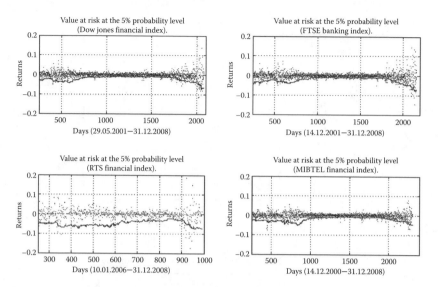

FIGURE 20.5   Estimated value at risk at the 5% probability level: English, Russian, and Italian financial sectors indexes.

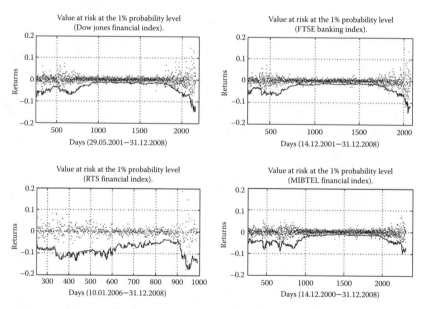

FIGURE 20.6   Estimated value at risk at the 1% probability level: English, Russian, and Italian financial sectors indexes.

20.3.3.2 Single Banks

After having examined the financial sectors, we now proceed to analyze four single banks, one for each country previously considered, that represent important cases due to their dimension and/or financial history:

1. *Citigroup* (*United States*): On 24 November 2008, the U.S. government announced a massive bailout of this bank, designed to rescue the company from bankruptcy while giving the government a major say in its operations. It was the largest bailout in financial history. As the subprime mortgage crisis began to unfold, heavy exposure to toxic mortgages in the form of collateralized debt obligations (CDOs), compounded by poor risk management, led the company into serious trouble. The Treasury has provided another $20 billion in TARP funds at the end of 2008, in addition to $25 billion given in October 2008. However, these measures seemed to be inadequate, and on 16 January, 2009, Citigroup announced that it would split into two businesses after reporting a fourth-quarter loss (in 2008) of $8.29 billion in a new dramatic move to survive. Citigroup lost $18.72 billion in 2008. It is clear that any further government intervention would bring this bank a step closer to nationalization.

2. *Royal Bank of Scotland* (*United Kingdom*): Despite heavy fundraising for many billions of Euros and amid the worsening of the 2008 global financial crisis, on 13 October 2008, the British Prime Minister, Gordon Brown, announced a U.K. Government bailout of the financial system, where RBS would be the main beneficiary. At the end of 2008, the U.K. Treasury held a near 58% controlling shareholding in the RBS, even though it stressed that it was not a "standard public ownership" and that the bank would return to private investors "at the right time."

3. *Unicredit* (*Italy*): It is the largest Italian bank and one of the top five banks in Europe. However, the global credit crisis has hit UniCredit harder than other Italian banks: "the share price collapsed (down 75% in 2008), and there was the revelation, buried in a prospectus released two days before Christmas, that clients of the bank could lose up to €805 million from the alleged fraud perpetrated by Bernard Madoff. As a consequence new capital has had to be raised: the bank is completing a €3 billion convertible bond issue and the

2008 dividend of about €3.6 billion will be paid in shares. Along with disposals of noncore assets at the end of last year, these measures have raised UniCredit's core tier 1 capital ratio—the key measure of a bank's balance sheet strength—to 6.7%. However, this may not be enough. After the forced recapitalizations of banks elsewhere in Europe as a result of the credit crisis, Italian banks have among the lowest capital ratios. Almost all of them are under 7%, and many are below 6%. Nevertheless, Henry MacNevin, chief analyst for Italian banks at Moody's Investors Service, said on January 14, 2009 on the FT that the Italian banks could get by with core tier 1 ratios of 7% and upward, compared with at least 8% for some European peers because of their relatively small exposure to structured products" (Boland, FT.com, 14 January 2009).

Similar to the Financial Sectors Indexes, we have employed AR(1)-T-GARCH(1,1) models with a Student's t distribution to model leptokurtosis in the price differences X_t and compute the ZPP for the considered banking stocks. Figure 20.7 shows the default probabilities computed by using the ZPP, again with a range between 0% and 100% for the last 500 trading days up to December 31, 2008.

The estimated ZPPs clearly show the difficult situation Citigroup and RBS are facing: the former has a 1-year-ahead default probability close to 50%, while the latter has already exceeded this limit in October 2008. Currently (January–February 2009), the RBS management and the U.K. ministers are discussing whether a new bailout is needed, while Obama's new administration is assessing whether a new intervention is required in the American banking sector. The main problem so far is that the amount of credit losses is still unclear and more probably has to come.

As for Unicredit and Sberbank, even though their risks of default have increased after the financial turmoil in October 2008 (close to 15% and 25%, respectively, for the 1-year-ahead ZPP), nevertheless, these risks have stabilized since then, differently from the previous American and English banks.

The previous conclusions are again confirmed by the VaR estimates at the 5% and 1% level reported in Figures 20.8 and 20.9, respectively, where the estimated VaR for the Citigroup and the RBS have become higher than those for Sberbank and Unicredit (while the reverse was true till August 2008).

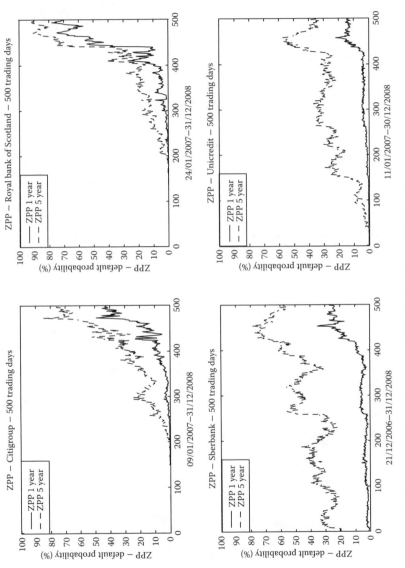

FIGURE 20.7   Estimated default probability: Citigroup, RBS, Sberbank, and Unicredit.

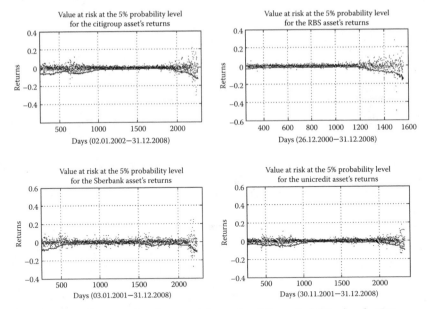

FIGURE 20.8 Estimated value at risk at the 5% probability level: Citigroup (American bank), RBS (English bank), Sberbank (Russian bank), and Unicredit (Italian bank).

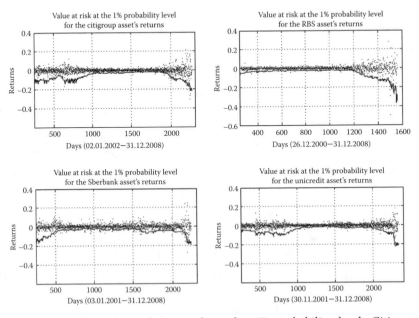

FIGURE 20.9 Estimated value at risk at the 1% probability level: Citigroup (American bank), RBS (English bank), Sberbank (Russian bank), and Unicredit (Italian bank).

20.4 WHAT IS NEXT?

20.4.1 Consolidation of Banking Sectors

The government has helped the state-controlled banks to buy banks in trouble. However, the current situation is also profitable for some large commercial banks, which can also acquire troubled banks. In general, in terms of stability of the banking system, the risks remain manageable. The top 20–30 Russian banks as well as the strongest regional banks are most likely to overcome the crisis: for them, the crisis is a good opportunity to raise their market share and to acquire small banks by negligible price. In fact, the required financial resources needed to save all troubled banks are much higher than the capacity of the available government financial funds, which thus may lead to the default of small- and medium-sized banks. On the contrary, state-controlled banks may enjoy low levels of risk, and the concentration of the loan portfolios in the largest banks may continue. Moreover, the liquidity provided by the state has become the determining factor in the banking system: accordingly, the CBR is likely to set some regulations on the banks, and those that do not cooperate may probably face significant regulatory risks (Figure 20.10).

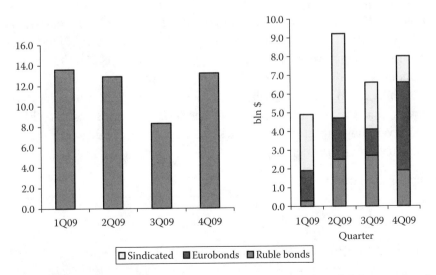

FIGURE 20.10 Russian external debt discharge (left) and total debt amount of Russian banks by bonds and syndicated credits (right). (From Central Bank of Russia, Troika Dialog.)

20.4.2 Dangers

The main risks that the Russian banking sector may face in the near future can be summarized as follows:

- A continued decrease in oil prices may determine a drop in economic growth, which may last much longer in the absence of cash and credits.

- Until we see some stabilization in the FOREX market, it is difficult to expect the recovery of the lending activity. Most likely, there will be further restructured loans and the declining quality of the collaterals, with an increase of bad loans in the banks' portfolios.

- Rising unemployment and deteriorating economic conditions can lead to loss of public confidence in the government.

Furthermore, reducing the rate of lending may cause a strong slowdown in demand and consumer loans, thus decreasing the production, and, in general, slowing the growth of the economy.

20.4.3 Opportunities

The Russian banks have used the current situation to try to optimize their expenses. For example, Sberbank announced the next target values for the main parameters of efficiency from a long-term perspective: expenses/revenue: 40%–45% (in a previous Sberbank report in 1H2008, this parameter was set to 51%), assets/employees—$3.3 billion (it was $0.8 billion), total number of employees—200–220 thousands (they were 258 thousands). Besides, VTB also announced a cut in its working staff.

Troika Dialog currently considers Sberbank and VTB as undervalued assets, and the recommendation for the long-run investors is to BUY. However, they have noted that much will depend on the degree of market devaluation and the behavior of the ruble.

REFERENCES

Biais, B., Glosten, L., and Spatt, C. (2005) Market microstructure: A survey of microfoundations, empirical results and policy implications. *Journal of Financial Markets*, 8(2): 217–264.

Boland, V., UniCredit chief scrutinised as bank struggles, FT.com, 14/01/2009, available at http://www.ft.com/cms/s/0/37fb4298-e262-11dd-b1dd-0000779fd2ac, dwp_uuid = 4da69efc-1b8f-11dd-9e58-0000779fd2ac.html.

Central bank review of Russian Banking system in Russian (2009) available at http://cbr.ru/analytics/bank_system/obs_081201.pdf.

Crosbie, P. and Bohn, J. (2001) *Modeling Default Risk*. KMV Technical document, New York.

Dekkers, A. L. M. and de Haan, L. (1989) On the estimation of the extreme-value index and large quantile estimation. *Annals of Statistics*, 17(4): 1795–1832.

Fantazzini, D. (2008) A new approach for firm value and default probability estimation beyond merton models. *Computational Economics*, 31(2): 161–180.

Fantazzini, D. (2009) Forecasting default probability without accounting data: Evidence from Russia, In: G. Gregoriou (ed.), *Stock Market Volatility*. Chapman Hall-CRC/Taylor and Francis: London, U.K.

Giacomini, R. and Komunjer, I. (2005) Evaluation and combination of conditional quantile forecasts, *Journal of Business and Economic Statistics*, 23(4): 416–431.

Glosten, L. R., Jaganathan, R., and Runkle, D. (1993) On the relation between the expected value and the volatility of the normal excess return on stocks. *Journal of Finance*, 48, 1779–1801.

Gonzalez-Rivera, G., Lee, T., and Santosh, M. (2004) Forecasting volatility: A reality check based on option pricing, utility function, value-at-risk, and predictive likelihood. *International Journal of Forecasting*, 20(4): 629–645.

Hasbrouck, J. (2007) *Empirical Market Microstructure*, Oxford University Press: Oxford, U.K.

Hao, H. (2006) Is the structural approach more accurate than the statistical approach in bankruptcy prediction? Working Paper, School of Business, Queen's University, Ontario, Canada.

Huisman, R., Koedijk, K. G., Kool, C., and Palm, F. (2001) Tail-index estimates in small samples. *Journal of Business and Economic Statistics*, 19(2): 208–216.

Jansen, D. and De Vries, C. G. (1991) On the frequency of large stock returns: Putting booms and busts into perspective. *Review of Economics and Statistics*, 73(1): 18–24.

Ketz, E. J. (2003) *Hidden Financial Risk. Understanding Off-Balance Sheet Accounting*. John Wiley & Sons: Hoboken, NJ.

Renaissance Capital report on Russian Banking system. (2008) Russian banking: Every day I love you less and less, December, New York.

Sberbank, available at http://en.wikipedia.org/wiki/Sberbank.

Troika Dialog Report on Russian Banking System in Russian. (2009) Russian banks: Credits by government, January, Moscow.

Troika Dialog Report on Russian Banking System in Russian. (2008) Russian banking sector: Why ruble is cheap, December, Moscow.

Vassalou, M. and Xing, Y. (2004) Default risk in equity returns. *Journal of Finance*, 59(2): 831–868.

CHAPTER 21

The Stability of the Australian Banking Sector

Philip A. Stork and Casper G. de Vries

CONTENTS

Australia's insulated geographical location and its Four Pillars policy have long sheltered its prosperous banking sector from external financial downturns. However, the 2007–2008 credit crisis uncovered its vulnerability to the most severe financial events. In this chapter, we use extreme value theory to examine the stability of the Australian banking system. We estimate the downside risks of the largest banks by calculating univariate extreme loss probabilities, which equal the inverse of the value-at-risk stress levels. The contagion risk within the banking sector is assessed by calculating multivariate co-crash probabilities. We find that spillover risk has increased markedly during the crisis.

21.1 INTRODUCTION

In this chapter, we focus on the Australian banking system because this region has a number of interesting characteristics. Geographically, it is relatively insulated and the Australian banks have historically had a strong domestic focus when compared to other regions. The Australian government has adopted a so-called Four Pillars policy, which prevents its four largest banks from merging with each other or with foreign banks. As the Australian economy has seen almost two decades of continuous growth, the banks have enjoyed years of uninterrupted strong profit growth with high, stable credit ratings, and impressive long-term shareholder returns. In ordinary times, the relatively insulated and healthy Australian banking system has worked very well. However, the 2007–2008 credit crisis has also affected the Australian financial system. In this chapter, we analyze the extent to which the contagion risk* has reached the Australian banks.

---

* Contagion risk refers to the transmission of an idiosyncratic shock, which affects an individual bank or possibly a set of banks, and its transmission to other banks, see *ECB Financial Stability Review*, December 2004. This transmission may take place through the interbank market, the payment system, contagious bank runs, or asset markets, see Gropp and Moerman (2004).

Finally, it should be noted that because New Zealand has no domestic exchange-listed banks, the four large Australian banks also dominate the New Zealand banking industry.* Therefore, the Australian banking industry findings in this chapter, generally speaking, also hold for the New Zealand banking sector.

21.1.1 Australian Banking System

Historically, the Australian banking industry has been relatively tightly regulated. Until the 1980s, Australian banks were divided into either savings banks or trading banks. The first category comprised banks that were restricted to providing only mortgages in their lending activities, while the latter category consisted mostly of merchant banks that provided no services to the general public. Prior to the 1980s the so-called non-bank financial institutions (NBFIs) flourished; these included building societies and credit unions. Prior to 1983, foreign banks played only a minor role in Australian banking; however, in that year the Australian financial industry was deregulated. The deregulation loosened some of the rules on banks, abolishing the distinction between savings banks and trading banks and allowing foreign banks an easier entrance into the Australian financial markets, including the opening of new branches. Many of the smaller NBFIs disappeared, while some of the largest transformed into banks. In the two ensuing decades, the financial system grew steadily and became quite profitable as compared to its international peers in terms of both returns on equity and in absolute amounts. Out of the many financial institutions, four major banks together have succeeded in dominating the Australian banking sector. These are Australia and New Zealand Banking Group, Commonwealth Bank of Australia, National Australian Bank, and Westpac Banking Corporation.

21.1.2 Four Pillars Policy

The Australian government has adopted the so-called Four Pillars policy that prevents mergers between these four banks. This policy, rather than formal regulation, has endured four successive prime ministers and seven successive treasurerships. In a recent statement† the Australian treasurer

---

* As of June 2008, the New Zealand assets of the four pillar banks equal 90% of the total assets of the New Zealand banking system, according to the Reserve Bank of New Zealand, Financial Stability Report, November 2008.

† See Media Release 02/06/2008 on www.treasurer.gov.au.

announced that "in the interest of stability and competition" the Rudd government will maintain the existing Four Pillars policy for the banking sector. Australia is considered to be best served by a stable banking system that can continue to draw on the strength and risk management skills of four major banks, and it has no wish to reduce this number. Clearly, the government favors the arguments based on competition, prudential risk, and labor market impact over those based on improving the ability (and size) of local Australian banks in order to compete against foreign banks. It has been said that none of the members of this well-run banking oligopoly step very far outside of the parameters set by the rest. And indeed, over the past couple of years, the big banks are not really getting a lot of traction in gaining market share from each other; overall, relative market shares have not changed to any great degree. Nevertheless, some smaller banks are attempting to acquire a bigger slice of the cake; these include Bendigo Bank, BankWest, Suncorp-Metway, and Bank of Queensland. The four pillar banks still remain significantly larger than their competitors and have been quite profitable while defending their position. Indeed, the number five bank, St. George Bank, announced a merger in 2008 with the much larger number four bank, WestPac Banking Corporation.

21.1.3 The Credit Crisis

The credit crisis originated in the United States. A strong deterioration in especially the residential property market combined with the strong growth of the subprime and alt-A mortgages has led to large losses for nearly all parties involved: the mortgage lenders, the home owners, the dealers in the securitized mortgages, the insurers, and so forth. The repackaging and further distribution of these risks by means of complex transactions has made it virtually impossible to determine either the location or the extent of the losses ensuing from the falling house prices, the subsequent payment arrears, and mandatory foreclosures. The very development that has added to the increased strength of the financial system to withstand a majority of events, through the wider dispersion of risk, has possibly now amplified the disruption caused by the credit crisis shock. As the market conditions worsened, the losses and negative surprises became more frequent and intense. Over the course of 2008, extreme events continued to occur in the United States, Europe, and other regions as well.

The Australian financial system has coped better with the credit crisis turmoil than have many others. The Australian banking system is soundly capitalized with an aggregate capital ratio* standing at 10.6% as of June 2008 and a Tier 1 ratio of 7.3%. It has only limited exposure to subprime-related assets and as of the third quarter of 2008 it continues to confirm excellent profitability with low levels of problem loans. All large Australian banks boast high credit ratings and have been able to access both domestic and offshore capital markets on a habitual basis. Credit standards in Australia have historically not been relaxed to the same extent as in the United States and the share of nonconforming loans is much lower. An additional factor that has kept the Australian banks in a good position throughout the current turmoil is that they have usually not relied greatly on income from trading activities for profitability. For the five largest banks, trading income accounted for only 5% of their total income in 2007. Furthermore, Australian banks have traditionally had only small unhedged positions in financial markets. Unlike many of their international peers, the Australian banks have not been forced to raise new capital to offset write-downs. Nevertheless, the Australian banking sector has not been completely insulated from the negative developments abroad.

21.1.4 Funding Costs and Funding Availability

In 2008, the Australian banks' funding costs increased strongly. Simultaneously, funding availability reduced markedly for a number of products and markets. The rising funding costs and the near disappearance of the offshore securitization funding sources are direct effects of the credit crisis and thus clearly affect the Australian banking sector. Conditions in the residential mortgage backed securities (RMBS) market have become increasingly difficult since mid-2007. As an illustration of the decreased volume, quarterly issuance averaged only $2 billion between mid-2007 and mid-2008 as compared to $18 billion over the previous year. Apparently, the offshore demand for Australian RMBSs has fallen dramatically. Funding mortgages by issuing RMBSs has become unprofitable for most types of loans, which has a noticeable impact on those lenders whose business models are centered on securitization. Australian banks have shifted a larger share of their short-term wholesale funding onshore, benefiting from a strong growth in deposits, particularly those from households.

---

* We refer to RBA's September 2008 *Financial Stability Review*, which we use extensively in this section.

21.1.5 U.S.$ Offshore Funding

Offshore markets have become increasingly more important to Australian banks over the past decade, with foreign liabilities around mid-2007 accounting for approximately 27% of the banks' total liabilities, as compared with 15% in the mid-1990s. This offshore funding dependence is one of the key transmission mechanisms through which the international credit crisis is contagious to the Australian banking system. Around two-thirds of offshore borrowing has been through the issuance of U.S.$ debt securities by the four pillar banks. Of these offshore debt securities, around 80% had been issued into the U.S. and U.K. markets. The funding spreads have gone up noticeably. This more difficult financial environment has made the banks more cautious in providing new loans.

In line with their international peers, the Australian bank share prices have shown increased volatility and a strong fall since their peak in mid-2007. This is discussed in more detail in Section 21.3 below. Consistent with the general deterioration in sentiment, credit default swap (CDS) premia on Australian banks during 2008 remained elevated relative to historical averages; these nevertheless remain lower than those for the largest U.S. financial institutions.

21.1.6 Australian Regulators

In order to deal with these difficulties in the markets, the Australian authorities have continued to examine crisis management arrangements.* During the weekend of October 13, 2008, the government followed international developments and guaranteed, for unlimited amounts and for a 3 year time horizon, all bank deposits as well as wholesale borrowings by Australian financial institutions. Evidently, in spite of its relatively strong position, the regulators felt compelled to act decisively in order to ensure the stability of the Australian banking system. In the next section, we will discuss the methodology applied to analyze the banking sector risks.

21.2 UNIVARIATE EXTREME VALUE THEORY

The common assumption in finance is that asset returns are normally distributed. The standard value-at-risk methodology assumes normality of the empirical distributions, and various other financial statistics are based on

---

* See www.rba.gov.au.

the same assumption. While this serves wonderfully well to address many finance-related questions, it turns out that the tails of the probability distribution functions are in practice much fatter (see for instance Jansen and de Vries (1991) and Poon et al. (2004)). As a result, large loss events occur more frequently.

Over the last years, the popularity of extreme value theory (EVT) to assess the risk of an extreme event has increased considerably. For example, EVT has been used to examine the severity of stock market crashes, the pricing of catastrophic loss risk in reinsurance, or the extent of operational risk in banks.* EVT is particularly suitable for the analysis of financial stability problems because widespread stability problems are extremely rare events. This means that the usual data sample sizes are far too small for determining the probability, extent, or cause of widespread crisis using common econometric techniques. The semi-parametric EVT approach is methodologically superior, as it exploits the functional regularities that probability distributions display far away from the center.

21.2.1 Univariate EVT and the Hill Estimator

EVT is a discipline in statistics that analyzes the behavior of the tails of statistical distributions. Naturally, the left tails or minima are the relevant areas to assess the distributions. Under rather weak assumptions, the tails of the probability distribution functions display certain regularities. Because the cumulative distribution function ranges from 0 to 1, the tails can only behave in certain ways. Given some mild regularity conditions, these are the exponential and power functions, if the support is unbounded. Hence, using a semi-parametric approach based on these two functional forms is the basis of the EVT approach and explains why it is superior to a non-parametric approach such as worst case analysis. The estimation of the tail behavior and, therefore, the assessment of the probability of a financial crisis, may be conducted even if such crises are either not present in the sample of empirical data or occur only very infrequently. The above-mentioned regularities with regard to the shape of the probability distribution function allows one to determine the tail in its entirety on the basis of a few observations only. Naturally, if fewer observations are available, the reliability of the estimation is reduced, as for instance expressed by an increasing width of the confidence intervals.

---

* Based on the ECB June 2006 *Financial Stability Review*.

The fatness of the tails of a probability distribution function can be measured by the tail probability $P(X > x) = 1 - F(x)$, where $F(x)$ is a cumulative distribution function. For large values of x, the left tail probability may be approximated by the Pareto distribution, if in the limit the following condition holds:

$$\lim_{t \to -\infty} \frac{F(tx)}{F(t)} = x^{-\alpha}, \quad \text{with } \alpha > 0 \tag{21.1}$$

For various distributions, including the popular Student-t, this condition holds, as the tail probability declines according to a power function. The parameter α represents the thickness of the tail of the distribution function and is commonly referred to as the tail index. If α is high, the tail is thin, whereas for low values of α, the tail is fat. Only the first $k < \alpha$ moments of the distribution exist; for $k \geq \alpha$, the moments are unbounded.

The Hill (1975) estimator is used to estimate $1/\hat{\alpha}$. Only the outer parts of the distribution are used in order to minimize estimation biases. Let x_i be the ith order statistic, with $x_i \geq x_{i-1}$ for $i = 2, \ldots, T$, representing a sample of T observations. The Hill estimator is given as

$$\frac{1}{\hat{\alpha}} = \frac{1}{k} \sum_{j=1}^{k} \ln\left(\frac{x_j}{x_{k+1}}\right) \tag{21.2}$$

In this formula, k equals the number of higher order statistics in the tail of the distributions used to estimate $\hat{\alpha}$. Although the concept and the calculation of the Hill estimator are straightforward, the choice of k is not. The optimum depends on the sample size T and the tail thickness α; the further one moves out into the tails, the better becomes the Pareto approximation of those tails. However, this reduces the number of observations available and increases the uncertainty of the estimate. In practice, one may also resort to visual inspection of the so-called Hill plots to determine the optimal k. In small samples, best practice is to plot the estimates as a function of the threshold, $\hat{\alpha}(k)$ and to select k in the region where $\hat{\alpha}(k)$ tends to be constant[*].

---

[*] See also Straetmans et al. (2008) or Slijkerman et al. (2005, p. 29).

In order to derive the probability that the daily stock return of a bank is lower than a prespecified level x_{var}, the inverse quantile estimator from De Haan, Jansen, and Koedijk (1994) may be used:

$$\hat{p} = \frac{k}{T} \left(\frac{x_{k+1}}{x_{var}} \right)^{1/\hat{\alpha}}$$

(21.3)

21.3 MULTIVARIATE EVT

In this section, we take univariate EVT one step further by conditioning bank stock price "crashes" on other banks' stock price crashes.* We focus on extreme co-movements in order to assess the risk of banking system instability through aggregate shocks. We estimate multivariate extreme spillover risk or bank contagion risk: This determines how likely it is that a bank's stock price declines dramatically if there is an extreme negative (systematic) shock in the sector.

21.3.1 Co-Crash Probabilities

Let X_{it} denote the log first differences of the price changes in bank stock i at time t, with $i = 1, ..., N$ and $t = 1, ..., T$. As is common practice in the literature, we multiply the stock returns by −1 in order to work with upper tail loss returns. We choose crisis levels or extreme quantiles[†] Q_i such that the tail probabilities are equal across all N banks, thus

$$P\{X_{1t} > Q_1\} = ... = P\{X_{it} > Q_i\} = ... = P\{X_{Nt} > Q_N\} = p \qquad (21.4)$$

Because we impose equal probability levels across the banks, the crisis levels Q_i will in general have different values, as the marginal probability distribution functions $P\{X_{it} > Q_i\} = 1 - F(Q_i)$ are bank-specific. On average, only once in every $1/p$ days will such a severe problem arise. We will estimate the conditional co-crash (CCC) probability of one Australian bank given that another bank crashes:

---

* We follow Hartmann et al. (2004).
[†] Percentiles refer to the probabilities of certain outcomes; here extreme negative returns and quantiles refer to their sizes. Once one of the two is fixed, the other follows from the observed or estimated distribution.

$$
\begin{aligned}
\text{CCC} &\equiv P\{X_{it} > Q_i(p) \mid X_{mt} > Q_m(p)\} \\
&= \frac{P\{X_{it} > Q_i(p), X_{mt} > Q_m(p)\}}{P\{X_{mt} > Q_m(p)\}} \\
&= \frac{P\{X_{it} > Q_i(p)\} + P\{X_{mt} > Q_m(p)\} - P\{X_{it} > Q_i(p), \text{ or } X_{mt} > Q_m(p)\}}{P\{X_{mt} > Q_m(p)\}} \\
&= \frac{2p - P\{X_{it} > Q_i(p), \text{ or } X_{mt} > Q_m(p)\}}{p}
\end{aligned}
\tag{21.5}
$$

The second step follows from Equation 21.4 where we fixed the marginal crash probability. In effect only the numerator has to be estimated. This is discussed in the next section.

21.3.2 Estimation

In order to estimate the CCC probabilities, we follow Straetmans et al. (2008). The CCC from Equation 21.5 may be estimated by a simple count measure:

$$
\text{CCC} = 2 - \frac{1}{k} \sum_{i=1}^{T} I\{X_{it} > x_{i,T-k}, \text{ or } X_{mt} > x_{m,T-k}\}
\tag{21.6}
$$

where I stands for the indicator function, $x_{i,T-k}$ and $x_{m,T-k}$ are the kth highest order statistics of bank i and bank m, respectively, and k reflects the number of extremes used in the estimation.[*] In general, the estimation boils down to counting the instances at which one or both stock price return series experience an extreme return over a given sample period. The level of threshold k should be chosen such that $\hat{p}(k)$ as a function of k tends to be constant.

21.4 DATA DESCRIPTION

First we determine the largest exchange-listed banks and insurers in Australia. We use the list of financials in the Datastream S&P/ASX 200 Financials–x–Property Index, which consists of 25 shares in total (Datastream mnemonic "LASG2XP10701"). Those companies that cannot be classified as either a bank or an insurance company are removed. For instance, neither the exchange itself nor investment management companies are included. Although the main focus of this study is on banking

---

[*] See also Coles et al. (1999, p. 346) for a similar approach.

risks, we include the largest listed insurance companies as well. This enables us to compare the downside risks between Australian banks and insurers, and enables us to analyze the codependence between the two sectors. Moreover, the use of derivatives and other financial innovations has blurred the distinction between the two sectors.

As the starting date, we take 08/10/1996 because for one insurer the prices are available only after this date. For all series, the end date is 13/10/2008. In order to prevent cross-sectional comparability problems due to scale differences, we use only those series for which the whole period between 08/10/1996 and 13/10/2008 is available, which amounts to a total of 3.135 observations per share. For one share, the available price series was shorter and we therefore removed it from our sample. After this selection a total of 11 shares remained, consisting of 8 banks and 3 insurance companies. There is some overlap of activities, as some of the banks are involved in insurance, and one of the insurance companies conducts banking activities as well. Because we study the stability of the largest Australian banks, and thus the overall systemic banking stability, the distinction between banks and insurers itself is of less importance. For this reason, we see no need to further investigate the precise functions of the financials. Three banking indices for Asia, Europe, and the United States are downloaded from Datastream as well and comprise the same period as the individual share price series.

21.4.1 Data Summary Statistics

In Table 21.1, we show the summary statistics of the 11 individual financials and the 3 indices.* The returns are calculated as the natural logarithms of the first differences of the original Datastream prices series.

Because banks are "long in the economy," the banks' average daily returns are all positive and within a limited range, varying between 0.034% for National Australia Bank and 0.062% for Macquarie Group. The minima and maxima cover a much wider range from −26.4% to 32.1% and even wider for QBE Insurance. The standard deviations show only limited divergence, ranging from the lowest volatility at 1.41% for Westpac banking to the highest at 2.32% for QBE Insurance Group. Interestingly, the skews have a less consistent pattern, because for only 4 out of the 11 series the skew is negative. For the other 7 series, the skew equals either zero or is positive. The kurtosis exceeds 3 for all series, which is evidence of the existence of fat tails in the return series.

---

* In Table 21.1, we abbreviate Australia as AU and New Zealand as NZ.

TABLE 21.1   Summary Statistics of Share Price Return Series

	Average	Minimum	Maximum	Standard Deviation	Skewness	Kurtosis
Banks						
AU and NZ Banking Group	0.047%	−11.54%	13.68%	1.53%	0.12	7.30
Bendigo and Adelaide Bank	0.051%	−10.38%	25.52%	1.73%	1.49	21.81
Bank of Queensland	0.043%	−19.97%	11.98%	1.53%	−0.58	13.88
Commonwealth Bank of AU	0.061%	−7.13%	7.71%	1.29%	0.00	4.38
Macquarie Group	0.062%	−26.38%	32.11%	2.09%	0.16	33.06
National AU Bank	0.034%	−14.46%	16.03%	1.52%	−0.44	12.59
Saint George Bank	0.055%	−9.28%	22.49%	1.42%	1.21	24.49
Westpac Banking	0.055%	−8.28%	8.59%	1.41%	0.01	3.18
Insurance companies						
Axa Asia Pacific Holding	0.041%	−10.59%	17.33%	1.97%	0.38	4.29
QBE Insurance Group	0.064%	−52.59%	41.88%	2.32%	−4.31	170.93
Suncorp-Metway	0.048%	−14.88%	9.86%	1.49%	−0.38	7.94
Bank indexes						
Asia Banks	−0.007%	−7.27%	9.97%	1.39%	0.23	4.54
Europe Banks	0.029%	−11.76%	15.71%	1.37%	−0.07	11.80
U.S. Banks	0.020%	−18.32%	16.86%	1.79%	0.16	13.48

Both the skewness and kurtosis estimates of QBE Insurance Group stand out. From a more detailed analysis of its return series, we glean that one extremely volatile episode in the time series of QBE Insurance Group is responsible for the relatively large standard deviation and kurtosis as well as the highly negative skew; between 14/9/2001 and 20/9/2001 the price dropped more than 64% and then recovered with 52% on 21/9/2001. These volatile share price movements were caused by the terrorist attacks on the World Trade Center in New York and investor concerns about QBE's exposure. The other share prices did react to the 11/9/2001 attacks, but not to the same extent.

21.5 EMPIRICAL RESULTS

In this section, we discuss the main empirical findings from the EVT analysis. We first study the individual return series of the 11 Australian financials. As we are interested in analyzing the risks—that is, the negative returns—we focus on the left tails of the distributions. The Hill estimator from Equation 21.2 is calculated in order to determine the fatness of the tails of the return distributions. From this, loss probabilities are derived in Section 21.5.1.

As discussed in Section 21.2, we construct Hill plots to determine the optimal number of order statistics k to estimate $\hat{\alpha}$ for the left tails. Figure 21.1a through d illustrate the Hill plots for four arbitrarily chosen banks. Evidently, a certain minimum number of order statistics is needed before a relatively stable Hill estimate is obtained. The figures show a declining trend in all four Hill estimates after including a certain number of order statistics, which reflects the bias due to increasing the number of nonextreme observations.

Table 21.2 shows the Hill estimates for the left tails of all series and for four different levels of the number of order statistics (k). For the remainder of this chapter, we use $k = 50$ for all series as at this level the Hill estimates turn out relatively stable.*

For $k = 50$ the Hill estimates vary between a minimum of 2.41 for the Macquarie Group and 4.62 for Westpac Banking. This range of estimates is in line with the results of Hartmann et al. (2006) and illustrates the nonnormality of bank stock returns. Macquarie Group's returns have the fattest left tail, even more than QBE Insurance Group (at 2.76), which has the highest standard deviation and kurtosis. The high Hill estimate for Westpac is in line with the low standard deviation and kurtosis shown in Table 21.1. Although not reported here in detail, we also calculated the Hill estimates for the right tails. The estimates vary between 2.60 for Macquarie Group and 3.88 for Axa Asia Pacific Holding. Apparently the right tail estimates are not too different from the left tail estimates. On average, the right tail Hill estimate equals 3.10 versus 3.43 for the left tail, which suggests that the right tail is slightly fatter.

21.5.1 Loss Probabilities

Next, we estimate the probability that the bank's daily stock price return is lower than a prespecified probability level x_{var} by using Equation 21.3. For a number of different loss levels, the probabilities are reported in Table 21.3.

---

* This threshold level compares nicely to the $k = 50$ used by Slijkerman et al. (2005) on a similar data sample size, as well as the 2% threshold level used by Poon et al. (2004, p. 593).

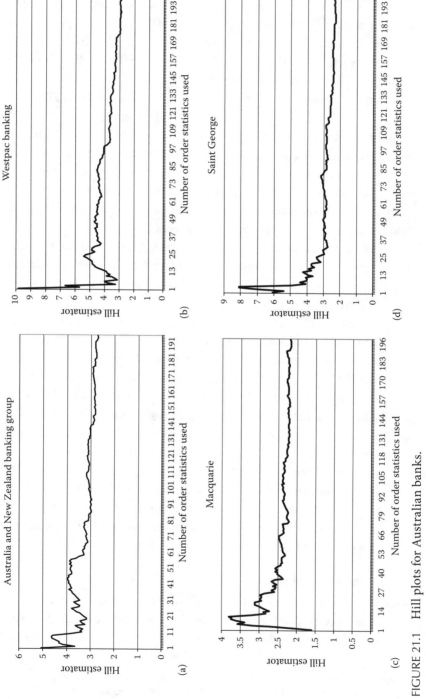

FIGURE 21.1   Hill plots for Australian banks.

TABLE 21.2   Hill Estimators of the Left Tails

	$k = 25$	$k = 50$	$k = 75$	$k = 100$
AU and NZ Banking Group	3.58	3.80	3.18	3.03
Bendigo and Adelaide Bank	4.17	3.78	3.34	3.17
Bank of Queensland	3.76	3.64	3.25	2.93
Commonwealth Bank of AU	3.69	2.91	3.03	2.88
Macquarie Group	2.95	2.41	2.25	2.38
National AU Bank	3.58	3.19	2.57	2.63
Saint George Bank	3.38	3.01	3.05	2.88
Westpac Banking	5.29	4.62	4.45	3.82
Axa Asia Pacific Holding	4.18	4.43	4.20	4.01
QBE Insurance Group	2.08	2.76	2.84	2.94
Suncorp-Metway	2.86	2.81	2.90	2.74
Asia Banks	3.91	4.11	3.65	3.52
Europe Banks	3.49	3.41	2.94	2.73
U.S. Banks	3.16	3.17	3.22	2.98

TABLE 21.3   Loss Probabilities Based on Univariate EVT

Threshold Loss Level	−20%	−10%	−5%
AU and NZ Banking Group	0.00%	0.04%	0.49%
Bendigo and Adelaide Bank	0.00%	0.05%	0.64%
Bank of Queensland	0.00%	0.05%	0.56%
Commonwealth Bank of AU	0.01%	0.04%	0.32%
Macquarie Group	0.05%	0.27%	1.43%
National AU Bank	0.01%	0.05%	0.63%
Saint George Bank	0.01%	0.01%	0.42%
Westpac Banking	0.00%	0.04%	0.24%
Axa Asia Pacific Holding	0.00%	0.11%	0.90%
QBE Insurance Group	0.02%	0.08%	0.78%
Suncorp-Metway	0.01%	0.04%	0.55%
Asia Banks	0.00%	0.02%	0.30%
European Banks	0.00%	0.10%	0.41%
U.S. Banks	0.01%	0.10%	0.87%
Average	0.01%	0.07%	0.61%
Median	0.01%	0.05%	0.56%

Depending on the context and purpose, risk managers or regulators will consider different levels of risk x_{var} as acceptable. To remain consistent with the estimation procedure of the Hill estimator, we set the number of order statistics k equal to 50.

Table 21.3 shows that Macquarie Group exhibits the highest probability of an extreme return, as compared with the other financials. The likelihood of a significant drop in the share price of 5% or more equals less than 1% for 10 of the 11 financials. A larger than 20% drop is seen as very unlikely with less than 0.05% probability for all series, except for Macquarie Group, and an average probability of 0.01%. For the average series, the estimates translate into a loss of more than 5%, 10%, and 20% occurring only once every 0.65, 5.81, and 43.75 years, respectively, or 1.54, 0.17, and 0.02 times per year.

Next, we compare the univariate EVT estimates in Table 21.3 to loss probability estimates that are based on the commonly used assumption of normality of the bank stock price returns. Hence, we use the best possible approximation by a normal distribution of the empirical distribution to measure the loss probabilities. For all banks, insurers, and bank indices shown in Table 21.3, we find a 0.00% probability of a greater than 20% loss. Even at 10% loss levels, the probabilities remain equal to 0.00%. Only after we further reduce the loss levels to 5% do we obtain nonzero probabilities, with an average probability of 0.24%*. Evidently the loss probability estimates under the normality assumption are much lower than those obtained by means of EVT.

21.5.2 Multivariate Extreme Value Theory Results

We subsequently calculate the CCC probabilities of observing joint meltdowns in bank stock prices. The estimated CCC probabilities are shown in Table 21.4.[†]

Table 21.4 shows that the average CCC probabilities are relatively high compared to prior studies on European and U.S. banking systems. For example, Slijkerman et al. (2005, p. 19) find in their sample of bank stock returns that the highest probability of a crash between two individual banks equals 37.5%, a finding that involves two Spanish banks. Hartmann et al. (2006) also find lower CCC probabilities, albeit for the largest European banks. The probability, for example, that the largest German bank faces

---

[*] For brevity reasons, we have not included detailed calculations. These are available from the author upon request.

[†] We note that the United States share market is the latest to close on any particular day, thus after the Australian, Asian, and European share markets. The United States share market typically leads the other international share markets. Following the approach used by Poon et al. (2004, p. 592), we use the previous day's United States returns whenever the calculation involves the United States banking index returns. We find higher CCC probabilities using the previous day's United States bank index returns than using the same day's returns.

TABLE 21.4   Multivariate Conditional Joint Meltdown Probabilities

	ANZ	Bend.	Qu.l.	CBA	Mac.	NAB	St.G.	W.pac	Axa	QBE
Bend.	36%									
Qu.l.	18%	28%								
CBA	44%	34%	20%							
Mac.	28%	34%	28%	36%						
NAB	44%	38%	24%	52%	38%					
St.G.	36%	32%	28%	42%	46%	46%				
W.pac	38%	28%	20%	28%	22%	36%	40%			
Axa	16%	16%	16%	18%	22%	20%	14%	12%		
QBE	28%	20%	16%	28%	24%	30%	24%	14%	14%	
Sunc.	30%	30%	14%	36%	34%	40%	32%	30%	18%	20%

an extreme spillover from the second largest German bank is estimated at 22.4%, while for Spain this probability is estimated at 11.2%.

One reason for the higher contagion risk could be that the Australian banking system is relatively more focused on its domestic market than the European ones. Thus Australian banks are exposed to the same macro risk drivers and are relatively under diversified. As a result, Australian banks may be more dependent on each other and have stronger links through, for instance, the interbank market, the syndicated loan market, and the payment system. This intuitive explanation would raise the level of the joint crash probabilities as compared to their European peers that have more open banking systems and better diversification against macro risks.

21.5.3 Individual Co-Crash Probability Estimates

Table 21.4 shows that the CCC probabilities vary strongly across the individual banks. The relatively small and locally orientated Queensland Bank has the lowest CCC probability average of 24%, whereas National Australia Bank has the highest of 40%. It should further be noted that Westpac Banking Corporation (Westpac) has relatively low CCC probabilities. Interestingly, Westpac's highest CCC probability of 40% is found to be with Saint George Bank, with which it has recently announced a merger. The high tail dependence is remarkable. Furthermore, the highest joint crash probability of all banks of 52% is found to exist between two of the pillars, namely, the National Australia Bank and the Commonwealth Bank of Australia. Should one of these two pillars see its demise, there is a higher than 50% likelihood that the other will fall as well. In such a dramatic scenario, in which two of the four pillars fall, other contagion effects are very likely to occur as well. The estimated level of this joint crash probability illustrates that at least two (and probably all) of the four pillars will in practice be seen as "too big to fail." Our analysis thus provides support for the idea that the Australian Central Bank or Government will likely consider the collapse of any of the four pillars as an unacceptable contagion risk. Some sort of preventative action is likely to be taken before any such scenario would unfold. The steps taken by the Australian Central Bank and Government in the fall of 2008 should also be interpreted in this light.

21.5.4 Has the Credit Crisis Raised the Co-Crash Probability Estimates?

Another intuitive reason for the elevated CCC probabilities may be that our sample includes the credit crisis effects, as opposed to the Hartmann et al. (2006) and Slijkerman et al. (2005) papers. One could easily argue that our relatively high CCC probability estimates at least partially reflect the credit crisis effects. In order to assess the validity of this line of reasoning, we reestimate CCC probabilities for a reduced sample period until May 1, 2007, excluding the credit crisis effects. Table 21.5 shows, per bank, the average CCC probability with all other banks, for both the full sample period and for the reduced sample period. The first row in Table 21.5 thus shows that the average probability of ANZ crashing, given that one of the other banks crashes, is estimated at 25% prior to the crisis and at 34% including the crisis data.

The results support our conjecture that the average CCC probability estimates have increased markedly during the credit crisis period, from an average of 25% to an average of 34%. A second observation from Table 21.4 is that

TABLE 21.5   Average Co-Crash Probability Estimates

	Excluding Crisis	Including Crisis
AU and NZ Banking Group	25%	34%
Bendigo and Adelaide Bank	21%	34%
Bank of Queensland	17%	24%
Commonwealth Bank of AU	31%	38%
Macquarie Group	24%	35%
National AU Bank	29%	40%
Saint George Bank	29%	38%
Westpac Banking	24%	30%
Average	25%	34%

the average joint crash probability for insurers equals 18%, thus lower than the 34% estimate found for banks. This finding contrasts with Slijkerman et al. (2005) who find similar joint crash probabilities and thus similar systemic risks for a sample of European insurers as compared to a sample of European banks. Interestingly, if we exclude the credit crisis period after May 1, 2007 from our sample and reestimate the CCC probabilities for the insurers, we find that the average decreases only slightly from 18% to 16%. This small difference suggests that the credit crisis did not strongly affect the market-perceived stability of the Australian insurers, which is in stark contrast to our findings for the Australian banking sector.

21.5.5 Tail Dependence of Australian Banks on the International Banking Sector

Finally, we calculate the CCC probabilities between the individual Australian banks and each of the three foreign banking indices. This enables us to assess the dependence of the Australian banking sector on foreign stress events. Table 21.6 shows that the contagion risk from the Asian and U.S. banks to the Australian banks is higher than that originating from the European banks.

In order to study the effects of the credit crisis on the international tail dependence, we reestimate the CCC probabilities with the shortened sample period ending May 1, 2007, thus excluding the credit crisis as in Section 21.5.4. These estimated averages are given in the bottom row of Table 21.6. One sees that during the credit crisis the tail dependence of the Australian banks on the European and especially the U.S. banks has

TABLE 21.6   International Tail Dependence

Probability	Asia	Europe	United States
AU and NZ Banking Group	22%	16%	22%
Bendigo and Adelaide Bank	28%	8%	18%
Bank of Queensland	12%	18%	16%
Commonwealth Bank of AU	24%	12%	16%
Macquarie Group	32%	18%	26%
National AU Bank	20%	12%	26%
Saint George Bank	18%	8%	20%
Westpac Banking	14%	14%	18%
Average incl. crisis	21%	13%	20%
Average excl. crisis	20%	9%	13%

increased in comparison to the period before the crisis. This is most likely because, in the case of the U.S. banks, they were the source of the crisis.

21.6 CONCLUSION

We apply EVT to analyze the stability of the Australian banking sector. We find that the Australian banks' share price return distribution functions exhibit fat tails. The risks thus exceed those indicated by the common assumption of normally distributed returns. We further find that the relatively high co-crash probabilities between the four pillar banks support the conjecture that these are too big to fail. Moreover, the co-crash probabilities between most Australian banks have increased during the 2007–2008 credit crisis. The tail dependence of the Australian banking sector on the American, Asian, and—to a lesser extent—European banking sectors was also boosted during the recent credit crisis.

REFERENCES

Coles, S., Heffernan, J., and Tawn, J. 1999 Dependence measures for extreme value analyses. *Extremes*, 2(4): 339–365.

de Haan, L., Jansen, D., and Koedijk, K.G. 1994 Safety first portfolio selection, extreme value theory and long run asset risks. In: J. Galambos (Ed.), *Extreme Value Theory and Applications*. Kluwer Publisher: Dordrecht, the Netherlands.

European Central Bank. 2004 *The Financial Stability Review*, December, 117–121.

European Central Bank. 2006 *The Financial Stability Review*, June, 159–164.

Gropp, R. and Moerman, G. 2004 Measurement of contagion in Banks' equity prices. *Journal of International Money and Finance*, 23(3): 405–459.

Hartmann, P., Straetmans, S., and de Vries, C.G. 2004 Asset market linkages in crisis periods. *The Review of Economics and Statistics*, 86(1): 313–326.

Hartmann, P., Straetmans, S., and de Vries, C.G. 2006 Banking system stability: A cross-Atlantic perspective. In: M. Carey and R.M. Stulz (Eds.), *The Risks of Financial Institutions*. The University of Chicago Press: Chicago, IL.

Hill, B. 1975 A simple general approach to inference about the tail of a distribution. *The Annals of Statistics*, 3(5): 1163–1173.

Jansen, D.W. and de Vries, C.G. 1991 On the frequency of large stock returns: Putting booms and busts into perspective. *The Review of Economics and Statistics*, 73(1): 19–24.

Poon, S.-H., Rockinger, M., and Tawn, J. 2004 Extreme value dependence in financial markets: Diagnostics, models, and financial implications. *Review of Financial Studies*, 17(2): 581–610.

Slijkerman, J.F., Schoenmaker, D., and de Vries, C.G. 2005 Risk diversification by European financial conglomerates, *Tinbergen Institute Discussion Paper*, No. 2005–110/2.

Straetmans, S., Verschoor, W., and Wolff, C. 2008 Extreme U.S. stock market fluctuations in the Wake of 9/11. *Journal of Applied Econometrics*, 23(1): 17–42.

CHAPTER 22

Why Have Australian Banks Survived the Recent Global Financial Crisis?

John Simpson and Jennifer Westaway

CONTENTS

Australia is a small developed country with a sophisticated, well-regulated banking and financial system. Because of its smaller stock market turnover, lower capitalization of its banking sector and smaller global interaction than other developed countries, the effect of the global financial crisis on the country has not been as great as in other developed economies. The banking regulatory framework has also had much to do with this insulation effect. This cross discipline study comments on the legal and institutional environment for banking in Australia and also reports empirical analysis of banking stock market data showing Australia's interaction and involvement in a global context.

22.1 INTRODUCTION

Australia's existing banking and financial sector regulatory and prudential framework has been under considerable local and international scrutiny, since the scope of the global financial crisis has become evident. The essential focus of that scrutiny has been on the effectiveness of the framework, and whether it was that framework that precluded Australia's banks and financial sector intermediaries being drawn into the financial "mire" of toxic debt securitized mortgage products that has so dramatically and critically impacted upon institutions in other parts of the world.

There is no doubt that the momentum of the financial crisis and the number of major financial institutions with significant levels of exposure to subprime debt took everyone by surprise, albeit arguably this should not have been the case. For many years, global financial markets had, as a result of innovative legal developments combined with technological changes, been operating in a financially liquid environment, where mortgages were able to be packaged and securitized into investment products, which were then marketed and found a high level of demand. This new "market" had one major downside—the opportunity for those marketing the products to make large sums of money, dependent of course on the number of mortgage loans that could be generated. The temptation therefore was in the following form: create and market as many mortgage loans as quickly as possible and in the process circumvent the need to retain the credit risk in relation to these mortgages.

The lure of short-term financial considerations, such as individual bonuses and institutional fees, as well as the corporate bottom line requirements of profit, were factors which drove the market into a frenzy of cross-institutional dependency and a domino-like existence. The role of credit rating agencies in this process is beyond the scope of this chapter, but this will doubtless be discussed at length and be the subject of many an academic paper. The extent of the spread of financial contagion that followed the collapse of the subprime mortgage market, which began with the collapse of Bear Stearns in June 2007, has focused attention on one word: "Why?" The answer, whilst it would appear to be simplistic, is far from being so, but it does provide a basis for the prudential framework within Australia and the Federal Government's response to the global financial crisis to be examined in terms of adequacy.

It could be argued that one answer or reason for what happened is that the global financial market did not operate according to the established market theories, and one could argue that a key aspect of this is the total failure or inadequacy of market regulators to do what they were required to do, or the absence of appropriate regulatory mechanisms to manage, scrutinize, or ensure the implementation of credit risk management practices. It may also be the case that established market theories were inappropriate for managing the global financial sector at the time that the crisis became apparent, and if so, then consideration of alternate theories is long overdue. However, that too is for another discussion.

There are two parts in this chapter. One part analysis and comments on Australia's positioning in the global banking system using banking stock market data prior to the global financial crisis beginning in late 2008. A comparison is therefore made with Australia's impact on and its degree of interdependence with the global banking system compared to that of larger developed economies and several developing economies. The other part makes comment on the Australian financial and prudential structure and comments on the importance of bank regulation in Australia and Australasia (which includes New Zealand) in combating the effects of global financial crisis. More comment follows on the Australian institutional and regulatory environment.

22.2 AUSTRALIAN FINANCIAL PRUDENTIAL STRUCTURE

The current regulatory structure of the Australian financial sector has its foundation in the recommendations of the 1997 Financial System Inquiry, otherwise known as the Wallis Inquiry. Up to this time, the Reserve Bank of Australia had responsibility for prudential supervision of banks as well as general regulatory powers. It should be noted that regulation of the banking and financial sector in Australia is a Commonwealth Government responsibility by virtue of the Constitutional power of the Commonwealth under s.51 (xiii). The Wallis Inquiry recommended that three new regulatory bodies be established—the Australian Securities and Investment

Commission (ASIC), which would have responsibility for the establishment and enforcement of rules of conduct and disclosure; the Australian Prudential Regulation Authority (APRA), which would take over the prudential supervisory functions of the Reserve Bank; and the Payments System Board, which would have responsibility for the payments system, ensuring stability, access to, and control of the system(Tyree, 2008).

The key responsibility of APRA is to encourage and to promote sound prudential practice methods. Under s.11B of the *Banking Act* 1959 (Cth) ("the Act"), APRA is instructed to collect and analyze information regarding prudential matters, and to evaluate the effectiveness and carrying out of practices relating to prudential matters. According to the s.5(1) of the Act "prudential matters" are those which relate to the conduct of an Authorised Deposit Taking Institution (ADI) or an nonbank "nonoperating holding company" (NOHC) of any of its affairs in such a way as to keep itself in a sound financial position, so as not to cause or promote instability within the financial system. Further, APRA is authorized by s.11A of the Act to develop and enforce prudential standards, known as the "harmonized" prudential standards, which apply to all ADIs. Importantly for the purposes of this discussion, APRA has developed standards relating to capital adequacy and risk management.

22.3 AUSTRALIAN PRUDENTIAL REGULATION AUTHORITY STANDARDS

Why are prudential standards in relation to capital adequacy and risk management so important? According to the 2008 Basel Committee on Banking Supervision's paper on *Liquidity Risk Management and Supervisory Challenges*, many banks had failed, in times of liquidity, to take account of a number of basic principles of liquidity risk management, thus leaving them in a position of significant vulnerability. The Committee highlighted that there were several key areas for concern and thus guidance which needed to be addressed, including the identification and measurement of the full range of liquidity risks, the maintenance of an adequate level of liquidity, and the need for a robust and operational contingency funding plan. Banks needed to have in place a risk management framework, which if inadequate given the bank's level of liquidity, could be reassessed by a supervisory body and steps implemented to rework the risk management framework.

APRA's Prudential Standards, specifically APS110-113, address these requirements and are therefore the platform from which it can be argued, prevented Australian ADIs from the types of collapse which other institutions around the world have experienced. Fundamentally, the Standards incorporate the basic principles of the Basel Capital Adequacy Framework, by addressing the form and quality of capital (APS 111); the credit risk associated with the on and off-balance-sheet exposures (APS 112); and, the market risk which arises from the trading activities of a particular ADI. On this last point, it is worth noting that the 2008 Basel paper mentioned previously, did note that the size, nature, and complexity of a bank's business needed to be taken into account when addressing how liquidity risk principles and supervision should be implemented. APRA's standards take into account these concerns of the Committee, partially as a result of the review of older prudential and monetary control standards, which operated in Australia, which were based largely on accounting standards, and which therefore did not allow for off-balance-sheet activities to be brought into the risk exposure calculations.

APRA, in APS 110, highlights the importance of ADIs taking responsibility for capital management, on the basis that capital forms the cornerstone of an ADIs financial strength, in that it is the buffer which can absorb unanticipated losses. According to this standard, the capital allows the ADI to operate in a "sound and viable manner in the event of problems as they are being addressed or resolved." It is the responsibility of the Board of Directors of the ADI to "ensure that the ADI maintains an appropriate level of capital commensurate with the level and extent of the risk to which the ADI is exposed from its activities."

The Standard goes on to state that an ADI must have in place an Internal Capital Adequacy Assessment Process (ICAAP) that has a minimum, firstly adequate systems and procedures to monitor, identify, measure, and manage those risks which arise on a continuous basis from the ADI's activities, ensuring that capital is held at a level consistent with the risk profile developed by the ADI; and secondly, a capital management plan, which is consistent with the overall business plan of the ADI for the management of the capital levels of the ADI on an ongoing basis. This capital management plan is required to set out the strategy of the ADI for maintaining adequate capital over time, including outlining its capital target for providing a buffer against identified risks, how that target will be met, and what means are available to source additional capital should it be required. The management plan is also required to specify what actions and procedures the ADI has in place to monitor its compliance with the minimum capital adequacy requirements, including "...the setting of trigger ratios to alert management to, and avert, potential breaches of these requirements."

APRA also requires, under this Standard that the ICAAP be subject to regular, effective, and comprehensive review, the frequency and size of the review being appropriate to the size, business mix, and complexity of the ADI's operations and the nature and extent of any change to its business profile and risk appetite. If there are significant changes to its ICAAP, then APRA can request that information be provided on these changes and how these changes will be managed in light of the requirement of compliance to the Standard.

What can therefore be seen from this standard is that ADIs have imposed upon them a comprehensive management process, which requires them to manage the business of the ADI in light of identified risks and to ensure that any changes in business activity within the ADI do not expose the strength and stability of the ADI to risks which are not calculated and which would pose a risk to the soundness and viability of the institution.

In addressing the issue of the capital adequacy of an ADI, APRA states in APS 111, that the capital base of an ADI should have four basic characteristics: (a) provide a permanent and unrestricted commitment of funds; (b) be freely available to absorb losses; (c) not impose any unavoidable servicing charge against earnings; and (d) rank behind the claims of depositors and other creditors in the event of winding up. Capital is divided into two tiers: "core" (Tier 1) and "supplementary" (Tier 2), with core capital meeting all four basic requirements, and Tier 2 capital meeting one or more, but still contributing to the overall financial strength of the institution as a going concern. The capital base of an ADI is, therefore, the sum of the Tier 1 and Tier 2 capital less itemized deductions allowed under the standard. The key to this standard is that at all times; an ADI will have permanent and unrestricted adequate funds in the form of core capital to meet any losses of the institution, thus ensuring the institution's strength and stability.

The focus of credit risk is addressed by ADS 112, with the aim of this standard being to ensure that all locally incorporated ADIs risk-weight their on-balance-sheet assets and their off-balance-sheet business according to certain risk categories. There are four categories of credit risk, with individual weights of 0%, 20%, 50%, and 100%. The category is determined by assessment of the likelihood of counterparty default, with higher risk counterparties being given a higher risk category. According to this standard, if in the opinion of APRA, an ADI fails to appropriately risk weight and exposure, then APRA has the power to determine what the risk weighted amount of a particular on-balance-sheet asset or off-balance-sheet exposure of the ADI shall be, and the ADI is then bound to that assessment. It is noted by APRA in this standard that the categorization used is a judgment of best practice by the regulator, but that at all times, the board of directors and the management of the ADI is ultimately responsible for assessing the risks of that ADI based on the activities and business exposure of the particular institution.

There are a two other key standards which also highlight the extent of the prudential scrutiny of ADIs in Australia. Firstly, there is APS 210 which addresses the issue of liquidity. The objective of this standard is stated to be "...to ensure that all ADIs have sufficient liquidity to meet obligations as they fall due across a wide range of operating circumstances." Importantly, this Standard imposes upon the board of directors and management of the ADI the responsibility for implementing, maintaining, and regularly reviewing its liquidity management strategy to ensure that at all times it is appropriate for the operations of the ADI, so that it has sufficient liquidity to meet its obligations at all times, taking into account any changing operating circumstances. If at any time the ADI has concerns about its current or future liquidity, then it is obligated to immediately advise APRA of those concerns and what its plans are to address those concerns.

The APRA key prudential role under this standard is to work with the ADI by reviewing and then agreeing with the ADI that its liquidity management strategy is both adequate and appropriate given its size and the nature of its operations. There are five key elements that must be addressed within the strategy, including a system for measuring, assessing, and reporting liquidity, clearly defined managerial roles and responsibilities, and a formal contingency plan for dealing with any liquidity crisis. An ADI is precluded from making any changes to this strategy, which applies to both the local and overseas operations of the ADI, until it has consulted APRA.

The other key prudential standard, for the purposes of this chapter, is APS 220 which addresses the issue of control of credit risk. This standard clearly identifies that the risk of counterparty default is the single largest risk facing an ADI and that an ADI must have a well-functioning credit risk management system, which addresses the particular type of credit risk undertaken by that ADI. Responsibility for the oversight of this system is placed on the Board of Directors and senior management, specifically imposing a responsibility to ensure that the ADI has in place:

1. Credit risk management policies, procedures, and controls appropriate to the complexity, scope, and scale of its business

2. Internal controls to consistently determine provisions and a general reserve for credit losses in accordance with the ADI's stated policies and procedures, applicable accounting framework, and the requirements of this prudential standard.

The significance of managing credit risk cannot be underestimated. Country and regional banking systems have become more interdependent driven by the availability of new securitized financial instruments and the growth in international trade between developed and developing countries. The free flow of financial services, the availability of capital, interbank lines of credit, and international lending are all the requirements of a globalized economy. However, interdependency has one critical flaw—the susceptibility to financial contagion and therefore to failure, especially in the absence of appropriate prudential supervision, as has been the case in the United States, where, it is clear there has been and still are uncoordinated and competitive regulatory agencies, revealing some serious gaps in regulatory reach. This leads to the possibility that the sale and issue of highly popular and heavily leverage financial products were essentially unregulated, even on prudential/systemic grounds, both within the domestic U.S. market and overseas.

There is a key relationship between APS 112 and APS 220, both of which, of course, address credit risk, but there is one other factor that ties the two closely together and that is the issue of counterparty risk. Counterparty risk can be defined as the risk of a party to a contract not meeting its contractual obligations—in other words, it is the risk that one party may "default." Interdependency of financial institutions naturally gives rise to counterparty risk, because the global financial system is dependent upon parties within that system being not only willing, but able to meet their contractual commitments when and as they arise. APS 220 is of key interest here.

As has been discussed above, APS 220 recognizes that counterparty default is the largest, and arguably, the most critical risk that an ADI will face, simply because default impacts directly upon the stability and soundness of the institution which is the nondefaulting party to the transaction in question. Counterparty relationships are based upon credit risk, as set out in APS 112. If the credit rating of a counterparty is high, then based upon that credit rating, the transaction may proceed. If the credit rating is low, then dependent upon the willingness of the institution to accept additional guarantees, for example, dependent upon the agreed financial management strategy of the institution, an Australian ADI operating under APRA, the transaction may still proceed. If however, the risk of default is assessed as too high, then an institution may find itself unable to secure or trade financial products, as was the case with both Lehman Brothers and Bear Sterns. This then directly impacts the ability of the unsuccessful counterparty to meet obligations to other counterparties with which it may have contractual obligations, and the result is that not only may that institution fail, but others contractually dependent upon it, may also fail. In other words, the domino effect prevails, trust evaporates, and market confidence collapses.

The direct flow-on effect of counterparty default within a market, and the resulting loss of trust between financial institutions, is exactly what we have seen across the globe—the availability or flow of credit diminishes as the cost of credit or borrowing increases. The London Interbank Offered Rate (LIBOR), which measures the interest rate at which banks are willing to lend to each other, can be said to be the "litmus test" of the health of counterparty relationships within the global financial system. If the LIBOR jumps, as it has done over recent months, then the market knows that the risk of interbank default within the system is high. This in itself may not mean that credit between institutions is unavailable per se, but it will mean that only those institutions able to meet the higher interest rates demanded will be able to borrow. If institutions are unable to borrow, then they are unable to lend, and the resulting economic, political, and social impact can be far reaching as, for example, has been the case in Iceland.

Returning then to prudential supervision. It is not being proposed here that prudential supervision will always provide the security against involvement in counterparty defaults. However, what is clear is that governments around the world have recognized that strong regulatory and prudential oversight is required of all financial markets. The G-20 Summit held in November 2008 in Washington to discuss the world economy and financial markets proposed a number of reforms for the financial markets, designed to strengthen regulatory regimes, prudential oversight and risk management, whilst not stifling innovation or the trade in financial products and services. The G-20 Summit was a clear indicator of the global scope and impact of the financial crisis, both of which could not have been anticipated. What has occurred, and arguably could be anticipated because of the existing regulatory and prudential supervisory structure is that no Australian bank has succumbed, and indeed profit margins, albeit downgraded are still high and yielding dividends for shareholders. The Australian government has also responded to the financial crisis similar to many other countries around the world by introducing further legislative measures designed to support the Australian financial system.

The Australian Government enacted the *Financial System Legislative Amendment (Financial Claims Scheme and Other Measures) Act* 2008 (Cth) (the Act) which implemented the financial claims scheme (FCS) designed to support a 3 year 100% Federal government guarantee of deposits with Australian ADIs. The Act also protects holders of claims in respect to the failure of a general insurer. The FCS is administered by APRA, with APRA making appropriate payments to eligible depositors in the event of an ADI going into administration or becoming insolvent. Funds paid out to eligible depositors can be recovered from the ADI concerned, and if this is not possible, a levy will be imposed on all ADIs to recover those funds. Another action taken by the Australian Federal Government has been to introduce a guarantee facility on term funding, so as to facilitate the ability of Australian banks to access funds in the local and international markets. The terms and conditions of the guarantee are strict and eligible ADIs will be required to meet specified set criteria before they receive coverage. The guarantee is also restricted to senior unsecured debt instruments which are denominated in all major currencies including the Euro, the Yen, the Hong Kong dollar, the U.S. dollar, and the British pound but only for a period of 60 months. Fees in relation to the guarantee will be assessed according to the credit rating of the eligible ADI.

Finally, the Australian Government has implemented an arrangement to support the mortgage market by directing the Australian Office of Financial Management to acquire up to A$8 billion of residential mortgage-backed securities from Australian mortgage lenders. Again, as with the guarantee, there are specific selection criteria: for example, all residential mortgage back securities should be rated at least AAA by two or more major credit rating agencies, and the maximum loan size is A$750,000, with low doc loans not exceeding 10% of the initial size of the pool.

22.4 LITERATURE AND EVIDENCE ON FINANCIAL CRISIS

The other part of this chapter deals with the evidence of the degree of interaction of the Australasian and Australian banking system with the global banking system in order to provide further reasons for the fact that Australia's financial system has, so far up to early March 2009, been less affected by the global financial crisis than other developed economies. From a financial economics view it may be good luck rather than good management that Australasia and Australia, though possessing developed banking systems that are well regulated, are also smaller systems compared to other developed countries. Australasian and Australian banking market capitalization is less and there is less interdependence with the larger developed country banking systems and with the global banking system. This chapter draws on some of the literature regarding financial contagion for its theoretical base for the financial economics analysis. For example, equity market literature and evidence in Baig and Goldfajn (1998), Forbes and Rigobon (1999), Dungey and Zhumabekova (2001), Caporale et al. (2003), Rigobon (2004), and also currency market literature in Ellis and Lewis (2000) have focused on the manifestation of financial contagion.

The position in this chapter is that real financial contagion exists when there is a "domino" or "chain reaction" effect of bank burials (not merely bank failures) and subsequent severe widespread global financial instability rather than when there is differentiation between changes in correlations of bank systemic returns over "crisis" and "noncrisis" periods. This chapter posits that real financial contagion has still not yet occurred, largely due to the impact of Basel Accord guidelines and capital adequacy adoption. This is not to say that there is no global financial crisis with most developed countries and other banking systems requiring central bank and government intervention. Universally, it seems the determined position of governments and central banks that banks will not be allowed to be buried. They will be recapitalized in some form. The approach in this chapter is to investigate the degree of financial integration, banking system interrelationships, and linkages and therefore the degree of interdependence between country and regional banking systems in equity prices prior to the 2008/2009 financial crisis. The specific area of interest is where the Australasian and Australian banking systems are positioned now and prior to the so-called financial meltdown. In this part of the chapter, unlagged daily banking stock index data are tested in single period correlation and regression analysis. The Australian and

Australasian banking systems are compared to other banking systems throughout the analysis.

22.5 BANKING SYSTEM ANALYSIS

This study specifies a basic linear market model to analyze unlagged banking stock price index data. The market model used is a simplified version of Sharpe's Capital Assets Pricing Model (Sharpe, 1964) as discussed and reported in Reilly and Brown (2003) who also feel that the analysis of indexed data is feasible in the study of risk/return relationships in stock markets, assuming the indices studied are representative. The indices used in this paper are taken from the commonly used Datastream database. The model for testing is as follows:

$$B_{i_t} = \alpha_t + \beta_t B_{w_t} + e_t \qquad (22.1)$$

where

B_{i_t} is the banking price index return for country i at time t

B_{w_t} is the world banking price index return at time t

α_t, β_t, and e_t are the regression intercept, coefficient, and error terms at time t, respectively

Daily time series banking price index data were collected for each country/region as well as a world banking price index from Datastream covering the period 31/12/1999 to 20/9/2004. Level and first difference data were analyzed. Preliminary analysis of the various time series was undertaken. Jarque Bera test statistics indicated that there were problems with skewness and kurtosis with each of the level and first differenced series for each country/region. An initial drawback of the analysis is that none of the series is normally distributed, which in the absence of any problem with sample size, provided an initial indication that the series was serially correlated.

With the country/regional banking price indices as the dependent variables and the world series as the independent variable, pairwise ordinary least squares (OLS) regression analysis was firstly undertaken to see how the unlagged level data behaved. The Durbin Watson (DW) test statistic detected significant serial correlation (i.e., the regressions were found to be spurious). White tests with no cross terms (as pairwise regressions were analyzed) revealed significant error term heteroskedasticity. First differences series were initially analyzed using pairwise OLS market models. First differencing removed the problem of serial correlation in the errors of the regressions. However, heteroskedasticity remained persistent. The OLS regression was respecified into a weighted least squares model to account for heteroskedasticity of an unknown form. Pairwise correlation analysis of the unlagged price index first differences also enabled an initial investigation of the degree of interdependence of the various country/regional banking systems with the world system.

22.6 FINDINGS: THE GLOBAL POSITIONING OF THE AUSTRALIAN BANKING SYSTEM

Pairwise correlation analysis of unlagged price index first differences data is a basic indicator of interrelationship and integration of country/regional banking systems in relation to the world banking system.

The ranking of the correlation coefficients provides an initial indication of the degree of systemic risk (that is the degree of interdependence of each country/regional system with the world system).

These relationships are confirmed when regressions are run and when the adjusted R square values and t statistics are considered. The initial indication of the strength and direction of the relationship is indicated in the correlation coefficients in Table 22.1 and the adjusted R square values in Table 22.2.

The ranking of the regression coefficients (Betas) indicates the level of market risk in each country/regional banking system considered pairwise with the world system. Care needs to be taken in the interpretation of these results. A low ranking of test statistics is also an indication of the lack of size and development of some banking systems. Australia has a small system but that system is developed, sophisticated, and comparatively well regulated.

In Table 22.2, it can be seen that each system shows explanatory power (as seen in adjusted R square values) on unlagged data and therefore the comparative strength of the regression coefficients in explaining banking system price index variation. The relationships are positive indicating that as the world banking–price index increases so does the respective banking market index.

From Tables 22.1 and 22.2 it can be seen that the developed banking systems in Europe, the United States, the United Kingdom, and Canada have strong interrelationships with the world banking system. The systems

TABLE 22.1   Correlation Analysis

Ranking according to Correlation	Banking System	Correlation Coefficient (Correlation with the World Banking Index)
1	Europe (excluding emerging systems)	0.8295
2	The Americas (including Canada and United States)	0.8108
3	Europe (excluding the United Kingdom)	0.7987
4	United States	0.7963
5	The EMU	0.7843
6	Europe (excluding the economic union)	0.6841
7	United Kingdom	0.6742
8	Canada	0.5301
9	The Pacific Basin	0.4346
10	The Pacific Basin (excluding Japan)	0.4335
11	Latin America	0.4281
12	Asia	0.4153
13	Asia (excluding Japan)	0.4064
14	The Far East	0.3975
15	Hong Kong	0.3627
16	Japan	0.3076
17	Singapore	0.2980
18	Australasia	0.2832
19	South East Asia	0.2822
20	South Korea	0.2363
21	Australia	0.1896
22	Taiwan	0.1699
23	Thailand	0.1542
24	Malaysia	0.0795
25	Indonesia	0.0571
26	The Philippines	0.0579
27	China	−0.0062

in the Pacific Basin, Latin America (which is strongly influenced by U.S. and Canadian systems), Asia, Hong Kong, the Far East, Japan, Singapore, Australasia, and South East Asia (including Singapore and Hong Kong) have a medium degree of integration and interdependence, while the smaller developed systems of Australia and Korea have a smaller degree of interaction with the world system along with the less developed banking

TABLE 22.2   Regression Analysis

Country and Regional Banking Price Index First Differences Regressed on the World Banking Price Index First Differences	Rank (Adjusted R Square Value and t Statistic Value)	Regression-Adjusted R Square Value	Regression Coefficient (Beta)	t Statistic for Regression Coefficient
Europe without emerging European systems	1	0.6881	0.8284	52.0867
Americas	2	0.6572	1.0544	48.57.7
Europe without United Kingdom	3	0.6378	0.7559	46.5411
United States	4	0.6340	1.4740	46.1639
EMU	5	0.6150	0.7003	44.3240
Europe without countries in the EMU	6	0.6103	1.2249	43.8934
Europe without the EU	7	0.4689	1.1482	32.8939
United Kingdom	8	0.4546	7.3415	32.0174
Canada	9	0.2796	0.7140	21.9207
Pacific Basin Countries without Japan	10	0.1879	0.2701	16.8739
Latin American Countries	11	0.1833	0.0339	16.6136
Asian countries including Japan	12	0.1720	0.4024	15.9998
Asia excluding Japan	13	0.1652	0.2821	15.6026
East Asia	14	0.1576	0.4279	15.1825
Hong Kong	15	0.1314	1.7724	13.6528
Japan	16	0.0938	0.1423	11.3225
Singapore	17	0.0887	0.1874	10.9430
Australasia	18	0.0796	0.1949	10.3611
South East Asia	19	0.0789	0.1110	10.3035
South Korea	20	0.0557	0.0305	8.5310
Australia	21	0.0354	0.2116	6.7778
Taiwan	22	0.0289	0.0370	6.0462
Thailand	23	0.0235	0.0617	5.4680
Malaysia	24	0.0061	0.0926	2.8017**
The Philippines	25	0.0014	0.0248	2.0180**
Indonesia	26	0.0017	0.0059	1.900**

Note: All *t* statistics are significant at the 1% level except those marked ** where significance is at the 5% level. The ranking is according to explanatory power in the adjusted *R* square value and the *t* statistic value. The results for China are not significant and not reported.

systems in Thailand, Taiwan, Malaysia, Indonesia, the Philippines, and China.

The objective for the purposes of this part of the chapter is to highlight the extent of global interaction of the Australasian banking system and the Australian banking system with the global banking system and the other major developed banking systems (such as those in Americas and in Europe) and also the banking systems in developing economies. Clearly there is substantially less global interaction with the global system in the cases of both Australasia and Australia than larger developed countries such as the United States and Canada.

22.7 DISCUSSION: AUSTRALIAN PRUDENTIAL AND REGULATORY CONDITIONS

The smaller market capitalization of Australia's banking system has much to do with its comparative global financial crisis insulation. However, the study now returns to a discussion on Australia's regulatory environment, which has provided further immunization from the global crisis. In October 2008, the International Monetary Fund released its *Global Financial Stability Report*. It stated when referring to restoring confidence in global markets:

> With financial markets worldwide facing growing turmoil, internationally coherent and decisive policy measures will required to restore confidence in the global financial system. Failure to do so could usher in a period in which the ongoing deleveraging process becomes increasingly disorderly and costly for the real economy. In any case, the process of restoring an orderly system will be challenging, as a significant deleveraging is both necessary and inevitable...

The IMF, in noting that the measures adopted would inevitably differ across countries, stated the five principles could be used as a guide to form the basis of restoration of confidence. These five principles were as follows:

1. The employment of comprehensive, timely, and clearly communicated measures, which will improve funding availability, cost, and maturity to stabilize balance sheets, as well as injecting capital, which is designed to support viable institutions. "...These measures should be clear and operational procedures transparent."

2. To aim for consistent and coherent policies across countries "...to stabilize the global financial system in order to maximize impact while avoiding adverse effects on other countries."

3. To ensure a rapid response on the basis of an early detection of strains but use of cross border cooperation and a framework which "...allows for decisive action by potentially difference sets of authorities."

4. To assure that all emergency government interventions are temporary and that taxpayer interests are protected, noting the importance of accountability of government and the minimization of moral hazard.

5. "Pursue the medium-term objective of a more sound, competitive and efficient financial system...by both an orderly resolution of non-viable financial institutions and a strengthening of the international macro financial stability framework to help improve supervision and regulation at the domestic and global levels, as well as to improve the effectiveness of market discipline."

In the opinion of the writers, the Australia's regulatory and prudential framework, together with the response of the Australian Federal Government represents the implementation of these principles. This is not to say that there is no work to be done, but Australian banks have survived so far and accordingly, Australia's regulatory and prudential structure should be seriously examined by governments around the world, seeking to implement or to modify their existing structures. The global financial and economic crisis is far from over and its effects will be long term. Cooperation, the sharing of information and experiences, and examination of alternatives cannot be too soon. The discussion now reverts to the contribution of the empirical study that also demonstrates the partial immunization of the Australian banking system from the global financial crisis.

22.8 CONCLUSION

The issues addressed in this study were as follows: What was the degree of financial integration, interdependence, and the threat of contagion in global banking systems and where was Australasia and Australia positioned prior to the 2008/2009 global financial crisis? What are the policy implications of financial integration for Australian banks and for regulators? How has Australia handled its prudential supervision and banking regulatory environment with the effect of reducing the impact of global financial crisis?

The financial economics part of this study makes the assumption that country, regional, and world-banking systems can be proxied by database time series of banking price indices. The study also takes the view that the real contagion refers to the "domino" effect melt down of financial systems and this has not to this point in time occurred. It is the continuing threat of contagion in the sense that is important (i.e., the research issue that requires ongoing examination). Such investigation can be undertaken effectively using correlation and regression analysis of unlagged banking stock market data.

Australasia and Australia, perhaps, by good luck rather than good management when financial economics issues are considered, has had a great deal less exposure to the global banking system than the larger developed economies. The Australian banking system is less capitalized and has had less of an involvement in, for example, the subprime mortgage market in the United States. Australian banks rely less on interbank lines of credit from the Eurocurrency market to bolster the liability and asset sides of their balance sheets. Generally speaking Australian banks have employed reasonable credit risk assessment and credit risk management guidelines. This is despite the fact that customer companies (e.g., mining companies) exposed to the global recession have suffered substantial loss of trade which have caused the Australian banks to incur significant but not impossibly problematic bad and doubtful debts. Australia is after all primarily a minerals exporter.

The continuing need is for global and Australian banks to refocus on investment, borrowing, and lending portfolio diversification, and for banking regulators to be vigilant in regulation of banking systems as the process of global financial services liberalization gathers pace. In respect of prudential supervision and regulation of the banking sector it must be said that the Australian authorities have been diligent and thorough and have assisted to reduce the impact of the global financial crisis on Australia. This regulatory side of the story indicates good management rather than good luck. Regulators will still need to closely monitor undiversified interdependence. A greater interaction between country central banks and a central global regulatory authority, such as the BIS may rapidly become desirable and necessary.

REFERENCES

Australian Prudential Regulation Authority, Australian prudential standards, January 2008, available at www.apra.gov.au.

Baig, T. and Goldfajn, I. 1998 Financial market contagion and the Asian crisis, IMF Working Paper, Washington, DC.

Basel Committee on Banking Supervision, 2008 Liquidity risk management and supervisory challenges, Bank for International Settlements, Basel, Switzerland.

Caporale, G. M., Cipollini, A., and Spagnolo, N. 2003 Testing for contagion: A conditional correlational analysis, Australasian meeting of the econometric society, October, Melbourne, Australia.

Dungey, M. and Zhumabekova, D. 2001 Testing for contagion using correlations: Some words of caution, Working Paper, Centre for Pacific Basin Monetary and Economic Studies, Economic Research Department, Federal Reserve Bank of San Francisco, CA.

Ellis, L. and Lewis, E. 2000 The response of financial markets in Australia and New Zealand to News about the Asian crisis, BIS Conference on International Financial Markets and the Implications for Monetary and Financial Stability, Basel, October 25–26, 1999.

Forbes, K. J. and Rigobon, R. 1999 No contagion, only interdependence: Measuring stock market co-movements, NBER Working Paper, Cambridge, MA.

International Monetary Fund, 2008 Global financial stability report–financial stress and deleveraging macro-financial implications and policy, International Monetary Fund, Washington, DC.

Reilly, F. K. and Brown, K. C. 2003 *Investment Analysis: Portfolio Management.* Thomson South Western: Florence, KY.

Rigobon, R. 2004 Identification through heteroskedasticity, *Review of Economics and Statistics*, 85 (3): 777–792.

Sharpe, W. F. 1964 Capital asset prices: A theory of market equilibrium under conditions of risk, *Journal of Finance*, 19 (3): 425–442.

Tyree, A. 2008 *Banking Law in Australia*. LexisNexis Butterworths: Syndey, Australia.

CHAPTER 23

Default Risk Codependence in the Global Financial System: Was the Bear Stearns Bailout Justified?

Jorge Antonio Chan-Lau*

CONTENTS

* The views expressed in this chapter do not reflect those of IMF or IMF policy. The author is solely responsible for any errors or omissions.

The current subprime crisis has encouraged renewed efforts to understand systemic risk. The assessment of systemic risk requires measuring default risk codependence, or in other words, how the default risk of a specific institution affects the conditional default risk of another institution. This chapter proposes a methodology for assessing default risk codependence among financial institutions based on quantile regression after accounting for the impact of fundamental and technical factors. Results are reported for a sample of 25 financial institutions in Europe, Japan, and the United States. Consistent with the earlier contagion literature, the results indicate that risk codependence is stronger during distress periods. The results also cast some light on the controversy surrounding the bailout of Bear Stearns and AIG, and the bankruptcy of Lehman Brothers and suggest that there is room for improving information disclosure and transparency in regard to policy decisions.

23.1 WHY RISK CODEPENDENCE MATTERS?

The increased globalization of financial markets, the deepening of financial integration, and the rapid pace of financial innovation have created a global financial system that transcends national boundaries. While the system has brought substantial benefits such as increased capital flows, lower borrowing costs, and better price discovery and risk diversification, the system has also strengthened and created news channels for the transmission of negative shocks across borders and across markets. The worldwide financial crisis that started in 2007 is a stark reminder of the risks posed by a more integrated and globalized financial system.

The 2007–2008 crisis has encouraged renewed efforts in four interconnected areas: the definition and measurement of systemic risk, the development of early warning indicators of financial crises, the enhancement of supervisory and regulatory frameworks, and the strengthening of current crisis management and bank resolution frameworks. Work in these four areas is currently under way in central banks and multilateral financial institutions as policy makers attempt to find an effective and prompt solution to the problems posed by the crisis and seek to avoid future crises.

This chapter contributes to work in the first area, the measurement of systemic risk, by proposing a methodology for assessing default risk codependence among financial institutions. Risk codependence is defined as the default risk of one institution conditional on the default risk of another institution after correcting for the effect of common observable fundamental and technical factors. Default risk codependence, therefore, is a manifestation of contagion that can be implied from observable security prices. Given observed default risk measures, it is possible to construct risk codependence measures using quantile regression.

Why does risk codependence matters? Risk codependence is closely linked to contagion and financial crises spillovers. In this context the measures of risk codependence derived in this chapter are associated with unobservable factors likely related to the interconnectedness among financial institutions. The interconnectedness arises from direct linkages, such as counterparty risk from interbank claims, or indirect linkages such as exposure to common risk factor such as reliance on wholesale markets for funding, and feedback effects from market volatility due to the adoption of similar risk management and accounting practices. When markets are efficient, these linkages are priced in by market participants and lead to comovements among institutions' security prices and risk measures.[*]

The results obtained from analyzing credit default swap (CDS) spreads for a sample of 25 financial institutions in Europe, Japan, and the United States suggest that risk codependence is stronger during distress periods, a finding consistent with the earlier empirical literature on contagion. More interestingly, the results also cast some light on the controversy surrounding the bailout of Bear Stearns and AIG and the bankruptcy of Lehman Brothers. Contagion effects, as implied from credit default swap spreads do not appear to justify the policy decision to bailout AIG and Bear Stearns instead of Lehman Brothers. In the latter case, the apparent contradiction between what markets imply and the policy decision suggests that credit derivatives markets exhibit, at best, semi-strong efficiency and do not incorporate private information.[†] It also suggests that there is room for improving the communication of private information relevant to policy decisions.

The remainder of this chapter is structured as follows. Section 23.2 explains why quantile regression is appropriate for measuring risk

---

[*] Common risk measures for financial institutions include distance-to-default, credit default swaps spreads, and the VaR of their trading portfolios. All these measures depend on market data and are considered forward-looking since current prices should reflect market expectations of earnings and discount rates.

[†] Fama (1970).

codependence among financial institutions. Section 23.3 describes in detail the specification of the econometric model and defines two measures of conditional risk codependence (CRC). Section 23.4 describes the data used to estimate the risk codependence measures and Section 23.5 presents and analyzes the results. Section 23.6 provides the conclusion.

23.2 MEASURING RISK CODEPENDENCE USING QUANTILE REGRESSIONS

For risk management, regulatory, and surveillance purposes, it is important to design and implement measures of risk codependence that account for large negative shocks. Accordingly, extreme value theory measures appear suitable for capturing the risk codependence of institutions subject to large negative shocks.[*] One common problem associated to extreme value measures, though, is that a significant amount of information from the whole data sample is lost as the analysis uses only the tail realizations of the series. This problem is more acute the shorter the length of the data sample is.

As an alternative to extreme value theory, this chapter introduces two conditional codependence measures that use all the information available in the sample. The measure is constructed using quantile regression, a simple approach that helps examining how risk codependence changes conditional on the level of risk of an individual institution.[†]

Quantile regression is a technique that expands on standard least squares estimation and provides an extensive analysis of the relationship among random variables. Quantile regression extends the notions of ordering, sorting, and ranking the data by ascending (or descending) order of univariate statistical analysis to regression models. As a result, it is possible to estimate the functional relationship among variables at different quantiles. In other words, in addition to computing the conditional mean function, as done through ordinary least squares, quantile regression also computes the full family of conditional quantile functions.

The use of conditional quantile functions in addition to the conditional mean function is important in the presence of regime switches since functional relationships may differ from one regime to the other. The existence of a multiregime approach is usually assumed formally or intuitively

---

[*] See among others Hartman et al. (2001); Chan-Lau et al. (2004); and references therein.

[†] We extend the analysis of Adrian and Brunnermeier (2008). For a detailed exposition of quantile regression techniques, see Koenker (2005), and for an intuitive exposition, Koenker and Hallock (2001).

when analyzing the impact of a financial crisis and the appropriate policy responses. In the context of the current crisis, moreover, there is a special emphasis on designing a regulatory framework where policy measures are contingent on whether the economy is in a precrisis, crisis, or postcrisis regime.[*]

In the empirical literature on contagion and financial crises, at least two different regimes are usually identified: a tranquil regime and a volatile regime. The volatile regime is usually characterized by rapidly falling security prices and a substantial increase of the risk of institutions. In the context of quantile regression, if risk measures or security prices are sorted by quantiles, the tranquil regime and volatile regimes are associated with the lower and higher conditional quantile functions, respectively.

Before describing the methodology in detail, we review briefly some other ways to measure systemic risk in the financial sector that complement the approach in this chapter. Equity price comovements could be used as a proxy for risk codependence in the financial system since they exhibit a high degree of commonality.[†] Furthermore, equity return data has also been used to estimate the probability of the simultaneous defaults of a group of banks.[‡] Similarly, the availability of prices for credit default swaps, which are insurance-like contracts, and estimates of asset return correlation derived from equity returns can be used to construct a default probability-based measure of systemic risk.[§]

23.3 MODEL SPECIFICATION AND RISK CODEPENDENCE MEASURES

As explained in Koenker and Hallock (2001), given N observations, the estimation of a quantile regression relies on the minimization of the sum of residuals. The residuals are weighted asymmetrically according to

$$\min_{\beta} \sum_{i}^{N} \rho_\tau (y_i - \xi(x_i, \beta)) \qquad (23.1)$$

where y is the dependent variable, $\xi(x_i, \beta)$ is a linear function of the parameters, and

---

[*] See Financial Stability Forum (2008) and the special issue of the *Journal of Financial Stability* edited by Goodhart (2008).

[†] Hawkesby et al. (2007).

[‡] Lehar (2005) and Allenspach and Monnin (2006).

[§] Huang et al. (2008).

$$\rho_\tau(u) = u(\tau - 1(u < 0)) \qquad (23.2)$$

is the residual weight function for the quantile τ. The minimization can be solved using standard linear programming techniques and the covariance matrices are usually estimated using bootstrap techniques which are valid even if the residuals and explanatory variables are not independent (Koenker, 2005).

When analyzing risk codependence, the relevant issue is to assess how the risk of institution i, denoted by $Risk_i$, is affected by the risk of institution j, denoted by $Risk_j$, after correcting for the effect of K aggregate risk factors, R_k. The aggregate risk factors account for the impact of fundamental factors such as the stage of the business cycle and technical factors such as investor sentiment. The minimization problem of interest, therefore, is given by

$$\min_\beta \sum_i^N \rho_\tau \left(Risk_i - \sum_k^K \beta_{k,\tau} R_k - \beta_{j,\tau} Risk_j \right) \qquad (23.3)$$

Equation 23.3 is solved for different quantiles τ in order to examine how the risk codependence between any two firms varies at different levels of risk. Indeed, the risk codependence is implicitly captured by the set of parameters $\beta_{j,\tau}$; therefore, we will refer to this coefficient as the risk codependence coefficient.

The behavior of $\beta_{j,\tau}$ as a function of τ illustrates how the risk of institution j affects institution i. An upward sloping pattern suggests that the riskier institution j is, the greater the impact is on the risk of institution i. Therefore, we can interpret the coefficient $\beta_{j,\tau}$ as a risk codependence measure. Furthermore, since the quantile regressions include aggregate observable risk factors, the coefficient $\beta_{j,\tau}$ can be interpreted as a proxy for the "frailty" effect, or an unobserved latent risk factor. Empirical studies using data on U.S. corporations suggest that the frailty effect is an important explanatory factor of correlated default in the sector.[*]

In addition to examining the behavior of $\beta_{j,\tau}$ as a function of τ, it is also possible to quantify the risk codependence between institutions i and j in terms of the conditional codependence function for a specific quantile, as

---

[*] See Das et al. (2006, 2007), Duffie et al. (2008), and Azizpour and Giesecke (2008).

first suggested by Adrian and Brunnermeier (2008). The CRC measure is defined as

$$CoRisk_{ij}(\tau) = \{Risk_i(\tau) \mid Risk_j(\tau), R_K\} \qquad (23.4)$$

where $Risk_i(\tau)$ is the level of risk of firm i for quantile τ. An alternative way to express the conditional risk measure is in terms of percent increase in the unconditional risk of firm i:

$$CoRisk_{ij}(\tau)(\%) = 100 \times \left(\frac{\{Risk_i(\tau) \mid Risk_j(\tau), R_K\}}{Risk_i(\tau)} - 1 \right) \qquad (23.5)$$

Section 23.4 describes the data used to estimate Equation 23.3 and its associated CRC coefficient $\beta_{j,\tau}$, and to construct the CRC measure.

23.4 DATA

The quantile regressions were estimated using daily market data for the period July 1, 2003–September 30, 2008. One advantage of using market data is that market data is available at high frequencies and on a timely fashion. Another advantage is that, provided market efficiency holds, market data is forward looking in general and captures market expectations on changes in the risk and performance of financial institutions.

Yields, spreads, and equity returns were obtained from Primark Datastream; credit default swap mid-price quotes were obtained from Bloomberg and Primark Datastream. The U.S. institutions included in this study are AIG, Bank of America, Bear Stearns, Citigroup, Goldman Sachs, JP Morgan, Lehman Brothers, Merrill Lynch, Morgan Stanley, Wachovia, and Wells Fargo; the European institutions are Fortis, Banque Nationale Paribas, Societe Generale, Deusche Bank, Commerzbank, BBVA, Santander, Credit Suisse, UBS, Barclays, and HSBC; and the Japanese institutions are Mitsubishi, Mizuho, and Sumitomo.

The risk measure chosen was the price of credit default swaps. The price of credit default swaps, which is quoted in basis points as spreads over Libor, is generally considered the best market proxy for default risk since they reference the credit risk of the issuer directly and because of its informational advantage vis-à-vis the prices of other securities.* Because of liquidity considerations, only 5 year contracts referencing senior debt were used.

The set of regressors could be classified into two categories, fundamental and market indicators. Fundamental indicators include the excess stock market return, the slope of the U.S. yield curve, and the corporate spread. The excess stock market return was set equal to the daily return of the S&P 500 index in excess of the U.S. 3 month Treasury bill. The inclusion of this variable attempted to capture overall market effects on the default risk of the financial institutions.[†] The slope of the U.S. yield curve, measured as the yield spread between the 10 year and the 3 month Treasury rates, was included as a business cycle indicator. Another business cycle leading indicator, the corporate spread, was constructed as the yield spread of BAA-rated corporate bonds and the 10 year Treasury bond.[‡]

Market indicators included the Libor spread, a liquidity spread, and a market sentiment indicator. The 1 year Libor spread over 1 year constant maturity U.S. Treasury yield was used as a proxy for systemic risk in the financial sector, as the Libor spread is usually regarded as a measure of the default risk in the interbank system. Liquidity shortages in the interbank and money markets may have distorted the information content of the Libor spread. To compensate for it, a short-term liquidity spread was included among the regressors. The liquidity spread series were constructed as the yield spread between the 3 month general collateral repo rate and the 3 month Treasury rate.[§] Finally, the implied volatility index (VIX) reported by the Chicago Board Options Exchange was used as a proxy for investor sentiment.

23.5 FINANCIAL INSTITUTIONS AS SOURCES OF RISK AND VULNERABILITY

Figure 23.1 summarizes the results corresponding to the estimation of Equation 23.3. The figure illustrates how the default risk of financial institutions in a specific region, or "source region," affects the default risk of financial

---

[*] Empirical studies suggest that CDS spreads have an informational lead over bond spreads and equity prices, especially for forecasting default events. See Hull et al. (2004), Blanco et al. (2005), Longstaff et al. (2005), and Chan-Lau and Kim (2005).

[†] At least for U.S. institutions, there appears to be a risk premium associated to default risk. See Vassalou and Xing (2004) and Chan-Lau (2007).

[‡] Empirical evidence supporting the business cycle properties of the corporate spread is presented in Chan-Lau and Ivaschenko (2001) among others.

[§] An alternative measure is the overnight index swap spread.

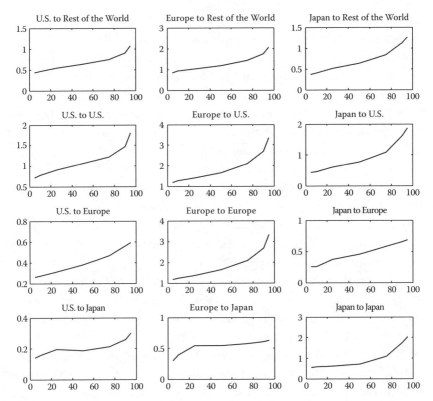

FIGURE 23.1   Risk codependence among financial institutions. The figure plots the average risk codependence coefficient for banks in the "recipient" region (vertical axis) as a function of the default risk of the financial institutions in the "source" region sorted by quantiles (horizontal axis). Higher quantiles correspond to higher levels of default risk.

institutions in other region or "recipient region." For a given level of default risk in the source region, denoted by the quantile τ, the default risk codependence arising from the source region is measured as the average value of the coefficient $\beta_{j,\tau}$ in Equation 23.3 estimated while setting the default risk of financial institutions in the "recipient region" as the dependent variable.*

The upward sloping pattern evident in Figure 23.1 suggests risk codependence is stronger in bad times (high quantiles) than in good times (low quantiles) regardless of what the source and recipient regions are. Although there is no underlying structural model, the results provide

---

* All coefficients are significant at the 95% confidence level. Detailed results are available upon request.

evidence supporting the existence of a "contagion" or transmission channel across financial institutions beyond that explained by the exposure to common aggregate shocks. The interconnectedness of the financial system arguably plays an important role supporting this transmission channel. The interconnectedness can arise through direct exposure and counterparty risk, as in the case of interbank lending. The interconnectedness can also arise indirectly from feedback mechanisms such as mark-to-market accounting and economic capital procyclicality driven by risk management systems or regulatory requirements.

Table 23.1 presents the results from a different angle using the CRC measure defined in Equation 23.5 above. The CRC measure assesses by how much the conditional default risk of an institution, when the default risk of another institution is at its 95th percent quantile, exceeds its unconditional risk at the 95th percent quantile.

In light of the debate surrounding the bankruptcy of Lehman Brothers, and the rescue of Bear Stearns and AIG, Table 23.1 presents two interesting results. First, Bear Stearns and AIG are among the institutions that are more vulnerable to default risk spillovers from other institutions. For all the institutions in the sample, the average conditional default risk increases by 40% relative to their unconditional value. For Bear Stearns, the average increase is of 134% and for AIG is 187%. For Lehman Brothers, while still vulnerable, the average increase is only 61%. If the averages are restricted only to risk spillovers from U.S. financial institutions, the average increase in conditional default risk is still substantial for Bear Stearns, (148%) and AIG (172%) relative to Lehman Brothers (55%). Among non-U.S. banks, risk spillovers appear to be modest, with HSBC, which reports an average increase of 44%, being the most vulnerable institution.

What makes these particular institutions stand out as the most vulnerable to the default risk of other institutions? Arguably, and partly with the benefit of hindsight, the vulnerability can be traced to the active participation of these institutions in the credit risk transfer market. For instance, a survey conducted by Fitch Ratings in mid-2007 reported that AIG was a large seller of protection through structured finance and corporate synthetic collateralized debt obligations (CDOs), with a gross sold position of $500 billion by end 2006. Bear Stearns and Lehman Brothers were ranked as the top twelfth and seventh counterparties by trade count in 2006.* The

---

* See Linnell et al. (2007).

TABLE 23.1　Conditional Risk Codependence among Financial Institutions Worldwide

Panel A	Bank of America	Bear Stearns	Citigroup	Goldman Sachs	JP Morgan
Bank of America	n.a.	28	9	9	11
Bear Stearns	154	n.a.	135	117	114
Citigroup	32	29	n.a.	12	15
Goldman Sachs	93	31	71	n.a.	50
JP Morgan	18	39	9	6	n.a.
Lehman	56	27	58	36	66
Merrill Lynch	20	13	16	17	25
Morgan Stanley	70	25	71	22	44
Wachovia	59	14	41	23	44
Wells Fargo	15	25	18	14	14
AIG	262	97	209	58	206
Fortis	31	43	32	23	17
Banque Nationale Paribas	41	46	25	23	28
Societe Generale	33	58	24	21	20
Deutsche Bank	39	40	26	19	27
Commerzbank	48	63	39	32	38
BBVA	36	47	22	22	21
Santander	38	45	25	23	26
Credit Suisse	43	49	28	25	28
UBS	40	50	19	14	16
Barclays	12	27	10	7	9
HSBC	69	41	53	46	52
Mitsubishi	50	50	40	33	34
Mizuho	32	−7	33	32	36
Sumitomo	47	28	40	32	39
Average	56	38	44	28	41
Minimum	12	−7	9	6	9
Maximum	262	97	209	117	206

Panel B	Lehman Brothers	Merrill Lynch	Morgan Stanley	Wachovia	Wells Fargo
Bank of America	12	13	13	16	8
Bear Stearns	158	176	163	180	142
Citigroup	24	36	19	36	31
Goldman Sachs	49	97	35	41	80
JP Morgan	17	22	18	18	17
Lehman	n.a.	61	68	38	53
Merrill Lynch	16	n.a.	20	20	20
Morgan Stanley	36	82	n.a.	40	48

(*continued*)

TABLE 23.1 (continued)   Conditional Risk Codependence among Financial
Institutions Worldwide

Panel B	Lehman Brothers	Merrill Lynch	Morgan Stanley	Wachovia	Wells Fargo
Wachovia	22	76	31	n.a.	39
Wells Fargo	15	21	18	17	n.a.
AIG	92	255	97	131	255
Fortis	34	37	29	37	27
Banque Nationale Paribas	31	47	32	41	37
Societe Generale	22	41	30	41	32
Deutsche Bank	40	43	34	38	31
Commerzbank	50	51	45	51	46
BBVA	34	37	33	38	30
Santander	34	44	35	42	32
Credit Suisse	40	50	41	48	39
UBS	36	46	26	41	34
Barclays	16	20	10	18	14
HSBC	64	76	59	74	63
Mitsubishi	33	52	43	36	34
Mizuho	14	29	30	28	33
Sumitomo	32	42	43	37	37
Average	38	61	41	46	49
Minimum	12	13	10	16	8
Maximum	158	255	163	180	255

Panel C	AIG	Fortis	Banque Nationale Paribas	Societe Generale	Deutsche Bank
Bank of America	13	19	26	25	22
Bear Stearns	248	129	119	148	99
Citigroup	30	14	27	24	19
Goldman Sachs	20	32	71	85	62
JP Morgan	12	13	20	18	20
Lehman	55	76	83	82	72
Merrill Lynch	25	31	45	40	33
Morgan Stanley	14	39	73	78	66
Wachovia	35	48	72	72	62
Wells Fargo	17	19	27	26	23
AIG	n.a.	136	246	237	219
Fortis	32	n.a.	14	14	15
Banque Nationale Paribas	34	20	n.a.	17	13
Societe Generale	32	12	6	n.a.	9
Deutsche Bank	32	23	17	25	n.a.

TABLE 23.1 (continued) Conditional Risk Codependence among Financial
Institutions Worldwide

Panel C	AIG	Fortis	Banque Nationale Paribas	Societe Generale	Deutsche Bank
Commerzbank	46	33	29	34	27
BBVA	27	14	16	15	13
Santander	28	16	17	21	15
Credit Suisse	37	26	15	24	13
UBS	39	16	30	36	18
Barclays	13	15	24	22	17
HSBC	66	42	28	41	31
Mitsubishi	36	38	24	30	28
Mizuho	32	33	36	33	33
Sumitomo	41	38	30	33	27
Average	40	37	46	49	40
Minimum	12	12	6	14	9
Maximum	248	136	246	237	219

Panel D	Commerz-bank	BBVA	Santander	Credit Suisse	UBS
Bank of America	24	18	14	26	22
Bear Stearns	96	113	116	104	141
Citigroup	11	17	20	25	19
Goldman Sachs	19	77	67	52	62
JP Morgan	14	16	14	20	12
Lehman	73	69	65	85	60
Merrill Lynch	43	28	28	42	34
Morgan Stanley	41	67	66	61	68
Wachovia	61	69	67	68	45
Wells Fargo	21	22	20	27	23
AIG	189	241	228	237	157
Fortis	−3	5	7	14	31
Banque Nationale Paribas	4	12	12	15	20
Societe Generale	1	7	10	14	24
Deutsche Bank	6	14	17	13	20
Commerzbank	n.a.	30	29	25	36
BBVA	8	n.a.	6	19	21
Santander	7	8	n.a.	21	24
Credit Suisse	7	20	21	n.a.	19
UBS	1	28	32	21	n.a.
Barclays	17	21	19	23	17
HSBC	21	30	30	34	51

(*continued*)

TABLE 23.1 (continued)   Conditional Risk Codependence among Financial
Institutions Worldwide

Panel D	Commerz-bank	BBVA	Santander	Credit Suisse	UBS
Mitsubishi	19	34	36	21	29
Mizuho	51	33	33	34	33
Sumitomo	22	32	36	24	32
Average	31	42	41	43	42
Minimum	−3	5	6	13	12
Maximum	189	241	228	237	157
Panel E	**Barclays**	**HSBC**	**Mitsubishi**	**Mizuho**	**Sumitomo**
Bank of America	19	19	25	31	25
Bear Stearns	159	80	129	147	153
Citigroup	26	16	14	12	10
Goldman Sachs	75	54	24	16	22
JP Morgan	22	15	17	18	13
Lehman	66	67	70	77	70
Merrill Lynch	26	37	39	46	40
Morgan Stanley	105	50	43	32	40
Wachovia	54	60	63	57	63
Wells Fargo	21	21	22	23	23
AIG	227	215	197	157	176
Fortis	53	0	17	14	13
Banque Nationale Paribas	37	12	18	19	16
Societe Generale	32	9	15	17	12
Deutsche Bank	31	12	20	27	21
Commerzbank	41	18	37	42	36
BBVA	27	5	17	25	20
Santander	33	7	21	27	23
Credit Suisse	39	19	22	30	21
UBS	33	17	8	7	5
Barclays	n.a.	20	16	19	15
HSBC	58	n.a.	53	49	48
Mitsubishi	46	31	n.a.	33	10
Mizuho	33	36	60	n.a.	60
Sumitomo	48	33	10	30	n.a.
Average	55	36	40	40	39
Minimum	19	0	8	7	5
Maximum	227	215	197	157	176

Note: Each cell reports the increase, in percent, of the default risk of the financial institution listed in the row relative to its unconditional 95th percent quantile when the
default risk of the institution listed in the column is at its 95th percent quantile holding the common explanatory factors constant.

high exposure of these institutions to the credit risk transfer market, and especially to the most leveraged and riskiest credit derivatives contracts, could also be judged from the relationship between their equity prices and the prices of synthetic CDO tranches.*

The analysis above considered financial institutions as "risk recipients." When the institutions are analyzed as "risk sources," the results indicate that Bear Stearns is not among the riskiest institutions. Among U.S. banks, it shares with Lehman Brothers the second lowest CRC of 38%. Only Goldman Sachs has a lower CRC of 28%. In the case of AIG, its average CRC is 40% and is not significantly different than the average CRC of Bear Stearns and Lehman Brothers. In contrast, the riskiest institutions are Merrill Lynch, Bank of America, and Barclays, with average CRCs of 61%, 56%, and 55%, respectively.

Under the assumption that systemic risk corresponds to a CRC of 50% or above, Bear Stearns posed systemic risk to AIG, Commerzbank, and UBS; AIG to Bear Stearns, HSBC, and Lehman Brothers; and Lehman Brothers to Bear Stearns, AIG, HSBC, and Commerzbank. These results suggest that AIG, Bear Stearns, Lehman Brothers in the United States, and Commerzbank, HSBC, and UBS could be grouped as one common systemic group of institutions.

23.6 CONCLUSION: WAS THE BAILOUT JUSTIFIED?

Assessing the risk codependence, or spillovers, among financial institutions is a major building block for monitoring financial stability in an increasingly integrated global financial system. This paper makes a contribution towards achieving this goal by proposing a simple methodology for measuring default risk codependence using quantile regression. The results indicate that default risk codependence is stronger during periods of distress, a finding that is consistent with and reinforces earlier results from the empirical contagion literature.

Furthermore, the results also cast some light on the controversy surrounding the bailout of Bear Stearns and AIG, and the bankruptcy of Lehman Brothers. Namely, publicly available price information from the credit derivatives market suggests that, from a financial stability perspective, Bear Stearns, AIG, and Lehman Brothers were among the most vulnerable institutions to an increase in the default risk of other financial institutions. On the other hand, the risk these institutions posed to the

---

* Chan-Lau and Ong (2007).

financial system was of similar magnitude. Therefore, either the government should have supported the three institutions or let both of them gone bankrupt.

The latter conclusion, however, relies exclusively on the analysis of market price data so some caveats apply. In the analysis, there is the implicit assumption that markets are efficient. This may not necessarily be the case, as anecdotal evidence, theoretical arguments, and empirical studies indicate that markets tend to overreact and/or underreact to new information. In particular, limits to arbitrage, herding behavior and investment constraints may prevent prices from adjusting rapidly to their fundamental prices.

It follows, then, that market prices are weakly or semistrongly efficient and only reflect the impact of past prices and past and current public information. In contrast, the decision to support a financial institution takes into account private information collected by regulators. This information is not usually available to market participants. Therefore, a definite conclusion derived from analyzing only available market prices may not be enough to judge the merits of the bailout of Bear Stearns and AIG and the bankruptcy of Lehman Brothers. It is important, therefore, for regulators to communicate clearly to the public the reasoning and more importantly, the information sources supporting policy decisions and to integrate quantitative and qualitative data carefully when building surveillance and monitoring systems.[*]

REFERENCES

Adrian, T. and Brunnermeier, M. K. (2008) CoVaR., Staff Report No. 348, Federal Reserve Bank of New York, New York.

Allenspach, N. and Monnin, P. (2006) International integration, common exposures and systemic risk in the banking sector: An empirical investigation, Working paper, Swiss National Bank, Zurich.

Azizpour, S. and Giesecke, K. (2008) Self-exciting corporate defaults: Contagion vs. frailty, Working paper, Stanford University, Palo Alto.

Blanco, R., Brennan, S., and Marsh, I. W. (2005) An empirical analysis of the dynamic relationship between investment grade bonds and credit default swaps, *Journal of Finance*, 60(5): 2255–2281.

Chan-Lau, J. A. (2007) Is systematic risk priced in equity returns? A cross-section analysis using credit derivatives prices, *ICFAI Journal of Derivatives Markets* 4 (1): 76–87.

---

[*] For instance, see Haldane et al. (2007); Huang et al. (2008); and Chan-Lau (2008).

Chan-Lau, J. A. (2008) Macro-finance: Old ideas in new clothes? Working paper, Tufts University, Medford, MA.

Chan-Lau, J. A. and Ivaschenko, I. (2001) *Corporate Bond Risk and Real Activity: An Empirical Analysis of Yield Spreads and their Systematic Components*, Working paper, International Monetary Fund, Washington D.C.

Chan-Lau, J. A. and Kim, Y. (2005) Equity prices, bond spreads, and credit default swaps in emerging markets, *ICFAI Journal of Derivatives Markets* 2(1): 7–23.

Chan-Lau, J. A. and Ong, L. L. (2007) Estimating the exposure of major financial institutions to the global credit risk transfer market: Are they slicing the risks or dicing with danger? *Journal of Fixed Income* 17(3): 90–98.

Chan-Lau, J. A., Mathieson, D. J., and Yao, J. Y. (2004) Extreme contagion, *IMF Staff Papers* 51(2): 386–408.

Das, S., Freed, L., Geng, G., and Kapadia, N. (2006) Correlated default risk, *Journal of Fixed Income* 16(2): 7–32.

Das, S., Duffie, D., Kapadia, N., and Saita, L. (2007) Common failings: How corporate defaults are correlated, *Journal of Finance* 62(1): 93–117.

Duffie, D., Eckener, A., Horel, G., and Saita, L. (2008) Frailty correlated default, forthcoming in *Journal of Finance*.

Fama, E. F. (1970) Efficient capital markets: A review of the theory and empirical work, *Journal of Finance* 25(2): 383–417.

Financial Stability Forum (2008) Report of the financial stability forum on enhancing market and institutional resilience, Basel.

Goodhart, C. A. E., ed. (2008) *Journal of Financial Stability Special Issue: Regulation and the Financial Crisis of 2007–08: Review and Analysis* 4.

Haldane, A., Hall, S., and Pezzini, S. (2007) Developing a framework for stress testing of financial stability, Financial Stability Paper No. 2, Bank of England, London, U.K.

Hartmann, P., Straetmans, S., and de Vries, C. G. (2001) Asset market linkages in crisis periods, *CEPR Discussion Paper No. 2916*, London, U.K.

Hawkesby, C., Marsh, I. W., and Stevens, I. (2007) Comovements in the equity prices of large complex financial institutions, *Journal of Financial Stability* 2(4): 391–411.

Huang, X., Zhou, H., and Zhu, H. (2008) *A Framework for Assessing the Systemic risk of Major Financial Institutions*, Working paper, Bank for International Settlements, Basel, Switzerland.

Hull, J., Predescu, M., and White, A. (2004) The relationship between credit default swap spreads, bond yields, and credit rating announcements, *Journal of Banking and Finance* 28(11): 2789–2811.

Koenker, R. (2005) *Quantile Regression*, Cambridge University Press, Cambridge.

Koenker, R. and Hallock, K. F. (2001) Quantile regression, *Journal of Economic Perspectives* 15(4): 143–156.

Lehar, A.(2005) Measuring systemic risk: A risk management approach, *Journal of Banking and Finance* 29(10): 2577–2603.

Linnell, I. K. R., Fahey, E., Burke, J., Abruzzo, T., Batterman, J., Merritt, R., and Rosenthal, E. (2007) CDX survey–Market volumes continue growing while new concerns emerge, Technical Report, FitchRatings, New York.

Longstaff, F., Mithal, S., and Neis, E. (2005) Corporate yield spreads: Default risk or liquidity? New evidence from the credit-default swap market, *Journal of Finance* 60(5): 2213–2253.

Vassalou, M. and Xing, Y. (2004) Default risk in equity returns, *Journal of Finance* 59(2): 831–868.

CHAPTER 24

The Implementation of MiFID in the Financial Crisis Context: An Ethnographic Research Conducted in Greece

Emmanuel Fragnière and Elena Grammenou

CONTENTS

One of the main objectives for the single European financial market represents the integration of markets in financial instruments directive (MiFID) to all European Union (EU) countries legislation. At the same time, the European financial market is crossing a very deep crisis. We can thus wonder how the first implementations of MiFID are impacted by the crisis. Consequently, we have recently conducted an ethnographic research to study the first impacts of MiFID implementation in the Greek financial market. Research findings and analyses indicate that the companies are not well prepared for the change and customers are lacking sufficient knowledge for properly managing this important issue. In addition, part of the staff of multiple enterprises is still ignorant about the subject and its relevant consequences in their everyday work life.

24.1 INTRODUCTION

Financial markets are pivotal to the functioning of modern economies. The more they are integrated, the more efficient the allocation of economic resources and long-run economic performance will be. Completing the single market in financial services is thus a crucial part of the Lisbon economic reform process and essential for EU's global competitiveness.

The objectives of the Commission's financial services policy over the next 5 years are to consolidate in order to obtain an integrated, open, inclusive, competitive, and economically efficient EU financial market. In addition, an important goal is to remove the remaining economically significant barriers, so financial services can be provided and capital can circulate freely throughout the EU to provide consumers with better benefits and protections. Finally, the main intention is to enhance supervisory cooperation and convergence in the EU, deepen relations with other global financial marketplaces, and strengthen European influence globally.

Dynamic consolidation is the *leitmotiv* of the Commission's approach, an approach that is practical, ambitious, and reflective of stakeholder sentiment, and the ultimate goal of financial services action plan (FSAP) initiative is this: single financial market (EU, 2005).

One of the main steps for the single European financial market was the integration of MiFID to all EU countries legislation. Based on an ethnographic study of the Greek financial industry, this chapter presents how effective was MiFID, at this early stage of its introduction, in meeting its primary goals in the Greek market, and how Greek companies took advantage of the opportunities offered by the directive in widening their scope of action.

This chapter is organized as follows: In Section 24.2, we briefly present the context of MiFID. In Section 24.3, we explain the ethnographic methodology employed to understand the first impacts of the MiFID implementation in Greece in the context of the current financial crisis. In Section 24.4, we highlight the main findings we have discovered through this ethnographic research. Finally, we conclude and give some perspectives regarding possible evolutions of MiFID.

24.2 CONTEXT

To understand the MiFID context, we propose a few extracts drawn from the final report of the committee of the wise men on the regulation of European securities markets (EU, 2001). In microeconomic terms, "An integrated European financial market should enable—subject to proper prudential safeguards and investor protection—capital and financial services to flow freely throughout the European Union" (EU, 2001, p. 9). Moreover, "European businesses, large and small, would be able to tap deep, liquid, innovative European capital pools centred around the euro for the financing they require to develop their business activities." Competition and choice would drive down the cost of capital. In short, the supply of European capital from European savings will be efficiently matched with the demand for capital from European businesses. Consumers would be better able to purchase financial services and securities from the best European suppliers of investment, insurance or pension funds, with net yields increasing as investment choice widens. Cross-border clearing and settlement should become cheaper" (EU, 2001, p. 9). On the other hand, "In macroeconomic terms, the productivity of capital and labour would increase, enhancing the potential for stronger GDP growth and job creation" (EU, 2001, p. 9). In the following subsections, we explain the MiFID in more detail.

24.2.1 Introduction to MiFID

The MiFID is one of the final steps in constructing an integrated, deep, liquid, and competitive EU capital market. It should intensify competition, improve standards of service to investors, and benefit companies by lowering the cost of capital. It would update the "single passport" for investment firms, allowing them to operate across the EU on the basis of an effective single authorization and across a wider range of financial instruments. Aside from stimulating cross-border competition, this would diversify the range of offerings that investors can access and markets that firms can tap into.

24.2.2 Scope of MiFID

MiFID would create the conditions, which will allow more integrated and competitive trading infrastructures to emerge. Transparency would be fuller and fairer.

As stated by Commissioner McCreevy (IIEA, 2006), "Investor protection rules will be harmonized at a high level so that investors can feel confident in using the services of investment firms wherever they are in Europe and wherever the investment firms come from in Europe."

In short, MiFID is intended to be an important step forward positioning Europe as a global leader in financial services, providing the necessary foundations for an enhanced overall economic performance. In this way, the European Commission could use financial integration as a stepping-stone to full political and economic integration of the EU (McCreevy, 2006).

MiFID is a wide-ranging legislation that introduces multiple changes; however, these are often summarized into eight major categories, portrayed in Figure 24.1, as follows:

MiFID shrouds all financial instruments except the foreign exchange spot transactions. In addition to the services covered by ISD, MiFID upgrades the consultancy that involves a personal recommendation to a substantial investment service, which can be transferred on a unique basis. Furthermore, MiFID clarifies that operating a MTF is covered by the licence—"passport"—that companies have and extends its scope, for the first time, to commodity and credit derivatives and financial contracts for differences (see Financial Services Authority, 2006, p. 35).

24.2.3 Picture of the Markets after MiFID

The benefits of MiFID can be distinguished to the direct, accrued to firms and consumers directly affected by MiFID, and the indirect ones, arisen to

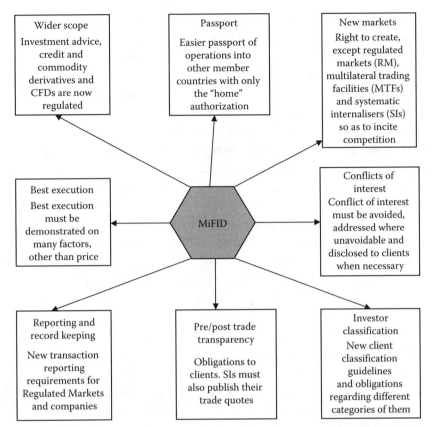

FIGURE 24.1   Key MiFID changes. (Adapted from Ahmad, A., Markets in financial instruments directive: Implications for the UK securities markets and the regulation of hedge funds, Working Paper, available at http://hdl.handle.net/1826/211, Cranfield University, U.K.)

EU economies, because MiFID may lead to a fall in the cost of capital and more efficiently directed investment, and hence to a rise in investment, reduced depth of recessions, increased sustainable growth rate of economies, and consequently higher GDP (Financial Services Authority, 2006).

The abolition of geopolitical barriers, thanks to the MiFID implementation, should encourage competition between trading venues directly by abolishing the special status of incumbent exchanges' prices and the concentration rules that in some European markets forced trading to take place on a particular exchange, levelling the playing field for MTFs and other alternative trading venues, such as dark pools, to compete.

It should support competition indirectly by compelling brokers and fund managers to communicate their execution policies to their clients, in which it is hard for them to make the case for ignoring the existence of multiple execution venues.

But while the brokerage community is all for breaking up the exchanges' monopoly pricing power, competition also raises the spectre of what is known as the "F" word in the equity market: fragmentation. Although brokers are keen to see lower exchange costs, which have spiralled over the past few years as electronic trading techniques have chopped up large trades into many small ones, fragmentation risks complicating the search for best execution and, some fear, the quality of price discovery.

"I think the real test will be to see what happens to total execution costs," says Rainer Reiss, managing director at Deutsche Börse. "Competition is good because it creates a more vibrant marketplace and lowers explicit transaction costs, but the risk is that liquidity fragmentation could lead to a rise in implicit costs, which are far more significant than the explicit costs."

"But the effects of fragmentation here won't be the same as it was in the United States because the market microstructure is different. The exchanges here are much better than they were in the United States when the order handling rules were introduced; their issue is more one of pricing than price formation. You have to consider both the competition effect and the fragmentation effect because the costs of fragmentation will hopefully be offset by the gains from competition. Where we've seen competition in the past the end user has benefited" (Koh, 2007, p. 135).

24.2.4 MiFID Business Opportunities

The new investment regulatory framework of Europe creates multiple opportunities for firms and exchanges to utilize and win a competitive advantage. Some pioneers have already moved on offering innovative services but there is still room for others. First, "As with all revolutions, the shock to the status quo will represent a profitable opportunity for those who are well-prepared—and a death sentence for those who cannot adapt to the new environment. The well prepared are the actors who in the post-MiFID world will generate higher revenue streams, steal market share from the less well-prepared, and begin to compete in areas lying outside their traditional scope of service provision—areas previously closed to them, or deemed to be unprofitable prior to MiFID" (Casey and Lannoo, 2006, p. 1).

Second, MiFID creates the opportunity for exchanges, MTFs, and third parties such as data vendors to compete as aggregators of post-trade and pre-trade data across all European equities. By comparison, the total revenue from data vending reported by the six largest EU exchanges was €458 million (2005), which is 12% of their total revenues. The most important are the Deutsche Börse and the London Stock Exchange, with about €130 million revenues each (Lannoo, 2007).

Third, exchanges can use their technology expertise, network, and trading systems to enable internalizers to provide this business model within the infrastructure of the exchange, as the cost of building such a facility in terms of technological infrastructure, risk management, and regulatory compliance is quite high. Thereby, internalizers are able to outsource the technological infrastructure, to transform fix costs into variable costs, and to use the established networks for routing, execution, and preferencing purposes. Exchanges, on the other hand, may generate revenues based on internalized trades that would be lost if executed within the internalizers' own facilities. Euronext has already developed a new facility that enables users to internalize their orders on the exchange itself using an algorithm that enables clients to give priority to any matching buy-and-sell orders they might have in the exchange's central order book. The innovative service will give users the benefits of internalization—lower clearing costs—without the development costs and obligation to become a systematic internalizer (Koh, 2007).

Furthermore, in the derivatives sphere, MiFID opens the door to direct competition between over-the-counter (OTC) derivatives platforms run by broker-dealers and exchanges—a fight that could get interesting if broker-dealers team up with clearinghouses to offer futures as well.

Finally, a lot of firms are also conducting a systematic review of their clients' profiles to determine if they sit within an investment advisory or execution-only portfolio. This reclassification of client profiles has seen the percentage of investment advisory clients' rise well above 50% at some firms, presenting an excellent opportunity for new advisory service offerings and fee structures.

The financial institutions making the biggest headway with their investment advisory offerings are using software that integrates their key front office activities such as profiling and planning, establishing objectives, setting asset allocation, generating investment proposals, implementing solutions and monitoring progress. In these instances, a single front office platform has become an enabler of product innovation (Global Investor, 2007).

24.3 RESEARCH DESIGN

The FSAP was designed for the creation of a single financial market in Europe and had three strategic objectives, namely, establishing a single market in wholesale financial services, making retail markets open and secure and strengthening the rules of prudential supervision. The financial integration of EU would be the first step for its political and economic integration, for which strives many years.

MiFID is a new regulatory framework within the general concept of FSAP and there is no extensive research on its implications, benefits, and opportunities or threats in the markets, firms, customers, and national economies. During the design process, everyone was very optimistic about it and awaited numerous positive changes.

Moreover, its implementation phase for all countries converged side by side with the recent global credit crisis and its implications on exchanges. The crisis has reminded market participants of the intrinsic qualities of trading on exchanges. It has reminded also of the value of regulated and transparent market environments. During this period of high volatility and uncertainty exchanges react differently, according to their power.

24.3.1 Research Question

The purpose of this study is to identify how effective was MiFID at this early stage of its introduction in meeting its primary goals in the Greek market in a context of global financial crisis, and how Greek companies took advantage of the opportunities offered by the directive in widening their scope of action. In this context, the research objectives can be formulated as follows:

- To identify the response of Greek firms to MiFID in meeting its requirements

- To obtain a deep understanding of the main strategic challenges and constraints of MiFID implementation

- To recommend ways on how Greek firms can utilize the new regulatory framework

Taking the research objectives into consideration, the research problem can be formulated as follows:

- How did the Greek financial market responded to the implementation of MiFID and how did they utilize the business opportunities that turned up?

In the context of the research problem, the study will also explore the main strategic challenges and opportunities for Greek companies. The research aims to enhance academic knowledge in the Greek capital and money market, its evolution through a pan-European directive, and its future prospects under the same or different structure.

24.3.2 Research Methodology

The philosophy of "interpretivism" was chosen as the most appropriate for the scope of this research. Its main objective is to understand how the market environment, has changed after the implementation of the directive, the perceptions of market participants under the new market conditions, and the way they took advantage of the emerged opportunities.

The research objectives—on a theme that it was very new, without much scientific research—were to identify the response of Greek companies to MiFID and to obtain a deep understanding of the main strategic challenges and constraints that its implementation has risen. The objectives required a deep understanding of the market conditions and the perceptions of different market participants within the new context, to be able to answer the research question, which could only be treated after the data collection. The inductive approach was preferred as the most suitable one for this research, taking into consideration the aforementioned elements.

The research strategy followed was the ethnography one. Its purpose is to describe and explain the social world that the research subjects inhabit in the way in which they would describe and explain it. It is a very appropriate strategy in business, if the researcher wishes to gain insights about a particular context and better understand and interpret it from the perspectives of those involved (Saunders et. al., 2007). This approach is well suited for situations facing deep structural changes like the current financial crisis.

This research constituted a vehicle for studying the new market that was formed, the opportunities that revealed for Greek firms and their new role in the new pan-European market. Greece is a country with several differences in the financial sector in relation with other EU countries and would be very interesting for a researcher to study how its financial sector's market participants would adopt, manage, and strategically benefit from the new rules of the game.

The data collection methods that were selected for the research included semistructured interviews with industry experts, unstructured interviews with customers and employees, as well as secondary data from several publications, reports, and special editions. Practically,

16 interviews of Greek banking experts, practitioners, and professional clients were conducted during summer and fall of the year 2008. The questions were open-ended, in order to allow interviewees to exhibit their views and perceptions and provide extensive answers. The set of questions was identical for all interviews, so as to be driven in comparable findings across interviews, with some additional questions based on the nature of interviewees.

24.4 FINDINGS AND DISCUSSIONS

In this section, we present and interpret the main findings of this empirical research. The different findings are grouped around the following themes: cost issues, client categorization, best execution, pre- and post-trading reporting, organizational adaptation of companies, competition in markets, and business opportunities.

24.4.1 Cost Issues

The findings of this research study provide evidence that the whole cost chain is increased and not decreased, as all the studies have reported. Investment companies, banks, and clients are complaining for the costs. The adaptation cost is quite high for all companies and most of them transfer the cost to clients. Clients, on the other hand, are complaining for the higher costs but they don't know that one of the main announced advantages of MiFID directive was cost reduction. They consider it as a logical outcome of the change. In addition, some firms not only try to cover their costs but also to regain the investment they have made for the implementation requirements.

Although all studies support the minimization of transaction and compliance costs—like Koh (2007) "MiFID lowers transaction costs," EU (2008) and McCreevy (2006), "MiFID will drive down cost of capital," FSA (2006) "…thus reducing costs for consumers… as a consequence, transactions costs may fall"—at this point of time, the study supports exactly the opposite. All the market participants are complaining for the costs and have not perceived any positive change on this issue.

24.4.2 Client Categorization

MiFID requires companies to categorize clients according to three different classifications: Retail clients, Professional clients, and Eligible Counterparties. Each category requires different level of protection and service. The findings of this research demonstrate many mistakes.

Although most of the firms have adopted developed software systems to do the categorization, some of them did it roughly or with wrong criteria and risk of damaging the firms' corporate brand and credibility, which are very important for this industry. Three out of five customers validated were very cautious about the future of their relationship with these companies. Some of them were more educated on dealing with such issues but others just changed firm, without even talking to them. The preservation of a company's brand image is really important and word of mouth expands a bad rumour twice as faster than a good one. Brand image is very important for companies under a high risky and volatile economic environment.

The client categorization is in line with the previous research findings—FSA (2005) "new client categorization rules with respect to opting up from certain categories also seek to increase investor protection especially for some instruments"—but theory has not referred on such issues. It is indeed important for the customers' protection but they have not estimated the mistakes presented, either on purpose or not, so as some companies to save costs of providing limited information to professional clients.

24.4.3 Best Execution

Best execution policies and procedures were an issue quite interesting to explore and had many differences across companies. Although MiFID requires companies to take into account "price, costs, speed, likelihood of execution and settlement, size, nature or any other consideration relevant to the execution of the order," when they executed orders, something relevant was not encountered. It is easily understood that best execution policies cannot be the same for equities, derivatives or foreign exchange transaction, but the research findings did not manage to confirm such a view. The experts, who were questioned, referred to only one best execution policy. Best execution has forced many companies to restructure their relationships with customers and design detailed procedures for their internal operation. In addition, there was a big confusion in Greek market about constant means of reporting transactions to customers, which made some firms to spend a lot of money on software that after a while proved to be worthless.

24.4.4 Pre- and Post-Trading Reporting

The finding of this research could not figure out what is happening in the Greek market about this issue. The OTC and SI transactions are not reported yet somewhere officially, and the market can have a view of these transactions only after the clearing phase. The transactions made to the Athens Stock Exchange (ASE) are reported automatically. Regarding all the other transactions, this research did not manage to decipher where they are reported. The only issue that came up is the re-registration of transactions to several information providers, which put their reports under question about their credibility and accuracy.

24.4.5 Organizational Adaptation of Companies

MiFID requires highly organized and transparent companies. The Greek interpretation of the directive forced mostly all companies to reapply for an operation licence, which created additional overload and strain to the industry participants. Most of them had to organize executive board meetings, change all their legal documentation, design new organizational structures, educate staff, implement new internal systems and procedures and be stricter on internal policies. In addition, one of the most important issues of the directive, transparency, was not supported by the findings. All markets participants have not sensed any increase in market and information transparency, but on the contrary they have all reported a decrease. Transparency affects the order flow and the price discovery process, but the proliferation of nontransparent venues and transactions may question or endanger the dominance and viability of transparent markets, one of the main goals of MiFID.

All companies have developed special departments—especially the bigger ones—or have hired competent people to deal with internal audit, compliance, anti-money laundering, and risk management issues. Some of the large companies have official operational procedures but some of the smaller ones are not ready yet on this aspect. They are based more on relationship building than official operations. Internal systems and controls are mostly in place for all companies and their record keeping policies are quite rigorous. The only element that was not supported by the research was the outsourcing of functions of investment services. Most companies are trying to preserve their market share and work with specialized providers, to secure the quality of services supplied, so as not to endanger their clientele and brand image.

Customers, on the other hand, did not have a hint about multiple trading venues and the accuracy of the data reporting they had. Even the most experienced ones knew nothing about Chi-X, Turquoize, or similar platforms and the issue of transparency. Transparency is more an issue for traders and brokers than customers, although it was designed for customer's benefit. Markets are still confused and companies have not yet found ways to solve the data vending, in order to have full view of markets and take advantage of the opportunities offered.

24.4.6 Competition in Markets

The research findings indicate that the industry expects competition to intensify and either they are afraid of the market conditions that will come and how they will survive or they consider it a plus for the market outlook and the products/services offered. On the retail part, the competition is greater because most of the firms have understood that their income will decrease and they are looking for partnerships, which will supply them with different customer basis, so as to keep their current cash flow. Another issue is the accessibility of small and medium players to the platforms and consortia that will be created. Whether these markets will be open or not is a very important issue for their development and the consequences it will have on investments. The more players are in a market the better outcomes can come up from their competition.

Finally, there is a debate about which markets the new directive will enhance, the big or the small markets, and what would the resulted reaction be. In big markets, the presence of many players is beneficial for all elements of competition—like price, cost, supply, etc.—but in small markets, the split of liquidity will might result in more nontransparent and maybe "dangerous" situations. On the other hand, the fragmentation caused in markets might jeopardize the quest for best execution and quality of price discovery or increase the implicit costs for clients. The research findings have shown that many of these alternative markets are not working yet and companies are still afraid to take advantage of "passporting" benefits. In addition, costs have raised and especially the intrinsic ones, while at the same time, the transparency of market is disputable. Companies have not experienced yet either the advantages or disadvantages of competition and for sure clients have not profited yet; on the contrary, they came across with the bad elements of competition, increase in costs and nontransparency of markets.

It is undeniable that vigorous competition is the lifeblood of strong and effective markets. Competition enables consumers get a profitable, full-informed deal. It encourages firms to innovate by reducing stagnancy, putting downward pressure on costs and providing incentives for the efficient organization of service production. The overall impact of increased competition should be an improvement in general economic prosperity. However, as the research findings have indicated, none of these elements are observable yet.

24.4.7 Business Opportunities

The MiFID directive is indeed a business challenge, but at the same time constitutes a cornerstone for the development of markets. The liberalization of markets and the abolition of borders create multiple advantages for market participants. The research findings of this study demonstrate that in Greece, although companies have identified some interesting opportunities, they have not taken advantage of them.

The first step was the creation of special departments for investment advisory and portfolio management with competent staff and luxurious physical evidence (e.g., nice offices and meeting rooms), to attract customers, who require such service and are very willing to pay more. It is the added value for companies on which they can invest for increasing their production and income generation. In addition, a big opportunity for Greek companies is the development of algorithmic trading and smart routing systems, to provide better service delivery and product variety. Such systems and software are not utilized yet by most of the companies, but the infrastructure is ready and waits for market participants to exploit it, before more informed and developed competitors enter the Greek market.

Moreover, all companies are much willing to participate in all alternative platforms, if there would be a chance for them and if the profit–cost analysis is positive. They are facing high costs and are welcoming any initiative of offering the same services with fewer costs. The only issue is that all these initiatives have not started yet, with the exception of two and it is difficult to judge the offers and benefits for market participants.

Most companies are more focused on the compliance part of MiFID rather than the business part. They have made some steps toward their business development but there are a lot more to be done, so as to be competent. Most of the time, opportunities come along with threats, which, if not confronted in the right way, might cause problems to the general economic environment. It is more than obvious that the issues that most likely create problems to companies are the conflicts of interest, best execution policy, and client categorization. Judicial conflicts will be raised most likely by these issues and companies have to be prepared for such contingencies. Although most of companies are trying to strictly follow the guidelines of directive, no one can be confident about the end results.

Furthermore, the time MiFID emerged in the markets was not the right one. The credit crisis has caused many problems and people confuse the outcomes of this crisis with the ones of MiFID. The uncertainty and volatility of markets make people insecure and afraid of investment services. Markets are questioned, have lost their value due to the lack of transparency and general financial conditions. Transparency—one of the core goals of MiFID—plays a very important role but the research findings could not support it. People need to be certain about their investment movements.

Most of the investment companies have to face higher operational risk. Client categorization, best execution policies, and internal systems and procedures increase the operational risk of companies together with the legal risk. Companies have to be very cautious about dealing with nonfinancial risk, which may cause serious problems. To conclude the discussion of research findings, MiFID directive in the Greek financial industry has a long way to go until it reaches its final goal: the creation of a pan-European open market, where competition, transparency, information distribution, costs and investment prospects will be better. Its implementation phase and adaptation would be better if the whole concept was built more on "managing the change" issues, rather than managing a compliance issue. It is not only about compliance; markets are transformed and market participants should accept, discuss and participate in this change, so as to raise the possibility of success and to accomplish the final goal, which puts aside many strategic, economic, and geopolitical interests of countries.

24.5 CONCLUSION

In the turbulence of the current financial crisis and the acquisition, conservatorship or bankruptcy of many major bank institutions and companies, even those supported or financed by governments, a concluding comment should be made about an important European directive concerning financial markets, MiFID.

MiFID is the cornerstone in the EU plan to become the most competitive market in the world by 2010 and, as such, is the broadest regulation to be introduced to Europe. MiFID aims at securing investor protection and at eliminating distortions of competition. Market infrastructure providers are expected to face competition from new trading venues and other third-party services. The aim of this research project was to identify how effective was MiFID at this early stage of its introduction in meeting its primary goals in the Greek market and how Greek companies took advantage of the opportunities supplied by the directive.

For the most part, the findings of this research have not established evidence for most of the main goals of the directive, either because of the general financial crisis in markets, or because the Greek financial industry have not managed the whole adaptation and implementation properly, by utilising all the related developments and opportunities. On the other hand, maybe, it's too soon to judge the effectiveness of such a huge change in the Greek market, taking into consideration the general global financial scene during the last year. A prediction for the future outlook of markets, in general, would be the one that follows.

The current crisis is the worst financial one since the Second World War, and it will have a definite effect. We've seen a substantial loss of wealth as a result of the decline in the value of property and bank shares. Clearly, the banks' balance sheets are going to have to contract. Less credit will be available to the real economy, and it will be dramatically harder to borrow than before. This is not just a sub-prime mortgage crisis, this is the crisis of an entire sub-prime financial system, and at the end of the day it will imply credit losses of at least $1 trillion and more likely $2 trillion. The rest of the world will not decouple from the US recession. Already 12 major economies are on the way to a recessionary hard landing. All of the G7 economies are now entering recession, while the rest of the world will experience a severe growth slowdown.

Both in Europe and in the United States, exchanges are consolidating, while the EU is looking for new ways to facilitate cheaper and better services, transcending national borders. There is no question that consolidation will continue with more linkages between exchanges. As we look to the future challenges ahead, it is clear that effective cross-border trading lies at the top of the agendas of most exchanges. The process of change has begun gradually, but it is now rapidly accelerating.

Assessing the impact of MiFID at this stage, one can conclude that it has definitely promoted new market activity. It remains to be seen whether it will promote genuine innovation, and increase competition and enhance market efficiency. What is undeniable is that MiFID will fragment Europe's markets and this may come at a cost. Fragmentation could lead to worse prices for the investors and lower the efficiency of raising capital in Europe. Fragmentation could also make it more difficult to supervise the markets and to fend off market abuse. Going forward, regulators and supervisors should definitely concentrate on the overall market quality.

If we look beyond these immediate new challenges, one thing becomes clear: the current credit crisis affecting world markets for over a year now has revealed the systemic importance of liquidity and transparent price formation processes. The structure of certain markets and the instruments they trade were at the origin of the crisis and augmented its effects (Capralos, 2007).

One of the most important long-term consequences of the current crisis will be the loss of investor confidence across all markets. This will inevitably spill over to the "real economy," hitting listed companies all over the world due to deteriorating market conditions. As with previous crises, investors' appetite for equity investments could diminish and impact overall capital formation and investments. This is why regulators across the globe should resolutely focus on improved and efficient investor protection. The EU in particular has traditionally focused on creating a level playing field and on increasing competition. It is now time to elevate investor protection to a policy priority alongside competition.

The supporters of capitalistic model and EU will now be very surprised with the states' interventionism, which we all witness every day. In the cradle of capitalism the government intervenes with a $700 billion proposal to support the financial system. In many countries of the EU, governments intervene with the same goal in mind, while it is reciprocal to the European regulations and the vision of Europe. Whether the bailouts will be better or not, it remains to be seen.

In addition to stricter regulations and intense state oversight, there was a constructive dialogue between EU and SEC for developing a greater transatlantic integration that may trigger multiple benefits, like price reductions, increased investment opportunities for both EU and U.S. investors, easier access to capital market for issuers on both sides of the Atlantic as well as cross-fertilization of market infrastructures. It remains to be seen how such a close cooperation and linkage will affect market operations during good and bad times, like the one we are experiencing today. Moreover, it is undeniable that in the long-run stock exchanges will operate 24/7, when there will be an online common clearing system—an initiative, which is under study and development from multiple market representatives.

The current credit crisis is a good opportunity for national governments, international institutions and market players to study their mistakes and build a new competitive framework, which will protect people from illegal activities and national economies from experiencing such hazards. The greed has never been a good advisor for designing strategies and proposing solutions. In other words, we borrow the earth from our children, and, at this moment, we are robbing them at all aspects. The world has to reflect and move on to big changes. Stronger rules, institutions, and ethics are essential for preventing business from cheating. Global income needs to be redistributed on a more equitable basis, and although it might be tough and demanding, the alternative offered is unthinkable. Responsible business attitude could be an initial step forward for ensuring that the current crisis was indeed not wasted.

REFERENCES

Ahmad A. (2007) Markets in financial instruments directive: Implications for the UK securities markets and the regulation of hedge funds, Working Paper, available at http://hdl.handle.net/1826/211, Cranfield University, U.K.

Capralos I. Spyros, (2008) Latest developments in the European exchanges industry, *5th International Capital Markets Conference*, September 2008, Thessaloniki, pp. 1–6, available at http://www.ase.gr/content/gr/announcements/Files/Spyros%20Capralos_speech_11092008_12.pdf.

Casey J.P., and Lannoo K. (2006) The MiFID revolution, policy brief, available at http://.shop.ceps.be.

European Commission (2001) Final report of the committee of the wise men on the regulation of European securities markets, available at http://ec.europa.eu/internal_market/securities/docs/lamfalussy/wisemen/final-report-wise-men_en.pdf, February.

European Commission (2005) White paper on financial services policy (2005–2010), December, available at http://ec.europa.eu/internal_market/finances/docs/white_paper/white_paper_en.pdf.

European Commission (2008) The final report of the committee of wise men, available at http://ec.europa.eu/internal_market/securities/docs/lamfalussy/wisemen/final-report-wise-men_en.pdf.

Financial Services Authority (FSA) (2005) Planning for MiFID available at http://www.fsa.gov.uk/pubs/international/planning_mifid.pdf.

Financial Services Authority (FSA) (2006) Overall impact of MiFID available at http://www.fsa.gov.uk/pubs/international/mifid_impact.pdf.

Global Investor (2007) The power of article 19, Global Investor, November, Issue 207, p. 5.

Koh P. (2007) Changing the rules of THE GAME, *Euromoney*, November, 38(463): 134–136.

IIEA–Events MiFID Conference (2006) Keynote by commissioner McCreevy, available at http://www.iiea.com/eventsxtest.php?event_id=179, June 30.

Lannoo K. (2007) Financial market data and MiFID, BM, available at http://shop.ceps.eu, March, 6.

Mccreevy C. (2006) Financial markets integration, Centre for European Reform, available at http://ec.europa.eu/, March.

Saunders, M., Lewis, P., and Thornhill, A., (2007) *Research Methods for Business Students*, 4th edn., Harlow Pearson Education, Upper Saddle River, NJ.

III

Preventing Banking Crises, Bank Runs, Regulation, and Bailouts

CHAPTER 25

Credit Derivatives and What Happened Next: Analysis and Recommendations

Bastian Breitenfellner and Niklas Wagner

CONTENTS

The 2007–2008 credit crisis not only vastly affected the financial system, but it also has severe ongoing effects on the global economy. Consequently, it is of particular importance to understand what actually triggered the financial collapse and how such a collapse could have been prevented. We take a view from the credit derivatives perspective and explain what led to the crisis. We describe the instruments fostering the instability of the financial system and show how the collapse was triggered. We then comment on the recent measures of short-term government intervention, which aim at limiting the acute damage to the financial and economic system. Finally, we suggest relevant areas for improved long-term financial regulation.

25.1 INTRODUCTION

The recent 2007–2008 credit crisis has been of economic policy concern at least since early spring 2007. At this point in time, several low-grade (subprime) lending institutions serving the U.S. housing market faced severe financial distress. The subprime crisis then spread throughout the global financial system in an unprecedented way. As of October 2008, the Bank of England (2008) reports the total volume of government support packages set up in order to support the distressed financial system amounts to approximately €5.55 trillion. This and the ongoing efforts of central banks around the world clearly reveal the vulnerability of the financial system in its current form. Hence, it is of importance to understand what actually triggered the collapse of the financial system, and how such a collapse might be prevented in the future.

Our purpose here is to explain what led to the current crisis, and which conclusions can be drawn from it. We describe the instruments fostering the instability of the financial system and show how the collapse of the financial system was eventually triggered. We then comment on the different possible means of government intervention. These short-term actions intend to limit the acute damage to the financial system. The crisis also calls for improved regulation. We suggest areas, which appear particularly relevant to long-run government intervention.

Several authors have addressed the 2007–2008 credit crisis so far. Shiller (2008) traces the roots of the crisis back to the bubble in the U.S. housing market. One of the consequences of the probable existence of bubbles in asset prices is the situation of severe misallocation in the economy. Corrections of such bubbles lead to costly financial and economic crises such as the current credit crisis. In line of such reasoning, we would argue that financial economists arrive at a very critical question: Can we trust market prices? Even less than 10 years ago, most experts would have claimed that we can, as markets were and still are supposed to be informationally efficient. Today, the unsatisfactory answer seems to be that—during some periods or in some states of the world—we would be better off not to trust market prices.

Apart from such fundamental considerations, it is obvious that short-run measures that adequately address the crisis are of foremost importance today. Despite this, the amount of literature commenting on acts of government intervention, including the recent global bailout programs, appears small. Three papers on this issue include Hoshi and Kashyap (2008), Bebchuk (2008), and Breitenfellner and Wagner (2009).

Hoshi and Kashyap (2008) investigate government intervention during the recent Japanese financial crisis. Given this experience, they draw conclusions for the design of the troubled asset relief program (TARP) in the United States. They argue that buying distressed assets is an appropriate way to recapitalize banks. Nevertheless, they conclude that the Japanese program lacks efficiency, as assets cannot be purchased for more than their economic value and hence, the total amount of assets purchased remains low. Therefore, no capital is rebuilt and the system remains undercapitalized. Hence, the authors propose that besides buying distressed assets, government assistance should also be conducted via direct equity injections. Bebchuk (2008) comments on the design of the TARP emergency legislation. He agrees that asset purchases are suitable to cope with the financial crisis, nevertheless he proposes a redesign of the legislation in order to achieve the targets of the program, that is, restoring stability in the financial system, while limiting costs to taxpayers. He argues that the possibility to overpay for certain assets is not in the interest of taxpayers. In order to prevent undercapitalization, he rather advocates allowing the purchase of securities newly issued by troubled institutions. Additionally, he argues that financial firms should be required to raise additional capital from their existing shareholders. Breitenfellner and Wagner (2009) suggest a simple model of government guarantees and asset purchases. They argue that insolvent banks should face bankruptcy even under a bailout program. In their model, only rescue packages including a purchase program for distressed assets create a setting where illiquid but otherwise solvent banks can be separated from insolvent banks. We will address some of their findings in Section 25.4.

The remainder of this chapter is structured as follows. Section 25.2 outlines recent developments in the market for credit risk, which eventually led to the crisis. In Section 25.3, we discuss the breakdown in the financial system. A discussion of some of the relevant methods of short-term government intervention is given in Section 25.4. Section 25.5 provides a brief conclusion.

25.2 THE MARKET FOR CREDIT RISK

The market for credit risk has grown very rapidly since the early 1990s. It seemed to be one of the biggest success stories in the history of financial intermediation. The new paradigm was based on the assumption that underwriting and bearing credit risk is perfectly separable. Banks could transfer credit risk with hardly any constraints to those seeking exposure in certain credit risks. On the other hand, any player in the financial system would be able to gain exposure in the credit risk of certain entities, without direct involvement with the respective entity or even without upfront capital outlays.

The tools for credit risk transfer are numerous, among which residential mortgage backed securities (RMBSs), credit default swaps (CDSs), and collateralized debt obligations (CDOs) are the most prominent. Details on credit derivatives products and markets can, for example, be found in Batten and Hogan (2002), Scheicher (2003), and Chaplin (2005). The reason behind risk transfer activities is quite obvious. Financial institutions are able to specialize on certain segments of the banking landscape. For example, institutions with no expertise in the lending business are able to gain exposure in any kind of credit risk. On the other hand, originators are able to eliminate large positions from their books, by passing them to other market participants. In turn, the relieved capital can be used to grant new loans. This development paves the way for new cash flows to credit markets, allowing the whole economy, as well as the public to profit from eased funding opportunities, which would not have existed without the risk transfer.

The traditional way of transferring credit risk is by means of securitization. In a typical securitization transaction, the originator of a credit portfolio sells his credit portfolio to a special purpose vehicle (SPV), which is refinanced via capital markets. Although the assets transferred to the SPV do no longer occur on the originators balance sheet, the ties between the originator and the SPV are manifold, for example, through swap agreements or guarantees. Securitization transactions have many advantages for the originator. The proceeds from selling the loan portfolio can readily be used to grant new loans. Therefore, securitization can be regarded as a form of refinancing. Among the other advantages are the transfer of credit risk to the SPV (and eventually to investors), and lower regulatory capital requirements for the originator. See, for example, Gorton and Souleles (2005) and Bluhm and Overbeck (2007) for more details on SPV transactions.

Overall, securitization transactions clearly augment lending capacities within the financial system and improve allocational efficiency of financial markets, concerning both funds and exposures. In turn, optimal allocation of resources helps to reduce the cost of credit (see, e.g., Duffie (2007)). Banks may use credit protection in order to hedge exposures in the credit market. Hedging by means of credit protection has an effect on the regulatory equity as required by the Basel II accord. Credit exposures which are hedged via credit protection transactions are no longer subject to regulatory capital requirements. Only the swap itself is accounted for, which reduces regulatory capital requirements as long as the swap carries a lower risk weighing than the underlying (see Basel Committee on Banking Supervision (2004)).

Unfortunately, the number of high-quality obligors in the system is limited. Hence, the overall proportion of low-quality creditors will increase with the volume of lending. At this point, a disadvantage of securitization becomes obvious. Since the originator may eliminate all the credit risk associated with the loan portfolio, he will not be overly concerned with the quality of his obligors. Consequently, there will be loans included in the portfolio, which would not have been granted by the originator, if he still had to account for them. This is a classical adverse selection problem. The deterioration of the average loan quality is further amplified by the incentive schemes within the lending business, where employee compensation largely depends on lending volume rather than risk-adjusted return, see, for example, Mills and Kiff (2007).

On the other side, the SPV tries to obtain maximal funding from selling securities on the capital market. In turn, the proceeds are transferred to the originator as a compensation for acquiring the loan portfolio. Hence, the SPV has an incentive to overstate the quality of the loan portfolio, a moral hazard problem, since the investors buying the SPV's bonds and commercial papers will typically have an information disadvantage concerning the quality of the loans contained in the portfolio. The complexity of many securitization transactions adds to this information disadvantage. The overall quality of the loans underlying the securitization transaction declines with every new transaction, as the amount of high-quality borrowers in the financial market is limited. However, the capital inflow due to the securitization transaction will tempt the originator to grant further loans, despite the lower quality of obligors seeking debt financing via loans. The overall situation clearly reveals problems of adverse selection and moral hazard on the sides of the loan originators.

It is notable that the empirical results on the pricing adequacy of credit derivatives pricing models appear to be relatively scarce in the literature so far. Additionally, pricing performance evidence even for simple CDS contracts appears to be mixed. This is supported, for example, by the empirical findings of Stewart and Wagner (2008) for the U.S. CDX market. One of the many reasons of partially weak pricing performance surely is that counterparty risk—as discussed, for example, in Läger et al. (2008)—was typically neglected. Studies on the empirical pricing results for models of complex products such as CDOs include Bourdoux et al. (2008), Leijdekker et al. (2008), Longstaff and Rajan (2008), and Moreno et al. (2008).

25.3 THE COLLAPSE OF CREDIT MARKETS

So far, we argued that the market for credit risk is to the benefit of the economy. So why do we experience such a devastating crisis? The answer is quite simple. There has been a lack of risk awareness among overconfident market participants. In mid-2007, market participants believed that advances in credit risk transfer would prevent severe losses on highly rated tranches and credit derivatives, see, for example, Mills and Kiff (2007).

As stated above, securitization transactions were widely used by originators to eliminate credit portfolio risk from their balance sheets. One hypothesis would be that the notion was that loans which are off-balance sheet do not contribute to the institution's risk profile. This proved to be rather myopic. While the number of securitization transactions increased, the quality of the loan portfolios declined. Those who invested in the tranches of the securitization transactions often were unaware of the inherent risk and relied on the external assessments of rating agencies, which in many cases were overly optimistic. After the burst of the housing bubble, more and more loans defaulted. Consequently, those who invested in securitization transactions incurred severe losses on their tranches, in particular on the first loss piece, and therefore had to write down their investments. This in turn put investors on the spot to liquidate their positions in order not to run into overindebtedness. Additionally, the growing uncertainty concerning the actual risk profile of the securitization tranches led to an erosion of liquidity in the secondary market, resulting in enormous discounts on the tranches, if they were sellable at all. This created a vicious circle in the market for securitization tranches, which consequently collapsed. This collapse led to further write-downs on financial institutions' assets, which absorbed much of the financial system's liquidity. So what happened as a consequence, although exposures due to the loan portfolios securitized were not on the banks' balance sheets in the first place, they finally got there through the back door. As a consequence of the rising illiquidity within the financial system, financial institutions ran into refinancing problems. The fact that many financial institutions heavily relied on short-term refinancing, while being highly leveraged, further boosted the crisis. What made things even worse was the psychological effect of a loss of trust in the banks and in the overall financial system. To avoid overindebtedness, all kinds of assets had to be liquidated and so the crisis spread throughout the system. The contagion effect was enormous, spreading throughout institutions, as well as markets and continents.

25.4 GOVERNMENT INTERVENTION

Given the above, several questions arise. What are the lessons learned from the current crisis? What has gone wrong and how can such failures be prevented in the future? Is there a need for stricter regulation in the financial system? It appears to us that the following areas preserve particular attention.

- The design of *rescue packages* as a means of short-term intervention
- Several means of long-term intervention including
  - An improved internal risk management
  - The regulation of securitization transactions
  - The creation of incentives for long-term profitability

We discuss these points in the sections below.

25.4.1 Rescue Packages

Rescue packages or government bailout programs raise an important question. Should financial institutions in distress be rescued by the government and consequently by tax payers? On the one hand, rescue measures seem appropriate in case the bankruptcy costs for the whole economy exceed the costs of the rescue. On the other hand, with a government as the lender of last resort, there is little incentive for financial institutions to pursue sophisticated risk management strategies. In contrast, a moral hazard problem arises. The incentive would be to increase the overall risk profile of the institution in order to obtain a higher expected payoff for shareholders. With a lender of last resort, shareholders are then equipped with a put option written by the government, generating an incentive to increase the risk profile of the firm at the cost of the government.

In this light, Breitenfellner and Wagner (2009) argue that state guarantees alone do not seem to be the appropriate measure to rescue banks. In case a rescue is inevitable it should be perused with the additional help of capital injections in order to avoid principal agent conflicts. In case the rescue of a certain financial institution is inevitable, the measures of assistance should come at a significant cost for the respective institution. Otherwise, the rescue packages could encourage institutions to rely on them as a cheap source of funding. Given this, a suitable way to support the financial system seems to be to recapitalize secondary markets for certain products such as RMBSs. This is done by "equipping" them with liquidity, that is, through asset purchases at a discount to the economic value of the products. As the economic value of the assets is supposedly above their current market value on average, this strategy has major advantages. First, only those financial institutions with severe liquidity problems will be willing to sell assets, which are currently undervalued. Second, the government may profit from the expected higher payoffs from those assets in the future. Third, only this setting allows to distinguish between illiquid but solvent and illiquid but insolvent financial institutions.

25.4.2 Internal Risk Management

The absence of proper risk awareness among market participants is surely among the basic causes of the current crisis. As argued above, market participants may assume that various risks can easily be eliminated through instruments like securitization and credit protection. What has been overseen is the fact that there are risks besides credit risk, which cannot be completely eliminated, including market risks, liquidity risks, and counterparty risks.

The amount of bad loans in the economy does by no means justify the enormous volume of financial products which are labeled as "toxic waste" during the recent period of market stress. According to Bank of England (2008) estimates, about 37% of the mark-to-market losses on U.S. subprime RMBSs can be attributed to discounts for illiquidity and uncertainty rather than actual credit risk. Hence, the essential problem is that there no longer exists a market for these products due to a lack of liquidity as well as due to a lack of trust. This fact, in connection with the fair value accounting principle, causes serious write-downs on investments in those assets, although these write-downs might only in part be driven by a lack of quality of the product itself. The possibility of such an erosion of secondary markets has obviously not been taken into account. This is clearly a failure of internal risk management within financial institutions. In this light, one has to discuss the question, whether or not this failure can be prevented by stricter regulation. A possible solution may be to require financial institutions to hold an increased equity cushion in order to absorb losses due to market and liquidity risk.

Admittedly, it appears impossible to hold equity cushions to absorb all potential losses due to market or liquidity risk. The October 2008 market meltdown clearly represents some sort of tail event, which cannot be fully absorbed. Nevertheless, it seems doubtful that we would have entered a state of market meltdown, if all market participants had at least provided enough capital to absorb moderate losses due to market and liquidity risk.

25.4.3 Securitization Transactions

In the public discussion, securitization is often blamed to be "the root of all evil," leading to the meltdown in the financial system. Given this, what consequences should the crisis have for the securitization market? Is there still room for securitization transactions, or do the drawbacks of securitization outweigh the advantages discussed above?

The advantages of securitization transactions do warrant the risks inherent in securitization transactions. However, the system has to undergo certain changes to avoid a collapse. As argued by the International Monetary Fund (2003) and Franke and Krahnen (2008), there currently is a misalignment of the incentives of the different counterparties in a typical securitization transaction. As the originator may eliminate the overall credit portfolio risk from her balance sheet, she has little incentive to assure certain minimum quality requirements for loans contained in the securitized portfolio. This can easily be prevented by requiring the originator to retain a certain share of the transaction, preferably a fraction of the first loss piece, on her books. This would ensure that no "toxic waste" is contained in the loan portfolio, since the originator is directly exposed to its inherent credit risk. Furthermore, the originator should be required to publicly declare the share and the tranches of the transaction he/she retains. This signal can be used by investors to assess the risk associated with a certain transaction. The optimal size of the share of the transaction the originator is required to hold is subject to further research. On the one hand, if the share is too small, the alignment of incentives is not accomplished. On the other hand, if the share is too large, the whole securitization transaction becomes unattractive for the issuer, since other forms of refinancing, for example, the issuance of covered bonds, become more rewarding. This terminates the positive effects of securitization transactions for the economy. Requiring the originator to retain a certain share of the transaction has another positive effect, as it also limits the overall volume of loans granted. In case the originator retains a certain share of the transaction, she expends her balance sheet, which would not be the case if she fully passes on all the tranches of the transaction. This automatically limits the number of loans he/she can grant as he/she cannot (at least she should not) exceed a certain level of leverage. This assures that the quality of obligors does not decrease arbitrarily.

Another important issue, which led to the financial crisis is the enormous complexity of certain products in the market for credit risk. This complexity not only hampered investors' assessment of the risk associated with those products, even rating agencies were not able to specify and measure the risk underlying certain transactions. Unfortunately, many investors relied on ratings which proved to be overly optimistic, an issue where rating agencies played a major role. Nevertheless, it is astonishing that investors solely relied on ratings and invested in products they obviously did not fully understand. The only explanation is that sophisticated risk assessment and management was sacrificed on the altar of irrational return expectations.

25.4.4 Long-Term Profitability

The above discussion leads us to another trigger of the financial crisis, namely a set of inappropriate management incentives. Performance figures such as the Return on Equity (ROE), among others, do not seem to be sound target figures for a financial institution. Instead of rewarding sophisticated risk management within the institution, they rather induce managers to increase leverage and to pursue a more risky business model. This strategy may yield sound performance figures in the short run, but does not necessarily promote the long-term stability of an institution. Unfortunately, the number of long-term investors seems to be steadily decreasing in the markets. Instead, investors with a short-term investment horizon, for example, hedge funds, own significant shares in many financial institutions. Their focus is clearly on short-term cash generation rather than sustainable growth.

This situation clearly plays a supporting role in a failure of internal control mechanisms of publicly listed companies. Rather than assuring that the management acts in the sake of the companies long-term stability, it is offered an incentive to increase short-term profitability via linking compensation to performance ratios such as ROE.

25.5 CONCLUSION

The current global financial crisis is unprecedented. It clearly highlights the need for structural changes in the financial services industry. In our view, stricter existing regulatory rules are not the sole answer to this problem. What is indeed needed are substantial improvements in global governance and regulation.

Financial institutions need to focus more on appropriate risk management and risk assessment instead of maximizing short-term profitability ratios. In this light, the recent government support packages granted have to be seen with caution as—in the long run—they do not support the development of sound risk management strategies. Rather than solely providing guarantees, government support should aim at appropriate capital ratios within the banking system. Recapitalization must come at a significant cost in order to separate sound from deficient risk management and in order to provide accurate incentives. A related question of interest is why many banks seemed to be reluctant to rely on recent government bailout programs. Most programs clearly allow financial institutions to lower their refinancing costs. Hence, it would seem reasonable to rely on government assistance from a shareholder value perspective. Nevertheless, some of the restrictive covenants of the packages (e.g., capped management salaries) seem to tempt some managers not to rely on such aid. It remains questionable whether market participants have really learned their lesson from the current crisis. Given the long-run development of treasury rates worldwide, it seems as if yet another bubble in financial markets is imminent.

REFERENCES

Bank of England (2008) Financial stability report October 2008. Bank of England financial stability reports, Bank of England, London, U.K.

Basel Committee on Banking Supervision (2004) International convergence of capital measurement and capital standards: A revised framework. Working Paper, Bank for International Settlements, Basel, Switzerland.

Batten, J. A. and Hogan, W. (2002) A perspective on credit derivatives. *International Review of Financial Analysis*, 11(3): 251–278.

Bebchuk, L. A. (2008) A plan for addressing the financial crisis. Discussion Paper Series, John M. Olin Center for Law, Economics, and Business, Harvard, MA.

Bluhm, C. and Overbeck, L. (2007) *Structured Credit Portfolio Analysis, Baskets and CDOs*, Chapman & Hall/CRC: Boca Raton, FL.

Bourdoux, J.-M., Hübner, G., and Sibille, J.-R. (2008) CDO Prices and risk management: A comparative study of alternative approaches for pricing iTraxx. In: Wagner, N. (ed.) *Credit Risk: Models, Derivatives and Management*, Chapman & Hall/CRC: Boca Raton, FL, pp. 511–525.

Breitenfellner, B. and Wagner, N. (2009) Government intervention in response to the subprime financial crisis: The good into the pot, the bad into the crop. Working Paper, Passau University, Passau, Germany.

Chaplin, G. (2005) *Credit Derivatives: Risk Management, Trading and Investing*, Wiley: Chichester, U.K.

Duffie, D. (2007) Innovations in credit risk transfer: Implications for financial stability. BIS Working Paper, Bank for International Settlements, Basel, Switzerland.

Franke, G. and Krahnen, J. P. (2008) The Future of Securitization. CFS Working Paper, Center for Financial Studies, Frankfurt, Germany.

Gorton, G. and Souleles, N. (2005) Special purpose vehicles and securitization. NBER Working Paper, National Bureau of Economic Research, Cambridge, MA.

Hoshi, T. and Kashyap, A. K. (2008) Will the U.S. bank recapitalization succeed? Lessons from Japan. NBER Working Paper, National Bureau of Economic Research, Cambridge, MA.

International Monetary Fund, IMF (2003) Financial asset price volatility: A source of instability? In: *Global Financial Stability Report*, September 2003, pp. 62–88.

Läger, V., Oehler, A., Rummer, M., and Schiefer, D. (2008) Valuation of credit derivatives with counterparty risk. In: Wagner, N. (ed.) *Credit Risk: Models, Derivatives, and Management*, Chapman & Hall/CRC: Boca Raton, FL, pp. 21–37.

Leijdekker, V., van der Voort, M., and Vorst, T. (2008) An empirical analysis of CDO data. In: Wagner, N. (ed.) *Credit Risk: Models, Derivatives, and Management*, Chapman & Hall/CRC: Boca Raton, FL, pp. 457–483.

Longstaff, F. and Rajan, A. (2008) An empirical analysis of the pricing of collateralized debt obligations. *Journal of Finance*, 63(2): 529–563.

Mills, P. and Kiff, J. (2007) Money for nothing and checks for free: Recent developments in U.S. subprime mortgage markets. IMF Working Paper, International Monetary Fund, Washington, DC.

Moreno, M., Peña, J., and Serrano, P. (2008) Pricing tranched credit products with generalized multifactor models. In: Wagner, N. (ed.) *Credit Risk: Models, Derivatives and Management*, Chapman & Hall/CRC: Boca Raton, FL, pp. 485–509.

Scheicher, M. (2003) Credit derivatives—Overview and implications for monetary policy and financial stability. In: *OeNB Financial Stability Report 5*, pp. 96–111.

Shiller, R. J. (2008) *The Subprime Solution: How Today's Global Financial Crisis Happened, and What to Do About It*, Princeton University Press: Princeton, NJ.

Stewart, C. and Wagner, N. (2008) Pricing CDX credit default swaps with CreditMetrics and Trinomial trees. In: Wagner, N. (ed.) *Credit Risk: Models, Derivatives and Management*, Chapman & Hall/CRC: Boca Raton, FL, pp. 181–196.

CHAPTER 26

Identifying Bank Run Signals through Sociological Factors: An Empirical Research in the Geneva Area

Giuseppe Catenazzo and Emmanuel Fragnière

CONTENTS

A "bank run" corresponds to the phenomenon where people run to their banks to withdraw all of their deposits. In 2008, the major Swiss private banks, the UBS and the Credit Suisse, have been deeply troubled because of the "subprime" crisis. In this chapter, we show the main findings of a survey conducted in May–June 2008 upon 363 people living in the Geneva area, designed to assess individuals' confidence toward Swiss banks. Also, we attempt to identify signals that would lead to a bank run. Results show that most people do not plan to change banks in the coming future. Geneva inhabitants still have confidence in their banks while carefully watching the evolution of the crisis. Moreover, we find that people who believe their bank savings are at risk are more likely to take part in a bank run than the others. Confidence toward Swiss banks among Geneva people seems to be a factor that might reduce the likelihood of a bank run in the given area. These findings have been put in perspective with the evolution of the financial crisis in Switzerland until mid-February 2009.

26.1 INTRODUCTION

Modern history has shown several episodes when people run on banks to withdraw all of their deposits. The years 1873, 1893, 1907 (Carlson, 2002), 1910–1928 (Wheelock and Wilson, 1995), 1927 (Yokoyama, 2007), 1929–1933 (Saunders, 1996), 1994 (Schumacher, 2000), and 2001 and 2004 (Ennis and Keister, 2007) recall famous bank runs over the last two centuries. In 2008, the subprime crisis on mortgages in the United States severely hit many worldwide financial institutions. Among them, the Northern Rock Bank collapsed after a run occurred in September 2007: the United Kingdom had not experienced such a phenomenon since 1866 (Schotter and Yorulmazer, 2003; Llewellyn, 2008; Yorulmazer, 2008).

Swiss financial institutions are renowned worldwide for their financial stability and the high quality of the services they provide. However, the current subprime crisis has heavily affected the two main Swiss banks, i.e., the Credit Suisse and the UBS. The UBS bank has registered remarkable losses, notoriously in its overseas investments. The price of its shares (UBS N. Zurich) has fallen by about 66% from July 2007 to July 2008 (source: SWX Swiss Exchange AG, http://www.swx.com). Among others, this has led to the resignation of the UBS President, Mr. Marcel Ospel. To keep Swiss households and small- and medium-size enterprises (SMEs) confident in the UBS, in May 2008, the bank sent letters (see Appendix 1) to 2.5 million of its Swiss customers (Gumy, 2008). The aim of these letters was to reassure customers and avoid the eventual risk of a bank run within the national borders. Credit Suisse shares (CSGN, Zurich) are currently quoted at approximately 36% less than 1 year ago (July 2007) (source: SWX Swiss Exchange AG, http://www.swx.com).

In contrast, the Raiffeisen (Switzerland) bank, a network of small cooperative banks, has counted 60,000 new customers between January and June 2008 (Source: *Télévision Suisse Romande,* 2008). In Switzerland, all bank deposits are insured up to CHF 30,000 per person (source: http://www.einlagensicherung.ch); better conditions are applied to cantonal banks and the Swiss Post deposits.

This difficult context might raise incertitude and an overall loss of confidence among the Swiss people toward the entire Swiss banking system. Also, this feeling might eventually lead the Swiss people to a bank run movement in the coming future. For this reason, we surveyed 363 individuals living and working in the Geneva area to assess the confidence people have in Swiss banks. Also, we attempt to identify whether bank run signals currently exist among Geneva inhabitants. Geneva is the second largest city of Switzerland and it hosts several local and worldwide financial institutions.

This empirical research evidences sociological factors connected with bank customers' attitudes and behaviors. Using a sociological approach, we wish to identify whether the risk of a run on banks exists in the selected area. We learn from the risk management theory (Tchankova, 2002; Fragnière and Sullivan, 2006) that risk identification is a first step that should lead to actions aiding efficient risk management. In particular, when dealing with services, we can manage risks through *ex-ante* (to anticipate) or *ex-post* (to reduce damages) actions and controls (Catenazzo and Fragnière, 2008b; Dubosson et al., 2008). Thus, we wish to point out some relevant elements of perception connected with this theme to drive advanced risk management measures.

This research has been run by the LEM, a laboratory of market research (LEM, Laboratoire d'Études de Marché) of the Haute École de Gestion of Geneva. Three years ago, the Haute École de Gestion of Geneva, Geneva School of Business Administration created the LEM with the aim of teaching students effective marketing survey techniques. Among the mandates already carried by the LEM, we quote, among others, "Information Overload: A Survey Research Conducted in the French Speaking Area of Switzerland" (Debély et al., 2006), "The Consequences of Information Overload in Knowledge Based Service Economies" (Debély et al., 2007), "Influences of Public Ecological Awareness and Price on Potable Water Consumption in the Geneva Area" (Catenazzo et al., 2008), and "Attitudes Regarding New Enterprise Risk and Control Regulations by the Active

Population of the Geneva Area" (Catenazzo and Fragnière, 2008a). The mandate under study in this chapter, "Identifying Bank Run Signals through Sociological Factors: An Empirical Research in the Geneva Area," was conducted from March to June 2008.

The main research questions underlying this empirical study are as follows:

1. What is the Geneva individuals' confidence toward Swiss banks?

2. What are the main attitudes individuals would present in case of a bank run?

To identify in detail individuals' attitudes within hypothetical scenarios, we have included contingent valuation methods (see, for example, Hoevenagel, 1994; Imandoust and Gadam, 2007). Thanks to this method borrowed from psychology, we attempt to assess individuals' typical attitudes and behaviors if overall panic spread and bank runs followed. Relationships between classes as well as relationships between variables have been investigated and analyzed in depth. Then, research hypotheses have been verified on the basis of nonparametric statistical tests.

This empirical research attempts to provide some elements of perception concerning the confidence Geneva inhabitants have in Swiss financial institutions. Also, we attempt to identify whether bank runs predictive signals exist within the social context examined in our research. Among others, this sociological perspective aims at providing financial institutions with elements leading to recommendations for bank governance and communication practices.

This short chapter is organized as follows. In Section 26.2, we present some of the existing academic literature related to our research. Following the literature review, we present the main descriptive statistics obtained by our survey. Finally, our hypotheses related to the theme retained for this chapter follow: Identifying bank run hints in Switzerland: An empirical research in the Geneva area. In conclusion, we summarize the contribution of this chapter and we put it in perspective with the evolution of the financial crisis in Switzerland from summer 2008 to February 2009.

26.2 LITERATURE REVIEW

Zhu (2001) identifies two broad classes for analyzing bank run movements: type-I gathers general panic-driven behaviors in a context with no objective problems and type-II occurs in periods of economic difficulties. To understand individuals' behaviors in bank run (or "bank panic") episodes, Carlson (Carlson, 2002) presents two theories: the "random withdrawal theory" and the "asymmetric information theory." The random withdrawal theory states that people rush to banks because they fear the banks liquidity is insufficient to face the demand for withdrawals. The asymmetric information theory stresses the lack of information; people do not know which financial institution is in trouble, and this pushes the population to withdraw from all banks within the concerned geographical area. These two theories have been tested in the Denver (Colorado) 1893 bank run crisis. For this case, the author (Carlson, 2002) retains both theories, even if the second one is considered the most apropos to explain the event.

The importance of the asymmetric information theory is also validated in the 1994 bank run in Argentina. In this case, a policy based on clear and copious information about the crisis has limited the risk of a run to all local banks (Schumacher, 2000). The 2007 bank crisis in the United Kingdom provides similar issues: as soon as the bailout announcement of the Northern Rock bank was made, all the other banks of the country could benefit from an overall regain in depositors' confidence. Therefore, it seems that individuals behave logically to market news (Yorulmazer, 2008). Imperfect information might be considered as the main cause of type-I (panic-driven) bank run events (Zhu, 2001).

Thus, the role of information seems to be strategic when facing bank runs. For example, disclosing information about the banks' reserves is an asset in avoiding runs (Selvaretnam, 2007). Also, if individuals are kept informed throughout the whole financial crisis, they tend to postpone withdrawals. This is one of the main findings of an experimental study conducted by the CESS (Center for Experimental Social Science) at the New York University (Schotter and Yorulmazer, 2008). In addition, the two authors (Schotter and Yorulmazer, 2008) argue that bank insiders are beneficial in avoiding bank runs.

Furthermore, deposit insurances seem to be successful disincentives to bank runs (Schumacher, 2000; Carlson, 2002; Chen and Hasan, 2005; Hall, 2008; Schotter and Yorulmazer, 2008). A full coverage on deposits should be guaranteed: partial protection, despite being useful, is not sufficient to avoid the risk of a bank run. However, this policy distorts bank market equilibrium since it reduces customers' ongoing control over the bank solidity (Wheelock and Wilson, 1995; Chen and Hasan, 2005; Llewellyn, 2008). Pursuing deposit insurance policies results in an imperfect *social optimum* equilibrium (Zhu, 2001). For instance, the Kansas bank crisis (1910–1928) displays that the banks participating in the state deposit insurance system were more likely to go out of business than the others. A larger concentration of high-risk portfolios has been evidenced among these banks (Wheelock and Wilson, 1995).

Finally, the recent run on the Northern Rock bank (2007) is an interesting source of study for several authors (see, for example, Hall, 2008; Keasey and Veronesi, 2008; Llewellyn, 2008; Yorulmazer, 2008). Among others, Hall (2008) and Yorulmazer (2008) present in detail the story of the collapse of the Northern Rock Bank, the United Kingdom's No. 5 mortgage bank. The bank's business model weaknesses have been noted among the main default causes (Keasy and Veronesi, 2008; Llewellyn, 2008). Failures in the public regulator control activity (Hall, 2008; Keasy and Veronesi, 2008), as well as the breaches within the United Kingdom banking system have also been highlighted (Hall, 2008). Nevertheless, this episode represents only one case with limited spillover effects to the country's banking system; so far the United Kingdom has not been highly affected by the U.S. subprime mortgage crisis (Yorulmazer, 2008).

This literature review, although not exhaustive, indicates that too little sociological knowledge is available to understand *ex-ante* individuals' behaviors and attitudes to identify and predict bank runs. In this study, we intend to discover some social patterns associated with these issues in Geneva. This city is the second largest Swiss financial centre and is situated only a few hundred miles from other European financial hubs such as Zurich, Paris, Frankfurt, and Milan.

26.3 METHODOLOGY

This empirical study attempts to identify the main clues connected with bank run likelihood in Switzerland, with a more precise social outlook focused on the Geneva area. This survey research was conducted from March to June 2008, and realized with the help of a group of students under the supervision of the LEM (Haute École de Gestion of Geneva Laboratory of Market Research) research staff. Our work encompassed the following steps: determination of the axes of investigation, design of the questionnaire, data collection, coding and statistical data processing, and, finally, communication of the results.

In the exploratory phase, we conducted around 40 in-depth interviews exploring this topic. Among several themes discussed, we have retained two main ideas: individuals are generally confident in the institution in which they deposit their money and, in case of a bank run, people would probably follow the stream by either changing banks or by withdrawing all of their deposits.

On the basis of the two identified research themes, we designed a questionnaire administered to the active population of the Geneva area. Respondents were selected on a random basis in the streets, open spaces, and other public places of the Geneva area. The questionnaire (the complete questionnaire, in French, is available on request) was made up of 20 close-ended questions attempting to identify individuals' opinions concerning their deposits, the choice of the financial institution to confer money, and possible behaviors in a bank run context. Hypothetical scenarios (Hoevenagel, 1994) have also been presented to provide further elements of analysis.

26.4 DESCRIPTIVE STATISTICS

363 people responded to our questionnaire: 56.7% men and 43.3% women. Generally, socioeconomic statistics of the sample are representative of the Geneva Canton population. Here are some of the main descriptive results obtained from the analysis of the questionnaires collected. First of all, we asked the respondents whether they are worried or not about their savings: 55.5% say "no," while this bothers 37.3% of them, and 7.2% do not know.

Then, we asked the interviewees if their money is deposited in one or more banks. Fifty percent (50.4%) affirm their money is split into several banks, while 46.5% have all their money in one bank. One percent (1.7%) provide alternative possibilities, and 1.4% do not know. Sixty-nine percent of the sample do not foresee a bank change in the short run. Fourteen percent (14.1%) would, and 16.9% cannot answer.

In case of a panic movement (i.e., a bank run), 40% affirm they would not follow the rush on bank for cash withdrawals, while 26.5% would. About a third of the sample (33.4%) cannot answer. Fifty percent (50.1%) of the respondents judge themselves as "average savers," 23.1% say they are "poor savers," 13.2% are "not savers at all," 11.8% are "high savers," and 1.7% do not know. Thirty-five percent (35.7%) of the interviewees affirm saving approximately 5% to 20% of their income. Thirty-three percent (33.1%) save less than 5%, 12% save more than 20%, and 9.7% do not save any of their income. Nine percent (9.5%) cannot answer this question. According to 41% of the sample, it is unlikely that one of the main Swiss banks defaults. Thirty-seven percent (37.2%) say this is likely to happen, and 21.8% cannot answer.

We asked respondents to assess their confidence vis-à-vis the Swiss banks through assigning them a mark. The average mark given is 6.88 on a scale spanning from 1 to 10 (10 being the best mark). The mode of the answers provided is 8, and the median is 7. Fifty-one percent of the respondents disagree with the statement, "the Swiss economy will threshold in a recession by the end of the year," 19% agree, and 30% cannot answer.

We enquired in which bank or financial institutions the respondents have deposited their money. The main Swiss private banks such as the UBS or the Credit Suisse gather about 65.8% of the respondents' deposits. Thirty-four percent (34.2%) have placed their money at public-driven financial institutions such as cantonal banks or the Swiss Post. National "retail" banks, such as Raiffeisen, Migros, and Coop banks are chosen by 17.1% of the sample followed by other Swiss private banks such as LODH, Pictet, or Sarasin that cover 9.6% of the choices. Eight percent (8.5%) have chosen other banks, mainly French banks (the French borders are less than 10 km away from the Geneva city centre), and 3.9% prefer not to answer this question. Respondents could choose more than one possibility; this is why the sum of the percentages provided above is over 100%.

Then, we asked the interviewees: If your bank defaulted, to which one of the following banks would you confer your money? The answers provided are as follows: 31.4% do not know what bank he/she would choose, 27.3% would choose Swiss private banks such as Raiffeisen, Migros, or Coop banks, 22% would redirect themselves toward cantonal banks or the Swiss Post, 12.1% would choose another institution, 10.5% would choose LODH, Pictet, or Sarasin banks, and only 9.9% would turn toward the main Swiss private banks like the UBS or the Credit Suisse.

Seventy-four percent (74.4%) of the respondents affirm they keep up-to-date regarding the current financial crisis (the subprime crisis on U.S. mortgages) through media communication means, 23.6% do not, and 1.9% do not know. Sixty-three (63.4%) of the interviewees affirm to have more confidence in Swiss banks than in foreign banks, 26.8% affirm their confidence in Swiss and foreign banks is alike, 5% rather prefer foreign banks, and 4.7% do not know.

Finally, the resignation of Mr. Marcel Ospel from his position of president at the UBS bank is considered "normal" by 44.2% of the sample and 27.1% "have no opinion over this matter." His choice is judged as "reassuring" by 13% of the interviewees, and "unacceptable" by 6.9%; 6.1% prefer not to answer, and 2.8% have another or a different opinion from the answer possibilities provided.

26.5 HYPOTHESIS TESTING

We now investigate some relationships among variables. As most of these variables are defined over qualitative scales, we employ nonparametrical tests to validate our hypotheses. Questionnaires have been coded through the SPSS 15 software and tests have been conducted according to the methods described by Bryman and Cramer (2006). We have identified two main relationships that we explore in more detail in this chapter. First of all, we have realized a connection between the fear individuals have of losing their savings and the likelihood that they would follow a bank run movement. To assess this link, we asked the sample the two following questions: "Are you worried about your savings?" Respondents had to choose among the following possible answers: "yes," "no," and "I do not know." The second question was, "In case of a bank run movement, would you follow the stream?" The provided answers people were asked to choose among were "yes," "no," and "I do not know." On the basis of these nominal variables, we have designed the following hypotheses scheme:

H_o: The fear of losing one's savings is not connected with individuals' behavior within a bank run scenario.

H_a: The fear of losing one's savings is connected with individuals' behavior within a bank run scenario.

We used the Pearson's Chi square test to study two independent nominal variables. We have retained a significance level of 5% that is the first-type error (or the risk to reject the null hypothesis when it is actually correct). The p-value of 0.000 indicates that we can reject the null hypothesis at the significance level of 5%. Therefore, we can conclude that there is a relationship between the individuals' fear of losing their savings and their possible behavior in following a bank run stream.

Also, we found a link between the individuals' opinion regarding the likeliness of default of one of the main Swiss banks and the possibility they would follow a bank run movement. To evidence this connection, we refer to two of the questions addressed to our sample: "Do you think possible that one of the main Swiss banks defaults?" "Yes," "no," and "I do not know" were the given answers that respondents had to choose. Then, we asked, "In case of a bank run movement, would you follow the stream?" The provided answers people were asked to choose were "yes," "no," and "I do not know." On the basis of these nominal variables, we have designed the following hypotheses scheme:

Ho: The perceived likelihood of default of one of the main Swiss banks is not connected with individuals' behavior within a bank run scenario.

Ha: The perceived likelihood of default of one of the main Swiss banks is connected with individuals' behavior within a bank run scenario.

We used the Pearson's Chi square test to study two independent nominal variables. We have retained a significance level of 5% that is the first-type error (or the risk to reject the null hypothesis when it is actually correct). The p-value of 0.000 indicates that we can reject the null hypothesis at the significance level of 5%. Thus, we can argue that whether people perceive the possibility of default of one of the main Swiss banks, the run stream would probably take them in.

26.6 CONCLUSIONS

Risk identification is a necessary step for efficient corporate risk management, both to financial and nonfinancial institutions. Within the banking environment, customers' run is a risk that only hits episodically, but its consequent damages are of high concern. In this chapter, we present the main findings of a research survey conducted in spring 2008 in the Geneva area, Switzerland. We have attempted to identify whether key elements leading to a bank run movement existed within the retained geographical area. Identifying individuals' perception toward the occurrence of a bank run and their possible behaviors provides useful information for keeping the overall banking system in good health. In fact, managements could implement in advance some shielding policies to avoid or keep under control the risk of a bank run.

Our sample counts 363 individuals living and working in the Geneva area. Among descriptive results, we have evidenced that most people keep up-to-date of the current financial (subprime on US mortgages) crisis, and the current confidence toward Swiss banks is generally high. Also, it seems that Geneva inhabitants do not plan to change banks in the coming future. This evidences a broad level of confidence, but, simultaneously, broad awareness and attention to this theme.

To address these concerns with practical recommendations, we have tested two hypotheses that highlight the following relationships: first, we can affirm that those people who believe their bank savings are at risk are more likely to take part in a rush to banks for cash withdrawals. This means that individuals' lack of confidence toward the institution they have conferred their money might push them to take part in a run to their banks. Second, confidence toward Swiss banks among the Geneva people reduces the likelihood of a rush to the city banks. Thus, clear, accurate, and faithful corporate communication of the whole Geneva banking system would likely help avoid panic-driven behaviors during difficult periods. Undertaking complete information policies to improve confidence between customers and their banks seems to be an efficient long-run strategy to follow.

This research was conducted during spring (data collection) and summer (hypotheses testing) of the year 2008. As the findings correspond to elements of "social perceptions," we might wonder whether they happen to be stable or not over time in regard of the evolution of the financial crisis in Switzerland or can be used as an anticipative tool for decision makers. On October 10, 2008, the two main Sunday newspapers of Switzerland, SonntagsBlick and Le Matin Dimanche, revealed the results of a national survey related to our topic. Their main question was roughly the same as the one that was asked in our survey, whose results appeared in the edition of June 27 of Le Temps. Indeed, 79% of the respondents said that they are worried about their savings (46% are not worried at all and 33% are rather not worried). Even though the scale employed in this survey is not the same as in ours (they are using a likert one and we used a dichotomous one), we can see that the perception regarding this matter of confidence seems to have remained stable over the period from April to September. On the opposite, many national indicators of other European countries were indicating a clear deterioration of trust during the same period.

From September 2008 on, the main attention of the Swiss public was astonishingly not that much focused on the UBS but rather on the Credit Suisse. Indeed, the Credit Suisse had sold for an amount of roughly 600 millions of Swiss francs of Lehman Brothers' financial structured products to its retail clientele. Many articles in the Swiss press and television programmes presented the cases of small investors having lost all their savings after Lehman Brothers defaulted. This issue was not covered by foreign newspapers. However, no official indicators have shown that the trust toward banks has lowered during the second half of 2008. So it seems that until fall 2008, the Swiss financial banking sector was still perceived as "reliable."

However, at the end of October 2008, everything changed regarding the perception of the Swiss banking system. The UBS received an unexpected (for the Swiss public) bailout of 60 billions of Swiss francs from the Swiss National Bank. On February 10, 2009, the UBS announced that it suffered an overall net loss for the year 2008, of 20 billions of Swiss francs. Moreover, the UBS had faced during the year 2008, an overall withdrawal of 226 billions of Swiss francs. Most of this money was reinjected in cantonal banks (state banks) and regional banks. Our survey had signaled this trend, which has been verified all along the last year by press releases of regional banks. Two-hundred and twenty-six billion Swiss francs are enormous when looking at the UBS balance sheet, but still we have not reached so far what we could call a "bank run." Given that the UBS has launched marketing campaigns to improve its perception among its Swiss clientele, because it is now convinced that this is the only way to avoid a bank run that could be fatal for the whole Swiss economy.

Nevertheless, our sociological survey has foreseen well in advance that clients of the two giant Swiss banks (the CS and the UBS) were withdrawing money to the profit of Swiss regional banks, as well as that confidence had remained still sufficiently high to avoid a disaster. We hope that this study, which has met a vivid interest in Switzerland (also among the banking sector, since after the publication of the above-mentioned article in Le Temps, both directions of the CS and the UBS have requested the full report of the study and have thanked us for its quality), will contribute to restore the solidity of the Swiss banking industry.

ACKNOWLEDGMENT

We would like to thank the students of Laboratoire d'Études de Marché of the Haute École de Gestion de Genève, Geneva School of Business Administration, who participated in the construction of the survey, the data collection, and the transcription of answers. Without them, this research would not have been possible.

REFERENCES

Bryman, A. and Cramer, D. (2006) *Quantitative Data Analysis for the Social Scientist with SPSS 15 & 16*. London, U.K.: Routledge.

Carlson, M. A. (2002) Causes of bank suspensions in the panic of 1893. Board of Governors, Federal Reserve System WP No. 2002-11. Available at SSRN: http://ssrn.com/abstract=301527.

Catenazzo, G., D'Urso, J., Fragnière, E., and Tuberosa, J. (2008) Influences of public ecological awareness and price on potable water consumption in the Geneva area. Available at SSRN: http://ssrn.com/abstract=1121664.

Catenazzo, G. and Fragnière, E. (2008a) Attitudes regarding new enterprise risk and control regulations by the active population of the Geneva area. Available at SSRN: http://ssrn.com/abstract=1247071.

Catenazzo, G. and Fragnière, E. (2008b) *La Gestion des Services*. Paris, France: Economica.

Chen, Y. and Hasan, I. (2005) The transparency of the banking industry and the efficiency of information-based bank runs. *Journal of Financial Intermediation*, 15(3): 307–331.

Debély, J., Derache, G., Fragnière, E., and Tuberosa, J. (2006) Information overload: A survey research conducted in the French speaking area of Switzerland. Available at SSRN: http://ssrn.com/abstract=945368.

Debély, J., Dubosson, M., and Fragnière, E. (2007) The consequences of information overload in knowledge based service economies. *Proceedings of the 2007 ESSHRA European Research Conference in Switzerland*, Kursaal, Berne, Switzerland, pp. 67–72, June 12–13, 2007. Available at SSRN: http://ssrn.com/abstract=999525.

Dubosson, M., Fragnière, E., and Milliet, B. (2008) A control system designed to address the intangible nature of service risks. *Proceedings of the 2006 IEEE International Conference on Service Operations and Logistics, and Informatics*, Shanghai, China, pp. 90–95, June 21–23, 2006. Available at SSRN: http://ssrn.com/abstract=1103110.

Ennis, H. M. and Keister, T. (2007) Commitment and equilibrium bank runs. FRB of New York Staff Report No. 274. Available at SSRN: http://ssrn.com/abstract=962714.

Fragnière, E. and Sullivan, G. (2006) *Risk Management, Safeguarding Company Assets*. Boston, MA: Crisp—Thomson NETg.

Gumy, P. (2008) UBS fait son acte de contrition dans une lettre adressée à 2,5 millions de clients, *Le Temps*, May 8, 2008.

Hall, J. B. (2008) The sub-prime crisis, the credit squeeze and northern rock: The lessons to be learned. *Journal of Financial Regulation and Compliance*, 16(1): 19–34.

Hoevenagel, R. (1994) An assessment of the contingent valuation method. In *Valuing the Environment: Methodological and Measurement Issues*, Pethig, R. (Ed.), Dordrecht, the Netherlands: Kluwer Academic Publishers.

Imandoust, S. B. and Gadam, S. N. (2007) Are people willing to pay for river water quality, contingent valuation. *International Journal of Environmental Science and Technology*, 4(3): 401–408.

Keasey, K. and Veronesi, G. (2008) Lessons from the northern rock affair. *Journal of Financial Regulation and Compliance*, 16(1): 8–18.

Llewellyn, D. T. (2008) The northern rock crisis: A multi-dimensional problem waiting to happen. *Journal of Financial Regulation and Compliance*, 16(1): 35–58.

Saunders, A. (1996) Contagious bank run: Evidence from the 1929–1933 period. *Journal of Financial Intermediation*, 5(4): 409–423.

Selvaretnam, G. (2007) Regulation of reserves and interest rates in a model of bank runs. Working Paper CDMA07/14, University of St Andrews, Scotland, U.K. Available at SSRN: http://ssrn.com/bstrct=1015234.

Schotter, A. and Yorulmazer, T. (2008) On the dynamics and severity of bank runs: An experimental study. Available at SSRN: http://ssrn.com/abstract=1090938.

Schumacher, L. (2000) Bank runs and currency run in a system without a safety net: Argentina and the "Tequila" shock. *Journal of Monetary Economics*, 46: 257–277.

Tchankova, L. (2002) Risk identification—Basic stage in risk management. *Environmental Management and Health*, 13(3): 290–297.

Télévision Suisse Romande (2008), Les banques Raiffeisen sont en verve, News July 6, 2008, http://www.tsr.ch.

Wheelock, D. C. and Wilson P. W. (1995) Explaining bank failures: Deposit insurance, regulation, and efficiency. *The Review of Economics and Statistics*, 77(4): 689–700.

Yokoyama, K. (2007) Too big to fail: The panic of 1927. Working Paper No. 465, Nagoya City University, Japan. Available at SSRN: http://ssrn.com/abstract=980879.

Yorulmazer, T. (2008) Liquidity, bank runs and bailouts: Spillover effects during the northern rock episode. Available at SSRN: http://ssrn.com/abstract=1107570.

Zhu, H. (2001) Bank runs without self-fulfilling prophecies. BIS Working Paper No. 106. Available at SSRN: http://ssrn.com/abstract=847445.

CHAPTER 27

Bank Default Risk in the United States and the United Kingdom

Robert Powell and David E. Allen

CONTENTS

The current financial crisis places the spotlight on the ability of banks to meet their financial obligations. This paper examines and compares changes in bank default risk in the United States and United Kingdom over time, including the current crisis period. A common approach used by banks to measure the probability of default (PD) among customers is the KMV/Merton structural model, which measures distance to default (DD). We use this same approach to measure the DD of the banks themselves. As a further measure of variation of bank risk over time, we use the value-at-risk (VaR) methodology to examine the banks' equity risk, as well as the increasingly popular conditional VaR (CVaR) methodology to measure their extreme equity risk. In addition, we incorporate CVaR techniques into structural modeling to measure extreme default risk. The study finds that the U.S. and U.K. banks are in an extremely precarious capital position based on market asset values, especially in the United Kingdom where the banks are more highly leveraged. We also find that the existing credit ratings of banks are much more favorable than what the default probabilities indicate they should be. Movements in market asset values are currently not factored into capital adequacy requirements, and based on our findings, recommendations are made for a revised capital adequacy framework.

27.1 INTRODUCTION

The banking industry has been at the forefront of the 2008/2009 financial crisis. Twenty-five banks failed in the United States in 2008 as compared to 3 in 2007 and none in the prior 2 years. Bank capital shortages have led to the need for government-led financial rescue packages in the United States and the United Kingdom and in many other countries.

The troubled asset relief program (TARP) was introduced by the U.S. Treasury in 2008 to purchase the troubled assets of up to $700 billion and to stabilize bank capital ratios. In the United Kingdom, a financial support package of approximately £500 billion was introduced in 2008 that included measures to recapitalize banks, provide government guarantees, and improve bank liquidity. Further rescue plans have been proposed in both the United States and United Kingdom.

U.S. Federal Reserve (2008) statistics for charge-off and delinquency rates approximately doubled from 2007 and trebled from 2006. In their 2008 third quarter report, Federal Deposit Insurance Corporation (FDIC) report mounting loan losses for banks and deposit institutions, with a trebling of loan loss provisions as compared to the prior year. 2009 is expected to see a continuation of high default rates and a need for capital at levels above current minimums, with FDIC introducing a range of measures to facilitate resolutions (MacDonald et al., 2009).

The Bank of England credit conditions survey (2008b) reports rising default rates, widening spreads, and reduced credit availability to households and businesses due to economic outlook concerns as well as the cost and availability of funds. The survey shows an expectation that these circumstances will continue going forward.

In their financial stability report (2008c), the Bank of England reports that "system-wide vulnerabilities were exposed...rooted in uncertainties about the value of banks assets...amplified by excessive leverage." The report expresses concern that the major banks had assets of just over £6 trillion and equity capital of around £200 billion, which could magnify default probabilities with increased volatility of asset returns. The example provided was that if the standard deviation of asset returns was, say, 1.5% pre crisis and increased to 3% post crisis, then PD would increase from around 1% to around 15%. As default probabilities rise, there is a greater likelihood of assets that need to be liquidated at market prices, and thus market participants need to revalue their assets with greater weight placed on mark to market values. This gives rise to reduced asset values and a need for increased capital.

U.S. banks typically operate on lower leverage (higher equity) ratios than the U.K. and European banks. The banks used in our study show equities (based on book value of assets and total liabilities) of 4.3% for U.K. banks as compared to 7.7% for U.S. banks. In terms of the Basel Accords, regulatory capital requirements are calculated as a percentage of risk-weighted assets. Tier 1 Capital ratios average 8.1% for the 4 largest U.S. banks and 7.8% for the largest 4 U.K. banks. Total Risk based capital ratios for the same banks are 11.3% (United States) and 11.7% (United Kingdom), which are significantly higher than based on the book values of total assets. The risk-based ratios are much closer between the U.S. and U.K. banks than the nonrisk weighted ratios, implying that a higher risk weighting is being applied to U.S. bank assets than is the case in the United Kingdom. High leverages and concerns over the Basel approach have led to the call for maximum bank leverage ratios from several quarters, including the Bank of England (2008c) and the Swiss National Bank (Hildebrand, 2008), based on capital to total assets that could run in parallel with Basel II standards.

In line with the concerns expressed by the Bank of England, the objective of this study is to examine increases in the PD of banks over the financial crisis period based on the market value of assets. As asset volatility is linked to movements in equity (the market value of assets equating to equity values plus liabilities), we commence by examining the volatility of bank equities using both VaR and CVaR approaches. The background to these approaches is discussed in Section 27.2. We then go on to examine PD using the Merton model. Based on CVaR approaches, we also include extreme risk into our default model and calculate the conditional PD (CPD). These approaches are also discussed in Section 27.2. The contribution and benefits of the study are discussed in Section 27.3, and our methodology is explained in Section 27.4. Results are presented in Section 27.5, which also includes an examination of whether current bank credit ratings are consistent with PD values. Conclusions are provided in Section 27.6, which also includes recommendations for a revised capital adequacy framework based on our findings.

27.2 BACKGROUND TO VAR, CVAR, AND PD MEASUREMENTS

VaR has become the recognized standard approach for market risk measurement since its introduction by JP Morgan in 1994 and subsequent adoption of VaR by the Basel Committee on Banking Supervision. VaR describes the loss that can occur over a given period, at a given confidence level, due to exposure to market risk.

A description of the various methodologies for the modeling of VaR can be seen at http://www.gloriamundi.org/. There is a comprehensive survey of the concept by Duffie and Pan (1997), and discussions in Jorion (1997), Pritsker (1997), RiskMetricsTM (J.P. Morgan and Reuters, 1996), Beder (1995), and Stambaugh (1996).

Although popular, VaR has shortcomings. It has certain undesirable mathematical properties; such as the lack of subadditivity, convexity, and monotonicity (Artzner et al., 1997; 1999). VaR says nothing about losses beyond VaR, and it is at times of extreme risk that businesses are most likely to fail. CVaR, which measures losses beyond VaR is becoming increasingly popular, as it not only measures losses beyond VaR, but also does not demonstrate the undesirable mathematical properties of VaR. Pflug (2000) proved that CVaR is a coherent risk measure with a number of desirable properties such as convexity and monotonicity, among other desirable characteristics. A number of papers apply CVaR to portfolio optimization problems; see for example Rockafeller and Uryasev (2002), Uryasev and Rockafeller (2000), Andersson et al. (2000), Alexander et al.,(2003), Alexander and Baptista (2003) and Rockafellar et al. (2006).

The Merton/KMV approach as described in Crosbie and Bohn (2003) and Saunders and Allen (2002) is based on the work of Black and Scholes in 1973. The model assumes that the firm has one single debt issue and one single equity issue. The debt (D) consists of a zero coupon bond that matures at time (T). The firm defaults if the debt obligation exceeds the asset value at T. At this stage, the bond owners take ownership of the firm and the shareholders get nothing. The Merton approach has been adopted and modified by KMV, and the KMV model is used widely by banks throughout the world to estimate the PD of corporate borrowers. KMV calculates DD values and then uses their extensive worldwide database of corporate defaulters to modify the PD values produced by the Merton model.

The examples of studies using structural methodology for varying aspects of credit risk, include asset correlation (Cespedes, 2002; Kealhofer and Bohn, 1993; Lopez, 2004; Vasicek, 1987; Zeng and Zhang, 2001), predictive value and validation (Bharath and Shumway, 2004; Stein, 2002), and fixed income modeling (D'Vari et al., 2003). The effect of default risk on equity returns has also been examined (Chan et al., 2008; Gharghori et al., 2007; Vassalou and Xing, 2002). These papers also examine PD as an extension to the Fama and French (1992; 1993) three factor view of asset pricing that includes the market, size, and book-to market. Ghargori et al. find that default risk is not priced in equity returns and that the Fama-French factors are not proxying for default risk. Vassalou and Xing find support for size and book to market as influences on default risk, but do not find strong linkage between default risk and return. Chan et al., using an extensive 30 year data sample of micro stocks, find significant linkage between default risk and returns. When conditioning for business cycles, they find that default risk premium is twice as high during expansions than during contractions. Some examples of studies linking the bank capital and the market values of banks include Saita (2007), who looks at improving bank capital allocation decisions using VaR and Bischel, and Blum (2004) who find a positive correlation between changes in capital and changes in risk but despite this, does not find a significant ratio between the default probability and the capital ratio. Allen and Powell (2007; 2009) have found a linkage between those industries that are risky from a credit perspective (lending to companies in those industries) and those industries that are risky from a market perspective (share price volatility). The authors incorporate industry VaR and CVaR techniques into structural PD models as well as transition matrix credit VaR models.

27.3 CONTRIBUTIONS AND BENEFITS OF THE STUDY

While most studies look at the default risk of borrowers of banks, our study looks at the default risk of the banks themselves. Importantly our study incorporates the current financial crisis period. A key point of difference is that, unlike most studies, we also incorporate CVaR techniques into the PD modeling of the banks. This is an important inclusion, as it is precisely at times of extreme risk that banks are most likely to fail.

Consistent with the concerns expressed by the Bank of England regarding the noninclusion of market values of assets into the capital adequacy assessment of banks, our study finds U.S. and, even more so, U.K. banks to be in a very precarious capital position based on market values. PD is found to be much higher than the current credit ratings of banks indicate.

The study provides contributions in many respects. Firstly, it provides a greater understanding of the default risk of banks over the crisis period. Secondly, our study builds on the concern expressed by the Bank of England regarding the lack of consideration of fluctuating market values in determining bank capital. Thirdly, we provide a unique point of difference to existing studies by incorporating CVaR into bank PD modeling over the crisis period. Fourth, we make recommendations on a revised bank capital adequacy framework, which incorporates the market values of bank assets.

The study can benefit banks, bank regulators, and investors through a better understanding of bank default probability, through a provision of new modeling techniques for incorporating CVaR into PD modeling, and by providing recommendations on a revised capital adequacy structure incorporating two new key factors, i.e., market fluctuations in asset values and extreme risk.

27.4 DATA AND METHODOLOGY

27.4.1 Data

Fifteen years of equity and balance sheet data were obtained from DataStream. The study includes entities listed as banks on the New York and London Stock exchanges for which equity prices and Worldscope balance sheet data are available in Datastream. This includes 52 U.S. banks with total assets of $7.8 trillion, and 8 U.K. banks with £5.4 trillion in assets. The data include all major banks and represent approximately 70% of the combined total asset values of all banks in the United States and United Kingdom as reported by FDIC (2008) and the Bank of England (2008a).

27.4.2 VaR and CVaR Methodologies

We follow the parametric VaR method used by RiskMetrics (J.P. Morgan and Reuters, 1996), who introduced and popularized VaR. This is the most commonly used VaR method. Daily equity returns are calculated for each of the 15 years using the logarithm of price relatives, which is obtained by the following calculation:

$$\ln\left(\frac{P_t}{P_{t-1}}\right) \tag{27.1}$$

i.e., the logarithm of the ratio between today's price and the previous price.

VaR is calculated for each bank at a 95% confidence level. Based on standard tables $VaR_x = 1.645 \times \sigma_x$

For investors holding a portfolio of bank assets, it would be usual practice to use a correlated approach to obtain the overall portfolio VaR. However, our study is not for investment purposes, and we do not need to show the effect of portfolio diversification. We, therefore, use an undiversified approach, whereby total VaR is the weighted average of the individual bank VaRs.

CVaR is calculated as the average of the worst 5% of returns, i.e., returns beyond VaR.

27.4.3 PD Methodology

We use the Merton approach to estimating default. DD and PD are measured as

$$DD = \frac{\ln(V/F) + (\mu - 0.5\sigma_V^2)T}{\sigma_V \sqrt{T}} \tag{27.2}$$

$$PD = N(-DD) \tag{27.3}$$

where
  V is the market value of firms debt
  F is the face value of firm's debt
  μ is an estimate of the annual return (drift) of the firm's assets
  N is the cumulative standard normal distribution function

To estimate the market value of assets, we follow approaches outlined by KMV (Crosbie and Bohn, 2003) and Bharath and Shumway (2004). Equity returns and their standard deviations are calculated exactly the same as for our market approach. Initial asset returns are estimated from our historical equity data using the following formula:

$$\sigma_V = \sigma_E \left(\frac{E}{E+F} \right) \tag{27.4}$$

These asset returns derived are applied to Equation 27.4 for estimating the market value of assets every day. The daily log return is calculated and new asset values are estimated. Following KMV, this process is repeated until asset returns converge (repeated until difference in adjacent σ's is less than 10^{-3}). These figures are then applied to the DD and PD calculations in Equations 27.2 and 27.3. We measure μ as the mean of the change in $\ln V$ as per Vassalou and Xing (2002). We also examine how current DD would change, based on historical asset volatility. The model measures the historical asset volatility using a combination of current balance sheet data and historical equity values, which are then used to estimate the historical asset values as described in earlier in this section. This allows us to examine how the current DD would change if asset volatilities reverted to historical levels. Anchoring the default variable allows the loss distribution to shift with changes in another variable, as is noted by Pesaran et al. (2003) whose credit risk model anchors default and determines loss distribution changes brought about by changes in macroeconomic factors. The authors note that "the problem is not properly identified if we allow both to be time varying."

Standard & Poor's (2005) provide default probabilities for each of their credit ratings. Once we have calculated PD values, we examine whether the current credit ratings of banks are consistent with our PD values.

27.4.4 CPD Calculation

For the purposes of this study, we define CPD as being PD on the condition that the standard deviation of asset returns exceeds the standard deviation at the 95% confidence level, i.e., the worst 5% of asset returns. We calculate the standard deviation of the worst 5% of daily asset returns for each period to obtain a conditional standard deviation (CStdev). We then substitute CStdev into the formula used to calculate DD in order to obtain a conditional DD (CDD). The CPD is calculated by substituting DD with CDD into the CPD formula:

$$CDD = \frac{\ln(V/F) + (\mu - 0.5\sigma_V^2)T}{CStdev_V \sqrt{T}} \qquad (27.5)$$

and

$$CPD = N(-CDD) \qquad (27.6)$$

27.5 RESULTS

Table 27.1 shows a very similar pattern for both the United States and the United Kingdom. Lower VaR is shown during the mid-1990's, increasing late 1990's and early 2000's with events such as the Asian financial crisis and the 2001 terrorist attacks in the United States. Through the mid-2000's VaR reduces, and then spikes dramatically during the current financial

TABLE 27.1   VaR and CVaR–Results Summary

	U.S. Banks		U.K. Banks	
	Daily VaR	**Daily CVaR**	**Daily VaR**	**Daily CVaR**
2008	0.095	0.141	0.075	0.117
2007	0.029	0.042	0.030	0.040
2006	0.015	0.021	0.017	0.024
2005	0.016	0.024	0.015	0.021
2004	0.017	0.022	0.017	0.023
2003	0.021	0.029	0.024	0.031
2002	0.041	0.058	0.046	0.051
2001	0.035	0.042	0.037	0.053
2000	0.044	0.058	0.038	0.053
1999	0.044	0.058	0.043	0.060
1998	0.045	0.061	0.049	0.065
1997	0.029	0.041	0.038	0.051
1996	0.026	0.036	0.024	0.032
1995	0.024	0.032	0.023	0.031
1994	0.023	0.031	0.025	0.032

Note: The table shows Daily VaR and CVaR. Annual VaR can be obtained by multiplying the Daily VaR by the square root of 250. Figures are undiversified and represent the weighted average of the individual bank VaRs. VaR is calculated at 95% confidence level, and CVaR is based on the worst 5% of returns.

crisis period. VaR in 2008 is 5 times greater than during 2004–2006, and more than double that seen in 2000–2001. This is illustrated in Figure 27.1. The U.S. and U.K. figures follow a very similar VaR pattern up to 2007. Both spike dramatically in 2008, but the United States more so than the United Kingdom. The differential is even greater for CVaR in 2008, with the United States at 0.141 and the United Kingdom at 0.117.

Figures 27.2 and 27.3 apply polynomial trend lines to VaR and CVaR. An interesting observation is that the differential between CVaR and VaR increases during the periods of higher volatility, which is especially notice-able in 2008. Thus, the importance of measuring tail risk (returns beyond VaR) increases during crisis periods.

The important consideration for this paper is how the trends above translate into default risk. DD and associated PD are shown in Table 27.2. Trend lines are shown in Figures 27.4 and 27.5.

From the results, we make five key observations. Firstly, as expected, the DD trend follows the VaR trend, with heightened risk in the early 2000's and during 2008, and significantly reduced DD during the mid-2000's.

Second, as shown in Table 27.2 and Figures 27.4 and 27.5, the DD and PD levels for the United Kingdom are higher than the United States. The two key components to DD are the equity of the firm (distance between the market values of assets and debts) and the asset volatility. It is in the

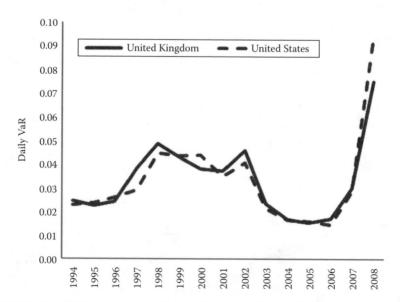

FIGURE 27.1   VaR comparison–United States and United Kingdom.

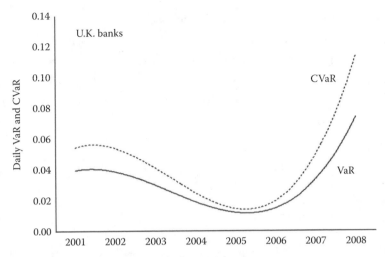

FIGURE 27.2   VaR and CVaR trends for U.K. banks.

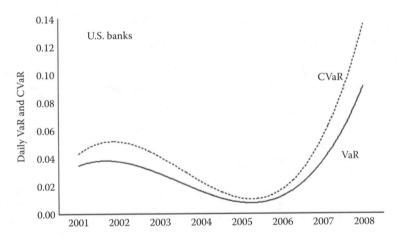

FIGURE 27.3   VaR and CVaR trends for U.S. banks.

first of these two components that the significant difference lies, due to the high leverage of U.K. banks as discussed in Section 27.1. While volatility is closely matched between the two banks, U.K. banks have a much shorter distance between the asset and debt values.

Third, the industry has almost breached the default line in 2008. While the graph only shows the total picture for the industry, our DD figures for individual entities show that approximately 27% of U.S. banks and 38% of U.K. banks have breached the default line in 2008.

TABLE 27.2   Distance to Default—Results Summary

	U.K. Banks				U.S. Banks			
	DD	PD	CDD	CPD	DD	PD	CDD	CPD
2008	0.48	0.3144	0.12	0.45	0.92	0.1778	0.23	0.41
2007	1.51	0.0658	0.41	0.34	3.21	0.0007	0.81	0.21
2006	2.51	0.0061	0.69	0.24	6.35	0.0000	1.73	0.04
2005	2.89	0.0019	0.83	0.20	6.27	0.0000	1.74	0.04
2004	2.57	0.0051	0.72	0.24	5.85	0.0000	1.66	0.05
2003	1.85	0.0319	0.53	0.30	4.55	0.0000	1.24	0.11
2002	1.02	0.1549	0.29	0.39	2.28	0.0113	0.63	0.27
2001	1.14	0.1263	0.31	0.38	3.10	0.0010	0.83	0.20
2000	1.14	0.1278	0.31	0.38	2.10	0.0178	0.59	0.28
1999	0.85	0.1988	0.17	0.43	2.02	0.0217	0.56	0.29
1998	0.75	0.2281	0.15	0.44	1.97	0.0246	0.54	0.29
1997	0.94	0.1726	0.20	0.42	3.02	0.0013	0.84	0.20
1996	1.50	0.0673	0.32	0.37	3.36	0.0004	0.91	0.18
1995	1.60	0.0550	0.33	0.37	3.70	0.0001	0.99	0.16
1994	1.48	0.0701	0.32	0.38	3.84	0.0001	1.06	0.14

Note: Calculations are described in Sections 27.4.3 and 27.4.4. DD shows the number of standard deviations to default, based on the market value of the assets. Prior year figures calculate the DD of current balance sheet values based on historical fluctuations in asset values. CDD and CPD are based on the worst 5% of asset returns.

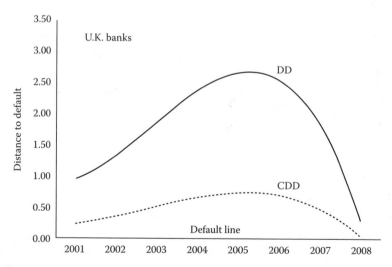

FIGURE 27.4   Distance to default. U.K. banks.

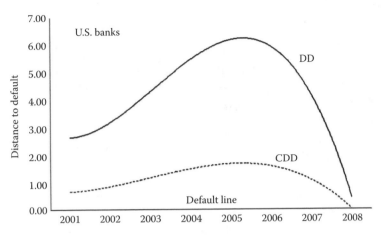

FIGURE 27.5   Distance to default. U.S. banks.

Fourth, the changes in VaR and their corresponding impacts on DD are magnified when translated into default probabilities. Default probabilities reach less than 1% for U.K. banks and less than 0.01% for U.S. banks based on volatility in the mid-2000's. This increases to 31% for U.K. banks and 18% for U.S. banks in 2008.

Fifth, there is a dramatic increase in risk when using CDD and CPD as opposed to DD and PD. In 2008, CPD is 45% for U.K. banks as compared to 31% PD. The CPD for U.S. banks is 41% as compared to 18% PD. The vast majority of banks in our study have a credit rating equating to a Standard & Poor's BBB+ and above. Yet, the PD's calculated value in this study equate to ratings far below this. Table 27.3 shows Standard & Poor's PD rates for each rating category. Based on our Merton model calculations, PD's calculated value in 2008 have increased to the extent that only 6% of banks would qualify for a rating of BBB upward. Thus, current credit ratings appear out of line with PD.

27.6 CONCLUSIONS AND RECOMMENDATIONS

Concerns expressed by the Bank of England regarding the PD based on asset values are well founded. Based on the market value of assets, the banking industry as a whole is precariously close to default and insufficiently capitalized to deal with the large fluctuations in asset values. PD's calculated value are much higher than the current credit ratings of banks

TABLE 27.3   Credit Rating Indicators Based on Default Probability

Rating	S&P PD Values	% Number of banks in Our Study Falling into This Category Based on Our 2008 PD Calculations
AAA	0.00%	—
AA	0.01%	—
A	0.04%	—
BBB	0.29%	6%
BB	1.20%	7%
B	5.71%	24%
CCC/C	28.83%	33%
D	Default	30%

Note: The first two columns in the table are based on Standard & Poor's global average one year transition rates, 1981–2004. This shows the probability of each category moving to a *D* (default rating). We map our PD's calculated value using the Merton model, to the S&P PD's for each rating category. Column three shows the percentage of banks which would fall into each rating category based on our results.

indicate. While U.K. banks show higher leveraging and default probability than U.S. banks, the situation in both the United States and the United Kingdom is highly concerning, particularly based on CPD calculations. Calls for a regulatory leverage ratio have merit in order to shore up bank capital. However, neither the current regulations nor a leverage ratio factors in the market fluctuations of asset values. We, therefore, recommend the consideration of a tri-part capital adequacy model, with each of the three components needing to be satisfied by banks. Part 1 to be the existing risk weighted assets approach. Part 2 to be a simple leverage ratio based on the book values of equities and assets. Part 3 to be based on the market value of assets.

To illustrate part 3 of the capital adequacy ratio, assume regulators set a maximum PD of 1% based on CStdev (the worst 5% of a specified period as per Section 27.4.4) of a bank's asset returns. Using standard normal distribution tables, we know 1% PD equates to an asset standard deviation of 2.33. In its simplest format, DD equates to capital/asset standard deviation. Assuming that a bank has a CStdev of 5% and asset values of $100 billion, it would be required to hold a capital of 2.33×5 billion = 11.67 billion (11.67% of assets).

In summary, we find the considerable erosion of the bank capital over the financial crisis period and recommend a market value based capital adequacy approach to counter this going forward.

REFERENCES

Alexander, G. J. and Baptista, A. M. (2003) CVaR as a measure of risk: Implications for portfolio selection. Working Paper, School of Management, University of Minnesota, Saint-Paul.

Alexander, S., Coleman, T. F., and Li, Y. (2003) Derivative portfolio hedging based on CVaR. In G. Szego (Ed.), *New Risk Measures in Investment and Regulation*, New York: John Wiley & Sons Ltd.

Allen, D. E. and Powell, R. (2007) Transitional credit modelling and its relationship to market at value at risk: An Australian sectoral perspective. Working Paper, Edith Cowan University, Perth, Australia.

Allen, D. E. and Powell, R. (2009) Structural credit modelling and its relationship to market value at risk: An Australian sectoral perspective. In G. N. Gregoriou (Ed.), *The VaR Implementation Handbook*, New York: McGraw Hill.

Andersson, F., Uryasev, S., Mausser, H., and Rosen, D. (2000) Credit risk optimization with conditional value-at risk criterion. *Mathematical Programming*, 89(2): 273–291.

Artzner, P., Delbaen, F., Eber, J. M., and Heath, D. (1997) Thinking coherently. *Risk*, 10, November: 68–71.

Artzner, P., Delbaen, F., Eber, J. M., and Heath, D. (1999) Coherent measures of risk. *Mathematical Finance*, 9(3): 203–228.

Bank of England. (2008a) *Bankstats*. Available at http://www.bankofengland.co.uk/statistics/.

Bank of England. (2008b) *Credit Conditions Survey Q4*. Available at www.bankofengland.co.uk/publications.

Bank of England. (2008c) *Financial Stability Report*, October. Available at www.bankofengland.co.uk/publications.

Beder, T. (1995) VaR: Seductive but dangerous. *Financial Analysts Journal*, 51(5): 12–25.

Bharath, S. T. and Shumway, T. (2004) *Forecasting Default with the KMV-Merton Model*. Available at http://w4.stern.nyu.edu/salomon/docs/Credit2006/shumway_kmvmerton1.pdf.

Bischel, R and Blum, J. (2004) The relationship between risk and capital in Swiss commercial banks. *Applied Financial Economics*, 14(8): 591–597.

Cespedes, J. C. G. (2002) Credit risk modelling and basel II. *ALGO Research Quarterly*, 5(1): 57–66.

Chan, H., Faff, R., and Koffman, P. (2008) Default risk, size and the business cycle: Three decades of Australian pricing evidence. Working Paper, Monash University, Melbourne. Available at http://ssrn.com/abstract=1097444.

Crosbie, P. and Bohn, J. (2003) Modelling default risk. Available at http://www.moodyskmv.com/research/files/wp/ModelingDefaultRisk.pdf.

Duffie, D. and Pan, J. (1997) An overview of value at risk. *Journal of Derivatives*, 4(3): 7–49.

D'Vari, R., Yalamanchili, K., and Bai, D. (2003) Application of quantitative credit risk models in fixed income portfolio management. Paper presented at *The Third International Workshop on Computational Intelligence in Economics and Finance (CIEF'2003)*. Available at http://www.rondvari.com/CIEF%20 2003_Final.pdf.

Fama, E. and French, K. (1992) The cross section of expected stock returns. *Journal of Finance*, 47,(2): 427–465.

Fama, E. and French, K. (1993) Common risk factors in the returns of stocks and bonds. *Journal of Financial Economics*, 33: 3–56.

Federal Deposit Insurance Corporation (2008) Homepage available at http:// www2.fdic.gov/SDI/.

Gharghori, P., Chan, H., and Faff, R. (2007) Are the Fama-French factors proxying default risk? *Australian Journal of Management*, 32(2): 223–250.

Hildebrand, P. (2008) Is Basel II enough? The benefits of a leverage ratio. Available at http://www.bis.org/review/r081216d.pdf.

J. P. Morgan and Reuters. (1996) RiskMetrics technical document. Available at http://www.riskmetrics.com/rmcovv.html.

Jorion, P. (1997) *Value at Risk: The New Benchmark for Managing Financial Risk*. New York: McGraw-Hill.

Kealhofer, S. and Bohn, J. R. (1993) Portfolio management of default risk. Available at http://www.moodyskmv.com/research/files/wp/Portfolio_Management_ of_Default_Risk.pdf.

Lopez, J. A. (2004) The empirical relationship between average asset correlation, firm probability of default and asset size. *Journal of Financial Intermediation*, 13(2): 265–283.

MacDonald, R., Kelly, C., Barragate, B., Orr, K., and Olsen, J. (2009) United States: Bank failures in 2008 and a look ahead to 2009. Available at JonesDay www. jonesday.com/files/Publication/758a02ac-078c-4715-89d6-02aacb1afac9/ Presentation.

Pesaran, M. H., Schuermann, T., Treutler, B.-J., and Weiner, S. M. (2003) Macro-economic dynamics and credit risk: A global perspective. Wharton Financial Institutions Center, Working Paper No. 03-13. Available at http://papers.ssrn. com/sol3/papers.cfm?abstract_id=432903.

Pflug, G. (2000) Some remarks on value-at-risk and conditional-value-at-risk. In R. Uryasev (Ed.), *Probabilistic Constrained Optimisation: Methodology and Applications*. Dordrecht, the Netherlands/Boston, MA: Kluwer Academic Publishers.

Pritsker, M. (1997) Evaluating value-at-risk methodologies: Accuracy versus computational time. *Journal of Financial Services Research*, 12(2/3): 201–242.

Rockafellar, R. T. and Uryasev, S. (2002) Conditional value-at-risk for general loss distributions. *Journal of Banking and Finance*, 26(7): 1443–1471.

Rockafellar, R. T., Uryasev, S., and Zabarankin, M. (2006) Master funds in portfolio analysis with general deviation measures. *Journal of Banking and Finance*, 30(2): 743–776.

Saita, F. (2007) *Value at Risk and Bank Capital Allocation*. Amsterdam, the Netherlands: Elsevier, Academic Press.

Saunders, A. and Allen, L. (2002) *Credit Risk Measurement*. New York: John Wiley & Sons, Inc.

Stambaugh, F. (1996) Risk and value at risk. *European Management Journal*, 14(6): 612–621.

Standard & Poor's. (April 2005) S&P quarterly default update & rating transitions. Global fixed income research. Available at http://www2.standardandpoors.com/spf/pdf/fixedincome/DefaultUpdate2005Q1.pdf.

Stein, R. M. (2002) Benchmarking default prediction models: Pitfalls and remedies in model validation. Available at http://www.moodyskmv.com/research/files/wp/BenchmarkingDefaultPredictionModels_TR030124.pdf.

Uryasev, S. and Rockafellar, R. T. (2000) Optimisation of conditional value-at-risk. *Journal of Risk*, 2(3): 21–41.

US Federal Reserve. (2008) Federal reserve statistical release. Charge-off and delinquency rates on loans and leases at commercial banks. Available at www.federalreserve.gov/.

Vasicek, O. A. (1987) Probability of loss on loan portfolio, KMV. Available at http://www.moodyskmv.com/research/files/wp/Probability_of_Loss_on_Loan_Portfolio.pdf.

Vassalou, M. and Xing, Y. (2002) Default risk in equity returns. *Journal of Finance*, 59(2): 831–868.

Zeng, B. and Zhang, J. (2001) An empirical assessment of asset correlation models. Moodys/K.M.V. Portfolio Credit Risk White Papers. Available at http://www.moodyskmv.com/research/files/wp/emp_assesment.pdf.

CHAPTER 28

Remuneration, Risk, and Financial Crisis

Guy W. Ford, Tyrone M. Carlin, and Nigel Finch

CONTENTS

The sharp deterioration in the subprime mortgage market in the United States in 2007 triggered a decline in confidence in global financial markets of unprecedented scale. The sizable losses that resulted in many financial institutions revealed significant weaknesses in basic market practices and principles—notably deteriorating lending and underwriting standards; excessive (undercapitalized) risk-taking; inadequate risk management processes; undue reliance on ratings of structured products; and an almost obsessive focus on actions driving short-term results with little consideration of long-term implications. In the aftermath, the attention of regulators has turned to the structure of remuneration in financial institutions, and in particular, the failure of these structures to adequately capture and adjust for the risk taken in order to earn returns. This chapter asserts that the efficacy of risk-adjusted remuneration structures relies on the resolution of three main factors: (1) risk appetite can be defined and translated into ex-ante risk measures; (2) agency problems that arise with the use of risk-adjusted performance measures can be resolved; and (3) managerial overconfidence in assessing risk can be taken into consideration in the design of remuneration structures.

28.1 INTRODUCTION

The sharp deterioration in the sub-prime mortgage market in the United States in 2007 triggered a decline in confidence in global financial markets of unprecedented scale. The sizable losses that resulted in many financial institutions revealed significant weaknesses in basic market practices and principles—notably deteriorating lending and underwriting standards; excessive (undercapitalized) risk-taking; inadequate risk management processes; undue reliance on ratings of structured products; and an almost obsessive focus on actions driving short-term results with little consideration of long-term implications. While this has been a familiar prescription over recent years in specific cases of large losses in financial institutions such as Barings Bank, Irish Allied Bank, and Societe Generale, losses in these institutions were largely linked to the actions of a single individual who was able to circumvent internal controls and take large market positions that eventually failed. Subsequent investigations into each case revealed a common thread: the existence of remuneration schemes that encouraged individuals to take large risks by rewarding them for short-term profit performance without appropriate adjustment or recognition of the risks taken in order to achieve these profits. In each case, regulators took little intervention, charging boards with the responsibility for designing and implementing appropriate remuneration structures.

In postmortems of the causes of the financial crisis, remuneration schemes have featured prominently—a problem that once appeared to exist in a limited number of investment banks has revealed itself to exist across a range of financial institutions, encompassing all aspects of loan origination, underwriting, and distribution. Somewhat surprising has been the propensity for many institutions to encourage imprudent behavior on the part of their sales teams by setting incentives that encouraged rapid growth in lending where capacity to repay must clearly have been in question—the propensity to front-load compensation and not link it to ex-ante risk measures or subsequent outcomes being a major contributing factor to the buildup in risk that precipitated the financial crisis.

This role of inappropriate remuneration structures in the crisis has not been lost on various regulatory and industry bodies, many of whom have subsequently issued position statements on bank remuneration policies. The Basel committee of banking supervision announced proposed enhancements to the Basel II framework, which set the expectation that banks establish appropriate incentives throughout the firm to reflect the long-term risks and rewards associated with their respective business models (Bank for International Settlements, 2009). The Australian Government, in consultation with the Australian Prudential Regulatory Authority, announced that it would examine what domestic policy actions on executive remuneration would be appropriate to avoid excessive risk-taking in Australian financial institutions, with focus on the structure of remuneration and the incentives built into such structures (Australian Prudential Regulatory Authority, 2008). The Financial Services Authority (FSA) in the United Kingdom states that the remuneration structure of firms "gave incentives for staff to pursue risky policies, undermining the impact of systems designed to control risk, to the detriment of shareholders and other stakeholders, including depositors, creditors and ultimately tax payers" (Financial Services Authority, 2008, p. 1). In a report on recommendations for market best practices, the Institute of International Finance (IIF) writes that market evolution related to the originate-to-distribute model and growth in structured products "led some firms to apply compensation incentives that exacerbated weaknesses that contributed to the market turmoil" (Institute of International Finance, 2008, p. 11). According to Baily et al. (2008), other bodies expressing similar views are the OECD (2008), the Counterparty Risk Management Policy Group III (2008), and the Brookings Institution (2008).

In light of the recommendations of various regulators and industry bodies that remuneration policies in financial institutions be reviewed and adjusted to reflect the risk taken in order to generate returns, with the potential for capital requirements to be formally linked to structure of remuneration, this chapter sets forth and examines three main propositions:

1. Remuneration structures in banks cannot be linked to risk unless the "risk appetite" of the bank can be clearly defined and "risk tolerances" effectively articulated. This, however, is fraught with difficulties as qualitative guidelines prove to be of little use when pricing business and measuring performance on a risk-adjusted basis, while quantitative measures—while preferred—cannot hope to capture the tail risks that characterize losses of the magnitude experienced in times of global financial crisis. Indeed, unless clarity is achieved on risk appetite, measures related to risk tolerance will prove largely useless.

2. The most effective remuneration structures should incorporate the ex-ante views of specialist managers regarding risk distributions into capital allocation and risk-adjusted performance methodologies given the notion of bank capital is to protect the bank against unexpected losses. Agency problems arise, however, when compensation links ex-post outcomes to ex-ante perspectives on risk. These in part reflect the nature of the risk measures that are used, and potential gaming on the part of managers. If risk tolerance is based on target credit ratings, which seems to be the market convention given economic capital in banking is based on a predetermined solvency standard, then agency problems are likely to thwart structures that link remuneration to risk.

3. If risk appetite and risk tolerance levels can be appropriately defined and measured, and agency problems prove unfounded, there exists a growing body of evidence that suggest even the best "behaved" managers suffer from overconfidence when it comes to identifying and assessing the risk associated with their decisions. Ratings agencies and contributors of bank capital can increasingly be expected to consider this when determining appropriate hurdle rates on bank investments.

This chapter proceeds by examining each of these propositions in detail. The final section concludes this chapter.

28.2 RISK APPETITE: ALLURING BUT ILLUSIVE

Understanding and articulating risk appetite is fundamental to the design of effective risk-adjusted remuneration structures. Put simply, how can a manager know the degree of risk that is acceptable when framing/pricing a new product or building business without a clear understanding of the risk appetite of the institution? Further, other than using a simple checklist approach, how can risk-adjusted performance be measured, and remuneration ultimately determined, without quantification of the risk taken in order to realize profits or other targets?

An increasing number of regulatory bodies require financial institutions to disclose statements on their risk appetite. The Basel Committee on Banking Supervision requires boards of directors and senior management to aggregate firm-wide exposure information in a timely manner using "easy to understand and multiple metrics" and to "define risk appetite in a manner than considers long-term performance over the cycle" and to "set clear incentives across the firm to control risk exposures and concentrations in accordance with the stated risk appetite" (Bank for International Settlements, 2009). The FSA requires firms under their purview to follow remuneration policies that are aligned with the "stated risk appetite" of the firm (Financial Services Authority, 2008). The IIF recommends that firms set basic goals for risk appetite, consider all types of risk when defining risk appetite, and involve finance and treasury functions as well as risk management in monitoring the overall risk of the firm (Institute of International Finance, 2008).

Let us consider definitions of risk appetite according to regulators and industry bodies in insurance and banking:

- The level of aggregate risk that a company can undertake and successfully manage over an extended period of time (Society of Actuaries)[*]

- A company's ability and/or willingness to absorb declines in the value of an asset, liability, trade, transaction, or portfolio (Committee of Chief Risk Officers)[†]

- The broad-based amount of risk a company or other entity is willing to accept in pursuit of its mission or vision (Basel Committee on Banking Supervision)[‡]

- The amount of risk than an organization is prepared to accept, tolerate, or be exposed to any point in time (HM Treasury)[§]

---

[*] Kamiya et al. (2007, p. 29).
[†] Ibid.
[‡] Ibid.
[§] HM Treasury (2006, p. 3).

Broad qualitative statements on risk appetite might conform to disclosure requirements but should provide little comfort to managers, boards, investors, or other stakeholders. As an example, consider the following disclosure statement on risk appetite released in December 2008 by an investment management business subject to consolidated supervision by the U.K. FSA in accordance with Pillar Three of the capital requirements directive:

> Martin Currie has an appetite for risk that varies from very low to medium depending on the nature of the risk concerned and how central it is to the Martin Currie culture. Generally speaking, it has a low or very low risk appetite in areas that are fundamental to the culture of Martin Currie and its business model – the investment process, client service, and regulatory compliance. It is more accepting of risks where – for example, risks to product development plans – where the overall effect, although potentially significant, is likely to be short to medium term, given its belief that continuing to focus on the fundamentals will deliver long-term success.*

The intention here is not to single out a particular firm, but rather, point out the vagueness that can arise when firms are required to articulate their risk appetite. Such qualitative statements should give little assurance to stakeholders who desire knowledge of how much risk an entity is prepared to take in order to achieve its business objectives. At the other extreme, risk appetite disclosures based exclusively on quantitative metrics can expose the institution to agency problems on the part of managers including gaming and risk arbitrage. This is discussed further in Section 28.3.

Requiring a financial institution such as a bank to articulate its risk appetite is a curious concept given the diverse interests of the multiple stakeholders that the bank represents. Specifically, in the event that the bank defaults on its debt, some stakeholders may be less concerned with the magnitude of losses than other stakeholders. The economic impact of default on bank shareholders and employees is likely to be largely invariant to the size of actual losses, with costs to these stakeholders a function of the event of default itself. Managers face loss of employment regardless of the size of default, while losses to owners are capped by the institution of limited liability. In contrast, the economic impact of default on

---

* "Capital Requirements Directive: Pillar Three Disclosure," Martin Currie Limited, December 2008.

regulators and creditors is more directly related to the size of losses in the event of default. For the design of remuneration structures, this means risk measures based on the probability of default are likely to be of more relevance to shareholders and employees, while measures linked to losses in the event of default may be more relevant to regulators and creditors. If a bank does represent stakeholders who carry different attitudes to risk or tolerance to unexpected losses, then this presents difficulties in terms of a consistent measurement of risk for remuneration structures. It places focus directly on the question of the appropriate risk appetite and tolerance levels for the governing body of a bank.

How then can risk tolerance levels be determined? The standard approach in banking is to relate the economic capital of a bank to some predetermined solvency standard (target credit rating). As discussed in Section 28.3, a bank targeting a specific credit rating is expected to hold capital sufficient to cover unexpected losses up to the solvency threshold—a target credit rating of AA, for example, suggests a risk tolerance of around 0.03%, meaning the bank would expect to hold sufficient capital to cover losses over one year with 99.97% probability. Other approaches might be based on earnings-at-risk or cash-flow at risk limits (avoid losing more than a defined percentage or multiple of earnings or cash over a given time horizon). The problem with approaches based on the probability of losses is they tend to embody a risk-neutral attitude to risk, which may not reflect the risk appetite of bank stakeholders who are averse to losses beyond some target threshold. Table 28.1 presents a simple example to support this point.

The data in Table 28.1 shows the distribution of losses for two portfolios and three risk measures that embody different concepts of risk. If the 99% solvency standard is applied, portfolio A has a 1% probability that losses will be $40m or more, while portfolio B has a 1% probability that losses will be

TABLE 28.1   Risk Measures

Loss ($m)	Portfolio A	Portfolio B
70		0.5%
40	1%	
10		0.5%
Solvency standard	40.00	10.00
Expected loss	0.40	0.40
Loss aversion	16.00	25.00

$10m or more—hence portfolio A carries the greatest risk. If investors are concerned with the full distribution of losses, both portfolios have the same expected loss of $0.40m and would be considered to carry the same risk. If, however, investors are "loss averse," meaning they are more concerned with larger losses carrying smaller probabilities than smaller losses carrying larger probabilities, then portfolio B may be considered to have more risk than portfolio A—here a simple squaring of the loss to embody greater aversion to larger losses than smaller losses captures the larger tail risk in portfolio B. We conclude that any attempt to articulate the risk appetite of a bank and include risk tolerance in remuneration structures must be accompanied by congruent risk measures.

It is also not uncommon to observe institutions express their risk appetite and tolerance levels in terms of constraints placed upon managers, such as concentration/exposure limits, underwriting limits, and lending restrictions. These metrics may reflect the risk appetite of the institution, but of themselves, do not determine the risk appetite of the institution. In this sense, risk tolerance may be better expressed in terms of an acceptable variation around the achievement of objectives. It may be also relevant to define the "acceptability" of risks for an institution—for a retail bank, credit and liquidity risk may be deemed an acceptable risk arising in the normal course of operations, whereas mismatching the duration of asset and liabilities in order to take gambles on interest rate movements may be deemed an unacceptable risk given the existence of interest rate swap markets for hedging such exposures.

28.3 RISK-ADJUSTED PERFORMANCE: AGENCY PROBLEMS

The industry standard in banking for risk-adjusted performance measurement is "return on risk-adjusted capital" (RORAC). This measure relates the profit from an activity to the capital allocated against the activity, where capital is determined such as to reduce the probability of failure to a desired confidence level, within a desired measurement horizon. This level of "economic capital" thus defines risk at a common point (confidence level) in the distribution, where the confidence level typically represents the target solvency standard (credit rating) of the entity. In defining risk in probabilistic terms, economic capital represents a common currency for risk that allows exposures related to credit risk, market risk, and operational risk to be directly compared across the entity. The solvency standard adopted by an entity forms the link between its internal assessment of risk and the capital structure of its balance sheet. The economic capital of the entity is attributed to the difference between the mean of its loss distribution—expected losses—and the designated confidence level. In this way, economic capital acts to protect the entity against unexpected losses, being downside variations in the expected loss rate.

Several authors advocate the use of RORAC for performance measurement in banks on the basis that it promotes goal congruence between the center of the bank and managers within the bank by aligning realized returns with the risks taken to achieve these returns. Mussman (1996, p. 7) argues that the methodology fosters consistency between the objectives of business units and that of central management by "applying a consistent risk/return criterion to business units and individual transactions as it does to the organisation as a whole." Punjabi and Dunsche (1998) state that RORAC measures provide scope not only to capture portfolio-level effects of transactions, but they also correctly motivate lending actions and relationship strategy.

We argue that RORAC models may not achieve these outcomes due to the existence of agency problems in banks. If the basis upon which risk is estimated for the denominator of RORAC is historical data on volatility, and managers have a good understanding of where historical estimates of volatility understate current or expected volatility, then it is possible for managers to exploit this information asymmetry and select investments where the risk estimate used for RORAC is less than the true estimate in the eyes of the manager. This will allow managers to evade risk limits and take on greater risk than permitted or desired by the center, and increase the potential to achieve a high RORAC and associated bonus. Managers may also desire to evade risk limits to exploit convexities in the compensation payment function or because they are gambling to resurrect a position that is incurring losses.

A possible solution to this problem would be for the center to base risk measures for RORAC on estimates of volatility provided by managers, rather than using data based on historical volatility. However, managers may have little incentive to reveal their private information, particularly when this information is used to derive a measure of risk that forms the basis upon which performance will later be judged. Further, and more concerning, managers may be incentivized to misrepresent this information in order to achieve a more favorable risk capital allocation and increase the size of any potential bonus. Such misrepresentation of information may lead to the bank being undercapitalized with respect to risk.

The value-at-risk (VaR) methodology has become the industry and regulatory standard for assessing risk in financial institutions. This can be traced to the endorsement of VaR for determining regulatory capital requirements for market risks in banks by the Basel Committee of the Bank for International Settlements in 1996 in the publication of the amendment of the capital accord (Bank for International Settlements, 1996). This amendment gave banks the option to determine regulatory capital requirements for market risks using VaR estimates derived from their own internal risk measurement models and, in doing so, established VaR as the preferred regulatory measure of market risk. The commitment to the VaR methodology was confirmed in 1999 when the Basel Committee and the International Organization of Securities Commissions (IOSCO) Technical Committee jointly issued a report calling for banks to publicly disclose summary VaR information in their annual reports (Bank for International Settlements, 1999).

While regulatory endorsement for VaR has centered on the assessment of market risks in banks, broader market endorsement has evolved on the part of ratings agencies, where the basis for determining the solvency standard of a bank is the probability associated with the bank defaulting on its senior debt securities over a specified time horizon. If a bank is targeting a specific credit rating, then it is expected to hold economic capital sufficient to cover unexpected losses up to the $(1 - \alpha)$ percent solvency threshold. This corresponds to a VaR risk assessment, given VaR models measure the loss that will be exceeded with a specified probability over a specified time horizon. This suggests a propensity for banks to use VaR methodologies in their internal models as the basis for determining economic capital requirements. If this is the case, then there is a high probability that RORAC measures in use within banks are based on a VaR-type risk measure in their denominator.

This use of VaR for internal risk measurement has potentially serious consequences. Basak and Shapira (2001) show, in a partial equilibrium framework, that an agent with his VaR capped optimally chooses to insure against intermediate loss states (those that occur with $(1 - \alpha)$ percent probability) while incurring losses that occur in the worst states. This suggests that agents ignore the (α) percent loss states that are not included in the computation of VaR. The intuition is that when agents are remunerated on the basis of a RORAC measure that uses VaR as the base measure for risk, it is optimal to incur losses in those states against which it is the most expensive to insure. Basak and Shapiro demonstrate that while the probability of a loss is fixed, the optimal behavior on the part of agents lead to a larger loss (when a large loss occurs) than would have resulted if the bank had not engaged in risk management based on VaR measures.

If the remuneration of a trader or asset manager is linked to a RORAC based on VaR in the denominator, it is possible for the individual to engage in positions that manipulate the VaR measure without a compensating reduction in underlying risk. This could be achieved by entering into positions that produce small gains with a reasonably high probability and a large loss with lower probability. This position will have a low VaR if the probability of a large loss is sufficiently low, increasing the RORAC on the position and the size of the bonus to the trader. At the same time, the bank is exposed to a large loss that is not captured in the VaR measure, and is thus undercapitalized with respect to risk.

Next consider gaming of the risk measure used in the RORAC by managers using their private information. This information may relate to knowledge of estimation errors in the risk measure when historical data is used to estimate future return distributions. While banks may use a variety of models to derive and estimate future distributions, all models tend to suffer from the small size of the data set used for the estimation. This arises because the extreme events that cause very large losses are by definition rare and will tend not to be included in the data. The result is that the risk estimate will largely reflect outcomes under normal market conditions, and not capture the potential for correlations across markets to increase significantly during extreme conditions. If a limited number of observations are made in the lower tail of the distribution, estimation errors may be large. If managers have a good understanding of the estimation errors in the risk estimate they can exploit any bias by selecting positions where the risk estimate is lower than the actual risk being taken. Given that the expected return on the position should be determined by the true risk being taken, the manager will on average expect to generate higher profits for a lower risk measure, providing compensation for the true risk being taken. The resulting higher RORAC measures will increase the size of the potential bonus to be received by the manager.

A manager may also have private information on the expected distribution of returns, based on a detailed knowledge of local market conditions or experience gained in managing specific bank asset portfolios. The potential for adverse outcomes depends to some extent on the process by which the center of the bank assigns capital to positions. In the scenarios discussed above, managers selected positions where they knew the capital assigned to a position would be less than reflected in the true risk of the position. It would generally be the case that capital is assigned prior to a position being originated so, for instance, pricing can be set to earn the required hurdle rate on the assigned capital. In this case, managers were assumed free to select the positions where true risk matched or was higher than that based on historical distributions. It may be the case, however, that a manager is more restricted in the selection of portfolios given a lower number of alternatives in a particular period or pressures to achieve targets with respect to business volumes, etc. Under these circumstances, managers are less free to choose among alternative investments, and indeed, may be forced to accept positions where they expect risk to be lower than that implied by historical distribution of returns. This means the expected RORAC will be lower than the "true" RORAC and the position may fail to earn the bonus that the manager believes should accrue to the position.

Let us consider how the manager may react, and the implications for the bank, under circumstances where expected risk is lower than that implied by historical data, and consequently the capital assignment is considered by the manager to be excessive relative to the underlying risk. We present four possible responses:

1. Reject the position

2. Price the position to incorporate the higher capital charge

3. Increase risk in the position or substitute with a riskier position (asset substitution moral hazard)

4. Reveal private information on the expected distribution of returns to the center of the bank

The first option may be limited by the number of investment alternatives available in the current period and/or pressures to meet sales targets. If the position is rejected on the basis of the expected low RORAC, then the bank will be underinvesting to the extent that a position expected to add value to the banking entity will not be pursued by managers. The second option will arise if the bank prices its business to earn a minimum hurdle rate on assigned capital. The higher capital charge will increase the profit required to achieve the minimum hurdle rate, and thus increase the price of the business to the customer of the bank. This option may be adopted by the manager, but the higher price may force the business to be lost to a competitor. If the manager absorbs the price into a lower margin, the RORAC and potential bonus will be lower.

The basis of the third option is that managers will believe that their ability to meet aspiration levels for RORAC will be restricted if the capital assignment is too large relative to the expected risk in the position. In order to make up for the overallocation of risk, and increase the expected RORAC in the position, the manager may be less conservative in the management of risk in the position or substitute it for a position that carries greater risk. The third option thus presents the most potential damage to the bank, subject to the degree of risk taken by the manager relative to the perceived excess capital assignment. If any change in risk is not adequately compensated for in the capital assignment, the bank may be undercapitalized with respect to risk.

If we consider the results of empirical studies on the behavior of individuals when they perceive themselves to be operating below aspiration levels, it is not unrealistic to assume that individuals will change their risk attitude in order to achieve their objectives. In Chapter 2 of this book, we reported the results of a number of empirical studies that conclude that individuals reverse their risk attitude from one of risk-aversion to one of risk-seeking when confronted with the likelihood of performance below aspiration. This is in keeping with basic premise of prospect theory, being that individuals are risk-averse in the domain of gains and risk-seeking in the domain of losses (Kahneman and Tversky, 1979). If managers face restricted choice on the range of investment alternatives and are assigned capital against positions that they believe is high relative to their expectations on risk, they may display behavior consistent with an S-shaped utility function, as described in Chapter 2 of this book. Despite the cushion associated with the overallocation of capital relative to risk expectations, this is an undesirable outcome in the sense that a gambling strategy on the part of managers may expose the bank to extreme losses that are not covered by its capital base.

Theoretical support for this proposition is provided by Berkelaar and Kouwenberg (2002). In a general equilibrium setting, they find that while in most cases VaR-based risk management tends to reduce stock return volatility, with the stock return distribution displaying a thin left tail and positive skewness, in very bad states managers switch to a gambling strategy that adversely amplifies default risk through a heavier left tail of the return distribution (Berkelaar and Kouwenberg, 2002, p. 141). In keeping with the basic propositions of prospect theory, they conclude that an agent working under a VaR-based capital constraint tries to limit losses most of the time, but starts taking risky bets once wealth drops below the reference point upon which gains and losses are measured. This closely resembles the optimal strategy of loss-averse agents with the utility function described by prospect theory.

We assert that the tendency for managers to select or manage positions such that the risk estimate in the capital assignment is less than the "true" risk will exist whenever the capital assignment is based on historical return distributions, independent of the risk measure being used to determine the capital assignment. The motivation for this behavior is the exploitation of convexities in the compensation payment function, in order to maximize the RORAC measure upon which bonuses (or resource allocation decisions) are based. If the range of investment opportunities in the current period is large, managers can use their private information regarding errors in the risk estimate to choose positions in which risk is understated relative to current volatility and expectations of future volatility. If the range of investment opportunities in the current period is small, there may be circumstances where managers take positions where the expected risk, according to their private information, is less than that embodied in the capital assignment. Regardless of the risk measure used for the capital assignment, managers in these positions may be encouraged to adopt a gambling strategy in order to increase the probability that aspirations levels for RORAC are achieved.

In circumstances where historical distributions are such that the capital assignment reflects a lower assessment of risk than that expected by managers, we can expect that managers will not signal this information to the center. If the true risk is higher than the capital assignment, the potential for achieving a high RORAC and consequent bonus is enhanced.

Stoughton and Zechner (1998) demonstrate that whenever there is asymmetric information about a business unit's investment opportunity set, the optimal capital allocation will embody a risk premium. They show that the existence of asymmetric information increases the price of risk to the business unit, making capital allocations more sensitive to risk-taking. They conclude that in the presence of asymmetric information, the center will assign more capital for a given position than in the presence of perfect information on the return distribution of the position. This, in turn, may encourage managers to use their private information to evade risk limits and increase underlying risk of the position, in order to achieve aspiration levels with respect to RORAC.

Hermalin (1993) shows that a manager may be incentivized to invest in highly volatile assets if the personal abilities of the manager are under scrutiny. The basis of this proposition is that highly volatile assets create more noise in the performance assessment and thus deflect focus away from the manager in periods where investments may be underperforming relative to market expectations.

This leads us to the fourth option, which is for managers to reveal to the center their private information with respect to expected risk in order to achieve a capital assignment that matches the true risk being taken. This could limit the adverse consequences of using historical data to assign capital to positions, and increase the returns to decentralization by incorporating the specialized information of managers with respect to expected risk. The key question, however, is how can managers *credibly* convey this information to the center, particularly when information asymmetries may be significant and information verification difficult? The center may expect managers to misrepresent their information on the expected distribution returns in order to achieve a "favorable" assignment of capital and increase the size of the RORAC upon which performance will be measured.

28.4 PROBLEM OF MANAGERIAL OVERCONFIDENCE

An additional factor that may stymie the veracity of risk-adjusted performance measures and remuneration structures is the phenomenon of managerial overconfidence. In the setting of this chapter, overconfidence does not refer to the tendency to make overly aggressive forecasts, but rather, refers to a miscalibration of beliefs—an overestimation of the reliability of one's own abilities and knowledge. In these terms, overconfident managers will tend to underestimate the variance of risky processes such that their subjective probability distributions regarding proposed investments will be too narrow. In a landmark survey of 6901 forecasts of stock returns by finance executives (mainly CFOs) in the United States covering the period June 2001 to June 2007, Ben-David et al. (2007) found that only realized S&P 500 market returns fall within the executives' 80% confidence intervals only 38% of the time. Furthermore, expected market returns and confidence bounds are influenced by recent historical market returns and returns of the firms in which the CFOs are employed. As a consequence, executives are more confident following periods of high market returns and less confident following periods of low market returns (Ben-David et al. 2007, p. 3). This fits neatly with the observation that risk-taking was excessive and undercapitalized in the buoyant period leading up to collapse of the sub-prime mortgage market in 2007. The authors go on to show that overconfident CFOs underestimate cash flow volatility, and as a consequence, use lower discount rates to value cash flows, as well as invest more, use more debt, use proportionally more long-term debt than short-term debt, are less likely to pay dividends, and are more likely to repurchase shares.

In a similarly compelling paper, Goel and Thakor (2008) present a model that suggests top executives should be expected to be overconfident because promotion in corporations is undeniably linked to recent historical performance, which ultimately is linked to the risk taken by executives. They show that overconfident managers underestimate investment risk and have the highest probability of being promoted to CEO because the variance of outcomes from their actions is greater, and when competing with otherwise rational managers, overconfident managers will be overrepresented by upper tail winners and are thus more likely to be recognized and promoted. In short, boards are likely to choose CEOs from pools dominated by overconfident managers. This also has a familiar tune when examining the apparent lack of prudence with respect to risk management on the part of senior executives in financial institutions in the lead up to the sub-prime mortgage market collapse.

28.5 CONCLUSIONS

It is clear that the remuneration structures that failed to appropriately adjust for risk created perverse incentives that became a major factor underlying the large-scale losses associated with the market turmoil of 2007 and 2008. In recognition of this, a number of regulatory and industry bodies have placed attention on the structure of remuneration in financial institutions, and issued guidelines and position statements requiring institutions to investigate and where necessary change their methods for determining remuneration to insure compensation is appropriately aligned with the risk. In this regard, this chapter draws four main conclusions.

First, risk-adjusted remuneration structures require clear intent regarding risk appetite and risk tolerance levels. Unfortunately, qualitative statements, while potentially useful in describing perceptions of risk, are of little use when it comes to guiding managers to make investment decisions that carry different risk distributions. Risk-based compensation requires clear and unbiased metrics if managers are to make decisions that are aligned with the interests of firm stakeholders. Yet herein we find a second fundamental problem. Different stakeholders in financial institutions are likely to carry different attitudes to risk, making it virtually impossible to arrive at a congruent measure of risk tolerance that can be applied consistently for ex-ante investment decision-making and ex-post performance measurement. We have shown using a simple example how risk measures provide inconsistent rankings of risk subject to the risk attitude implicit in the metric.

Second, agency problems may create an environment where managers have little incentive to reveal their private information on the expected distribution of returns. If managers have a good understanding of where historical estimates of volatility understate current or future volatility, and this private information is difficult for the center of the bank to screen, then it is possible for managers to exploit this information asymmetry in order to achieve favorable outcomes with respect to resource allocation decisions and bonus payments. If the bank assigns risk capital to positions on the basis of historical volatility and this exceeds managers' estimates of current or future volatility, managers may reject profitable investment opportunities or increase risk-taking in order to achieve aspiration levels for the RORAC. If managers believe the capital assignment does not reflect current or future volatility, they are unlikely to reveal this information because a higher capital charge will reduce the potential RORAC. In either case, managers' desires to achieve high bonuses may lead the bank to be undercapitalized with respect to risk. Exacerbating this problem is the fact that mangers may be overconfident with respect to their abilities, leading them to underestimate the variance of risky processes.

Third, if risk capital is assigned to positions on the basis of achieving a target solvency standard for the bank, which in turn is based on a predetermined probability of default, the optimal behavior of managers is to incur losses in those states against which it is the most expensive for them to insure. This has been apparent in the financial meltdown of 2007–2008, where lack of attention to tail risks resulted in large losses across the financial services sector.

REFERENCES

Australian Prudential Regulatory Authority (2008), APRA outlines approach on executive remuneration, Press Release, 08.32, December 9, 2008.

Baily, M., Litan, R., and Johnson, M., (2008), The origins of the financial crisis. Working Paper, Brookings Institution, Washington, DC.

Bank for International Settlements (1996), Amendment to the capital accord to incorporate market risks, Basel Committee on Banking Supervision.

Bank for International Settlements (1999), Recommendations for public disclosure of derivatives activities of banks and securities firms, Joint Report by the Basel Committee on Banking Supervision and the Technical Committee of the International Organization of Securities Commissions (IOSCO).

Bank for International Settlements (2009), Proposed enhancements to the Basel II framework, Basel Committee on Banking Supervision, January.

Basak, S. and Shapiro, A., (2001), Value-at-risk-based risk management: Optimal policies and asset prices, *The Review of Financial Studies,* 14(2): 371–405.

Ben-David, I., Graham, J., and Harvey, C., (2007), Managerial overconfidence and corporate policies. Working Paper 13711, National Bureau of Economics Research, Cambridge, MA.

Berkelaar, A. and Kouwenberg, R., (2002), The effect of VaR-based risk management on asset prices and the volatility smile. *European Financial Management,* 8(2): 139–164.

Financial Services Authority (2008), Remuneration policies. October, pp. 1–5.

Goel, A. and Thakor, A., (2008), Overconfidence, CEO selection and corporate governance. *Journal of Finance,* 63(6): 2737–2784.

Hermalin, B., (1993), Managerial preferences concerning risky projects. *Journal of Law, Economics & Organisation,* 9(1): 127–135.

HM Treasury (2006), Thinking about risk, November, pp. 1–33.

Institute of International Finance (2008), Final report of the IIF committee on market best practices: Principles of conduct and best practice recommendations, IIF, Washington, DC, July, pp. 1–174.

Kahneman, D. and Tversky, A., (1979), Prospect theory: An analysis of decision under risk. *Econometrica,* 47(2): 263–291.

Kamiya, S., Shi, P., Schmidt, J., and Rosenberg, M., (2007), Risk management terms. Working Paper, University of Wisconsin-Madison, Madison, WI.

Mussman, M., (1996), In praise of RAROC. *Balance Sheet,* 4(4): 7–10.

Punjabi, S. and Dunsche, O., (1998), Effective risk-adjusted performance measurement for greater shareholder value. *Journal of Lending and Credit Risk Management,* 18(2): 18–24, October.

Stoughton, N. and Zechner, J., (1998), Optimal capital allocation using RAROC and EVA. Working Paper, University of California, Irvine, CA.

CHAPTER 29

Some Overlooked Ethical Aspects of Bailing Out Banks and the Philosophy of Frederic Bastiat

Robert W. McGee

CONTENTS

When banks get into financial trouble, it poses a threat to the entire economy, both nationally and internationally. Government intervention is often called for to reduce the adverse effects that would otherwise occur. Governments and the economists who work for them estimate the adverse effects that would ensue in the absence of intervention. Multiplier theory is often employed to show the secondary effects that are expected to ripple through the economy.

The problem with this approach is that policy makers focus only on the losses incurred by the banks and the adverse ripple effects that are caused by the problems in the banking sector. A good utilitarian analysis would examine the effects a policy has on all groups, both long term and short term. Rights issues are often ignored, since utilitarian analyses almost uniformly disregard the existence of rights. This chapter will examine the current banking crisis and will apply both the utilitarian ethics and rights theory of Frederic Bastiat to determine when, and under what circumstances, government intervention in financial markets can be ethically justified.

29.1 INTRODUCTION

As this chapter was written, a number of banks have become insolvent and more are in danger of bankruptcy. The mortgage market is in the dumps, housing values have declined, employers are cutting back on expenses by laying off staff, leading to an increase in unemployment. The economy appears to be in a downward spiral. The President and the Congress have passed a number of bailout bills that are intended to pump money into the economy with the view that doing so will save us from even worse disaster. According to one television report, the supply of money increased by 70% between October, 2008 and January, 2009. Yet many prices have either declined or remained constant and only a few people are talking about the inflation that some say is just around the corner. Monetary theorists predict a round of inflation, since they assert there is a direct relationship between the general price level and the amount of money in circulation. Keynesians would disagree, and it is Keynesians who are currently in charge of policy in Washington.

In order to solve a problem, it is generally a good idea to first determine the cause, so that the prescription for a cure will be on the mark. But there is no consensus about what caused the problem. Politicians and pundits are blaming greed on Wall Street. Business executives are blaming government regulation and intervention in capital markets. Nearly everyone agrees that something needs to be done and that it is the government that needs to act. The voices of those who disagree that the government is the solution are drowned out by the vast majority of politicians, news commentators, and even business people who think that further intervention can make things better. If anyone discusses ethics at all, it is the perceived lack of ethics on Wall Street. Practically no one applies ethical theory to the current situation or raises the question of whether it is ethical for government to bail out the banking industry, or any industry, for that matter.

This chapter is intended to fill the gap in that debate and to enhance the scant literature that addresses ethical issues in connection with government bailouts. The ethical philosophy of Frederic Bastiat has been chosen as the benchmark, since Bastiat's philosophy takes a comprehensive and unique perspective. Bastiat (1801–1850) was a French economist and journalist who could properly be labeled as within the French Liberal School (Salerno, 1978; Liggio, 2001). Like most economists, he applied utilitarian theory to solve problems. But unlike many utilitarian theorists, he really made an effort to examine the effects a policy would have on all affected groups, both in the short-run and in the long-run. This approach separates him from the Keynesians, who focus on the short-run and who do not include some very important groups in their utilitarian calculus.

But Bastiat's ethical analysis goes beyond utilitarianism and into rights theory. This approach is one thing that separates Bastiat from most other theorists. Most utilitarians ignore rights issues. Indeed, Jeremy Bentham, one of the founders of the utilitarian philosophy, actually stated that rights are nonsense on stilts (Waldron, 1987; Bentham, 1988). Bastiat would disagree.

29.2 LITERATURE REVIEW

A number of reasons have been given for the causes of the banking crisis and a number of disparate solutions have also been offered. Analyses have been made of a number of banking crises in several countries and regions, including Argentina (del Negro and Kay, 2002), Asia (Pyun, 1999; Rahman et al., 2004), Indonesia (Fane and McLeod, 2002; Frecaut, 2004), Japan (Hoshi and Kashyap, 2004), Jamaica (Kirkpatrick and Tennant, 2002), Mexico (McQuerry, 1999; Castañeda, 2007), and Russia (Aleksashenko et al., 2001; Thiessen, 2005). One thing these studies all have in common is their approach. The underlying premise is utilitarianism, which is understandable, given the fact that the vast majority of economists are utilitarian. They are looking for positive-sum games or loss minimization. They seldom, if ever, discuss anyone's property rights and almost never consider the effects that pulling the funds out of one sector will have in their utilitarian calculus.

Accounting has been blamed, either because the accounting rules are inadequate or insufficiently transparent or misleading or because the

542 ■ The Banking Crisis Handbook

accounting rules were not followed. Allen and Carletti (2008) blame the mark-to-market accounting rule, which requires companies to list their financial assets at market value, causes companies to appear to be insolvent when in fact they are not. They recommend historical cost as a better alternative. Sapra (2008) examined the Allen and Carletti model, criticized it, and offered an interesting extension to it.

Others (Anonymous, 2009) have also focused on mark-to-market accounting as a possible cause. Black (1997a,b) discusses this issue as well as the argument that mark-to-market accounting, if it had been used instead of the historical cost model, could have prevented the savings and loan crisis. Black (1997b) also points out that it was the Securities and Exchange Commission that pushed for mark-to-market accounting, which means that if mark-to-market accounting were the cause of the present financial crisis it is government that is to blame, not the private sector accounting standard setters.

Davis and Hill (1993) discuss a study that concluded that saving and loan institutions adopting mark-to-market accounting were less likely to fail than were non-adopters but that the surviving adopters increased their high-risk investments more so than did the surviving non-adopters. King (2009) discusses the problems that result when companies are forced to report temporary declines in financial asset market values but cannot reflect gains, which is what the lower of cost or market method requires.

Related to the mark-to-market argument is the argument that derivatives, or more precisely, the accounting for derivatives, are responsible for the financial crisis. Studies by Sapra (2008) and Sawyer (2008) dispute that claim.

Some studies place the blame on government intervention in financial markets (Vasquez, 1998). Bailing out any industry causes what economists call a moral hazard. Several studies have applied the concept of moral hazard to the bank bailout (Anonymous, 2004; Lee and Shin, 2008). Studies suggest that bailing out the financial sector when they make risky investments will lead to more risky investments (Dell'Ariccia et al., 2006). Burnside et al., (2004) conclude that government guarantees lead to banking crises and that banking crises could not occur in the absence of government guarantees, since banks would not have incentives to engage in risky behavior in the absence of government guarantees.

Government regulation has also been blamed for the financial crisis. Although the Federal Reserve Board is nominally a private institution, all its members are appointed by government, and board members are frequently summoned to testify before Congress. There is also pressure on the Fed to adopt a monetary policy that helps achieve federal government goals through monetary policy. The Fed has been blamed for helping to cause the crisis by keeping interest rates abnormally low (Vadum, 2008).

Another regulation that has been blamed for causing the financial crisis is mortgage lending legislation that pressures banks to lend at below-market rates to customers who are not creditworthy. The Community Reinvestment Act, for example, pressured banks to lend to poor customers who want to purchase homes in poor, minority neighborhoods when normal credit analysis would classify them as high risk. Vadum (2008) cites this act as one of the primary causes of the financial crisis. Vitaliano and Stella (2006, 2007) estimated the cost of compliance with the act. Minton (2008) suggests a better approach would be to deregulate credit markets and make the act voluntary. Others have called for the act's repeal (Moroney, 2008).

Acharya and Yorulmazer (2008) point out that liquidity problems can result when the assets of too many failed banks are put on the market in a too-short period of time and suggest that government should inject liquidity into the non-failed banks so that they can have the funds needed to acquire the assets of the failed banks. The suggestion has been made that a bailout is necessary because of some "too big to fail theory" (Mayer, 1999; Stern and Feldman, 2004; Ennis, 2005) or that the government should be the lender of last resort (Garuda, 1998). These are scary ideas indeed, since adopting them would increase moral hazard and cause the economy to work even less efficiently.

29.3 BASTIAT'S UTILITARIAN ETHICS

Before one can do justice to Bastiat's approach to utilitarian ethics, one must first say a few words about the brand of utilitarian ethics that is used as a tool of economic analysis by most economists and political theorists. Jeremy Bentham (1748–1831) and John Stuart Mill (1806–1873) are generally credited with being the founders of utilitarian ethics (Bentham, 1988; Mill, 1993), although utilitarianism has been traced back to Plato's *Republic* and some of his other dialogs (Barrow, 1975).

There are several ways to look at utilitarian ethics (Goodin, 1995; Quinton, 1988; Shaw, 1999). One way to quickly summarize the philosophy is to say that it advocates the greatest good for the greatest number. Economists would say that a policy is good if the result is a positive-sum game, meaning there are more winners than losers. Superficially, that sounds plausible, but when one digs beneath the surface the philosophy quickly begins to unravel.

One fatal flaw with utilitarian ethics is that it is mathematically impossible to maximize two variables at the same time (von Neumann and Morgenstern, 1947: 11; Hardin, 1968). Another fatal flaw is that it is not possible to measure utility (Rothbard, 2004). The best one can do is rank preferences. Adding another nail to the coffin, one might point out that it is impossible to compare interpersonal utilities (Rothbard, 2004). And economists sometimes rightly raise the question of how to deal with situations where a few gain much while the many lose a little, which is always the case with rent seeking. But let us not quibble with the deficiencies of the utilitarian calculus for now. Economists, political theorists, and news commentators continue to use it as a frame of reference, so we cannot ignore utilitarian ethics, since doing so would result in an incomplete analysis.

Bastiat does a better-than-average job of using utilitarian ethics to make a point. His classic essay on utilitarian ethics is *What Is Seen and What Is Not Seen* (1850; 1964: 1–50). The first part of this essay illustrates what has come to be called the *Broken Window Fallacy*. He starts by pointing out that the only difference between a good economist and a bad economist is that a bad economist confines his analysis to visible effects, whereas a good economist also includes the effects that must be foreseen. Henry Hazlitt (1946; 1979) expands on this point by stating that a good economist looks at the effects an action will have on all groups in both the short-run and the long-run whereas a bad economist does not.

In his example, Bastiat relates the story of a father whose son breaks a pane of glass. Onlookers start to gather and discuss what has happened. Upon reflection, they begin to see a bright side to the incident. If it costs six francs to replace the window, some glazier will be six francs richer because his business will prosper. If one were to stop the analysis at that point, one might conclude that destroying things is good for the economy because it creates employment. This is the line of reasoning some economists use to conclude that war is good for the economy.

But their analysis is incomplete. As Bastiat would be quick to point out, what is seen is prosperity for the glazier. What is not seen is that there are two losers as well as one winner. The father loses because now he has just a window, whereas before the window was broken he had both a window and six francs.

But I just said there are two losers, yet I have identified only one. Who is the second loser? No one knows. In Bastiat's example, the cobbler or the local book seller are also losers because if the window had not been broken the father might have used the six francs to buy some shoes or add a book to his library. No one knows who the second loser will be because the father never had the opportunity to decide what he would do with his six francs.

However, it is not absolutely necessary to identify all losers in order to arrive at a policy decision. In Bastiat's example, all we know is that there is one winner and two losers, even though we can identify only one of the losers. If we assume that interpersonal utilities are equal, then we can reasonably conclude that the losers exceed the winners, and that the act cannot be justified on utilitarian grounds. Let us forget about the son's intent because that just complicates matters. In order to add an ethical dimension to the analysis, there must be intent.

That being the case, let us take an example that has intent. Let us say that some banks have failed and that the government decides to inject billions or trillions of dollars into the banking system to bail it out. Those same bystanders in Bastiat's example would be quick to point out that injecting funds into the banking system would lead to prosperity and they would be right, to a point. But that is not the end of the analysis. It is only the beginning.

On the surface, everyone is a winner. A strong banking system benefits everyone and there are no losers … until one asks, "Where is the government going to get the money?" There are a few places the government can get the funds needed to bail out the banks. It could levy taxes, in which case taxpayers would have less to spend. Or it could borrow the money, in which case the taxpayers would have to pay interest on the new debt and perhaps repay the principal at some point. If that solution is not palatable, the third way to get the funds is to simply create them through the banking system. But if the government resorts to this third method, the result will be an increase in the money supply, which will cause inflation and a declining purchasing power for anyone who has cash or financial assets, such as workers, people on pensions, and just about everyone else. In other words, someone is going to have to pay one way or the other. There is no such thing as a free lunch (Friedman, 1975).

So far in this analysis we have identified one group of winners–the banks–and one group of losers–taxpayers. If that were the end of the story, it might appear to be a zero-sum game, since there is one group of winners and one group of losers. But, as in the Bastiat example, that is not the case. If the taxpayers were not forced to bail out the banking system, they would have money to spend on other things, like shoes, books, and video games, or education, housing, automobiles, and a wide assortment of other goods and services. It is impossible to tell in advance where their money would flow but it certainly would flow somewhere.

The problem with Keynesian economics is that the analysis is confined to the effects that can be seen and identified. If the government pumps $1 trillion into the economy and the multiplier is five, then GDP will increase by $5 trillion. But it never works out that way (Hazlitt, 1959, 1960; Hutt, 1963; Terborgh, 1968; Skousen, 1992), and for good reason. Keynesians, like the spectators in Bastiat's example, are overlooking some affected groups. Keynesian apologists attribute this failure to live up to projections to leakages (Hansen, 1953)

In Chapter 10 of *The General Theory of Employment, Interest and Money* (1936), Keynes (1883–1946) discusses his multiplier theory. According to this theory, if the government pumps money into the economy, GDP and employment will expand by some multiple of whatever is pumped into the economy. That is what is seen. But what is not seen is where the government gets the money to pump into the economy. Government has no resources of its own. Whatever it has it must first take from somebody. Another fact that is almost universally overlooked is that the sector receiving the funds is always going to be less productive than the sectors that are forced to come up with the funding. Were that not the case there would be no need for a bailout.

Let us take an example to see how this pump priming would work in practice. Let us say that the government wants to pump $1 billion into the economy and that the multiplier is five. Let us further assume that the sector receiving the funds has a 1% rate of return while the sectors providing the funds have a 5% rate of return. In other words, the sector receiving the funds can expect to earn 1% and the sectors providing the funds will forfeit a 5% rate of return. The sectors providing the funding will shrink by 5% while the sector receiving the funds expands by 1%.

This scenario is actually optimistic, since the sector receiving the funds actually has a negative rate of return. Otherwise there would be no need for a bailout. But let us assume the targeted sector has a slightly positive return to keep the math simple. Let us further assume that the economy's annual GDP before the bailout is $10 billion. Here is what the outcome would look like (Table 29.1).

As can be seen from the example, the bailout would result in a net loss to the economy because the more efficient sectors are forced to subsidize

TABLE 29.1   GDP before the Bailout

GDP before the bailout		10,000,000,000
Increase in GDP resulting from pumping money into the targeted sector [$1,000,000,000 × 5]	5,000,000,000	
Decrease in GDP resulting from sucking money out of other sectors of the economy [$1,000,000,000 × 5]	(5,000,000,000)	0
Profits earned by the targeted sector [$5,000,000,000 × 1%]	50,000,000	
Reduced profits in other sectors [$5,000,000,000 × 5%]	(250,000,000)	(200,000,000)
GDP after the bailout		$9,800,000,000

a less efficient sector. In the above example, the funds injected into the economy totaled $1 billion. If the funds had been $1 trillion instead, the loss to the economy would be $200 billion instead of $200 million.

The above example estimates the total dollar losses to the economy. What the example does not show is the employment losses. Let us estimate what those losses would be. In a speech President Obama made a few days before I started writing this chapter, he said he expects the bailout to create 4 million jobs. At the time he made that statement, one estimate was that the bailout would cost $850 billion. Both of those numbers were changing on a daily or weekly basis, but let us take those numbers for purposes of this example. If those numbers are accurate, then pumping $850 billion into the economy would create 4 million jobs, which comes to about $212,500 per job. That does not seem like a good investment, given the fact that the average worker might earn $25,000 or $30,000. Let us ignore that fact for the moment and focus on the total effects that the bailout would have on employment.

If the total cost of the bailout comes to $1 trillion, which seems like a conservative estimate, and if the total losses from the bailout are $200 billion, which is what could be expected based on the above calculations, then there would be a net loss of 941,176 jobs if we use the cost per job of $212,500. As can be seen, failing to take all affected groups into account leads to bad policy decisions, which Bastiat pointed out in the 1840s, a generation before Keynes was born. What is curious from the perspective of the history of economic thought is that Keynes was aware of Bastiat and his theories, yet he persisted in pushing his multiplier theory anyway.

29.4 BASTIAT'S PROPERTY RIGHTS ETHICS

Bastiat's full and comprehensive approach to utilitarian ethics serves as a better model than the Keynesian model being used in Washington. But Bastiat goes beyond utilitarian ethics and into rights theory. Very few economists delve into the area of rights theory, since rights cannot be quantified. Most of them are also unprepared by training to apply rights-based ethics to economic problems.

Basically, rights ethics holds that an act is automatically unethical if performing it results in the violation of someone's rights. Some economists would suggest paying compensation to people whose rights are violated as a means of solving this rights problem. For example, when government seizes property in an eminent domain proceeding it pays compensation to those who have had their property taken from them without their consent.

But this solution suffers from several weaknesses. For one, the government probably does not pay sufficient compensation. If it did, the property owners would have sold their property voluntarily instead of resisting. But there is another facet to the problem that almost universally gets ignored by economists. The funds the government uses to pay the people whose property has been confiscated had to come from somewhere. It does not fall from the sky. So taxpayers are forced to part with their money in order to pay the targets of the eminent domain. So there are at least two groups of losers, the people targeted by eminent domain proceedings and the taxpayers who must pay them. And then there are the businesses that are not able to sell their products and services to those taxpayers, since the taxpayers no longer have money to do business with them.

But the line of reasoning above is getting us back into a utilitarian analysis of the issue. Let us look at what Bastiat has to say about property rights. The fullest exposition of Bastiat's rights theory is in *The Law*, first published in 1850 and later included in his collected works and translated into English and several other languages.

Under the pretense of organization, regulation, protection, or encouragement, the law takes property from one person and gives it to another; the law takes the wealth of all and gives it to a few—whether farmers, manufacturers, shipowners, artists, or comedians (Bastiat, 1968: 17).

One might add bankers to Bastiat's list. Bastiat referred to this process as legal plunder. According to Bastiat's view of the law, the government cannot do anything that citizens acting as individuals cannot do on their own. Individuals merely delegate some of the powers they have to the government. For example, people have the right to defend themselves, and since they possess this right, they can delegate this function to the government, which can form a police force or army in order to defend the people more efficiently.

But people cannot band together to take property that belongs to some people and give it to other people. If the government engages in such activity, it is legal plunder.

But how is this legal plunder to be identified? Quite simply. See if the law takes from some persons what belongs to them, and gives it to other persons to whom it does not belong. See if the law benefits one citizen at the expense of another by doing what the citizen himself cannot do without committing a crime. (Bastiat, 1968: 21)

He recommends abolishing such laws immediately, before they can multiply and grow into a system. That would be difficult to do in the twenty-first century America, since the system of redistribution is firmly established in both Washington and most of the states. But basically the bailout is nothing more than a redistribution scheme. Money is being transferred from the general population to a special interest group—bankers. Supposedly, the money will trickle down to the people who gave it, but if that were really the case there would be no need to give the money in the first place, since the ultimate recipients—the general citizenry—already have that money in their pockets.

29.5 CONCLUDING COMMENTS

Bailing out banks does not pass either the utilitarian ethics test or the property rights test. Sucking money out of the relatively productive sectors of the economy to subsidize failed banks results in a negative-sum game because one must take funds out of the more efficient sectors of the economy before they can be invested in the less-productive sectors. Furthermore, subsidizing failed banks sends the wrong signal. It is a basic law of economics that if you subsidize something you will get more of it. If banks or other industries think the government will bail them out if they fail they will be encouraged to engage in risky activities. An economist would say that the profits are privatized and the losses are socialized. Bailing out any industry causes what economists call a moral hazard. Several studies have applied the concept of moral hazard to the bank bailout (Anonymous, 2004; Lee and Shin, 2008). Studies suggest that bailing out the financial sector when they make risky investments will lead to more risky investments (Dell'Ariccia et al., 2006).

Bailing out failed banks also does not meet the property rights test of ethical behavior because people must have their property confiscated from them in order for the government to have the funds to turn over to the banks. The confiscation can take the form of taxes or inflation. Either way their purchasing power decreases. The result is not much different if the bailout is financed by debt, since the same people who would pay under either of the first two alternatives would be forced to pay the interest on the debt and perhaps the principal as well.

The only ethical solution to the banking crisis is to let the bad banks fail. Their assets can be sold to healthy banks that are better able to manage them. Any other solution necessarily violates either property rights or the principles of utilitarian ethics. The argument that more jobs will be lost if we do not bail out the banks does not hold up to analysis. As was shown above, more jobs will be lost if the bailout goes through than if the idea is abandoned. Government cannot create jobs anyway. Government does not create wealth; it redistributes it.

REFERENCES

Acharya, V. V. and Yorulmazer, T. (2008) Cash-in-the-market pricing and optimal resolution of bank failures. *The Review of Financial Studies*, 21(6): 2705–2742.

Aleksashenko, S., Astapovich, A., Klepach, A., and Lepetikov, D. (2001) Russian banks after the crisis. *Problems of Economic Transition*, 44(3): 27–49.

Allen, F. and Carletti, E. (2008) Mark-to-market accounting and liquidity pricing. *Journal of Accounting and Economics*, 45: 358–378.

Anonymous (2004) Bank bailout policy: The problems of time inconsistency and moral hazards. *I.D.E. Occasional Papers Series*, 39: 76–93.

Anonymous (2009) The role of mark-to-market accounting in the financial crisis. *The CPA Journal*, 79(1): 20–24.

Barrow, R. (1975) *Plato, Utilitarianism and Education*. London: Routledge, Kegan & Paul.

Bastiat, F. (1850; 1964). *Selected Essays on Political Economy*. Irvington-on-Hudson, NY: Foundation for Economic Education. Reprinted in *The Bastiat Collection*, vol. 1. Auburn, AL: The Ludwig von Mises Institute, 2007.

Bastiat, F. (1850; 1968). *The Law*. First published as *La Loi*, reprinted in *Sophismes Économiques*, vol. 1, *Oeuvres Complètes de Frédéric Bastiat*, 4th edn. Paris: Guillaumin et Cie, 1878, pp. 343–394. Reprinted in English in 1968 by the Foundation for Economic Education in Irvington-on-Hudson, NY.

Bentham, J. (1988). *The Principles of Morals and Legislation*. Amherst, NY: Prometheus.

Black, R. J. (1997a) Market value accounting: Panacea or poison for the banking industry? (Part 1). *The Journal of Bank Cost & Management Accounting*, 10(2): 49–66.

Black, R. J. (1997b) Market value accounting: Panacea or poison for the banking industry? (Part 2). *The Journal of Bank Cost & Management Accounting*, 10(3): 53–70.

Burnside, C., Eichenbaum, M., and Rebelo, S. (2004) Government guarantees and self-fulfilling speculative attacks. *Journal of Economic Theory*, 119(1): 31–63.

Castañeda, G. (2007) Business groups and internal capital markets: The recovery of the Mexican economy in the aftermath of the 1995 crisis. *Industrial and Corporate Change*, 16(3): 427–454.

Davis, H. and Hill, J. W. (1993) The association between S&Ls' deviation from GAAP and their survival. *Journal of Accounting and Public Policy*, 12(1): 65–83.

Del Negro, M. and Kay, S. J. (2002) Global banks, local crises: Bad news from Argentina. *Economic Review—Federal Reserve Bank of Atlanta*, 87(3): 89–106.

Dell'Ariccia, G., Schnabel, I., and Zettelmeyer, J. (2006) How do official bailouts affect the risk of investing in emerging markets? *Journal of Money, Credit, and Banking*, 38(7): 1689–1714.

Ennis, H. M. and Malek, H. S. (2005) Bank risk of failure and the too-big-to-fail policy *Economic Quarterly—The Federal Reserve Bank of Richmond*, 91(2): 21–44.

Fane, G. and McLeod, R. H. (2002) Banking collapse and restructuring in Indonesia, 1997–2001. *Cato Journal*, 22(2): 277–295.

Frecaut, O. (2004) Indonesia's banking crisis: A new perspective on $50 billion of losses. *Bulletin of Indonesian Economic Studies*, 40(1): 37.

Friedman, M. (1975) *There's No Such Thing as a Free Lunch: Essays on Public Policy.* LaSalle, IL: Open Court.

Garuda, G. (1998) Lender of last resort: Rethinking IMF conditionality. *Harvard International Review*, 20(3): 36–39.

Goodin, R. E. (1995) *Utilitarianism as a Public Philosophy.* Cambridge & New York: Cambridge University Press.

Hansen, A. H. (1953) *A Guide to Keynes.* New York: McGraw-Hill.

Hardin, G. (1968) The tragedy of the commons. *Science*, 162(3859): 1243–1248.

Hazlitt, H. (1946; 1979) *Economics in One Lesson.* New York: Harper & Brothers, 1946. Reprinted by Arlington House Publishers in New York in 1979.

Hazlitt, H. (1959) *The Failure of the "New Economics."* Princeton, NJ: D. Van Nostrand Company.

Hazlitt, H. (Ed.) (1960) *The Critics of Keynesian Economics.* Princeton, NJ: D. Van Nostrand Company.

Hoshi, T. and Kashyap, A. K. (2004) Japan's financial crisis and economic stagnation. *Journal of Economic Prospects*, 18(1): 3–26.

Hutt, W. H. (1963) *Keynesianism—Retrospect and Prospect: A Critical Restatement of Basic Economic Principles.* Chicago, IL: Henry Regnery Company.

King, A. M. (2009) Determining fair value. *Strategic Finance*, 90(7): 27–32.

Keynes, J. M. (1936) *The General Theory of Employment, Interest and Money.* New York: Harcourt Brace & World.

Kirkpatrick, C. and Tennant, D. (2002) Responding to financial crisis: The case of Jamaica. *World Development*, 30(11): 1933–1950.

Lee, J. and Shin, K. (2008) IMF bailouts and moral hazard. *Journal of International Money and Finance*, 27(5): 816–830.

Liggio, L. (2001) Bastiat and the French school of laissez-faire. *Journal des Economistes et des Etudes Humaines*, 11(2/3): 495–506.

Mayer, M. (1999) Is everything too big to fail? *The International Economy*, 13(1): 24–27.

McQuerry, E. (1999) The banking sector rescue in Mexico. *Economic Review—Federal Reserve Bank of Atlanta*, 84(3): 14–29.

Mill, J. S. (1993) *On Liberty and Utilitarianism*. New York: Bantam Books.

Minton, M. (2008) The community reinvestment act's harmful legacy: How it hampers access to credit. *CEI On Point*, No. 132 (March 20). Washington, DC: Competitive Enterprise Institute.

Moroney, A. (2008) Government must look in mirror when fixing blame. *Wall Street Journal*, Eastern edition (September 24): A28.

Pyun, C. S. (1999) Roles of the IMF in the Asian financial turmoil. *Multinational Business Review*, 7(2): 68–72.

Quinton, A. (1988) *Utilitarian Ethics*. LaSalle, IL: Open Court.

Rahman, S., Tan, L. H., Hew, O. L., and Tan, Y. S. (2004) Identifying financial distress indicators of selected banks in Asia. *Asian Economic Journal*, 18(1): 45–57.

Rothbard, M. N. (2004) *Man, Economy, & State*. Scholar's Edition. Auburn, AL: The Ludwig von Mises Institute.

Salerno, J. T. (1978) A comment on the French liberal school. *Journal of Libertarian Studies*, 2(1): 65–68.

Sapra, H. (2008) Do accounting measurement regimes matter? A discussion of mark-to-market accounting and liquidity pricing. *Journal of Accounting & Economics*, 45: 379–387.

Sawyer, N. (2008) Derivatives are not to blame. *Risk*, 21(10): 16.

Shaw, W. H. (1999) *Contemporary Ethics: Taking Account of Utilitarianism*. Malden, MA and Oxford: Blackwell.

Skousen, M. (Ed.) (1992) *Dissent on Keynes: A Critical Appraisal of Keynesian Economics*. New York, Westport, CT and London: Praeger.

Stern, G. H. and Feldman, R. J. (2004) *Too Big To Fail: The Hazards of Bank Bailouts*. Washington, DC: Brookings Institution Press.

Terborgh, G. (1968) *The New Economics*. Washington, DC: Machinery and Allied Products Institute.

Thiessen, U. (2005) Banking crises, regulation, and growth: The case of Russia. *Applied Economics*, 37(19): 219–231.

Vadum, M. (2008) Financial affirmative action. *The American Spectator*, online edition (September 29).

Vasquez, I. (1998) The Asian crisis: Why the IMF should not intervene. *Vital Speeches of the Day*, 63(13): 399–401, April 15.

Vitaliano, D. F. and Stella, G. P. (2006) The cost of corporate social responsibility: The case of the Community Reinvestment Act. *Journal of Productivity Analysis*, 26(3): 235–244.

Vitaliano, D. F. and Stella, G. P. (2007) How increased enforcement can offset statutory deregulation: The case of the Community Reinvestment Act. *Journal of Financial Regulation and Compliance*, 15(3): 262–274.

von Neumann, J. and Morgenstern, O. (1947) *Theory of Games and Economic Behavior*. Princeton, NJ: Princeton University Press.

Waldron, J. (Ed.) (1987) *Nonsense upon Stilts: Bentham, Burke and Marx on the Rights of Man*. London & New York: Methuen.

Index